THE BIOLOGY OF CHOLINESTERASES

NORTH-HOLLAND RESEARCH MONOGRAPHS

FRONTIERS OF BIOLOGY

VOLUME 36

Under the General Editorship of

A. NEUBERGER

London

and

E. L. TATUM
New York

NORTH-HOLLAND PUBLISHING COMPANY
AMSTERDAM · OXFORD

THE BIOLOGY OF CHOLINESTERASES

ANN SILVER

Agricultural Research Council Institute of Animal Physiology,
Babraham, Cambridge, England

1974

NORTH-HOLLAND PUBLISHING COMPANY – AMSTERDAM · OXFORD
AMERICAN ELSEVIER PUBLISHING COMPANY, INC. – NEW YORK

Library of Congress Catalog Card Number: 74–79240

North-Holland ISBN for this series: 0 7204 7100 1
North-Holland ISBN for this Volume: 0 7204 7140 0
American Elsevier ISBN: 0 444 10652 9

PUBLISHERS:
NORTH-HOLLAND PUBLISHING COMPANY – AMSTERDAM
NORTH-HOLLAND PUBLISHING COMPANY, LTD. – OXFORD

SOLE DISTRIBUTORS FOR THE U.S.A. AND CANADA:
AMERICAN ELSEVIER PUBLISHING COMPANY, INC.
52 VANDERBILT AVENUE, NEW YORK, N.Y. 10017

PRINTED IN THE NETHERLANDS

Editors' preface

The aim of the publication of this series of monographs, known under the collective title of *'Frontiers of Biology'*, is to present coherent and up-to-date views of the fundamental concepts which dominate modern biology.

Biology in its widest sense has made very great advances during the past decade, and the rate of progress has been steadily accelerating. Undoubtedly important factors in this acceleration have been the effective use by biologists of new techniques, including electron microscopy, isotopic labels, and a great variety of physical and chemical techniques, especially those with varying degrees of automation. In addition, scientists with partly physical or chemical backgrounds have become interested in the great variety of problems presented by living organisms. Most significant, however, increasing interest in and understanding of the biology of the cell, especially in regard to the molecular events involved in genetic phenomena and in metabolism and its control, have led to the recognition of patterns common to all forms of life from bacteria to man. These factors and unifying concepts have led to a situation in which the sharp boundaries between the various classical biological disciplines are rapidly disappearing.

Thus, while scientists are becoming increasingly specialized in their techniques, to an increasing extent they need an intellectual and conceptual approach on a wide and non-specialized basis. It is with these considerations and needs in mind that this series of monographs, *'Frontiers of Biology'* has been conceived.

The advances in various areas of biology, including microbiology, biochemistry, genetics, cytology, and cell structure and function in general will be presented by authors who have themselves contributed significantly to these developments. They will have, in this series, the opportunity of bringing together, from diverse sources, theories and experimental data, and of integrating these into a more general conceptual framework. It is unavoidable, and probably even desirable, that the special bias of the individual authors will become evident in their contributions. Scope will also be given for presentation of new and challenging ideas and hypotheses for which complete evidence is at present lacking. However, the main emphasis will be on fairly complete and objective presentation of the more important and more rapidly advancing aspects of biology. The level will be advanced, directed primarily to the needs of the graduate student and research worker.

Most monographs in this series will be in the range of 200–300 pages, but on occasion a collective work of major importance may be included somewhat exceeding this figure. The intent of the publishers is to bring out these books promptly and in fairly quick succession.

It is on the basis of all these various considerations that we welcome the opportunity of supporting the publication of the series *'Frontiers of Biology'* by North-Holland Publishing Company.

E.L. TATUM
A. NEUBERGER, *Editors*

Author's preface

> 'Begin at the beginning and go on
> till you come to the end: then stop'.
>
> Lewis Carrol, *Alice in Wonderland*

The King of Hearts' instructions to the White Rabbit are relevant to many tasks but not, unfortunately, to the discussion of cholinesterases. In a field with so many facets it is difficult to recognize the beginning and there is, as yet, no end. What, then, is the best arrangement of the chapters? In trying to decide this, I came to the conclusion that the precise order is relatively unimportant since no one is likely to sit down and read the book from start to finish; more probably, odd bits will be read at odd times. With this in mind I have used cross-references and have included a comprehensive list of abbreviations and a glossary at the end of the book. Such a design means that some repetition is inevitable, and for this I apologize.

An enjoyable part of writing this book is to thank all the people who have helped in its preparation. I want to express my gratitude to the many friends and colleagues, at Babraham and elsewhere, who have generously spent time answering my questions; to Mr. D.W. Butcher and Miss Wendy Reynolds of the Institute ibrary for tracing and obtaining papers; to Mr. A.L. Gallup and his staff for all the photography, and to Mrs. Patricia Munn, Janet Silver and Alastair Silver for help with references.

I owe much to all those who have typed for me but particularly to Mrs. Sally Rochead and Mrs. Janet Hood. Special thanks are due to Mrs. Jennifer Reed for invaluable work on the tables, figures and bibliography; to Drs. J. Bligh and J.S. Perry and Mr. M.L. Ryder for expert help with proof-reading and corrections; to Dr. E.A. Munn for fig. 4.1; to Dr. P.F.V. Ward for advice on the compilation of tables 11.1 and 2 and to Dr. D. Morris for help with the glossary.

Chapters or sections were read by the following to whom I am most grateful. Drs. W.N. Aldridge, J.B. Furness, P.S. Guth, Catherine Hebb, G.W. Kreutzberg, V. Lew, P.R. Lewis, D. Morris, J.S. Perry, C.C.D. Shute and V.P. Whittaker. I am also indebted to all who have given me permission to use their material, both published and unpublished, but most especially to Drs. P. Kása, G.W. Kreutzberg and L. Tóth for sending me electronmicrographs.

Finally I would like to thank Dr. R.D. Keynes (then Director of this Institute) for permission to accept the invitation to write the book.

ANN SILVER

Acknowledgement

Several illustrations and diagrams in this volume have been obtained from other publications. Some of the original figures have been slightly modified. In all cases reference is made to the original publications in the figure caption. The full sources can be found in the reference lists at the end of each chapter. The permission for the reproduction of this material is gratefully acknowledged.

Contents

CHAPTER 1

Introduction

1.1 Cholinesterases: definition and nomenclature

1.1.1 What cholinesterases are

During the 40 odd years since their discovery cholinesterases have been described in a number of ways (see Mounter and Whittaker 1950; Augustinsson 1963, 1971). A composite definition which incorporates their main characteristics is that cholinesterases are hydrolases which, under optimal conditions, catalyse the hydrolysis of choline esters at a higher rate than that of other esters and which are inhibited by low concentrations (10^{-5} M or less) of physostigmine and organophosphorus compounds. As with most definitions there are exceptions, more particularly with regard to susceptibility to inhibitors. In frog brain, for example, the hydrolase present appears to fulfil the other criteria for a cholinesterase but it is not totally inhibited by physostigmine even at a concentration of 10^{-4} M (Hawkins and Mendel 1946). This lack of conformity underlines an important point: cholinesterases are not identical in all species. They constitute a family of enzymes which fall broadly into two types, those which preferentially hydrolyse acetyl esters such as acetylcholine (ACh) and those with a preference for other types of esters such as butyrylcholine (BuCh). Again this division is not absolute and is more applicable to mammalian than to non-mammalian species.

1.1.2 Nomenclature

The naming of cholinesterases represents a long-standing controversy. When Stedman et al. (1932) isolated an ACh-splitting enzyme from horse serum they called it 'choline-esterase'. Mendel and Rudney (1943a) showed later that the serum contained two enzymes which were capable of hydrolysing ACh. The greater part of the activity was due to an enzyme which was non-specific in the sense that it could split not only choline esters but also non-choline esters such as tributyrin. The remainder of the activity was attributable to a second enzyme, thought to act exclusively on esters of choline. This latter enzyme resembled the 'specific' enzyme which they themselves had found in brain and which Alles and Hawes (1940) had found in erythrocytes. Mendel and Rudney (1943a, b) suggested that this enzyme should be called cholinesterase or true cholinesterase while the other, the non-specific enzyme, should provisionally be called pseudocholinesterase.

Over the years other names have been introduced. True cholinesterase has been variously termed ChEI, specific, erythrocyte-, E-, aceto- or acetylcholinesterase. In 1964 the Enzyme Commission proposed acetylcholine hydrolase (EC 3.1.1.7) as the systematic name with acetylcholinesterase as the trivial name. This latter name, or its abbreviation AChE will in general be used in this book.

Although Mendel and Rudney's suggested term, pseudocholinesterase, has been very widely used for the non-specific types of enzyme it has nevertheless been under fire almost since its inception. Glick (1945) pointed out that 'pseudo' usually implied 'false' and this, in his opinion, made the term unacceptable. Even now, nearly 30 years later, there is still no really satisfactory terminology. Names which have been used include ChEII, non-specific, serum-, ψ-, S-, butyro- or butyrylcholinesterase. The Enzyme Commission (1964) recommended acylcholine acylhydrolase (EC 3.1.1.8) as the systematic name and simply 'cholinesterase' as the trivial name. The use of 'cholinesterase' for the non-specific enzyme in contradistinction to 'acetylcholinesterase' for the specific type follows the scheme proposed in 1949 by Augustinsson and Nachmansohn but in my view it has a major drawback. If the name cholinesterase is reserved solely for the non-specific enzyme we are bereft of a simple term applicable to an enzyme of unestablished specificity or to the group of enzymes as a whole. The word cholinesterase has long been used in this latter way, covering both acetyl- and pseudocholinesterases and habit dies hard. Moreover, it seems inappropriate that if we talk of 'cholinesterases' we are by definition excluding the acetylcholinesterases which are probably of greater functional significance than any other type. In what follows I propose to continue the use of 'cholinesterase' or 'cholinesterases' (ChE) to cover the enzymes as a whole. The non-specific types will be called 'pseudocholinesterases' (pseudoChE) or will be named more explicitly according to their substrate preferences as, for example, butyryl- (BuChE) or propionylcholinesterase (PrChE) (see also ch. 10).

A particular problem of classification is posed by the so-called myosincholinesterases (for review see Varga and Szöőr 1971). It has been found that L-meromyosin behaves like a cholinesterase but the pattern of activity shown towards various substrates depends on the source of the myosin. Myosin from rabbit striated muscle and cardiac muscle have different substrate preferences and in neither case does the pattern fit that of rabbit AChE or BuChE (Kövér et al. 1957; Kövér and Kovács 1957). The relation of myosincholinesterases to other types of cholinesterase is not clear and they will receive little further attention.

1.1.3 The meaning of 'cholinergic'

With the development of theories of chemical transmission Dale (1934) recognized the need to classify cells on a functional rather than on an anatomical basis. He therefore proposed that neurones which release ACh from their terminals should be called cholinergic and those which release adrenaline, adrenergic. The transmission occurring at synapses formed by such cells would also be termed, respectively, cholinergic and adrenergic. Some years later Dale introduced a further term, cholinoceptive, to describe the cells which respond to ACh (Dale 1954).

Initially 'cholinergic' was used exactly in accordance with the original definition but gradually the sense began to change. Dale regretted the tendency to widen the meaning and in 1953 wrote that although 'transference of epithet' could be attractive in a poem 'for a scientific term it is destructive of its only value, precision'. Despite his plea the application has continued to broaden. In pharmacology, for instance, it is used to describe any drug capable of affecting ACh receptors. It is also common for the word to be applied to a cell not to denote any functional characteristic but to indicate its AChE content. Because proven cholinergic cells give a strong histochemical reaction for AChE, authors have come to use 'cholinergic' as an epithet for any AChE-containing cell regardless of whether or not it is known to release ACh. To use 'cholinergic' as a synonym for 'acetylcholinesterase-containing' may be convenient but it can be misleading: although all cells known to use ACh as their transmitter contain AChE, not all AChE-containing cells are functionally cholinergic. As far as possible I shall avoid using 'cholinergic' in any but its functional meaning.

1.2 The history of cholinesterases

Cholinesterases have an interesting history. Their existence was predicted by Dale in 1914, some 15 years before he and Dudley showed acetylcholine to be a natural constituent of animal tissues. Dale made his prediction on the basis of the evanescence of the effect produced when ACh was injected into an animal. He wrote, 'In the blood at body temperature it seems not improbable that an esterase contributes to the removal of the active ester from the circulation and the restoration of the original condition of sensitiveness'. Loewi and Navratil (1926) provided experimental support for this view when they showed that if eserine was present the effect of ACh on frog heart was prolonged. They attributed this result to the inhibition, by eserine, of the enzyme normally responsible for destroying ACh. It was another 6 years, however, before Stedman et al. (1932) prepared the first crude extract of cholinesterases from horse serum. Three years later Stedman and Stedman (1935) found cholinesterase in the cat brain and reported that the concentration in the basal ganglia (see ch. 7) was twice that in the cerebral cortex.

During this same period physiologists, notably Dale and his colleagues (see Dale 1953) were trying to establish whether ACh acted as a neurotransmitter. A necessary condition for the theory of cholinergic transmission was that the ACh liberated as a transmitter should be destroyed during the brief refractory period. While it was tempting to assume that cholinesterase fulfilled this function there was initially no evidence that the enzyme was present at the relevant sites in an adequate concentration. In 1937, however, Marnay and Nachmansohn working on frog muscle showed that almost all the cholinesterase present in the tissue was localized at nerve endings (Marnay and Nachmansohn 1937a, b). A year later they published figures indicating that the concentration was quite adequate to hydrolyse ACh with the rapidity demanded by the theory (Marnay and Nachmansohn 1938).

During the second world war cholinesterases acquired a special significance in the context of chemical warfare. Very many of the agents developed as potential war gases were powerful inhibitors of these enzymes (see ch. 11). Soon after the war the study of cholinesterases received a new stimulus with the introduction of a histochemical method (Koelle and Friedenwald 1949; Koelle 1950) for their detection. This enabled correlations to be made between biochemical, physiological and morphological findings. In the last decade techniques used to examine the enzyme have included electron microscopy, radiochemistry and nuclear magnetic resonance. These methods are providing the answers to some questions but at the same time they are introducing new ones. The varying data given in the next chapter, particularly in the sections on molecular weight and isozymes, illustrate that we are still a long way from knowing the physical properties of cholinesterases, let alone understanding their physiological subtleties.

CHAPTER 2

Some enzymological aspects of acetylcholinesterase

2.1 Introduction

Enzymologists and physiologists – and it is at physiologists that this book is primarily aimed – tend to view cholinesterases rather differently. The enzymologist is interested in the enzymes themselves. He wants to know the protein structure and molecular weight, to identify the active sites and isozymes, and to determine substrate specificity and the various kinetic constants which characterize the hydrolytic potential of an enzyme in vitro. The physiologist, in contrast, may have little interest in the enzymes per se. For him it may be enough to know that in vivo AChE is responsible for hydrolysing ACh at cholinergic synapses and that the presence of AChE may help in the identification of cholinergic pathways. In other words, he is mainly concerned with what the enzyme does rather than how it does it. Furthermore, the capabilities and characteristics established for a purified enzyme preparation in the test-tube, although important academically, may be of limited relevance to the enzyme which is performing its normal physiological function in intact nervous tissue. This point will be discussed more fully in § 2.5.3. While the physiologist can argue that an enzymologist's treatment of biochemical aspects may be too esoteric, the enzymologist can argue with equal justification that a physiologist should have some basic understanding of AChE if he is to invoke its action to explain his experimental findings.

This chapter attempts to outline some of the established properties of cholinesterases, more particularly those with a bearing on the interpretation of results from physiological experiments. In addition, it deals with certain of the still controversial aspects of enzyme action at the molecular level. Most of the emphasis is on acetylcholinesterase since pseudocholinesterase is considered separately in ch. 10.

2.2 Some characteristics

In the introductory chapter cholinesterases were defined as enzymes which, under conditions that are optimal for each substrate, hydrolyse choline esters faster than other esters. Another distinctive feature of cholinesterases in most species (but see

Hawkins and Mendel 1946 re resistance of frog brain enzyme) is their susceptibility to inhibition by physostigmine (eserine) in concentrations of 10^{-5} M or less. Both the arylesterases (EC 3.1.1.2; previously A-esterases) and the carboxylesterases (EC 3.1.1.1; previously B- or aliesterases) are resistant to much higher concentrations of eserine. Cholinesterases are also inhibited by organophosphorus compounds but this property is not exclusive to these enzymes, being shared with the carboxylesterases.

It was mentioned in ch. 1 that a distinction is made between acetylcholinesterases (EC 3.1.1.7) and non-specific or pseudocholinesterases (EC 3.1.1.8). The criteria on which this division is based include differences in substrate and inhibitor specificities (table 2.1) and in kinetic constants. These dissimilarities reflect differences in the molecular structure of the various types of cholinesterase, more particularly differences in the active centre (§2.4.2; see also ch. 10). To interpret the characteristics of each cholinesterase in these terms it is obviously necessary to have some understanding of basic enzyme kinetics and of events which occur at the active centre. Both of these aspects, enzyme kinetics and the nature of the active centre, will be tackled later in the chapter; the purpose of this section is merely to catalogue certain properties of AChE and thus to provide the data for subsequent discussion. It must be emphasized here that the properties of acetylcholinesterase from different species are not necessarily identical (see Augustinsson 1948, 1963; Ciliv and Özand 1972; Heller and Hanahan 1972).

2.2.1 Substrate specificity

The natural substrate for vertebrate AChE is assumed to be acetylcholine (ACh) which is the predominant choline ester found in the tissues. In the test-tube too, preparations of AChE generally exhibit maximal activity towards ACh but with some species, propionylcholine, another choline ester present in certain animals (see Banister et al. 1953) is hydrolysed as fast or even faster. This is particularly true of the enzyme in avian brain tissue (Myers 1953). Acetyl-β-methylcholine (MeCh) is split considerably more slowly than ACh but it is more specific in the sense that in most species except birds it is not appreciably hydrolysed by other types of cholinesterase (see Mendel et al. 1943). Because of this it is the substrate of choice in specificity studies (see ch. 3). It has recently been shown by Hosein et al. (1970) that bull erythrocyte AChE splits acetylcarnitylcholine but horse serum pseudocholinesterase does not. AChE activity towards BuCh is very low compared with that towards ACh; nevertheless the enzyme in certain species including rat (see Bayliss and Todrick 1956) will split BuCh to a significant extent. Benzoylcholine, on the other hand, is hydrolysed only slightly by mammalian AChE and has been used as a specific substrate for pseudocholinesterase (ch. 3). As will be described in §2.4.2, part of the active site of AChE is negatively charged and while it is the presence of this charge which enables the enzyme to accept cationic (positively charged) substrates such as esters of choline and thiocholine the enzyme is also

TABLE 2.1
Properties of AChE compared with pseudocholinesterase.

Property	AChE (EC 3.1.1.7)	Pseudocholinesterase (EC 3.1.1.8)	Comment
Substrate specificity			
Preferred substrate	ACh	BuCh or PrCh	This is the most general pattern of specificity but there are some species differences
Activity towards:			
MeCh	+	−	
BzCh	−	+	
Acetylcarnitylcholine	+	−	
Carbamylcholine (carbachol)	−	−	
Inhibitors			
Physostigmine	* Inhibited by 10^{-5} M	Inhibited by 10^{-5} M	Aryl- and carboxylesterases are resistant
DFP	Inhibitory conc: $10^{-5} - 10^{-6}$ M	Inhibitory conc: $10^{-7} - 10^{-8}$ M	Carboxylesterases susceptible, arylesterases resistant
Iso-OMPA	Resistant	Susceptible	
Ethopropazine	Resistant	Susceptible	See ch. 11 for details of degree of resistance
BW 284c51	Susceptible	Resistant	
Effect of ions			
Ni & Zn	Strong inhibition	Weak or no inhibition	
Mn & Mg	Mg the more effective activator	Mn the more effective activator	Cier et al. (1970)
Inhibition by excess substrate	Yes. Most active towards substrate in low concentration	No. Most active towards substrate in high concentration	

* Eserine-resistant AChE found in frog (Hawkins and Mendel 1946).

TABLE 2.2
Human erythrocyte cholinesterase. Activity towards choline and non-choline esters.
(From Adams 1949.)

Substrate	Rate of hydrolysis	Substrate	Rate of hydrolysis
Acetyl-β-methylcholine (0.03 M)	100	3:3-Dimethyl butyl acetate	180
Acetylcholine (0.C06 M)	200	Benzyl acetate	68
Acetylcholine (?M) (estimated peak) $\simeq$	300	Ethyl acetoacetate	6
Benzoylcholine (0.03 M)	1.5	Monoacetin	58
Butyrylcholine (0.03 M)	5	Triacetin	125
Ethyl acetate	6.2	Ethyl propionate	2
n-Propyl acetate	29	*n*-Propyl propionate	9
n-Butyl acetate	49	*n*-Butyl propionate	16
n-Amyl acetate	45	*n*-Amyl propionate	5.7
n-Hexyl acetate	22	*iso*-Amyl propionate	29
iso-Butyl acetate	20	*n*-Butyl formate	10
iso-Amyl acetate	72	*iso*-Amyl formate	10
1:3 Dimethyl butyl acetate	26	*iso*-Amyl butyrate	$\simeq$ 2
2-Ethyl butyl acetate	65	Tributyrin	$\simeq$ 2
2-Ethyl hexyl acetate	23		

Rates expressed as percentages of the rate of hydrolysis of acetyl-β-methylcholine. Concentration of substrates 0.1 molar unless otherwise stated. No hydrolysis could be detected with the *n*-alkyl butyrates from methyl to amyl.

capable of hydrolysing uncharged substrates. For example, it rapidly hydrolyses the carbon analogue of ACh, 3,3-dimethylbutyl acetate in which the quaternary nitrogen is replaced by an uncharged carbon atom. This ability of the enzyme to act on non-choline esters is sometimes overlooked although it is implicit in the definition given above. Adams and Whittaker (1948) and Adams (1949, see also Whittaker 1951), using human erythrocyte AChE showed that with aliphatic esters, as with choline esters, the nature of the acyl group determines the rate of hydrolysis. Acetates give the optimal reaction, propionates and formates are hydrolysed more slowly and butyrates scarcely at all. This point is illustrated in table 2.2 which lists a number of substrates including fatty acid esters. Aromatic or arylesters such as phenyl acetate and napthyl acetate are also hydrolysed by AChE (see Augustinsson 1963) and activity towards acetylaneurin (a thiamine derivative) has been demonstrated with AChE from guinea-pig erythrocytes and snake venom (Augustinsson 1948, 1951). Further data on the hydrolysis of different substrates by AChE are given in §2.5.2. but the special substrates developed for use in the histochemical methods are listed in ch. 3.

2.2.2 Inhibitors and activators

As mentioned already, susceptibility to inhibition by physostigmine and organo-phosphorus compounds is one of the distinctive characteristics of cholinesterases. These inhibitors will be considered in detail in ch. 11 (see also ch. 3) and all that need be said here is that AChE can be distinguished from pseudocholinesterases by the use of so-called 'specific' inhibitors. Some of these, for instance BW284C51 (see table 11.1 for formula) inhibit AChE but not pseudocholinesterase while for others such as ethopropazine hydrochloride (table 11.1) the reverse is true: AChE is resist-ant and pseudocholinesterase susceptible. 'Differential' inhibitors like di*iso*propyl-phosphorofluoridate (DFP) affect both types of enzyme but at different concentra-tions. AChE requires a higher concentration for effective inhibition than does pseudocholinesterase in all the species so far tested. In addition to the carbamate and organophosphorus inhibitors, which are effective in very low concentrations, many other agents will reduce enzyme activity. Among these are acetone (see Augustinsson 1948), zinc and nickel (see for instance Cier et al. 1970) and various anions including fluoride. Copper is also inhibitory and at one time it was thought that AChE activity was dependent on intact -SH- groups but other reagents which react with -SH- groups, such as iodoacetate and oxidized glutathione, do not produce significant inhibition (see Mounter and Whittaker 1953). Various other compounds which reduce AChE activity are given in the more comprehensive list in ch. 11 §4.

Substances which activate AChE are less numerous but include cadmium (Pham-Huu-Chanh and Plancade 1971), magnesium (see Cier et al. 1970), lithium, potas-sium and sodium (for references see Maheshwari et al. 1971). Magnesium seems to be the most potent but, as Cohen and Oosterbaan (1963) pointed out, the effects of ions vary greatly with other conditions of the experiment. This is particularly true in the case of sodium and potassium because they not only activate the enzyme but also increase the value for the optimum substrate concentration. Mendel and Rudney (1945) examined mouse brain homogenized in 25 mM sodium bicarbonate. When the substrate concentration was low the addition of KCl or NaCl reduced activity but when the substrate concentration was high the addition of salt postponed substrate inhibition (see below); the optimum concentration was shifted to a higher value and at this concentration the rate of hydrolysis greatly exceeded the optimum achieved in the absence of KCl or NaCl (fig. 2.1). Mendel and Rudney suggested that this effect of K^+ could be of some significance in the physiological regulation of AChE activity towards ACh released from nerve endings. Calcium, which is an activator of pseudocholinesterase, has little if any effect on AChE of horse erythrocytes (Cier et al. 1970) but it activates human erythrocyte (Heller and Hanahan 1972), bovine erythrocyte (Dawson and Crone 1973) and electric eel electroplaque AChE (Nachmansohn 1940a; Gridelet et al. 1970).

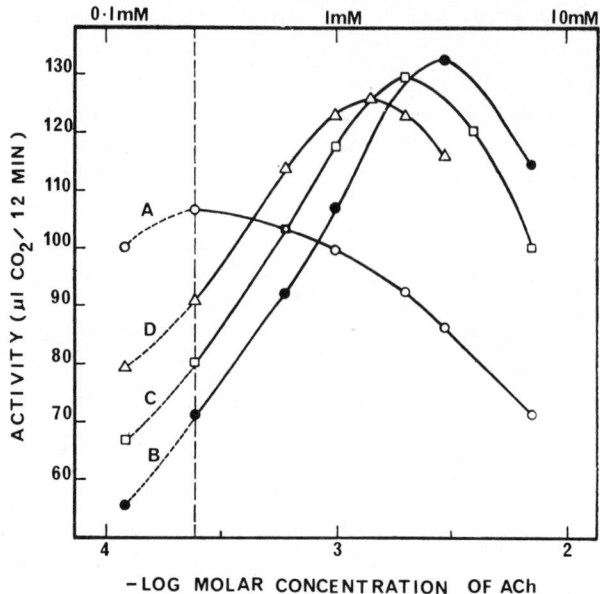

Fig. 2.1. The effect of substrate concentration and of salts on the activity of mouse brain AChE towards ACh. The vertical line marks the 0.25 mM (4 mg%) acetylcholine level. A) 25 mM $NaHCO_3$; B) 25 mM $NaHCO_3$ + 160 mM KCl; C) 25 mM $NaHCO_3$ + 80 mM KCl; D) 25 mM $NaHCO_3$ + 40 mM KCl. (Modified from Mendel and Rudney 1945.)

2.2.3 Substrate inhibition, pH and temperature

An important characteristic of AChE is its susceptibility to inhibition when substrate is present in excess. This inhibition of AChE by excess substrate is one of the features which distinguishes it from pseudocholinesterase. Whereas AChE is most active against low concentrations of ACh, and is inhibited by high concentrations, pseudocholinesterase is relatively ineffective against low concentrations of substrate but its activity increases as the concentration rises. The phenomenon of substrate inhibition will be discussed in detail in §§2.5.2, 3.

As is to be expected of an enzyme, the activity of AChE is affected by temperature and pH. The optimum temperature for AChE from mammalian tissues is between 37 and 40°C (see Augustinsson 1948). According to Shukuya (1953a; see also Wilson and Cabib 1956) the K_M (see glossary) is independent of temperature over the range of 15–30°C. Inactivation of enzyme in mammalian tissues seems to begin at about 56°C (Plattner and Hintner 1930) and all cholinesterase activity is lost at 70°C (see ch. 5 regarding enzyme in invertebrates). Freeze-drying does not reduce cholinesterase activity, a point which has useful practical implications (see ch. 3). The optimum pH varies with the source of AChE but for most preparations with ACh as the substrate, it is in the range of 8.0–8.5 (see Cohen and Oosterbaan 1963). As will be discussed in §2.5.2, the bell-shaped curve (fig. 2.2) obtained when the rate of activity towards ACh is plotted against pH has been taken as evidence that the esteratic site in the active centre contains both a basic group and an acidic group. A shift in pH away from the optimum will alter the degree of ionization of one group or the other. Bergmann et al. (1958) working with electric

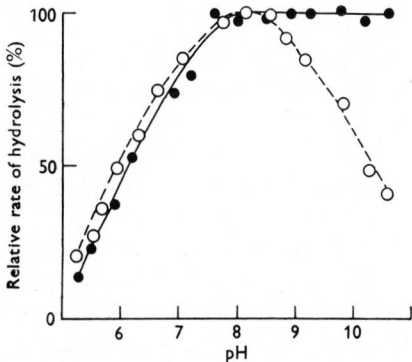

Fig. 2.2. The effect of pH on the activity of electric eel AChE towards ACh and AcThCh. Enzyme dilution, 1 : 1500; ○, acetylcholine, 5 mM; ●, acetylthiocholine, 6 mM. Temperature, 27°C. (From Bergmann et al. 1958.)

eel AChE showed that with certain substrates including acetylthiocholine and phenyl acetate, activity is pH dependent only on the acid side: activity increases with pH up to 7.0 but thereafter, as fig. 2.2 shows, it is unaffected by any further rise in pH. The isoelectric point (see glossary) varies from species to species. A value of 5.1–5.2 has been given for electric eel AChE (Hargreaves 1961) while for horse erythrocyte enzyme the value found by Augustinsson (1948) was 4.65–4.70.

2.2.4 Catalytic potential

As befits an enzyme involved in the very rapid process of nervous transmission, AChE reacts extremely fast, its rate of activity being among the highest for any enzyme. Nachmansohn (1972) has calculated that the enzyme associated with 1 g of the excitable membrane of eel electroplax can hydrolyse more than 30 kg ACh per hr. Activity is usually expressed in terms of molecules of substrate hydrolysed per min per active centre or per molecule of enzyme. The older term 'turnover' covered both expressions. The specific activity of an enzyme preparation denotes the amount of substrate hydrolysed in unit time (min or hr) per mg protein. Since enzyme activity is influenced by pH, temperature and substrate concentration these conditions must be noted in any comparison of results obtained from different preparations and species. Some of the values in the literature are included in tables 2.4a and b. These tables also give, in some instances, 'turnover time' which is the time taken by one active site to hydrolyse one molecule of substrate.

2.3 The acetylcholinesterase molecule

2.3.1 Molecular weight

Although more is known about the composition of the active site (§ 2.4.2) than about the chemical nature of the rest of the molecule, some data are available for the enzyme as a whole. Table 2.3 lists the results of amino acid analysis of purified electric eel AChE (Leuzinger and Baker 1967). As judged from the results of disk electrophoresis and analytical ultracentrifugation the preparation appeared to be a homogeneous protein.

Early estimates of the molecular weight of eel AChE were in the order of millions (Rothenberg and Nachmansohn 1947; Michel and Krop 1951). Lawler (1963) extracted an AChE polymer with a specific activity of 425 mmoles ACh/mg protein/hr and calculated molecular weights of 25,000,000 and 31,300,000 from viscosity and from light-scattering measurements, respectively. Another preparation, with a specific activity of 275, gave a molecular weight of 13,400,000. In one experiment a smaller molecule (mol. wt. 330,000) was obtained. This had a lower specific activity and Lawler suggested it might be the monomer. As methods of purification have improved, estimates of molecular weight have decreased and the

TABLE 2.3
Amino acid analysis of electric eel AChE. (From Leuzinger & Baker 1967.)

	I	II		I	II
Lysine	8 (0.210)	8 (0.186)	Alanine	10 (0.2658)	10 (0.235)
Histidine	4 (0.1082)	4 (0.0984)	Half cystine	2 (0.053)	2 (0.0475)
Arginine	10 (0.255)	10 (0.239)	Valine	12 (0.338)	12 (0.295)
Tryptophan	4 (0.100)	4 (0.090)	Methionine	5 (0.141)	5 (0.133)
Aspartic acid	20 (0.455)	21 (0.528)	Isoleucine	6 (0.178)	6 (0.159)
Threonine	8 (0.211)	8 (0.183)	Leucine	16 (0.432)	16 (0.386)
Serine	12 (0.334)	12 (0.293)	Tyrosine	7 (0.185)	7 (0.168)
Glutamic acid	16 (0.451)	16 (0.405)	Phenylalanine	10 (0.250)	10 (0.235)
Proline	14 (0.395)	14 (0.345)	Hexosamine	3 (0.078)	5 (0.118)
Glycine	14 (0.379)	13 (0.328)	Ammonia	20 (0.554)	18 (0.450)

Data are given for two different preparations of enzyme (I & II), both with a specific activity of 730 mmoles/mg/hr. For details of purification see Leuzinger & Baker (1967). The integers of amino acid residues are based on the assumption of 4 histidine residues per minimum molecular weight. The analytical data (μmoles per sample) are given in brackets.

TABLE 2.4a

Determination of molecular weight and other properties of vertebrate AChE.

Source	Purification	Specific activity μmole/hr/mg protein	Molecular weight
1) Bovine caudate nucleus (calf). Two preparations, A & B, with sedimentation coefficients of 11.7S and 6.65S respectively	30–35 ×	1,110 (pooled preparation)	A: 284,000–360,000 B: 161,400–204,600
2) Bovine caudate nucleus	5,000 ×	≏ 100,000	Not determined
3) Bovine caudate nucleus. Two peaks of activity, A & B, separable by gel filtration	≏ 400 ×	4,600 13,500	A: 430,000 B: 291,000
4) Bovine caudate nucleus. After an initial 700-fold purification by affinity chromatography, chromatography on Sephadex G-200 gave 3 peaks, A, B & C	≏ 23,000 ×	575,000 (average)	A: 390,000 B: 270,000 C: 130,000
5) Bovine caudate nucleus. In addition to the particulate enzyme AChE was also detected in a soluble form	Not cited	Not determined	80,000 On storage this form aggregated to give forms with molecular weights of 250,000; 510,000 and 1 million or more
6) Rat Brain. Preparations extracted by Triton X-100 (A) differed from those solubilized with venom or bacterial protease (B)	A ≏ 10 × B ≏ 150 ×	60 880	> 200,000 100,000

TABLE 2.4a (continued)

K_M	Additional data and comment	Reference
1) 1.38×10^{-4} M (ACh) (at 37°C; pH 8.0)	For combined preparation: Substrate optimum: 7.5×10^{-4} M Substrate inhibition: 1.04×10^{-2} M pH optimum: 8.0	Jackson & Aprison (1966a)
2) 7.0×10^{-4} M (ACh) 15.0×10^{-4} M (MeCh) (at 30°C; pH 7.5)	pH optimum, 7.5–8.5; substrate optimum, 3.3×10^{-3} M	Kaplay & Jagannathan (1970)
3) 0.87×10^{-4} M (AcThCh) 1.40×10^{-4} M (AcThCh) (at 25°C; pH 8.0)	pH optimum, 8.2; substrate optimum, $1.25–1.5 \times 10^{-3}$ M pH optimum, 7.8; substrate optimum, $1.0 - 1.25 \times 10^{-3}$ M	Chan et al. (1972a)
4) 1.4×10^{-4} M (AcThCh) 1.5×10^{-4} M (AcThCh) 2.3×10^{-4} M (AcThCh) (at 25°C; pH 8.0)	pH optimum, 8.1; substrate optimum, 1.0×10^{-3} M pH optimum, 8.0; substrate optimum, 1.0×10^{-3} M pH optimum, 7.5–8.0; substrate optimum, 5.0×10^{-4} M	Chan et al. (1972b)
5) Not determined	Attempts to purify the low molecular weight preparation caused aggregation to the higher molecular weight forms. Significance of soluble enzyme in cytosol discussed	Hollunger & Niklasson (1973)
6) 1.02×10^{-4} M (AcThCh) 1.05×10^{-4} M (AcThCh) (at 37°C; pH 8.0)	Substrate inhibition: 1.20×10^{-2} M Substrate inhibition: 1.28×10^{-2} M Authors suggest A to be precursor of B	Ho & Ellman (1969)

TABLE 2.4a (continued)

Source	Purification	Specific activity μmole/hr/mg protein	Molecular weight
7) Human erythrocytes. Two peaks of activity, A and B, separable by column chromatography	A 200 × B 50 ×		420,000 420,000
8) *Electrophorus* electroplax	330 ×	660,000	230,000
9) *Electrophorus* electroplax	720 ×	730,000	260,000 ± 10,000
10) *Electrophorus* electroplax		13,500	66,000 ± 1,000
11) *Electrophorus* electroplax	(in situ)	290	70,000
12) *Electrophorus* electroplax	commercial preparation	≏ 100,000	270,000

TABLE 2.4a (continued)

K_M	Additional data and comment	Reference
7) 1.25×10^{-4} M (AcThCh) 1.70×10^{-4} M (AcThCh) (at 37°C; pH 8.0)	Neither preparation showed substrate inhibition with AcThCh in concentrations up to 25×10^{-2} M. If Triton X-100 was omitted from gel filtration stage, mol. wt. > 2,000,000	Shafai & Cortner (1971a)
8)	Activity per mole active site = 6.1×10^5 μmole/min. Calculations suggest 4 active sites per mole	Kremzner and Wilson (1964)
9)	See also Leuzinger & Baker (1967)	Leuzinger et al. (1969)
10) Not determined	Authors suggest this represents α subunit (Leuzinger et al. 1969). Values for equivalent subunit from human erythrocytes, *Torpedo* and plaice enzymes were respectively 76,500, 74,000 and 76,000	Gentinetta & Brodbeck (1972)
11) Not determined	Assayed by irradiation inactivation	Levinson & Ellory (personal communication)
12) Not determined		

specific activity has increased. Kremzner and Wilson (1964) reporting a 'partial characterization' of electric eel AChE found a molecular weight of 230,000 for material with a specific activity of 660 mmoles ACh/mg protein/hr. A value of the same order (260,000 ± 10,000) was obtained by Leuzinger et al. (1969) for a crystalline, electrophoretically homogeneous preparation with a specific activity of 730. Froede and Wilson (1970) used a preparation of only 10% purity but since it was labelled with [^{14}C]-DFP its sedimentation velocity could be measured. By extrapolation from data for catalase (mol. wt. 250,000) they estimated the molecular weight to be 224,000. Table 2. 4a gives some of the molecular weights determined for enzyme preparations obtained from both electric eel and mammalian tissues.

Experiments to determine the molecular weight have indicated that the AChE molecule can be broken down into subunits but it is not clear what these represent in terms of enzyme function. Leuzinger et al. (1969; see also Leuzinger 1969, 1971a,b) suggested that AChE was a dimer formed from 2 protomers each of which was composed of two non-identical polypeptide chains. As table 2.5 shows, the molecular weights of the subunits are the same, each equalling approximately a quarter of the total molecular weight of 260,000. According to Leuzinger's group (see Leuzinger 1971a) each molecule of AChE possesses only 2 active sites and they proposed two possible schemes for relating molecular structure to function. In the first scheme each active centre would be built up from the α-chain which provides the serine for the esteratic site, and the β-chain which provides the anionic site (§ 2.4.2). As an alternative they suggested that only the α-chain contributed to the active centre, providing both the anionic and the esteratic site, while the β-chain was of unknown function but could represent the ACh receptor (§2.6.). This latter view has been reiterated by Leuzinger and Schneider (1972). Kremzner and Wilson (1964) believed AChE to contain at least 4 active sites but they pointed out that, at that time, there was no evidence that the enzyme could be dissociated into 4 subunits nor was there evidence that such units would be independently active. According to Wilson (1971) the early data of Michel and Krop (1951) which indicated 48 active sites for a preparation of molecular weight of 3,000,000 would support an estimate of 4 active sites per molecule of weight 260,000 (but see below re calculation of Millar and Grafius 1970). Froede and Wilson (1970), like Leuzinger et al. (1969), were able to break down *Electrophorus* AChE into four subunits. They started with a preparation with a molecular weight of 225,000 and a sedimentation coefficient ($s_{20,w}$) of 10.5 S. On guanidine treatment this produced two subunits each with a molecular weight of 102,000 and a sedimentation coefficient of 6.2 S. Further treatment with guanidine and 2-mercaptoethanol yielded subunits of about 49,000 mol. wt. and a sedimentation coefficient of 3.8 S. These figures suggest the active enzyme may be composed of 4 subunits of more or less equal weight but whether each subunit carries an active centre remains in doubt. Froede and Wilson subscribe to the view of 4 active centres per molecule but neither the

The biology of cholinesterases

TABLE 2.4b

Source	Turnover number moles/min/active centre	Turnover time μsec	Reference
Erythrocytes			
Dog	50,000		Berry (1951)
Horse	13,000		Berry (1951)
Man	171,000		Berry (1951)
Rat	10,000		Berry (1951)
Ox	295,000	*203	Cohen & Warringa (1953)
Ox	278.000		Cohen et al. (1955)
Caudate nuclei			
Man	2,061,000	29.12	Jackson & Aprison (1966a)
Calf	436,000	137.8	Jackson & Aprison (1966a)
Sheep	406,000	147.8	Jackson & Aprison (1966a)
End-plate membranes			
Rat	161,000	372.0	Namba & Grob (1968)
Electroplax			
Electric eel	740,000		Wilson & Harrison (1961)
Electric eel		30–60	Lawler (1961)

* Calculated by Lawler (1961) from data of Cohen & Warringa (1953).

TABLE 2.5
Properties of the subunits of AChE from electric eel. (From Leuzinger 1971a.)

Property	Subunits	
Molecular weight	64,000	64,000
Isoelectric point	pH 4.8	pH 7.0
Relative electrophoretic mobility	1.6	1.0
Subunit label	* DFP	** TDF
C-terminal amino acid	serine	glycine

Isoelectric points are given for carboxamidomethylated AChE-subunits.
* DFP; ** TDF: see abbreviations list.

large subunits (102,000 mol. wt.) nor the small (49,000 mol. wt.) were enzymically active. Various possibilities might account for these findings: treatment needed to disrupt the enzyme may inhibit the active centres, the molecule may split in such a way that no subunit carries a complete active centre or the active centres may be complete on each subunit but can function only in relation to the more complete molecule. Disulphide links which will be disrupted by treatment with 2-mercaptoethanol could in some way be necessary for activity. A suggestion that AChE contains not 2 or 4 active sites, but 6 has come from Millar and Grafius (1970). Their preparation of *Electrophorus* AChE had a molecular weight of 259,000 which is in good agreement with Leuzinger's figure. Guanidine treatment plus dithiothrei-tol at pH 7.0 and at pH 2.0 gave products with molecular weights of 42,200 and 21,500 respectively. The authors tentatively suggest that the enzyme has 6 subunits with molecular weights of 42,000 and that these subunits are composed of 2 chains, each of which has a molecular weight of 21,500. They accept the possibility that the number of active sites may be less than the number of subunits but point out that the early data of Michel and Krop (1951) could be recalculated to give 6 binding sites for a molecular weight of 260,000. As mentioned above, Wilson (1971) invoked the same data to support his idea of 4 sites.

2.3.2 Isozymes

The fact that an enzyme may exist in more than one molecular form has been accepted for a number of years. In his review of isozymes of lactic dehydrogenase, Markert (1968) discusses the general question of heterogeneity of enzymes and provides some useful definitions. The existence of multiple molecular forms of enzymes was first detected by electrophoresis. When run on starch gels, prepara-tions of enzyme separated into a number of bands with different electrophoretic mobilities but which were catalytically equipotent. The various molecular forms of enzyme thus revealed were termed isozymes. For a time isozymes were considered to be identical in all respects except electrophoretic behaviour. Subsequently, how-ever, heterogeneity was revealed by many other tests and gradually the definition of an isozyme has become broader (see also Wieland 1968; Shaw 1969). Some am-biguities have resulted but, according to Markert (1968), these have stemmed from failure to define the enzyme rather than the isozyme. He writes 'once the accepted criteria for defining a collection of molecules as an enzyme have been successfully applied, and if these molecules by any means (electrophoretic, chromatographic, solubility, immunochemical, etc) can be separated into distinguishable types, then these types represent isozymic forms of the enzyme'. As Markert says, difficulties can arise if the definition of the 'enzyme' is in doubt. Some authors (see for example Eldefrawi et al. 1970) are at pains to distinguish 'isozymes' which repre-sent *different* molecular forms, from variably sized aggregates of a *single* molecular form. Others (see ch. 10 §3.3) do not make this distinction and include as iso-zymes interconvertible agglomerations of a common form. It has been shown by Grafius and Millar (1967) that the formation of such reversible aggregates in prepa-rations of *Electrophorus* AChE is influenced by experimental conditions of pH and ionic strength. Whether the phenomenon of aggregation seen in the test-tube is relevant to the situation in vivo is not clear (see below) but the degree of aggrega-tion must be taken into consideration in kinetic studies in vitro. Thus Main (1969), working with human and horse serum pseudocholinesterase and bovine erythrocyte AChE, found evidence that the kinetics of reactions with inhibitors could be affected by the state of agglomeration. The multiple molecular forms of cholinesterases have been studied most intensively in the case of serum pseudocholinesterases (see ch. 10) but data about isozymes of AChE are accumulating. Some of these findings are indicated in table 2.6.

The isozymic pattern of some enzymes, e.g. lactic dehydrogenase, changes during development. This may reflect differences in the degree of oxygenation in embryonic and adult tissues (see Markert 1965; Masters and Holmes 1972). Using electrophoresis, Maynard (1966) showed that the patterns of AChE isozymes in brain and muscle of chick differed from those in adult hens (see table 2.6). Some indications of developmental changes in the nature of cholinesterases as judged by substrate specificity have been detected histochemically in brain tissue of a number of species including cat (Kása and Csillik 1965b; Krnjević and Silver 1966). Taylor and Anderson (1973) found a change in the specificity of the ChE present in the atrioventricular node and bundle in foetal human heart (see also Finlay 1972 re rat). Similarly, Gerebtzoff (1959) reported evidence of a change in the enzyme in human thymus during post-natal development. Whether such changes represent alterations in isozymes is doubtful; more probably one type of enzyme merely replaces another. Niemierko and Skangiel-Kramska (personal communication) have examined the isozymes of AChE in the proximal and distal stumps of dog sciatic nerve within 6 to 24 hr after transection. They found that the proportion of the two isozymes, I and II, is different from the proportion in control nerves. In sectioned nerves the relative amount of the electrophoretically faster form, isozyme I, is increased and this form accumulates to a greater extent than does isozyme II. The authors suggest that the latter isozyme is more firmly bound to the axon.

The functional significance of the different isozymic forms of AChE is unclear. Grafius and Millar (1967), in discussing how the aggregation of AChE molecules is influenced by pH and ionic strength, suggested very tentatively that this phenomenon could be of relevance in the control of the permeability cycle in membranes but they stressed that the aggregates observed in the test tube may not exist in that form in vivo. The occurrence of AChE in various states of solubility is of some practical importance in histochemical experiments and may explain why some methods which successfully demonstrate AChE at end-plates are not so satisfactory for ganglionic tissue and are even less satisfactory for the CNS (see Koelle et al. 1970, and ch. 3 §3.2.3).

2.4 Mechanism of action

2.4.1 Introduction

The primary function of acetylcholinesterase in animal tissue appears to be to

Isozymes of AChE
(Vertebrates and invertebrates)

Source	Detection method	Result	Comment	Reference
Human caudate nucleus and putamen	Starch gel electrophoresis of supernatant from saline homogenates	AChE associated with 2 major bands, A & D, and a lighter band, C	See also Barron et al. (1963); Bernsohn et al. (1963); Barron & Bernsohn (1965)	Bernsohn et al. (1962)
Human brain (developing and mature)	Starch gel electrophoresis of saline and Triton X-100 extracts	In adult, AcThCh gives 2 major bands in caudate, cerebellar grey and white matter and brain stem. In cerebral grey matter, bands fainter. Faster band sometimes absent. With BuThCh, 3 bands. For developing brain, patterns with AcThCh and BuThCh became identical with those of adult brain between 1 and 4 months (grey matter matured faster)	Other esterases also studied	Barron & Bernsohn (1968)
Human erythrocytes, normoblasts and erythroblasts	Electronmicroscope histochemistry (copper thiocholine) and spectrophotometric (Ellman et al. 1961)	The AChE present in mature erythrocytes does not give a histochemical reaction product because it is inhibited by fixation, low pH and Cu^{2+}. The form in red cell precursors is relatively insensitive to these factors and will react histochemically	Author suggests that isozyme demonstrable in the ER and perinuclear cisternae of immature cells could be the precursor of the more sensitive isozyme associated with the erythrocyte membrane. For further discussion see ch. 3	Skaer (1973)
Bovine caudate nucleus	Polyacrylamide gel electrophoresis of enzyme solubilized in sucrose and EDTA and purified chromatographically on DEAE cellulose and Sephadex G-200	Two types of AChE with different molecular weights and small differences in K_M and substrate and pH optima (table 2.4a)	See also Jackson & Aprison (1966a) (table 2.4a)	Chan et al. (1972a)
Bovine caudate nucleus	Polyacrylamide gel electrophoresis. Solubilization as above followed by affinity gel and Sephadex chromatography	Three types of AChE with different molecular weights and small differences in K_M and substrate and pH optima (table 2.4a)		Chan et al. (1972b)

TABLE 2.6 (continued)

Source	Detection method	Result	Comment	Reference
Rat brain	Agar gel electrophoresis of supernatants from homogenates in barbitone buffer	Two forms of AChE and one form of BuChE. AChE isozymes differ in K_M and pH optima as well as in electrophoretic mobilities	Authors suggest rat isozymes are unlike those in man which seem to differ only in electrophoretic mobility	Bajgar & Žižkovský (1971)
Rat superior cervical and spinal ganglia	Starch gel electrophoresis of lyo- (soluble) and desmo- (bound) esterases extracted from tissue. Histochemical examination of enzyme in sections before and after extraction with water	Sympathetic ganglia contained more desmo- (bound) & spinal ganglia more lyo- (soluble) cholinesterase	Other esterases studied as well. For practical implications see ch. 3 § 3.1. See also Berry & Rutland (1971) and Hollunger & Niklasson (1973) re soluble AChE in, respectively, rat diaphragm and calf caudate nucleus	Eränkö et al. (1964)
Rat retina (also brain and muscle)	Polyacrylamide gel electrophoresis of enzyme solubilized by detergent Myrj 53 and Triton X-100	10 AChE bands in retina. Brain and muscle were not examined in detail but gave similar patterns	Half-life of isozymes measured. One retinal band had very rapid turnover with a half-life of 3 hr	Davis & Agranoff (1968)
Retina, various species			See Table 6.32.1	Esila (1963)
Cat ganglia, cat and mouse intercostal muscle	Histochemical demonstration (thiocholine technique, details in paper) that AChE diffuses more readily from some ganglia than from end-plates	Treatment of ganglia with saline or Triton X-100 caused diffusion or loss of demonstrable AChE. Fixation minimized this effect. AChE at end-plate did not diffuse on treatment with saline or Triton X-100; on the contrary staining was enhanced. Following fixation the intensification was less but still apparent	Authors suggest AChE in ganglia is in the lyoenzyme (soluble) form and that at end-plates is in desmoenzyme (bound) form (see Eränkö et al. 1964)	Koelle et al. (1970)

TABLE 2.6 (continued)

Source	Detection method	Result	Comment	Reference
Rabbit sciatic nerve	Acrylamide gel electrophoresis of Triton X-100 solubilized enzyme	Fast band, mol. wt. 170,000 and slow band, mol. wt. 316,000	Authors suggest slow band is dimer of fast band	Skangiel-Kramska & Niemierko (1971)
Chicken brain and muscle (embryonic and mature tissue)	Acrylamide gel electrophoresis of aqueous homogenates	Embryonic muscle and brain have 3 AChE bands. Activity varies in different brain areas. In adult brain, band 1 less active relative to 2 and 3 compared with 18-day embryo. Adult muscle lacks band 1	See also Wilson et al. (1968) who found embryonic isozyme pattern in muscles from hens with muscular dystrophy	Maynard (1966)
Goldfish brain	Polyacrylamide gel electrophoresis of enzyme solubilized by the detergents Myrj 53 and Triton X-100	5 AChE bands detected. Patterns differ in different brain regions. Half-life (detected with DFP) varied from band to band (whole brain)	'Training' of Goldfish did not affect AChE significantly (cf. cockroach)	Lim et al. (1971)
Torpedo electric organ	Various extraction methods including toluene treatment. Sucrose gradient separation	Three main forms, A, C and D. B appears with autolysis or proteolytic enzymes	See also Massoulié et al. (1970a, b*) and Bauman et al. 1972) Papers in French with English summaries (*excepted)	Massoulié & Rieger (1969) Paper in French with English summary
Electrophorus electroplax	Sucrose gradient separation of homogenates of toluene-treated tissue	Four forms A, B, C and D. A lost on purification by affinity chromatography	Brief report. Molecular weight ratio suggests forms are aggregates built of 2, 3 or 4 protomers	Wermuth & Brodbeck (1972)
Brain of housefly; also cockroach and southern army worm (Prodenia eridania)	Polyacrylamide gel electrophoresis of supernatant of aqueous homogenate	In housefly preparation supernatant contained particulate and soluble enzyme. The latter was electrophoretically separated into 4 bands. Mol. wt. < 500,000	Cockroach has 2 bands, army worm 1; both have particulate form. Authors consider that the isozymes are separate molecular forms and not aggregates of a single form	Eldefrawi et al. (1970)

TABLE 2.6 (continued)

Source	Detection method	Result	Comment	Reference
Cockroach ganglia	Polyacrylamide gel electrophoresis of supernatants from homogenates in phosphate buffer. Also sucrose gradient separation	Two major bands. These differ in K_M, substrate optimum and mol. wt.	'Training' of cockroach caused 70% decrease in activity of major isozyme due to an increase in K_M (no change in V_{max})	Kerkut et al. (1972)
Lobster (*Panulirus argus, Panulirus guttatus & Homarus americanus*) brain, ganglia & nerve cord	Acrylamide gel and starch gel electrophoresis of aqueous homogenates	Two bands for AChE. In CNS the fast band, and in peripheral nerve the slow band, predominates		Maynard (1964)
Sea anemone, liver fluke, *Ascaris*, earthworm and king prawn	Polyacrylamide gel electrophoresis of aqueous homogenates	Anemone & liver fluke gave 5 bands, and earthworm and king prawn 2, bands for AChE. King prawn gave, in addition, 2 bands with equal affinity for acetyl and butyryl substrates. *Ascaris* gave 3 bands with preference for butyryl substrates	Carboxyl- and aryl-esterases also studied	Haites et al. (1972)
Nematode (*Nippostrongylus brasiliensis*)	Polyacrylamide gel electrophoresis of aqueous homogenates	Three isozymes of AChE present; proportions change when host (rat) develops immunity following infection		Edwards et al. (1971)

catalyse the hydrolysis of acetylcholine.

Acetylcholine   Choline   Acetic acid

To understand the intermediate steps in the reaction it is necessary to know something of the nature of the active centre of the enzyme. The active centre (also called the active site) of an enzyme has been defined by Dixon and Webb (1964) as that part of the protein structure which combines with the substrate and is responsible for the enzymic properties of the molecule (see Koshland 1963 for a discussion of general principles). Thanks to the work of Wilson and his colleagues (see Wilson 1951) the fundamental structure of the active site of AChE has long been known although, as will become clear, many of the details are still to be filled in. As fig. 2.3 shows, the active centre has two subsites, a negatively charged anionic site and an esteratic site containing both an acidic (electrophilic) and a basic (nucleophilic) group (respectively, H and G). At the simplest level, the events occurring in hydrolysis may be described as follows: i) the quaternary nitrogen of acetylcholine is bound by coulombic and Van der Waals' forces (see glossary) to the negatively charged anionic site and so orientated that the carbonyl group is presented to the esteratic site and an enzyme-substrate complex is formed; ii) electron shifts occur at the esteratic site with the result that choline is liberated leaving an acyl-enzyme; iii) the acylated enzyme is hydrolysed very rapidly yielding acetic acid and resulting in the restoration of the esteratic site to its inactive state. These basic steps are indicated in the scheme produced by Wilson et al. (1950) and re-drawn by Nachmansohn (1962):

Although the diagram shown in fig. 2.3 has proved very useful in attempts to relate this reaction to structure, the catalytic process is more complicated than the

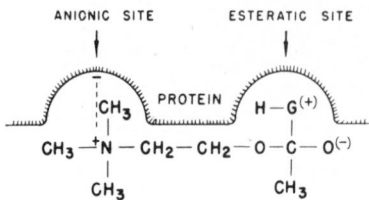

Fig. 2.3. Schematic presentation of interaction between the active groups of acetylcholinesterase and its substrate. Early version, cf. fig. 2.4. For explanation see text. (Redrawn by Nachmansohn 1962 from Bergmann et al. 1950.)

two-dimensional representation suggests. To clarify the mechanism by which AChE effects the hydrolysis of ACh, not in the test-tube but within intact living tissue, we need the answers to several interrelated questions. We must know first the precise nature and orientation of the chemical groups in the esteratic site; second, the number, nature and sphere of influence of other binding sites in the enzyme, which, while not located in the active centre, are nevertheless capable of affecting the catalytic reaction, and third, the type and extent of conformational changes which result from reactions at the active and allosteric sites (see §2.7). Next, all the data on these individual points must be collated so that the components of the catalytic and allosteric sites can be related to the three-dimensional structure of the protein molecule as a whole. The final problem is to decide how, if at all, the enzyme, AChE, is related to the postulated receptor for acetylcholine in post-synaptic membranes.

2.4.2 Structure of the active site

Various estimates have been given for the dimensions of the active site of AChE. The absolute figures may still be in doubt but it seems clear that the dimensions are not the same in all species. O'Brien (1963) found an intersite distance of 4.5 Å for bovine erythrocyte AChE and of 4.5–5.9 Å for flyhead AChE (see also ch. 5). Krupka and Laidler (1961) who also studied bovine erythrocyte enzyme suggested that the distances between the anionic site and the acidic and basic groups in the esteratic site are of the order of 2.5 and 5.0 Å respectively.

In fig. 2.3 the anionic site is represented as a negative charge. It is the possession of this charged site which gives cholinesterases their specificity for cationic substrates and distinguishes them from other esterases. Engelhard et al. (1967) have suggested that the charge is carried by an ω-carboxyl group, possibly of glutamic acid. Other details about the anionic site are sparse but some findings of Beddoe and Smith (1971) indicate that groups in its immediate vicinity are not involved in alkylation reactions. O'Brien (1969) voiced what he termed the heretical thought that a true anionic site might not exist. He pointed out, among other things, that certain uncharged substrates have a lower K_M than does ACh.

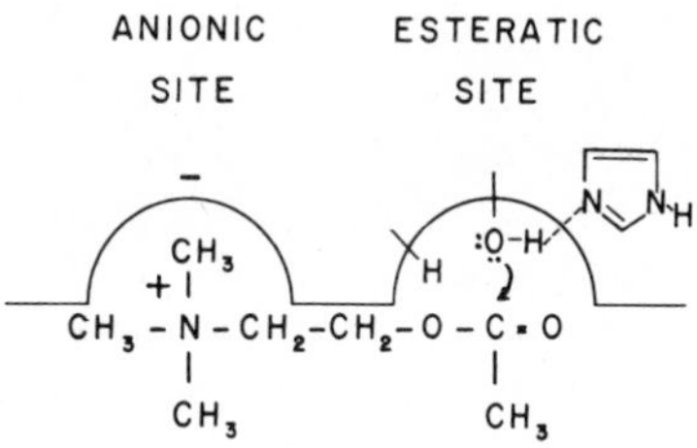

Fig. 2.4. Schematic presentation of interaction between active groups of acetylcholinesterase and substrate. Later version, cf. fig. 2.3. (From Wilson 1967.)

In the esteratic site 'H' represents an acidic group as yet unidentified but presumed to be an acidic side-chain of an amino acid residue (see Kienhuis 1964); the early suggestion that it could be tyrosine (Bergmann et al. 1956, 1958) has not apparently been refuted. The nature of 'G', the basic group, has been debated over the years but it now seems very likely that it is the hydroxyl group of serine. Many mammalian esterases contain serine in the esteratic site and this appears to be true of both AChE and pseudocholinesterase (see ch. 10). The amino acid sequence at the esteratic site of AChE is glutamic acid–serine–alanine. These data have been obtained from studies of the ^{32}P-labelled peptides that can be recovered from enzyme which has reacted with ^{32}P-labelled DFP (see Cohen and Oosterbaan 1963). The technique is not infallible since the label may move during degradation to a group other than that to which it was bound initially. Despite this and other reservations it is fairly widely held that the serine hydroxyl group does constitute

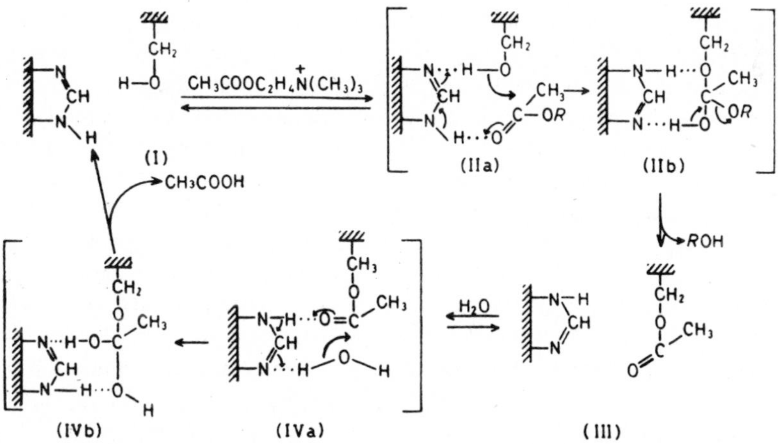

Fig. 2.5. To illustrate the proposed role of imidazole in the esteratic site. For explanation, see text. (From Brestkin and Rozengart 1965.)

the nucleophilic acyl acceptor (see glossary). One of the difficulties of casting serine in this role has been that the *pK* of the hydroxyl group is 11.79 (Merck Index) while that predicted for the basic group, as derived from studies of the effect of pH on enzyme activity is 5.8–7.0 (see Augustinsson 1963). Since the imidazole ring of histidine is the only group in proteins which has a *pK* in the required range it has often been considered to be a better candidate than serine as the acyl acceptor. It now seems likely that although it is the serine hydroxyl group which functions as the ultimate acyl acceptor the histidine imidazole group plays an important part in the catalytic action. It has been suggested (Cunningham 1957) that imidazole acts as a general base catalyst and enhances the nucleophilicity of the serine oxygen by forming a hydrogen bond with the hydroxyl group. This is indicated in fig. 2.4 (from Wilson 1967). Brestkin and Rozengart (1965) envisage a rather more active participation for imidazole (fig 2.5). They suggest that the first reaction at the esteratic site consists of the formation of a hydrogen bond between the carbonyl oxygen of ACh and the iminonitrogen (see glossary for terms) of imidazole. This bond increases the nucleophilicity of the second nitrogen, the azolic nitrogen, and brings about a rearrangement of the active centre. This results in the formation of the hydrogen bond between the azolic nitrogen and serine hydroxyl group which, as indicated earlier, enhances the nucleophilicity of the serine oxygen and allows its interaction with the carboxyl carbon of ACh. The subsequent steps, elimination of choline and the hydrolysis of the acyl-enzyme, can be followed in the figure. It was suggested by Brestkin and Rozengart that the acidic group of the esteratic site makes possible the transfer of protons from the imidazole ring to choline (IIb) and, in the restoration of the active site, from imidazole to the serine oxygen (IVb).

   This scheme embodies an important point: that the active site cannot be regarded as a static arrangement of a number of independent reactive groups. Not only may the reactivity of one group be affected by its proximity to another but the degree of this interaction may alter during the course of catalysis as a result of conformational changes which occur in the enzyme molecule. Krupka (1966b) put forward the theory that the catalytic groups involved in acylation and deacylation are different and have different *pK*s. He suggested that after the acyl-enzyme complex has been formed a conformational change shifts the acyl group out of the influence of the basic group responsible for acylation into the ambience of another which is responsible for deacylation. His idea is indicated in fig. 2.6. The components of the active site which are involved in Krupka's scheme are shown in (a); COO^- is the anionic group, and B_1 is an adjacent basic group. B_2, AH and OH are respectively the basic group, acidic group, and serine hydroxyl of the esteratic site. Krupka's main concern was with mechanisms rather than with chemical identification of specific groups but he suggested that both B_1 with a *pK* of 6.3 and B_2 with a *pK* of between 5.2 and 5.5 could be imidazole residues. The acidic group, AH, has a *pK* of 9.2. When substrate and enzyme react the position of ACh as determined by the anionic site enables B_2 and AH to catalyse the acylation of the serine hydroxyl to produce the enzyme-substrate complex (fig. 2.6a). As choline splits off, leaving an acyl-enzyme, conformational changes occur which have the effect of bringing the acyl-serine group closer to the anionic site and into the influence of the adjacent basic group, B_1 (fig. 2.6b). It is this basic group which is responsible for the deacylation (hydrolysis). Krupka points out that factors favouring acylation tend to disfavour deacylation. For example, the proximity of groups which increase the nucleophilicity of serine hydroxyl will promote the formation of the acyl complex but will hinder its hydrolysis. It is thus theoretically very reasonable to postulate that conditions operating during deacylation are likely to be different from the initial conditions of acylation. What has to be established is whether a conformational change in the active site would alter circumstances sufficiently to enable the *same* group to perform both tasks (Brestkin and Rozengart; see also Hillman and Mautner 1970) or whether a second group is involved (Krupka 1966b). In a modification of his earlier scheme Krupka (1967) proposed something of a compromise: B_1 is involved not only in deacylation but also in a fast step in acylation. The role of B_2 (the *pK* of which is 5.3) is confined to the slow rate-limiting step in acylation. Reiner and Aldridge (1967) discussed the difficulties in trying to identify individual amino acids from *pK* measurements. Since the substitution of an amino acid into a protein may alter its *pK*, as may conformational changes, they point out

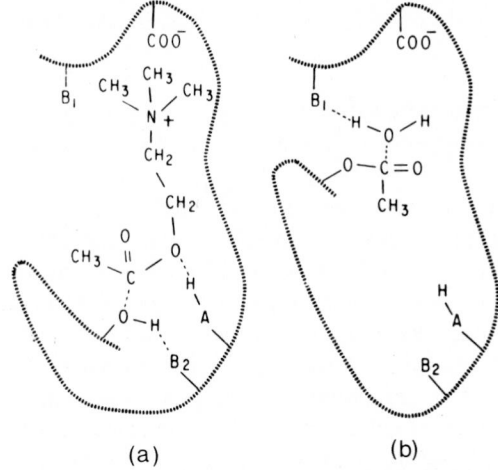

(a)   (b)

Fig. 2.6a, b. Krupka's representation of the active centre of AChE. B_1 and B_2 are the basic groups of pK 6.3 and 5.5, respectively; AH the acidic group of pK 9.2, and OH and COO⁻, the serine hydroxyl and the anionic site, respectively. Fig. (a) shows the enzyme-substrate complex for ACh. While the substituted ammonium ion is held at the anionic site, B_2 and the acidic group catalyse the transfer of the substrate acetyl group to the serine hydroxyl (acetylation) Fig. (b) shows the product of this reaction, the acetyl enzyme (EA). As the result of a conformational change, the acetyl residue has been brought near B_1, which catalyses hydrolysis of EA (deacetylation). (From Krupka 1966b.)

that differences in *pK* for acylation and deacylation do not necessarily imply that different groups are involved in the two reactions.

2.5 Kinetics

2.5.1 Introductory theory

The field or enzyme kinetics lies largely outside the area of this book but since catalytic properties of enzymes are usually described in kinetic terms some explanation of basic concepts may be useful in following the literature. The two terms most frequently encountered are V_{max} and K_M. V_{max} defines the maximum or limiting velocity attained by the enzyme reacting in the presence of excess substrate and is expressed as the amount of substrate hydrolysed per unit time per weight of enzyme preparation. K_M specifies the molar concentration of substrate at which the velocity of the reaction reaches half the maximum value. The theory behind this relationship is outlined below.

It has been known for a long time (see Segal 1959 for the development of this knowledge) that during hydrolysis the enzyme and substrate form a complex (the Michaelis–Menten complex) which on reaction with water breaks down to yield enzyme plus product. In the simplest case this may be explained as

$$\text{E} + \text{S} \rightleftharpoons \text{ES} \qquad\qquad\qquad\qquad \text{I}$$

$$\text{ES} + \text{H}_2\text{O} \rightarrow \text{E} + \text{P}_1 + \text{P}_2 \qquad\qquad\qquad \text{II}$$

where E and S are respectively enzyme and substrate, ES is the Michaelis–Menten complex and P_1 and P_2 are products.

Each step in the reaction occurs at a rate determined by the concentration (see below re pH effect) and a velocity constant, k. At any moment a certain amount of enzyme will be bound up as ES so if the total molar concentration of enzyme is $[\text{E}_0]$ the concentration of free enzyme will be $[\text{E}_0] - [\text{ES}]$. Because of the discrepancy between the molecular weights of substrate and enzyme the molar concentration of the substrate, $[\text{S}]$, is large compared with that of the enzyme and the alteration in $[\text{S}]$ due to formation of ES can be neglected. If k_1 is the velocity constant for the forward reaction in equation I, the rate v is given by

$$[\text{S}]\,([\text{E}_0] - [\text{ES}])\,k_1 \qquad\qquad\qquad\qquad \text{III}$$

similarly, for the reverse reaction

$$v = [\text{ES}]\,k_2 \qquad\qquad\qquad\qquad\qquad \text{IV}$$

and for the breakdown of the Michaelis–Menten complex (equation II)

$$v = [\text{ES}]\,[\text{H}_2\text{O}]\,k_3 \qquad\qquad\qquad\qquad \text{V}$$

where k_2 and k_3 are the respective velocity constants. In practice $[\text{H}_2\text{O}]$ is ignored since it remains constant. In the steady state, which is reached very quickly, the rate of formation and of breakdown will be equal; that is $[\text{ES}]$ will be constant, therefore

$$[S]([E_o] - [ES])k_1 = [ES]k_2 + [ES]k_3 \qquad \text{VI}$$

Hence

$$\frac{[S]([E_o] - [ES])}{[ES]} = \frac{k_2 + k_3}{k_1} \qquad \text{VII}$$

It is the term $(k_2 + k_3)/k_1$ which is the constant K_M defined above. K_M is also called the Michaelis constant. To the uninitiated it may seem strange that K_M is expressed as a molarity although it is derived from velocity constants. This is explicable as follows. With a large excess of substrate almost all the enzyme $[E_o]$ will exist in the combined form so that $[E_o] = [ES]$ and the velocity, v, of reaction will reach the limiting value V_{max}. In this special case equation V, $v = [ES]k_3$, can be written as

$$V_{max} = [E_o]k_3 \qquad \text{VIII}$$

Thus

$$\frac{v}{V_{max}} = \frac{[ES]}{[E_o]} \qquad \text{IX}$$

hence

$$V_{max} = \frac{[E_o]\,v}{[ES]} \qquad \text{X}$$

Rewriting equation VII in two stages gives

$$\frac{[S][E_o] - [S][ES]}{[ES]} = K_M \qquad \text{(a)}$$

$$\frac{[S][E_o]}{[ES]} = K_M + [S] \qquad \text{(b)}$$

Substitution of terms of equation X in (b) gives

$$\frac{[S]\,V_{max}}{v} = K_M + [S]$$

$$v = \frac{[S]\,V_{max}}{K_M + [S]} \qquad \text{XI}$$

In the particular case where the velocity, v, is half the maximum velocity i.e. $v = V_{max}/2$ substitution gives

$$\frac{V_{max}}{2} = \frac{[S]\,V_{max}}{K_M + [S]} \qquad \text{XII}$$

$$\frac{1}{2} = \frac{[S]}{K_M + [S]} \qquad \text{XIII}$$

$$K_M + [S] = 2[S] \qquad \text{XIV}$$

Thus $K_M = [S]$ \qquad XV

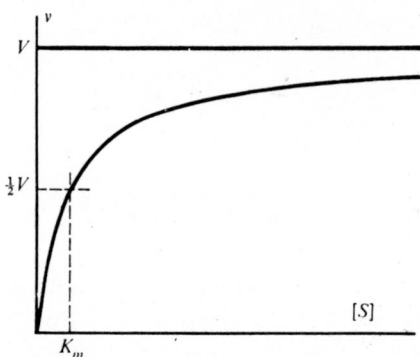

Fig. 2.7. Determination of the Michaelis constant, K_M. Ordinate, the initial reaction velocity, v. Abscissa, the concentration of substrate [S]. This gives the theoretical curve for the Michaelis (or Briggs and Haldane) equation, $v = V[S]/[S] + K_M$ where v is the limiting velocity. (From Baldwin 1967.)

The value for K_M in a particular reaction can be derived from a graph in which the velocity of reaction, v, is plotted against substrate concentration [S] (fig. 2.7).

K_M has now been defined, but what is its significance? In a reaction where the rate of formation of ES is high compared with that of its breakdown to products, the term k_3 contributes little to the equation (VII) $K_M = k_2 + k_3/k_1$ hence $K_M = k_2/k_1$ and is a measure of the affinity of the enzyme for the substrate. The greater the affinity, the greater the rate of reaction to form ES, and the smaller the value for K_M. In other types of reaction, where product formation is particularly fast, k_3 becomes a significant component (for a fuller description see Morrison 1965) hence K_M is no longer such a direct measure of affinity. Nevertheless, it is a useful index for comparing overall activity towards different substrates and, furthermore, the separate velocity constants for the different steps can be derived from experimental data and mathematical extrapolation. By comparison of such values for different substrates it is possible to determine which is the rate-limiting step for each one and in this way to gain some clues about the likely composition of the active centre (§ 2.4.2) of an enzyme.

2.5.2 Kinetics of AChE

It is generally accepted (see §2.4.2) that in the reaction between AChE and acyl-containing substrates, the enzyme-substrate complex which is formed undergoes a two-stage process. First the complex gives rise to an acylated enzyme and then this acylated form is hydrolysed. This means that equations I and II given in § 2.5.1 are an over-simplification. Wilson and Cabib (1956) have expressed the reaction as:

$$EH + S \underset{k_2}{\overset{k_1}{\rightleftharpoons}} EHS \overset{k_3}{\longrightarrow} ES' + ROH$$

$$ES' + H_2O \overset{k_4'}{\longrightarrow} EH + CH_3COOH$$

where EH is the enzyme, S the substrate, EHS the Michaelis–Menten complex, ES' the acylated enzyme and ROH the amino alcohol (in the case of ACh this is choline) which is split off.

Breaking down the hydrolysis of EHS into two steps introduces an extra kinetic constant, k_4'. K_M is equal to

$$\frac{k_2 + k_3}{k_1} \bigg/ \frac{1 + k_3}{k_4'}$$

Wilson and Cabib were able to derive values for both k_3 and k_4' for *Electrophorus* AChE reacting with a number of different acetyl esters of ethanolamine, including ACh. From their data they could compare the 'suitability' of each substrate as indicated by the speed with which it acylated the enzyme. With rapidly hydrolysed substrates such as ACh and AcThCh the rate-limiting step is deacylation ($k_4 < k_3$) but with poor substrates acylation may be slower than deacylation ($k_3 < k_4$). Wilson (1967) quotes a figure of 100 μsec for deacylation with ACh as substrate. Fig. 2.8, taken from Ecobichon and Israël (1967), shows examples of the different rates of activity of *Electrophorus* AChE towards a variety of substrates. The values for V_{max} and K_M calculated from this data by means of Lineweaver–Burk plots (see glossary) are given in table 2.7.

Detailed analyses of the effect of different substituent groups in substrates, and also of the effects of alterations in pH and other experimental conditions, on the kinetic constants of AChE reactions have enabled enzymologists to build up a picture of the catalytic centre of the enzyme in terms of the nature and interrela-

TABLE 2.7

V_{max} and K_M for electric eel AChE with different substrates. (From Ecobichon & Israël 1967.)

Substrate	V_{max}*	K_M
Acetylcholine chloride	545	1.4×10^{-4} M
Acetylthiocholine iodide	462	1.2×10^{-4} M
Propionylcholine iodide	429	2.3×10^{-4} M
Indoxyl acetate	198	0.7×10^{-4} M ·
Butyrylcholine iodide	66	0.4×10^{-4} M

* V_{max} represents the μmoles of substrate hydrolysed per hour per mg of enzyme protein, as determined by the method of Lineweaver and Burk (1934). Measurements made at pH 7.6.

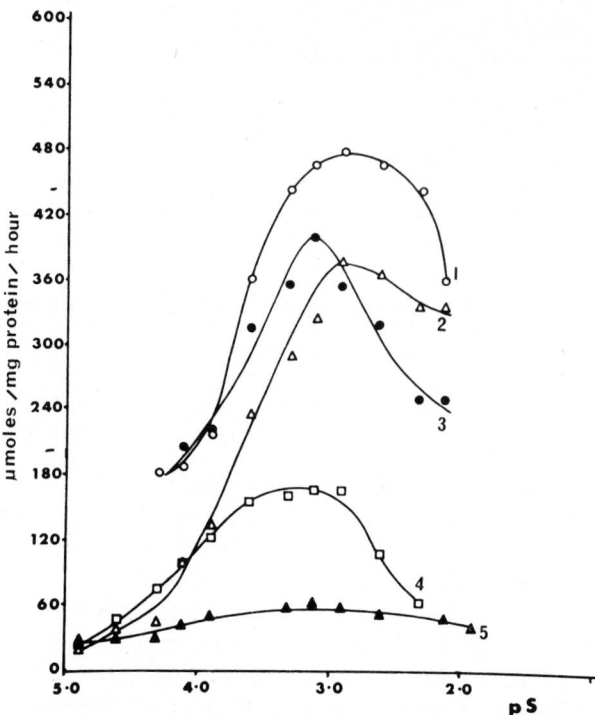

Fig. 2.8. The rate of hydrolysis of various esters by the water-soluble esterases from electric tissue of *Electrophorus electricus* at pH 7.6. 1) acetylcholine; 2) propionylcholine; 3) acetylthiocholine; 4) indoxyl acetate; 5) butyrylcholine. pS, $-\log$ molar concentration. (Modified from Ecobichon and Israël 1967.)

tions of the active groups. As more is known about the enzyme, early ideas initially very useful as working hypotheses have become inadequate. Wilson and Cabib's scheme is almost certainly too simple a treatment for the function of AChE and alternatives have been proposed. Rabin (1967), for example, has suggested a model which takes account of the conformational changes which occur in the enzyme when it reacts with substrate (see §2.4.2). Krupka (1966a, b) includes the pH dependence of V_{max} and K_M in his formulation. The pH factor was omitted by Wilson and Cabib (1956) on the grounds that it was not significant under their experimental conditions at pH 7.

The influence of pH on V_{max} and K_M stems from the fact that the reactions occurring in catalysis involve ionized groups in enzyme and substrate and the degree of ionization (dissociation) is pH-dependent. The relation between the dissociation constant K_a (more usually expressed as its negative log, pK_a) and pH is very clearly explained by O'Brien (1967). More complex treatments, both in general terms and in the case of cholinesterases, are given by Wilson and Bergmann (1950), Laidler

(1955a, b, c) and by Krupka and Laidler (1960). Fig. 2.2 showed a typical plot of the effect of pH on hydrolysis of ACh by a purified preparation of AChE from the electric organ of *Electrophorus*. The maximum velocity is attained at a pH of about 8.5. The bell-shaped curve for pH dependence was taken by Wilson and Bergmann as evidence for the existence of acidic and basic groups in the active site, a point mentioned already (§2.2.3). Reiner and Aldridge (1967) examined the influence of pH on *pK* in reactions with inhibitors rather than substrates but extended the discussion of their results to the question of acylation and deacylation in enzyme-substrate reactions. They criticized previous work in which pH effects were studied in relation either to percentage inhibition produced by a fixed concentration of inhibitor or to concentration of inhibitor required to give 50% inhibition. In their view the correct figures to which pH should be related are the *separate* rate constants for inhibition and for reactivation.

A phenomenon which is seen with a number of enzymes but which is of particular interest in the case of cholinesterases is substrate inhibition (for a general treatment of the subject see van Rossum 1964). Fig. 2.9 shows that a plot of velocity of reaction against the negative log of substrate concentration gives a bell-shaped curve with a maximum in the case of AChE (a) but a sigmoid curve in the case of pseudocholinesterase (b). Thus, as discussed earlier, AChE is inhibited by a large excess of substrate but pseudocholinesterase is not. A point to note is that AChE gives a bell-shaped curve whether the abcissa represents substrate concentration [S], or the negative log of concentration, −log [S]. In the latter case, however, the bell is symmetrical as shown here and the axis of symmetry corresponds to the optimal substrate concentration which for most preparations is 2.5–3.0 × 10^{-3} M ACh.

Various explanations have been proposed to account for this difference in the behaviour of AChE and pseudoChE towards excess substrate. One early theory was that it reflected a difference in the number of subsites in the active centre of the two enzymes. Zeller and Bissegger (1943) suggested that the active centre of pseudoChE consisted only of an esteratic subsite whereas AChE possessed both an anionic and an esteratic subsite. Fig. 2.10 shows Zeller and Bissegger's scheme for explaining substrate inhibition on the basis of the type of subsites present. For normal catalysis (a), both of the subsites of AChE are involved in the binding of a single molecule but if substrate is in excess (b), separate molecules are bound individually to each of the subsites thereby interfering with hydrolysis. Pseudocholinesterase, with only one type of subsite is not affected in this way (c). Subsequent evidence indicated that pseudoChE, like AChE, does possess an anionic site (but see ch. 10 for further discussion) and Zeller and Bissegger's theory was superseded by the suggestion (Bergmann 1955, 1958) that whereas pseudocholinesterase had one anionic site for each esteratic site, AChE had two. Bergmann postulated that the binding of separate molecules of excess substrate to each of these anionic sites would prevent either molecule combining with the esteratic site. The idea of two

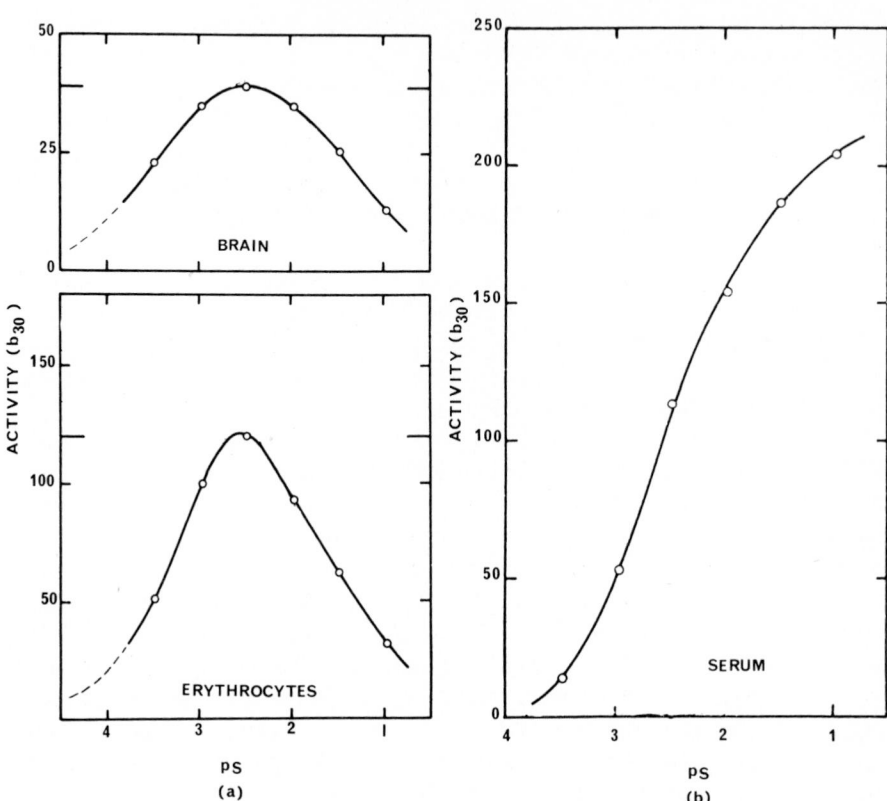

Fig. 2.9. The effect of substrate concentration on the velocity of hydrolysis of ACh by AChE and pseudocholinesterase. a) Activity-pS curves for hydrolysis of ACh by AChE of dog brain and cow erythrocytes. Bell-shaped curve indicates inhibition of the enzyme by excess substrate. b) activity-pS curve for hydrolysis of ACh by pseudocholinesterase from horse serum. Sigmoid curve indicates lack of inhibition by excess substrate. pS, $-$log molar concentration of ACh; b_{30}, enzyme activity (total hydrolysis minus non-enzymic hydrolysis) expressed as μl CO_2 evolved/30 min. (Modified from Augustinsson 1948.)

anionic sites in the active centre has not been confirmed and more recent theories of substrate inhibition invoke different factors. Krupka (1963, 1964) concluded that when present in excess, substrate molecules form a complex with the acylated enzyme by binding at the anionic site which is unoccupied at this stage (see § 2.4.2). This complex is deacylated only very slowly hence the inhibition. Krupka, while suggesting this as the mechanism for substrate inhibition of AChE, does not go on to speculate why pseudocholinesterase is immune from the effect. It seems likely that the reason must lie in fine distinctions in the arrangement and nature of groups in and around the anionic site. It should be pointed out here (but see also

Erythrocyte enzyme   Serum enzyme

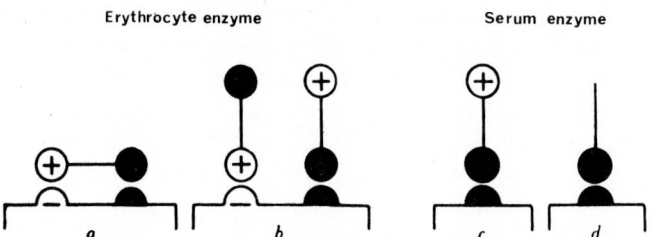

Fig. 2.10. Scheme proposed by Zeller and Bissegger to account for inhibition of AChE by excess substrate. Black circles, ester group of substrate; black semicircles, corresponding group on enzyme. White circles and semicircles, the respective positively and negatively charged groups of substrate and enzyme. a) normal binding of ACh with erythrocyte cholinesterase; b) binding conditions for erythrocyte enzyme with excess substrate; c) binding of ACh with serum cholinesterase; d) binding of aliphatic esters (e.g. methylbutyrate) with serum cholinesterase. (From Zeller and Bissegger 1943.)

ch. 10) that binding at the anionic site of pseudocholinesterase appears to be by Van der Waals' forces which, unlike the coulombic forces operating at the anionic site of AChE, do not favour formation of complexes. As discussed in §2.4.2, Krupka (1966b) has evidence that deacylation of AChE is controlled by a basic group in the vicinity of the anionic site. In pseudocholinesterase the activity of this group would not be so liable to interference by excess substrate molecules if these do not readily complex at the anionic site. Alternatively deacylation of pseudocholinesterase may not involve an equivalent basic group and the mechanism could be quite different. Brestkin et al. (1965, 1966) have a theory to account for substrate inhibition of AChE which differs from Krupka's idea. In their view substrate inhibition does not stem from the formation of a stable acylenzyme–substrate complex but from a decrease in the catalytic potency of the enzyme as a result of conformational changes induced by the presence of excess substrate. Other authors (for example Changeux 1966) have suggested that the binding of excess substrate molecules to allosteric sites (§2.7) could induce conformational changes sufficient to prevent normal catalysis. Kuhnen (1972) tentatively proposed that such an effect might be additional to any inhibition produced by an interaction of extra molecules at the active site. The involvement of allosteric sites in the mechanism of substrate inhibition is also discussed by Kato et al. (1972) and Kato (1972b).

2.5.3 Kinetics of AChE in situ

The question that must be asked is how closely the results obtained from purified preparations in vitro are related to the physiological situation. Substrate inhibition is obviously a very real phenomenon in the test-tube but what is its significance in terms of normal function? In 1960 Fehér and Bokri perfused the cat superior cervical ganglion with Locke's solution containing different concentrations of either ACh or acetyl-β-methylcholine (MeCh). They determined the rate of hydrolysis of the substrates by measuring the arteriovenous difference in substrate concentration by Hestrin's method or by bioassay. The results were compared with those for homogenized ganglia. With ACh as substrate the perfused ganglia gave rather scattered results while the homogenates gave the expected bell-shaped curve. MeCh also gave a bell-shaped curve with homogenates but in the intact tissue the curve was sigmoid like that for pseudocholinesterase, suggesting no substrate inhibiton was occurring. The authors took their results to indicate that the kinetics of AChE in vivo were not the same as those in vitro. There is, however, still a difference between their experimental 'in vivo' situation, in which the substrates are reaching the enzyme via the blood vessels, and the true physiological situation where ACh is released in the immediate vicinity of the enzyme. Some factor other than a difference in kinetics, for instance membrane permeability, may be responsible for the findings. Working with intact slices of rat brain, Hobbiger and Lancaster (1971) showed that the AChE activity increased with increasing substrate concentration whereas the activity of homogenized slices decreased over the same concentration range (0.5–8.0 mM). They concluded that the depth to which substrate penetrated the tissue, and hence the number of sites involved in the hydrolysis, was a function of substrate concentration.

Welsch and Dettbarn (1972b) found that AChE in the intact rat diaphragm did not show substrate inhibition even with a concentration of ACh of 5×10^{-2} M. Their suggestion was that substrate inhibition could be masked by slow but nevertheless significant hydrolysis by butyrylcholinesterase which is also present in the tissue. Some support for this idea comes from the finding that even with homogenized diaphragms substrate inhibition was not evident. This latter observation is, however, at variance with that of Goyer (1968) who demonstrated substrate inhibition with homogenates of both rat diaphragm and anterior tibial muscle. This discrepancy is a warning that differences in experimental design may affect kinetic values and in consequence it may prove extremely difficult to get a clear picture of the true physiological situation. One factor which may have a significant effect on results both in the test-tube and in tissue is the ionic strength. The effect of ions on substrate inhibition was mentioned earlier (§2.2.2). Mendel and Rudney (1945) showed that substrate inhibition of AChE extracted from mouse brain was partly prevented by K^+ and Na^+ (fig. 2.1). Results obtained by Gridelet et al. (1970) on AChE extracted from eel electroplax suggest these findings are not necessarily applicable to all types of AChE. Under the conditions of their experiments ACh in a concentration range of 1×10^{-3} to 5×10^{-2} M produced only slight inhibition in the absence of cations but with KCl in excess of 20 mM inhibition was marked and it was even more marked in the presence of 10 mM $CaCl_2$. Whether the difference in the results obtained by the two groups is due to experimental or to species differences is not clear. Results reported by Buckley and Heaton (1970) would seem to indicate that in intact tissue susceptibility of AChE to substrate inhibition does vary with species. They found that cholinesterase in end-plates from posterior latissimus dorsi muscle of chick suffered inhibition whether the preparation was intact or homogenized; with rat gastrocnemius muscle, on the other hand, substrate inhibition occurred only in homogenates. A point worth considering here is whether substrate inhibition is *likely* to occur under physiological conditions. On theoretical grounds it would seem a rather undesirable attribute of the enzyme. If the purpose of the enzyme is to prevent accumulation of transmitter the need would seem to be for greater rather than less efficiency when ACh is released in large quantities. Hellenbrand and Krupka (1970) pointed out that substrate inhibition might be an accident – a chemical consequence of the fact that once the rapid acylation step has occurred and choline has been released, the anionic site remains vacant during the slow deacylation phase and can accept cations. In invertebrates, the situation seems different from that in the vertebrates. Hellenbrand and Krupka found that for fly head AChE the slow step is acylation rather than deacylation. This means that the anionic site is not so readily available to excess substrate and, furthermore, evidence suggested that any cations which were bound to the acylated enzyme did not apparently prevent deacylation. Although both these factors argue against the likelihood of substrate inhibition of fly head AChE, Hellenbrand and Krupka showed it occurred strongly in their preparation. From this they inferred the existence of a second anionic site which binds excess cations and thereby interferes with the acylation step of catalysis. They went on to speculate that the apparent evolution of two separate mechanisms, one in invertebrates and the other in vertebrates, to mediate substrate inhibition might imply that it was not an accidental consequence of some other necessary attribute of the enzyme but served a definite purpose. It is still difficult to imagine what that purpose might be.

Barnard and his colleagues (Barnard and Rogers 1967; Rogers et al. 1969; Barnard et al. 1971a, b) have measured the amount of AChE at individual neuromuscular junctions in various muscles from different species. For mouse diaphragm, Barnard and Rogers took a figure of approximately 4,400 molecules AChE per μ^2 post-junctional membrane surface and calculated this to be equivalent to a maximum effective AChE concentration within the synaptic folds of 3.7×10^{-4} M. On the tentative assumption that values obtained for rat diaphragm were applicable to mouse, they used Krnjević and Mitchell's (1961) figure of 6×10^6 molecules as the amount of ACh released per impulse (this figure could be too high according to Mitchell and Silver 1963). After allowing for the restriction of the enzyme to the junctional folds they calculated the molar concentration of ACh in the vicinity of the AChE to be about 1.6×10^{-5}. Using this value, together with a K_M of 5×10^{-4} M, Barnard and Rogers estimated that 95% of the ACh released would be hydrolysed by AChE in 1 msec. In tonic muscles the number of molecules of enzyme per end-plate is smaller than in twitch muscles (Barnard et al. 1971a) and this difference in density of enzyme fits with the observation that the transmitter persists for some time in the more slowly acting tonic fibres. The calculations made by Barnard's group seem to indicate that the amount of AChE at neuromuscular junctions is only just adequate to cope with the ACh released by each impulse — there appears to be no surplus — and we come back to the question of whether substrate inhibition is liable to occur. In vitro, substrate inhibition does not become significant until concentration exceeds about 3×10^{-3} M and, according to the calculations above, the molar concentration at the junctional folds is appreciably lower than this, about 1.6×10^{-5} M. Of necessity, a large number of assumptions have to be made in calculations of this sort and it is dangerous to speculate from theoretical grounds about the possible build-up of ACh to inhibitory concentrations. One factor which might well upset any such estimation is that pseudocholinesterases (which do not suffer substrate inhibition) are also present at many neuromuscular junctions and may contribute to the hydrolysis of ACh. Barnard et al. (1971a) considered that at the end-plates in the diaphragm of the rat any part played by the pseudocholinesterase would be relatively minor but, as mentioned earlier, Welsch and Dettbarn (1972b) believed it could be significant. In histochemical experiments on the chicken (Silver 1963), I was unable to detect pseudoChE activity in end-plates of certain muscles although in other muscles the end-plates reacted strongly. Muscles which did not react included posterior latissimus dorsi; as mentioned above Buckley and Heaton (1970) found that substrate inhibition was demonstrable in this muscle while it was not demonstrable in intact rat gastrocnemius. In trying to decide whether or not pseudoChE masks substrate inhibition it might be useful to compare, under the conditions used by Welsch and Dettbarn (1972b), the reaction to excess substrate of those chicken muscles whose end-plates contain only AChE and those which contain both AChE and pseudocholinesterase. In such a comparison inhibitors would not be needed to distinguish between the activity of pseudo-cholinesterase and acetylcholinesterase. This would be an advantage because if inhibitors have to be used it is not entirely safe to assume that the non-inhibited enzyme is behaving completely normally (see ch. 3). Pseudocholinesterase occurs together with AChE not only at some end-plates but also in certain areas of the CNS (see ch. 10). Its role there is as obscure as it is elsewhere; the possibility that it takes over should AChE suffer substrate inhibition has been mooted but not tested (see Silver 1967).

The weight of evidence suggests that in the particular case of substrate inhibition the results of in vitro experiments on purified preparations may not be a good indication of the situation in intact tissue. It seems important to extend this discussion to other aspects of enzyme behaviour. A vast literature is available on the reaction of purified AChE towards substrates and inhibitors under a wide variety of in vitro conditions. This knowledge has helped to provide a picture of the structure and potential of the enzyme but we must not lose sight of the fact that very many of these data refer to substrates and conditions which could not normally be encountered by the native enzyme. Kinetic data from intact tissues are sparse compared with those for purified preparations but it is on such systems that experiments must be done if the physiological rather than the enzymological properties are to be elucidated.

Namba and Grob (1968) prepared isolated enzyme-bearing membranes from rat intercostal muscles and found the pH optimum to be 8, the turnover number (molecules ACh hydrolysed per min per active site) 1.61×10^5 and the K_M 3.1×10^{-3} M. This last figure is an order of 10 higher than the figures generally found for purified enzyme but has been borne out by subsequent work. Mittag et al. (1971a, c) working on intact rat diaphragm obtained values of from 4.7×10^{-4} M to 1.54×10^{-3} M and of 1.05×10^{-3} M respectively; the value for intact guinea-pig ileum measured by a similar radiometric method (Ehrenpreis et al. 1970) was 1.8×10^{-3} M (Mittag et al. 1971b). These workers (see Ehrenpreis et al. 1971) also found other differences between tissue-bound AChE and AChE in solution. Some of these are listed in table 2.8.

These findings led to the conclusion (Mittag et al. 1971a, b; Ehrenpreis et al. 1971) that there are inherent differences between the kinetics of AChE in situ and in solution; Furthermore, extra-enzymic factors that are not operative in solution may contribute to these differences. In the guinea-pig ileum, for instance (Mittag et al. 1971b), a jump in temperature from 30 to 37°C increased AChE activity by 60%

TABLE 2.8

Comparison of some properties of rat diaphragm acetylcholinesterase in solution and in situ. (From Ehrenpreis et al. 1971.)

	AChE in solution (I)	AChE in situ (II)
K_M (acetylcholine)	$3–5 \times 10^{-4}$	$2–3 \times 10^{-3}$
Effect of edrophonium	Competitive inhibitor	Non-competitive inhibitor
Effect of neostigmine	Competitive inhibitor plus carbamylating agent	Non-competitive inhibitor
Rate of phosphorylation by *DFP	I is 10–40 times greater than II	
Rate of reactivation by *PAM	I is several times faster than II	
Effect of d-tubocurarine	Inhibitor (allosteric)	Activator
Effect of decamethonium	Inhibitor (allosteric)	Activator plus inhibitor (incomplete)
Effect of succinylcholine	Inhibitor	Activator plus inhibitor (incomplete)
Effect of *TRIEG	Inhibitor (allosteric)	Inhibitor
Effect of ambenonium	Inhibitor	Inhibitor (incomplete)

* See abbreviations list.

but a sudden decrease from 37 to 30°C caused a reduction of only 6%. Mittag et al. suggested this might be due to alterations in the membrane (perhaps by allosteric effects) which occur quickly when temperature goes up but which are not rapidly reversed by a decrease. Recently Young (1973) has published figures indicating that the AChE on membranes exposed to fluctuations in pressure (as in the heart) could be subject to conformational changes of a magnitude sufficient to alter hydrolytic activity. Using AChE attached to artificial membranes, Silman (1969) found the curve for activity versus pH was S-shaped and not bell-shaped. He suggested this was a result of local decreases of pH in the vicinity of the membrane-bound enzyme, the decreases being caused by the enzyme activity (see also Katchalski et al. 1971).

Species differences are another factor to be kept in mind when considering the relevance of data obtained in vitro to the physiological behaviour of the enzyme in an experimental animal. Electroplax from electric fish are a useful source of AChE for purification and study but some characteristics of this enzyme may differ from those in mammals. For instance, Alid and Orrego (1972) obtained kinetic data indicating differences in the conformation of the active site of AChE from rat brain and from *Electrophorus* electroplax. Similarly Blaber and Cuthbert (1962) suggested that mammalian and avian AChE might differ in the anionic site since the chicken enzyme did not react with bisquaternary substrates and mammalian enzyme did. The findings of Lee and Pickering (1967) that brain cholinesterase of goose, duck and hen all have dissimilar rate constants for reactivation following inhibition by Haloxon (see ch. 11) are an indication that even within a class the active site of the enzyme may be different. In discussing differences in the behaviour of AChE of chicken and frog brain towards various inhibitors Andersen et al. (1972) advanced a number of alternative explanations including the possibility that frog brain enzyme is not subject to the same conformational changes as is chicken brain AChE. Not surprisingly, there is evidence (see ch. 5) of considerable differences between the acetylcholinesterase of vertebrates and invertebrates. Welsch and Dettbarn (1972b) concluded from experiments with inhibitors that the nature of the binding at the esteratic site of AChE from rat diaphragm is not the same as that of AChE from lobster peripheral nerve. Taken together these observations indicate that any physiologist interested in the possibility of cholinergic function in a particular tissue of a particular species requires data on the behaviour of AChE in that specific situation and that he would be especially unwise to extrapolate from data obtained in vitro from a different tissue and species.

2.6 Relation of AChE to the acetylcholine receptor

One of the most active lines of recent research in molecular pharmacology has been the quest for the acetylcholine receptor with the attendant controversy over its relation to AChE (see Watkins 1965; Karlin 1967, 1969; Namba and Grob 1967;

Changeux et al. 1969; Belleau 1970; Garland and Durell 1970; Smythies 1970; Changeux et al. 1971; O'Brien et al. 1972).

Various methods have been tried. Some workers have used the technique of affinity labelling (see Wofsy et al. 1962 for the principle) to study the properties and number of cholinergic binding sites in excitable membranes from electroplax and muscle. The method employs compounds such as p-(trimethylammonium) benzenediazonium fluoborate (TDF, see Wofsy and Michaeli 1967), decamethonium and d-tubocurarine which have a high affinity for cholinergic sites. If these are isotopically labelled their binding to membranes can be directly followed under different experimental conditions; indirect methods are also available for tracing unlabelled compounds. Papers dealing with the affinity labelling technique include those of Hassón-Voloch (1968); Takagi and Takahashi (1968); Changeux et al. (1969); Karlin (1970); Podleski and Changeux (1970); and O'Brien et al. (1971). A disadvantage of the method is its lack of specificity for, as Changeux et al. (1970a) have pointed out, several classes of sites in excitable membranes will bind these types of compound. An approach often used in conjunction with affinity labelling has been the attempted extraction of the receptor protein from excitable membranes (see for example Chagas et al. 1958; Ehrenpreis 1959; Namba and Grob 1967; Changeux et al. 1970a; De Robertis et al. 1970; La Torre et al. 1970; Lunt et al. 1971; Miledi et al. 1971). Although proteolipid preparations have been obtained which have a capacity to bind cholinergic agonists and antagonists it is very far from certain that these preparations are in any way equivalent to cholinergic binding sites in situ (see for instance Levinson and Keynes 1972).

It is not difficult to visualize a finite protein component of the excitable membrane which on reacting with ACh undergoes conformational changes thereby altering membrane permeability. Permeability changes are measurable and can be defined in physical terms but what they represent in terms of structure is conjectural. Among various theories (see for example Changeux 1969; Changeux et al. 1969; Smythies 1970) are those invoking little pores or gates, the ionophores, which are in close spatial relation to the receptor molecule and which open or shut depending on the conformation of the 'receptor'. Speculation along these lines is very tempting but it is important to recognize that, as yet, the receptor is an idea rather than an entity. Heilbronn made this point in 1970(b) at Skokloster when she said 'I think we should try to find out whether there is really a thing called the receptor that can be isolated or whether that is totally impossible. I believe it may turn out to be a word used to explain something that happens, but not a substance that can be isolated, or at least it may be difficult to identify isolated pieces of tissue as parts of a receptor'. Cavallito (1970) also speaking at Skokloster sounded much the same note of caution, saying 'To attempt isolation of something that has subtleties through which it can undergo minor changes and induce tremendous responses and to expect to tear this from its roots and cast it into a very strange environment and expect it to behave in the same fashion, I think is a more difficult project'.

Although, as Heilbronn suggests, 'receptors' may be concepts rather than sub-stances, receptor theory must influence our thinking about the role of AChE at a cholinergic synapse. In the very simplest terms we think of acetylcholine crossing from pre-synaptic endings to the post-synaptic membrane where it reacts with the receptor and induces permeability changes. When these changes have occurred it is imperative that the ACh is removed from the receptor so that the status quo can be restored and the synapse made ready to accept further impulses. It is assumed that the eventual destruction of ACh depends on AChE activity but how is acetylcholine disengaged from the receptor? When we look at the simple diagram depicted in fig. 2.3 we are considering only two things — the active surface of the enzyme and its substrate. What happens if the substrate is already attached to a third component, namely the cholinergic receptor? One possibility which has been mooted is that there is no third component, AChE itself being the receptor. This idea has a long history. As early as 1937 Roepke proposed that cholinesterase could be used as a convenient model for all receptors. Roepke is often quoted in the literature as suggesting *AChE* as the ACh receptor but this is inaccurate: AChE had not been distinguished from pseudocholinesterase at the time of Roepke's proposal and in any case his suggestion stemmed from work on cholinesterase in serum. In 1953 Župančič extended the idea suggesting that the receptor protein for ACh is not merely similar to but is identical with AChE. Since both the receptor and the active centre of AChE must on current theory be capable of binding acetylcholine it seems logical to expect some degree of structural similarity. It is this very point, their similarity, which has made it so difficult to establish what is the precise relation of receptor to enzyme. Some authors, for example Župančič (1967) who also cites Karassik (1946), Ehrenpreis (1967), Leuzinger and Schneider (1972) and a number of others (see Karlin 1967) have subscribed to the view that while the esteratic site of AChE is not involved in receptor activity some other part of the macromolecule, possibly the anionic site, may be common to enzyme and receptor. Despite some evidence in favour of the hypothesis (see below and also discussion by Changeux et al. 1969) current opinion tends to reject the idea that the AChE macromolecule is both enzyme and receptor. Evidence cited against the possibility (see also Belleau 1970; Garland and Durell 1970) includes the following findings. One of the toxins (the α-bungarotoxin) from the venom of the snake *Bungarus multicinctus* is a specific irreversible blocker of the acetylcholine receptor in certain muscles (Chang and Lee 1963) and in the electroplax of *Electrophorus electricus* (see Changeux et al. 1971), but it is without effect on the activity of purified AChE (Kasai and Changeux 1971). This means that the anionic site of the enzyme is still available to the substrate at a time when the binding site of the receptor is irreversi-bly blocked. In other words they do not seem to be the same thing. Further support for their separate identity comes from the finding that receptor activity, as judged by a response to applied ACh, is retained in muscles after AChE activity at the end-plate has been destroyed by proteases (Albuquerque et al. 1968). This latter evidence could be refuted, however, on the grounds that the loss of hydrolytic activity could be due to changes restricted to the esteratic site. Kinetic constants appear to be different for the ACh-receptor complex and AChE; in electroplax the dissociation constant of the former is very much below the K_M for AChE (see for instance, Nachmansohn 1964). Furthermore, neither the K_M nor the V_{max} of electroplax AChE is altered by dithiothreitol or p-chloromercuribenzoate both of which abolish the electrical response to ACh (Karlin 1967). Both Bartels (1968) and Podleski (1969) tested various compounds for their ability to activate receptors and to inhibit AChE activity in electroplaque membranes and they found the potency differed in the two systems. For example, Bartels using a series of neostigmine analogues reported an inverse relation between their effectiveness as receptor activators and enzyme inhibitors. Another of her findings which favours the view that enzyme and receptor action involve different sites was that acetyl-β-methylcholine, while highly stereospecific for AChE, did not depolarize the membrane.

De Robertis and his co-workers (see De Robertis et al. 1970) are among the many engaged in attempts to isolate the ACh receptor from membranes of electroplax. They have shown that on fractionation of membranes the AChE activity and the binding capacity (as measured with $[^{14}C]$-methylhexamethonium) parallel each other to some extent but there are also differences. For instance, in sucrose gradients the AChE activity in 1.0 M sucrose is 4.8 times lower than that in 0.8 M sucrose but the binding is only 1.7 times lower. In addition, when the AChE was solubilized and all but 10% removed from the membranes their binding capacity was equal to that of AChE-containing control membranes. Miledi et al. (1971) have studied receptor protein obtained from electroplax of *Torpedo marmorata* by solubilization with the non-ionic detergent Triton X-100. They found that although the number of α-bungarotoxin-binding sites per gram equalled the number of active sites of AChE (a similar equality has been reported in *Electrophorus* by Changeux et al. 1970b and in mouse diaphragm by Barnard et al. 1971b) enzyme and binding protein were chemically separable. Furthermore, studies on the molecular weight do not indicate a common subunit.

Taken together all these findings provide increasing support for the view that although AChE and the ACh receptor may be in a very close spatial relation to each other, perhaps forming a mosaic in the post-synaptic membrane (Barnard et al. 1971b) they are, nevertheless, separate macromolecules. What of the contrary view (Župančič 1970; Leuzinger 1971a; Leuzinger and Schneider 1972; Štalc and Župančič 1972; Župančič et al. 1972), that the ACh receptor site and the ACh hydrolysing site are present on the same macromolecule but on different parts, possibly different polypeptide chains? Štalc and Župančič (1972) have evidence that if α-bungarotoxin is tested on AChE bound to end-plates instead of on the purified preparation then an effect on AChE, an increase in activity, is detectable. They invoke this to support their contention that the enzyme is present on the same macromolecule as the receptor. Furthermore, they suggest that during purification the responsiveness of AChE to α-bungarotoxin is lost because its allosteric properties (see §2.7) are altered. This experiment re-emphasizes the difficulty of trying to deduce from work in vitro the exact situation in vivo. In view of this, it may still be premature to accept unreservedly that the ACh receptor and AChE are totally separate macromolecules.

If, as a working hypothesis, we do take it that AChE and the ACh receptor are not part of the same macromolecule what mechanisms might operate to ensure hydrolysis of ACh after it has effected changes in membrane permeability? First, the ACh-receptor complex might break down spontaneously and the free ACh could then present itself to the enzyme for hydrolysis. According to this scheme ACh would be involved in four separate steps during the excitatory process: binding to the receptor with initiation of permeability changes, dissociation from it, attachment to the enzyme and hydrolysis. As Župančič (1953) pointed out, the time-course of such a sequence would be slow and this was one of the considerations which led him to suggest that AChE itself could be the receptor.

Another, faster, sequence could be visualized if one takes account of the evidence that receptor-substrate-enzyme interactions are likely to result in appreciable conformational changes not only in 'ionophores' but in many other molecules. One could speculate that when the ACh interacts with the receptor the conformational changes responsible for opening the pores simultaneously orientate the ACh molecule in such a way that groups not bound to the receptor are presented to AChE occupying the adjacent part of the membrane. The feasibility of this sort of scheme hinges on the shape the ACh molecule assumes when part is bound to the receptor.

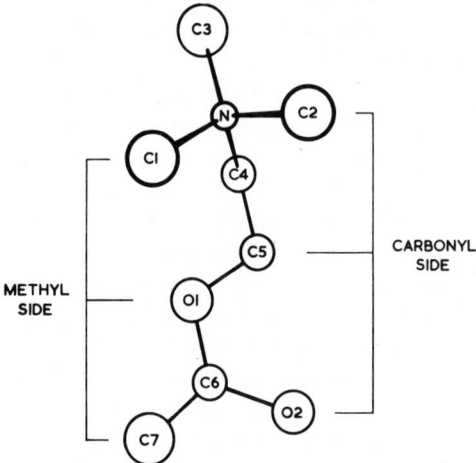

Fig. 2.11. Acetylcholine in the conformation relevant to muscarinic and nicotinic receptors in nervous tissue. (From Chothia 1970.)

Can it still fit the active surface of the enzyme? Evidence available suggests there could be difficulties in this respect. Chothia and his co-workers (see Chothia and Pauling 1970; Pauling 1970) have studied the crystal structures of various 'cholinergic' compounds and have concluded that the conformation of acetylcholine relevant to its reaction with AChE (Chothia and Pauling 1969) is not the same as that relevant to its reaction at muscarinic and nicotinic sites. Furthermore, although the conformation of ACh at nicotinic and muscarinic receptors seems to be the same (Chothia 1970) different *groups* on the ACh molecule are apparently involved at the two types of site. Chothia recognizes a 'methyl' side and a 'carbonyl' side to the ACh molecule (fig. 2.11) and by comparing the structures of specific agonists concluded that it is the methyl side which activates muscarinic receptors and the carbonyl side which activates nicotinic ones. Smythies (1974) has specified the possible molecular structure of cholinergic receptors in terms of the amino acid sequences concerned in 4 strands of the receptor protein: by using his models he has independently reached much the same conclusion as Chothia, that when bound to a nicotinic receptor ACh is 'upside down' compared with its position at the muscarinic receptor. Given this situation it is difficult to see how AChE could 'prise' the substrate off the receptor binding site, at least by a mechanism common to both types of receptor. Does this mean that we have to return to the idea of the 'slow' scheme by which the ACh-receptor complex must break down spontaneously before hydrolysis can occur? Obviously, much more must be done before the problems can be fully understood and this includes the need to establish beyond doubt whether or not the receptor is totally independent of the macromolecule of AChE.

Before leaving the subject of the receptor one other point must be raised, the question of its evolution. Work which has suggested the separateness of receptor and enzyme has also revealed similarities (see Belleau 1970). The possible evolutionary basis for this situation has been briefly discussed by Belleau and Di Tullio (1970). They postulated that during the development and specialization of the cholinergic nervous system gene duplication and mutation may have operated so that part of the enzyme became modified in such a way that the esteratic site was lost but the part capable of binding acetylcholine by cationic—anionic interaction was retained. This seems an acceptable speculation: if it is assumed that the primitive organism was already endowed with an efficient means of binding ACh during hydrolysis, appropriation and modification of this mechanism, rather than the de novo formation of a totally different ACh-binding receptor, would seem a likely step in development. This is line with Pantin's suggestion (1956, see ch. 5) that during the evolution of the nervous system the chemicals which acquired functional importance were ones which happened to be present in the cells and were not, as had been suggested by Bullock and Nachmansohn (1942), specially developed for the purpose.

2.7 Allosteric effects

In discussing the reaction between AChE and ACh the only binding sites so far considered have been the anionic and esteratic subsites of the active centre. It has, however, become increasingly clear over the last decade that the capacity of enzymes to bind compounds (or 'ligands', see glossary for definition) is not solely confined to the active site. Largely as a result of pioneering work at the Institut Pasteur it is now established that enzymes possess so-called 'allosteric' sites. The general principles of allosterism have been broadly discussed in two papers by Monod et al. (1963, 1965) and the basic concept is given very clearly in the former. 'These [controlling] proteins are assumed to possess two, or at least two stereo-specifically different, non-overlapping receptor sites. One of these, the *active site*, binds the substrate and is responsible for the biological activity of the protein. The other, *allosteric site* is complimentary to the structure of another metabolite, the *allosteric effector*, which it binds specifically and reversibly. The formation of the enzyme-allosteric effector complex does not activate a reaction involving the effector itself: it is assumed only to bring about a discrete reversible alteration of the molecular structure of the protein or *allosteric transition*, which modifies the properties of the active site changing one or several of the kinetic parameters which characterize the biological activity of the protein'. Monod et al. (1963) consider in some detail the need for a mechanism which can modify the activity of an enzyme according to the overall requirements of the organism. They point out that neither the operation of the law of mass action nor mere competition for common substrates is adequate to regulate enzyme activity in response to changing demands. They go on to say that 'for the chemical activities of a cell to be precisely adjusted to its own requirements, adapted to the environment and directed towards the performance of a particular function, the specific activity of those proteins which are responsible for critical metabolic steps must be altered electively in response to the presence of certain metabolites playing the role, not of substrates for the reaction in question, but of chemical signals'.

In essence then, an allosteric site is a binding site which is distinct from the active centre. Interaction between this site and the relevant 'regulator' induces conformational changes in the tertiary structure of the enzyme which in turn alter the conformation and hence the performance of the active site. Wilson (1971) argued that since all proteins undergo some conformational changes on interacting with a small molecule, every protein could be regarded as allosteric. In practice, however, the allosteric effects may be minimal in terms of the ultimate physiological or pharmacological performance of the enzyme and the question we have to consider is whether or not any allosterism shown by AChE is essential to its function. Firm evidence that AChE possesses more than one type of binding site came initially from the work of Changeux (1966) on extracts of AChE from *Torpedo marmorata*. Changeux studied the effect on ACh hydrolysis of different neuromuscular blocking agents believed to inhibit AChE by competition for the anionic site. Drugs tested included the 'leptocurares' (see glossary) such as decamethonium and 'pachycurares' e.g. d-tubocurarine and gallamine (Flaxedil). He found that the degree of inhibition was affected by salt concentration and that while pachycurares decreased the affinity of the enzyme for some anionic site inhibitors, they increased the affinity for others. These and other results indicated that the pachycurares were binding to some site other than that to which ACh bound. In contrast, decamethonium gave results suggesting that it did bind to the same site as ACh, i.e. to the anionic site in the active centre, but it also bound (decamethonium is bivalent) to another, separate, anionic site. Although the effects of salt concentration on the sedimentation characteristics of AChE showed that the enzyme was capable of considerable conformational change, Changeux was hesitant about attributing differences in enzyme activity under various conditions to such alterations in structure. Nevertheless, he pointed out how the allosteric model (Monod et al. 1965) could be applied to AChE. Results produced by the different blocking agents and also the phenomenon of substrate inhibition (see §2.5.2) could be explained if it were assumed that the enzyme exists in at least two conformational states in equilibrium and that these states differ in affinity for quaternary groups.

Kitz et al. (1970) have done somewhat similar experiments to examine the effect of neuromuscular blockers on the rate of inhibition of *Electrophorus* AChE by carbamates such as physostigmine (ch. 11). They showed that while some blocking agents increased the rate of inhibition and also of spontaneous recovery of AChE activity, other agents lowered these rates. Like Changeux, Kitz et al. concluded that their observations could be best explained by the presence of an allosteric site which was capable of binding curare-like blockers and of inducing conformational changes at the active centre. Changeux's original premise did not specify the number of allosteric sites related to each active site. Recent experiments by Roufogalis and Quist (1972) have demonstrated the existence of at least 3 bindings sites on the surface of bovine erythrocyte AChE. From analyses of the kinetics of interactions between various ligands they suggest that some compounds, e.g. tetramethylammonium, bind exclusively to the catalytic anionic site which they designate α. Decamethonium (and probably tetraethylammonium) bind to the α site and, in addition, to an allosteric site, β. Calcium also acts at β and it is there that it competes with decamethonium. On the other hand, since neither calcium nor tetraethylammonium nor tetramethylammonium interfere with gallamine binding it is suggested that the latter must react at a second allosteric site, γ. Occupation of this site by gallamine brings about conformational changes which affect conditions both at the catalytic site and at the other allosteric site, β.

In addition to these types of experiment (see also Patočka and Bajgar 1971; Wilson 1971; Kuhnen 1972) in which the results of ligand interactions are interpreted in terms of allosterically induced structural changes, other experiments have given more direct evidence that AChE is capable of alterations in conformation. Kitz and Kremzner (1968) have studied a purified preparation of *Electrophorus*

AChE with a spectropolarimeter. Their data indicated that the AChE macromolecule exists mainly as a right-handed helix but that the conformation altered when the enzyme was heated and also when it reacted with a substrate (acetylhomocholine chloride), anticholinesterase agents (TEPP and Tensilon, see ch. 11) or 0.1 N KOH. Results from nuclear magnetic resonance studies (NMR) can also be explained on the basis of binding sites additional to those in the active centre. Kato et al. (1970, see also Kato 1972a, b) looked at the changes in NMR produced when AChE from squid head ganglion reacted with atropine and eserine. Their evidence was that atropine, a very weak inhibitor of AChE, was bound at a site other than the catalytic site where eserine interacted. Further work (Kato and Yung 1971) with atropine and some of its analogues gave a clue to the nature of this non-catalytic site. It would appear to have 3 subsites: an electrostatic binding site, a hydrophobic site and a hydrogen-bonding site. Kabachnik et al. (1970) have suggested that the number and dimensions of hydrophobic sites in AChE and pseudoChE are different (see ch. 10).

This body of evidence seems to make a convincing case for the classification of AChE as an allosteric protein. That being so it is logical to invoke the 'regulatory' attributes of allosterism (see Monod et al. 1963, 1965) when considering the physiological role of AChE. One might assume that enzyme activity in vivo is regulated according to need by some sort of feedback via the allosteric sites. The difficulty about making such an assumption is that most of the data on allosterism have come from in vitro studies on enzymes of some degree of purity. More work must be done to determine whether observations made in vitro have a relevance to the situation in vivo. Some findings already available suggest that, even in vitro, membrane-bound and solubilized AChE may behave differently. Thus, whereas Changeux (1966) working with purified soluble AChE from *Torpedo*, and Roufogalis and Quist (1972) using purified bovine erythrocyte AChE, found that calcium interfered competitively with binding of decamethonium, no such antagonism was found by Wins et al. (1970) with membrane-bound AChE of *Torpedo*. It remains to be seen whether this discrepancy represents a true difference in behaviour between the two types of preparation or whether it stems from differences in experimental conditions. In the physiological situation enzymes are not working alone but interdependently with a number of other systems. Disagreement between results in vitro and in vivo might be explained in a number of ways, i) the mechanism of enzyme action is different in the two cases; ii) the basic mechanism is the same in both cases but conditions in intact tissue affect it in a way which is not mimicked in the test-tube; iii) the mechanism itself may be acting similarly under the in vitro and in vivo conditions but because, in vivo, additional systems are simultaneously reacting, the end result may be different. Reference was made earlier (§ 2.5.3) to the finding by Welsch and Dettbarn (1972b) that substrate inhibition of AChE could not be demonstrated in the intact diaphragm. A possible explanation given by the authors is that BuChE, also present in the tissue, participates in the hydrolysis at high substrate concentrations and thereby masks any inhibiton of AChE. Here is but one example of the complications introduced when studies are made on intact systems.

CHAPTER 3

Methods for the study of cholinesterases

3.1 Introduction

Cholinesterases impinge on a range of disciplines which include, among others, biochemistry, physiology, pharmacology, medicine and anatomy. In consequence, they have been studied from many angles and by many methods, hence this chapter must be regarded simply as a guide to some of the techniques available; it cannot claim to be a comprehensive catalogue. Old methods as well as new are included since this information can be helpful in the interpretation of earlier papers.

In 1967 Glick asked that something should be done to discourage the uncritical use by authors of the terms 'biochemical', 'histochemical' and 'staining'. He made the point that since histochemistry is a form of biochemistry, phrases such as 'histochemical evidence was confirmed biochemically' are unacceptable. He also deplored the tendency of authors to use the word 'histochemistry' solely in the sense of 'tissue staining' and to ignore its equal relevance to techniques such as X-ray absorption which also allow a correlation between histology and chemistry. Glick's protest was completely justified but it came too late to oust what he described as 'the already entrenched jargon'. Wrong though it is, the common practice is still to use 'biochemical' for quantitative studies, including the measurement of enzyme activity in intact tissue, and to use 'histochemical' for techniques which involve a staining reaction. On this basis, the chapter is divided into two main parts. The first deals with the estimation of cholinesterase activity and the study of its properties by quantitative biochemistry. The second part describes histochemical methods for the demonstration of the enzymes at the optical and electron microscopical level. The appendix includes practical details of some histochemical techniques and indicates methods whereby the examination of cholinesterases can be combined with the examination of catecholamines or choline acetyltransferase.

3.1.1 Choice of substrates and inhibitors

3.1.1.1 Choice of substrate   The final, choice of substrate depends to some extent on the purpose of the study and on the method. Requirements for the approximate estimation of the total cholinesterase activity in a tissue homogenate can be

TABLE 3.1

Type of substrate	Type of inhibitor	Conclusions from	
		A) Positive result	B) Negative result
1 'Specific' for cholinesterases	None	One or both types ChE probably present	ChE's probably absent
2 " "	10^{-5} M eserine	Non-specific esterases present, specificity of substrate therefore, in doubt	Confirmation of conclusion 1A
3 'Specific' for AChE	None	ChE's present, probably AChE	AChE probably absent
4 " "	Anti-BuChE	AChE very probably present	AChE probably absent but if 3 gave a positive result the specificity of inhibitor (or substrate) is to be doubted
5 " "	Anti-AChE	Potency or specificity of inhibitor (or specificity of substrate) is to be doubted	Confirmation of conclusion 4A
6 'Specific' for BuChE	None	ChE's present; probably BuChE	BuChE probably absent
7 " "	Anti-AChE	BuChE very probably present	BuChE probably absent but if 6 gave a positive result the specificity of inhibitor (or substrate) is to be doubted
8 " "	Anti-BuChE	Potency or specificity of inhibitor (or specificity of substrate) is to be doubted	Confirmation of conclusion 7A

different from those for a precise study of enzyme kinetics or for histochemistry. Whatever the problem, however, and whatever the method, it is necessary to establish the specificity of the enzyme or enzymes under examination. Cholinesterases must be distinguished from non-specific esterases, and acetylcholinesterases from butyryl- or other pseudocholinesterases. These distinctions are made by the use of the appropriate combination of so-called 'specific' substrates and 'specific' inhibitors in the correct concentrations (table 3.1). The important point to realize is that the specificity of these compounds may be relative rather than absolute, particularly in some species.

Acetylcholine, which is considered to be the main, but not necessarily the sole, natural substrate for AChE, is hydrolysed by BuChE as well; it cannot therefore be used in the biochemical estimation of AChE activity unless any BuChE present has been selectively inhibited (see below). It is the substrate of choice, however, for the measurement of total cholinesterase levels and it should be used as the standard of comparison when 'specific' substrates are used; Burgen and Chipman (1951; see also Sperti et al. 1960a) give the equations needed for making such a comparison (tables 6.9.1 and 6.15.1). ACh is also ideal for testing the properties of purified AChE but it is not suitable for histochemistry since its hydrolysis products are soluble. In some species including birds (Augustinsson 1948; Shute and Lewis 1963; see also ch. 5 re insects and mites) the enzyme in nervous tissue and erythrocytes, that is the enzyme which by analogy with other species would be considered as a true or acetylcholinesterase, hydrolyses propionylcholine (PrCh) as fast as ACh, or faster.

The 'specific' substrate most generally used for biochemical measurements of AChE is acetyl-β-methylcholine (MeCh; Mechoyl, Merck Ltd), which is a synthetic substance. It was first used by Mendel et al. (1943) who showed that whereas the 'true' cholinesterase of mammalian erythrocytes and brain hydrolysed it very rapidly, mammalian pseudocholinesterase and simple esterases did not attack it. Unfortunately this high degree of specificity does not hold for the cholinesterases of all species. In the chicken, for example, the serum pseudocholinesterase (identified by its action on butyrylcholine, BuCh) hydrolyses it very actively (Earl and Thompson 1952a; Myers 1953; Blaber and Cuthbert 1962).

The two substrates most commonly used in biochemical work on the pseudocholinesterases are butyryl- and benzoylcholine (BzCh). Neither is absolutely specific since both may be attacked to a greater or lesser extent by AChE (Adams 1949; Ord and Thompson 1952; Bayliss and Todrick 1956) and also by eserine-resistant esterases (Myers 1953). In addition, species variation in specificity patterns for pseudocholinesterase is marked (see ch. 10). Ruminant serum pseudocholinesterase, for example, will hydrolyse butyryl- but not benzoylcholine (see Mendel and Myers 1955) and in some invertebrates (ch. 5) activity towards the two may be very different.

The rate at which the cholinesterases hydrolyse specific substrates is usually lower than that for ACh; there are, however, some exceptions to this generalization,

particularly among the invertebrates. The thioesters of choline which were originally introduced as histochemical substrates but which are used biochemically as well,
are, in contrast, hydrolysed faster than is ACh (see for example Hobbiger and
Lancaster 1971). Since experimental conditions influence the absolute rate of
hydrolysis by a given amount of enzyme (see Burgen and Chipman 1951), the
direct comparison of results obtained by different methods and with different
substrates is rarely justifiable.

As yet, no really specific substrate has been developed for the histochemical
demonstration of AChE, the hydrolysis product of MeCh being soluble like that of
ACh. Acetylthiocholine (AcThCh) is relatively specific for cholinesterases as distinct from simple esterases and in some species hydrolysis by pseudoChE is slight.
Under the conditions which Hobbiger and Lancaster (1971) used for investigating
cholinesterases in rat brain homogenates, less than 1% of the hydrolysis of AcThCh
was due to pseudocholinesterase. In other species, for example the cat, the substrate specificity is rather less good. The preferred substrate for pseudocholinesterase in histochemistry is butyrylthiocholine (BuThCh). In some species (e.g. rat) it
does not appear to be attacked by AChE (see Shute and Lewis 1963; Klingman et
al. 1968) but as a precaution it should be tested with an inhibitor of AChE to make
sure that this holds for the particular species under investigation. Other histochemical substrates, including thiolacetic acid, are considered further in §3.3.2.5.

3.1.1.2 Inhibitors   It has been emphasized already that because no substrates are
absolutely specific for either AChE or pseudocholinesterase, inhibitors must be
used to aid identification. The inhibitor normally used to distinguish cholinesterases
from the other esterases is physostigmine (eserine, see ch. 11). While cholinesterases
in most species (again the invertebrates may provide exceptions, see ch. 5) are
totally inhibited by a concentration of 10^{-5} M, the concentration needed to inhibit
non-choline esterases is about 1000 times higher (Easson and Stedman 1937;
Mendel et al. 1943).

In the early days of histochemistry DFP (di*iso*propylphosphorofluoridate, see
ch. 11) was commonly used to distinguish between AChE and BuChE. In most
species BuChE is at least 10 times more sensitive than AChE to inhibition by DFP
but the sensitivities may overlap and the concentration needed to inhibit BuChE
completely may cause some inhibition of AChE as well (see fig 11.1). Species vary
considerably in their sensitivity to DFP. Ecobichon and Comeau (1973) showed an
800-fold difference in the I_{50} value (see glossary) for the serum cholinesterase of
man and goat; furthermore their value for the latter is more than 2000 times that
reported by Mendel and Myers (1955) for pseudocholinesterase in the spleen of ox,
another ruminant. DFP is a very toxic irreversible inhibitor and since it is volatile
and can be absorbed through the intact skin it represents a very real hazard in the
laboratory unless handled correctly. Because of this its use should be avoided if at
all possible.

BuChE can be more selectively inhibited by a number of compounds (ch. 11). Among the irreversible organophosphorus inhibitors the most widely used are iso-OMPA (tetramono*iso*propylpyrophosphortetramide) and Mipafox (*N,N'*-di*iso*-propylphosphorodiamidic fluoride). Holmstedt (1957a) found the latter to be excellent as a specific inhibitor of pseudocholinesterase in human serum (see fig. 11.1) and in cat brain, but on rat enzyme it seems less effective (Pepler and Pearse 1957; Naik 1963). Ethopropazine methosulphate or hydrochloride (the latter is sold as Lysivane, 10-(2-diethylaminopropyl)phenothiazine hydrochloride) are potent reversible inhibitors and are much safer to use than the organophosphorus compounds. Klingman et al. (1968) have reported slight inhibition of AChE by Lysivane at a concentration which gives 100% inhibition of BuChE in rat sympathetic ganglia, but Bayliss and Todrick (1956) found that a concentration which gave 90% inhibition of the pseudocholinesterase (BzCh as substrate) in rat intestinal mucosa actually caused a slight activation of AChE in the brain. A disadvantage of both Lysivane and the ethopropazine methosulphate is their tendency to produce cloudiness in the presence of substrate. This is more pronounced with the latter salt and can generally be avoided with Lysivane if the solution is freshly prepared for each experiment and is kept in the dark until used. As always, species differences must be considered; for example, C. Hebb (personal communication) has found the rabbit enzyme to be more resistant than that of rat or cat.

The most satisfactory of the various 'selective' inhibitors of AChE (see Fulton and Mogey 1954; Augustinsson 1963) are probably BW284C51 [1:5-bis (4-allyl-dimethylammoniumphenyl)-pentan-3-one-dibromide] or its corresponding diiodide BW297C50, and BW62C47 [1:5-bis (4-trimethylammoniumphenyl)-pentan-3-one-diiodide]. Again, these are not entirely selective; Holmstedt (1957a) found that BW284C51 in a concentration of 10^{-5} M produced slight inhibition of purified human serum cholinesterase (see fig 11.1) but in contrast to this, Bayliss and Todrick (1956) observed slight activation of crude pseudocholinesterase in rat intestinal mucosa (BuCh as substrate) by BW284C51, and also by BW62C47. These discrepancies may well be further examples of species differences or they could reflect a difference in the behaviour of purified versus crude enzymes (see Cohen and Warringa 1953; Mittag et al. 1971a; also ch. 2 §5.3). Further details of these and some other inhibitors are given in ch. 11.

When the substrate is present in the medium, the enzyme will to some extent be protected from these inhibitors which are essentially competitive in action. For inhibition to be maximally effective preparations should be preincubated with inhibitor for 20–30 min before the substrate is added. Another factor which influences the efficiency of the inhibitor is the pH (Heilbronn 1954; Lewis 1961). Holmstedt (1957a) advises that the concentration of inhibitor necessary for any histochemical experiment should be determined from biochemical controls run at the same pH as the histochemical medium. In my own experiments the pH of the histochemical medium is generally 5.5 which is well below the optimum for inhibition by DFP. Hence, whenever DFP is used the buffer for preincubation is at a pH of 6.5.

3.1.1.3 Combination of substrate and inhibitor   In summary, the choice of substrate and inhibitor is influenced by the object of the experiment, the method (whether biochemical or histochemical) and in particular, by the pH involved. Even when these considerations are met it is still not safe to assume that a particular combination of selective substrate and inhibitor is necessarily appropriate for the enzyme in an unfamiliar species. When a new species is first studied, preliminary experiments should be designed according to table 3.1 so that the various substrates are used alone as well as in combination with the relevant inhibitors. A series of experiments of this type (see for example Koelle 1955) can show up discrepancies which might not be apparent otherwise. An illustration of this sort of anomaly is given in ch. 5. No histochemical reaction occurred in the leech nervous system with BuThCh as substrate and BW284C51 as inhibitor and it would seem reasonable to conclude that the tissue lacks pseudocholinesterase. In the absence of inhibitor, however, the tissue gave a positive reaction to BuThCh which suggested that a pseudocholinesterase might be present. There are at least two explanations for the inconsistency; either BW284C51 is non-selective in the leech and inhibits pseudocholinesterase as well as AChE or, alternatively, the specificity of BW284C51-sensitive 'AChE' is broad and the enzyme actively hydrolyses BuThCh as well as AcThCh. The findings illustrate the errors which may arise if what appears to be an appropriate combination of substrate and inhibitor is used uncritically. For work on tissue cultures the choice of inhibitor needs special care. A compound such as OMPA (Schradan, ch. 11) which is relatively inactive until metabolized in the liver (Dubois et al. 1950) is inappropriate as an inhibitor in cultures of neural tissue.

In certain histochemical experiments (for example Osen and Roth 1969) substrate is omitted from the incubation medium as an additional control for the specificity of the enzyme reaction (see appendix §3.4.4 re other adjustments to the medium).

3.2 Biochemical methods

3.2.1 Preparation of material for purification and assay

It is simple to make a crude extract of acetylcholinesterase suitable for measurement of activity by the assay methods described below; preparation of the purified enzyme for characterization of the protein, including its molecular weight, is more difficult. In general, tissues are best homogenized in the medium to be used in the analytical procedures. This is most often bicarbonate or phosphate buffer but if necessary sucrose can be used in certain methods (e.g. the Hestrin method, the radiochemical method and that of Ellman et al., see below). Sucrose is useful where purification techniques involve differential centrifugation or where preparations are to be analysed for other components such as choline acetyltransferase. Since AChE is not easily solubilized, measurements of tissue activity are generally made on the whole homogenate, care being taken to ensure uniform sampling. For simple assay work, as opposed to subcellular fractionation or purification studies, centrifugation should be avoided but when centrifugation is used the medium must be carefully chosen. Ord and Thompson (1951) showed that with rat brain homogenates in 0.025 M sodium bicarbonate activity in the supernatant was very low, but with dilute aqueous homogenates up to 90% of the total activity was present in the supernatant; recovery fell off sharply, however, if the tissue concentration was increased.

The AChE in the supernatant is in a lipoprotein complex; unlike pseudocholinesterase it is not in true solution. Many reagents have been used to disrupt this complex as a starting point for purification and these include organic solvents, detergents, snake venoms, lipases and bacterial proteases (for references see Jackson and Aprison 1966a; Ho and Ellman 1969; Chan et al. 1972a).

3.2.1.1 AChE from electroplax and brain tissue Because purification methods producing the highest specific activity have tended to give poor yields, the most popular starting material has been the electric organs of *Electrophorus electricus* and *Torpedo marmorata* which are very rich in enzyme but unfortunately some of the methods suitable for electric tissues are not equally satisfactory for mammalian brain (see Lawler 1964). The method for electric eel used by Rothenberg and Nachmansohn (1947) consisted of removal of mucin and solubilization of the enzyme with toluene. Ammonium sulphate fractionation followed by ultracentrifugation resulted in a preparation with a specific activity of about 400 mmole ACh/mg protein/hr; the yield was approximately 15%. Kremzner and Wilson (1963, 1964) used essentially the same initial steps followed by chromatographic purification on a series of columns including DEAE-cellulose and Sephadex G-200. The yield was only 9% but the specific activity was 660 mmole ACh/mg protein/hr. Leuzinger and Baker (1967) modified the column procedure to make the method more suitable for large-scale extraction. About 10 kg of toluene-treated electric organ can be processed to yield 60 mg of enzyme with a specific activity of 750 mmole/mg protein/hr and a percentage recovery of between 8 and 12. A further step involving sucrose gradient centrifugation was introduced by Leuzinger et al. (1969). Leuzinger et al. (1968) obtained crystals of AChE when the concentrated enzyme was kept for 2 to 4 days in 35% ammonium sulphate at 4°C. AChE has been extracted from the electric organ of *Torpedo* and *Gymnotes* by Massoulié and Rieger (1969). In their method the enzyme in the exudate produced by tissue autolysis under toluene is solubilized with the detergent sodium desoxycholate.

References to some of the more recent methods for purifying AChE from

mammalian brain were included in tables 2.4a and 2.6. Jackson and Aprison
(1966a) used freeeze-dried ox caudate nuclei treated with n-butanol and ether to
remove lipids; the enzyme was extracted with glycine-sodium hydroxide buffer.
Further steps involved ammonium sulphate fractionation, gel filtration on Sephadex G-200 and sodium chloride fractionation. This was followed by desalting on a
Sephadex column, dialysis against buffered Dextran and, finally, zone electrophoresis. The yield was only 4—6% but the specific activity was 1110 μmole
ACh/mg protein/hr. Ho and Ellman (1969) obtained a much higher yield, about
60—80%, but the specific activity was lower (880 μmole ACh/mg protein/hr). In
their method the enzyme is solubilized with Triton X-100 and then run on a
DEAE-Sephadex column. The effluent is dialysed against buffer and freeze-dried.
The resulting material is taken up in buffered Triton X-100 and centrifuged, the
enzyme being concentrated in the supernatant. Chan et al. (1972b) solubilized the
AChE in bovine caudate nucleus by repeated homogenization and centrifugation in
0.32 M sucrose plus EDTA. Following ammonium sulphate fractionation, the
preparation was purified 700-fold by agarose affinity-gel column chromatography.
Subsequent filtration on Sephadex G-200 yielded three peaks with an average
specific activity of 575 mmole ACh/mg protein/hr. The method produced an
overall purification of about 23,000.

Before leaving the purification methods which have been applied to nervous
tissue I should make some comment about Triton X-100 (for studies on some other
surface-active agents including Myrj and Tween see Jackson and Aprison 1966b).
Table 2.4a in the previous chapter indicated that the method of preparation of a
purified enzyme could influence its behaviour. For example, Ho and Ellman (1969)
found that enzyme solubilized with snake venom or bacterial protease was more
water-soluble than the form which resulted from extraction by Triton X-100.
Shafai and Cortner (1971b) have suggested that Triton-solubilized enzyme (designated AChE-3) is a dimeric hybrid formed of two equal-sized but dissimilar components α and β. When the Triton-solubilized enzyme is run on DEAE-Sephadex
dissociation occurs and each component is eluted separately. Dimerization of like
components follows, yielding two modified forms, AChE-1 (α_2) and AChE-2
(β_2). Srinivasan et al. (1972) have recently found evidence that Triton X-100 not
only solubilizes the enzyme but enhances its catalytic potential. They suggest that
it causes conformational changes which may affect the catalytic or allosteric sites
(ch. 2).

Other preparative procedures which are useful in the study of AChE in nervous
tissue include the methods for the isolation of single cells (see Giacobini 1969), the
microscale technique described by Giacobini et al. (1971) which allows subcellular
fractionation of individual autonomic ganglia, and zonal centrifugation for the
isolation of enzyme-rich membranes (Kuenzle et al. 1972).

3.2.1.2 Erythrocytic AChE   The enzyme in erythrocytes, like that in brain, is not easily solubilized but freeze-drying followed by butanol extraction is effective. A method for the rapid purification of commercial preparations of bovine erythrocyte AChE by affinity-gel chromatography on a d-tubocurarine-Sepharose column with elution by NaCl has been described by Jung and Belleau (1972). Berman and Young (1971) employed a similar principle, but used different inhibitors, in their purification of AChE from bovine erythrocytes and *Electrophorus* electroplax.

Various workers have found that whereas AChE can be released from bovine erythrocytes by hypotonic salt solution, the AChE on human erythrocytes is not solubilized by this procedure. Ciliv and Özand (1972) achieved a 2537-fold purification of AChE from human erythrocytes as follows. The enzyme was extracted from the membrane fragments with Triton X-100 and after centrifugation the supernatant was run on DEAE cellulose. Following dialysis, the preparation was chromatographed on calcium phosphate gel. The most active samples, representing a 5% yield, were pooled, lyophilized, washed with toluene to remove Triton X-100 and centrifuged. Finally the toluene was evaporated off under vacuum.

3.2.1.3 Pseudocholinesterase

Because pseudocholinesterase is soluble there is less difficulty about its purification. Surgenor and Ellis (1954) attained a 3400-fold purification of human serum pseudocholinesterase by an ethanol fractionation procedure. A further purification of the resulting preparation was achieved by Malmström et al. (1956) with chromatography on calcium phosphate gel and Dowex-2 anion exchange resin. Das and Liddell (1970), who provide a useful list of various additional methods, used a three-stage procedure consisting of chromatography on DEAE-cellulose, electrofocussing and gel filtration on Sephadex G-200. The percentage yield was 54 and the purification factor 13,000.

In the case of tissue pseudocholinesterase, Tucci and Seifter (1969) reported a 3200-fold purification of BuChE from the pig parotid gland by ammonium sulphate fractionation followed by sequential chromatography on DEAE-Sephadex, carboxymethyl cellulose and Sephadex G-200. The overall yield was 5–10% and the specific activity 74 mmoles ACh/mg protein/hr.

3.2.2 Methods of measurement

3.2.2.1 General considerations and sources of variation

With the exception of the radiochemical methods, the basic techniques for measuring cholinesterase activity have been used for a very long time. In 1957 Augustinsson published a very excellent sort of 'consumers guide' to all the main methods giving precise technical details and authoritative comments on the advantages, disadvantages and scope of each. Much of what he wrote then is still applicable but he has recently produced an updated version (Augustinsson 1971a) covering the improvements which have stemmed from the increasing sophistication of apparatus. Anyone embarking on the determination of cholinesterase activity could do no better than to consult these two reviews. For this reason I do not give many technical details in what follows; instead my aim has been to outline the principles.

The conditions to which the enzyme is subjected are not the same in all methods hence results obtained in different ways are not directly comparable. Nevertheless, as Augustinsson pointed out, values are more meaningful if they can be expressed in some sort of absolute unit, such as μmole ACh hydrolysed or μl CO_2 evolved, rather than as changes in pH or in light absorption. Factors which influence the activity include pH, the concentration and nature of the substrate, the type of buffer and the temperature. The nature of the enzyme preparation is also important. Not only may the characteristics of the enzyme itself be different in the purified and the crude state but non-enzymic components of the crude preparation, such as salts, may affect the activity directly or may interfere with the recording and so cause anomalous results. In all methods precautions must therefore be taken to minimize and to allow for any non-enzymic chemical or physical effects of the preparation as well as for non-enzymic hydrolysis of the substrate. Even with adequate controls, all that can be measured is the activity of the enzyme under the particular conditions of the experiment. The results do not necessarily indicate the activity in vivo. It is possible that even the mode of death may influence results and while this factor does not seem to have been studied systematically it is worth bearing in mind where studies which are similar in all other respects give divergent results. On theoretical grounds Urethane (ethyl carbamate) should not be used to kill animals since it is a weak anticholinesterase (Augustinsson 1948, see also ch. 11 §4). However, Souza et al. (1970) found that values for the total ChE activity in blood of rats anaesthetized with Urethane or Halothane were less variable than those from animals anaesthetized with other agents including ether, chloroform, chloral hydrate and thiopentone (see ch. 9 §4 re delayed effects of ether and pentobarbitone). I routinely use sodium pentobarbitone (Nembutal, Abbott Laboratories Ltd) and the values for AChE in the cerebellum of rats killed with an overdose are similar to those from decapitated animals (P. Kása personal communication).

3.2.2.2 Gasometric methods   The Warburg method is suitable for cholinesterases and has been widely used. As a result of enzymic hydrolysis of the substrate acid is released into the bicarbonate-Ringer's solution and the CO_2 evolved is measured manometrically.

If the total volume of CO_2 produced in 30 min is a_{30} and the volume due to spontaneous hydrolysis is s_{30}, the activity of the enzyme is expressed as

$$\frac{b_{30}}{22.4} \mu\text{mole substrate/30 min}$$

where $b_{30} = a_{30} - s_{30}$ and 22.4 is the normal molar volume (see glossary).

The Warburg apparatus is relatively expensive and a certain expertise is needed to operate it but in practised hands the method is reliable. It can be used on all types of preparation including spots of blood on filter paper and it is not affected by the colour or consistency of the enzyme preparation. A disadvantage is that it cannot be used to test pH effects because of the limitations imposed by the bicarbonate-CO_2 system; there is evidence too (Smallman and Wolfe 1954) that bicarbonate may itself lower activity and may to some extent also reduce the activation produced by added salts.

The Cartesian diver (Linderstrøm-Lang and Glick 1938) and the magnetic diver (Brzin et al. 1964; Brzin and Zeuthen 1964) employ the gasometric principle but on the ultramicroscale. The methods can be used for measuring enzyme in single cells (see Giacobini 1956b, 1969) and in axonal membranes (Brzin et al. 1965). The lower limit of sensitivity of the Cartesian diver is 5×10^{-6} μmole ACh/hr or 1×10^{-4} μl CO_2/sample; the sensitivity of recently developed models of the magnetic diver balance is considerably greater (Augustinsson 1971a). Both types of diver can be used for μg quantities of tissue but since they are difficult to use and the introduction of inhibitors is complex, they must be regarded as a specialist, rather than a routine, tool.

3.2.2.3 Methods depending on a pH change   a) *Electrometric recording.* The acid produced by hydrolysis of the substrate can be recorded as a pH change. According to Michel's (1949) original electrometric version, which he used for erythrocytes and serum, the enzyme preparation is equilibrated for 10 min with sodium barbital buffer and the pH (pH_1) is then read with a meter sensitive to 0.01 pH units. ACh is added and after $1-1\frac{1}{2}$ hr the pH (pH_2) is re-read.

The unit activity is Δ pH/hr which is calculated as

$$\frac{(pH_1 - pH_2 - b)}{t} f$$

where t is time and b and f are the correction factors given for pH_2 in Michel's table; b allows for spontaneous hydrolysis of the substrate and f for the effect of pH on enzyme activity. The method is versatile and although less accurate than the Warburg technique it is still popular for the routine measurement of ChE in blood. Augustinsson points out, however, that care is needed in the choice of buffer. The buffering must not be so good that there is no measurable fall in pH but on the other hand it must be good enough to prevent the pH falling so far that enzyme activity is severely reduced. Michel's choice of sodium barbital was a good one because as pH falls there is a reduction in its buffering capacity and this happens to parallel the fall in enzyme activity. There is a risk that buffering by the enzyme preparation itself may influence the measurement but this difficulty can be overcome if the enzyme can be diluted sufficiently. By fitting the pH meter with a suitable recorder which can be calibrated with acetic acid (Tammelin and Löw 1951; Tammelin and Strindberg 1952) the values for pH/t can be converted to μmole acetic acid released/min. The calibration must obviously be carried out under the conditions of the particular experiment but, in addition, because of the buffering effect discussed above, recalibration is necessary for each concentration of enzyme.

b) *Photometric recording*. pH changes can be followed with an indicator such as phenol red or bromothymol blue and recorded photometrically rather than with a pH meter. The pH shift represented by the change in colour between the start and the end of incubation is read from a standard chart. This graph, prepared in a preliminary experiment with substrate, heat-inactivated enzyme and acid of varying pH, also allows for non-enzymic hydrolysis. The value of pH/t is then calculated from Michel's formula. According to Augustinsson the figures given by the electrometric and indicator methods are nearly identical. Indicator methods are suitable for use with the Auto-Analyzer and Augustinsson (1971a) gives references to the application of this technique to the routine screening of anticholinesterase residues on plants.

3.2.2.4 Titrimetric methods   The principle of these methods is that the acid liberated by hydrolysis is titrated with alkali and the enzyme activity is expressed in ml alkali/hr. The alkali can either be added continuously to keep the incubate at the initial pH or it can be added after a fixed time to restore the pH to its initial value. Augustinsson (1957) gives long lists of variants in which indicators are used to determine the end-point of the titration and he recommends their use in preliminary procedures and in experiments where accuracy in not vital. The disadvantage of indicators is that colour matching is at best imprecise and may be made more difficult by any colour in the enzyme preparation. Furthermore, there is a possibility that the indicator may affect the enzyme or, conversely, the colour may itself be affected by inhibitors or other components added to the incubation medium. These last considerations are also relevant to the methods discussed in §3.2.2.3.

Greater precision is possible when the end-point of the titration is signalled electrometrically; Augustinsson (1971a) cites a number of such methods. The development of the pH-stat (Jacobsen et al. 1957) has made the titrimetric technique one of the most accurate methods available and according to Augustinsson (1971a) its applications appear virtually unlimited. The pH-stat consists of a reference electrode (e.g. calomel) and a hydrogen-ion electrode (e.g. glass or antimony) which is immersed in the reaction mixture. The temperature of the reaction vessel is thermostatically controlled and its contents can be stirred. Any deviation from a preset value in the pH of the reactant is detected by the electrodes and is signalled via the pH meter to an automatic burette which then delivers whatever amount of alkali is needed to restore the pH. The additions of reagent over a given time are recorded on a chart and the slope of the trace is proportional to the enzyme activity.

A great advantage of the method is that the enzyme is functioning at a virtually constant pH; this is particularly important in kinetic studies. As with any method, account must be taken of hydrolysis of the substrate in the absence of enzyme (see Delaunois 1962) and of any release of acid from the tissue samples; Jensen-Holm et al. (1959) found that unless tissues were cooled rapidly after removal from the body they liberated sufficient acid to introduce appreciable errors in the results. More recently Ballantyne (1968) has shown that absorption of acidic components from air inside reaction vessels can cause an added error and he advocated filling them with nitrogen.

3.2.2.5 Colorimetric and spectrophotometric methods Augustinsson (1957) mentions two colorimetric techniques, one with ACh as substrate and the other, specific for pseudocholinesterase, with β-carbonaphthoxycholine. Subsequently Ellman et al. (1961) published a spectrophotometric method in which either acetyl- or butyrylthiocholine is used. The method which is simple, accurate and flexible has been widely used. Various modifications which have been introduced over the years are listed by Augustinsson (1971a). In principle, the thiocholine released by hydrolysis reacts with 5:5-dithiobis-2-nitrobenzoate (DTNB) to give the yellow anion of 5-thio-2-nitrobenzoic acid. Each mole of anion produced represents hydrolysis of 1 mole of substrate. The rate of production of the coloured ion is recorded at 412 nm on a spectrophotometer; corrections can be made for any absorbance by the tissue, for the spontaneous hydrolysis of substrate and, as may be necessary in the case of a suspension, for the release of thiol groups from the tissue. Ellman et al. express results in moles hydrolysed/min/g tissue but it is rather more common to use nmole/min/mg and when comparing the values given by different authors it is vital to note the units. The explanation of the calculation provided by Ellman et al. is a little brief so the steps involved in expressing activity in nmole substrate hydrolysed per minute per mg of tissue are given here in some detail.

1) The change in absorbance/min, ΔA, is measured from the slope of the spectrophotometric record.

2) The extinction coefficient (see glossary) for a molar solution (1 mole/litre) of coloured anion is 1.36×10^4 (Ellman 1959) therefore the extinction coefficient for 1 nmole in a cuvette of volume V ml =

$$\frac{1.36 \times 10^{-2}}{V}$$

3) The rate of hydrolysis of substrate is therefore

$$\frac{\Delta A \times V}{1.36 \times 10^{-2}} \text{ nmole/min}$$

4) If the concentration of the enzyme preparation is C mg/ml and the volume used is Y ml the rate of hydrolysis in nmole/min/mg is

$$\frac{\Delta A \times V}{1.36 \times 10^{-2}} \times \frac{1}{C \times Y}$$

Ellman et al. suggest that the rate of hydrolysis given by this method with AcThCh as substrate can be taken as a good indication of the rate of hydrolysis of ACh. Within limits this is a justifiable approximation but as the authors themselves show (table 3.2) the kinetic constants obtained with the two substrates are not identical and this must be recognized in any comparison with results from other methods.

Recently van Hooidonk et al. (1972) have described the synthesis of 1-[2-thiazolylazo]-2-acetoxybenzene derivatives which they recommend for use as substrates in another type of spectrophotometric assay. The method depends on the colour difference between the ester and the hydrolysis product. One advantage of these compounds is that the products absorb at long wavelengths (485–585 nm) thus interference from protein absorption is eliminated.

3.2.2.6 Unreacted substrate   Hestrin's (1949) method involves a colour reaction but the principle is different from the colorimetric methods just discussed. The enzyme is allowed to hydrolyse ACh for a given time after which hydroxylamine is added. This will react with any residual ACh to form acethydroxamic acid and on the addition of ferric chloride and HCl a red-purple complex is formed. The intensity of the colour, which can be measured with a spectrophotometer, is proportional to the concentration of ACh present, thus the absolute value for the unhydrolysed substrate can be obtained from a standard curve.

The method is versatile with regard to pH and concentration of enzyme and substrate and it requires only small amounts of enzyme. It is not as accurate as the manometric or titrimetric methods and is unsuitable for kinetic studies but is convenient for routine analysis and is sensitive down to 4×10^{-2} μmole ACh/ml final solution. A microscale version of the Hestrin technique was developed by Bonting and Featherstone (1956); the method allows the removal of an aliquot of medium from each tube before the addition of ferric chloride and this can be analysed for protein. The original Hestrin method has been variously modified (see

TABLE 3.2

Kinetic constants of bovine erythrocyte cholinesterase given by the substrates acetylthiocholine and acetylcholine. (From Ellman et al. 1961.)

Constant	Acetylthiocholine	Acetylcholine	Reference
Michaelis (K_M)	1.4×10^{-4}	2.0×10^{-4}	Wright and Sabine (1948)
Inhibition (K_i)			
Substrate	2.9×10^{-2}	1.5×10^{-2}	Wright and Sabine (1948)
Decamethonium	6.1×10^{-5}	2.5×10^{-5}	Bergmann et al. (1950)
Physostigmine	2.3×10^{-8}	6.1×10^{-8}	Nachmansohn and Wilson (1951)
Quinidine	8.7×10^{-4}	9.6×10^{-4}	Wright and Sabine (1948)

Augustinsson 1971a) particularly by those using it for studies of organophosphorus residues. It can now be used for automated assays and other improvements include the stabilization of the colour by substituting HNO_3 for HCl; in the original method fading was a problem.

3.2.2.7 Radiometric methods   During the last decade a number of radiochemical methods have been introduced for cholinesterase measurements. In all the methods $[^{14}C]$-ACh is used as substrate. At the end of the incubation the labelled acetate formed by hydrolysis is separated from unreacted substrate and either one or the other, depending on the particular method, is counted. The various methods differ from each other mainly in the procedures for this separation. In the original technique (Winteringham and Disney 1964) the $[^{14}C]$-acetate was allowed to volatilize and the residual $[^{14}C]$ activity present in the unhydrolysed substrate was counted. In most other versions it is the acetate which is counted after separation from ACh. In Potter's (1967) method the acetate is extracted with pentanol-toluene which gives a recovery of 80–90%. Higher recoveries are achieved in the method of McCaman et al. (1968) in which the ACh is removed by reinekate precipitation. Fonnum (1969) has used sodium tetraphenylboron to remove unhydrolysed ACh; he found the method less laborious than reinekate precipitation and the acetate recovery was still high.

Koslow and Giacobini (1969) developed a microscale version of the method of McCaman et al. for use on single cells. For this technique the concentration of $[^{14}C]$-ACh is reduced to avoid the substrate inhibition which is liable to occur with the very small amounts of enzyme involved. Tests have shown that under the conditions of the experiment ACh is able to penetrate intact cells. Fifty to 150 samples can be dealt with in a single experiment and the limit of detectable hydrolysis is about 7 picomole ACh/hr. A microscale technique employing racemic $[^{14}C]$-acetyl-β-methylcholine as a substrate has been described by Hoskin et al. (1969). This method allows the simultaneous measurement of both the labelled unhydrolysed substrate and the labelled hydrolysis product.

Radiochemical methods have proved suitable for the study of AChE in intact tissue. Buckley and Heaton (1968) used a modification of Potter's method to assay AChE in motor end-plates which had been stained histochemically and Ehrenpreis et al. (1967, 1970) applied a radiometric method to pieces of intact guinea-pig ileum. Fonnum's method was used by Storm-Mathisen (1970) to measure AChE in samples of tissue dissected from freeze-dried sections of rat hippocampus.

3.2.2.8 Miscellaneous methods   Prince (1966) introduced a fluorimetric method in which the substrate is 1-methyl-7-acetoxyquinolinium iodide. On hydrolysis this gives the highly fluorescent compound 1-methyl-7-hydroxyquinolinium iodide. The intensity of fluorescence can be measured with a fluorimeter and its development plotted against time. A calibration curve prepared from standard concentrations of the fluorescent compound is then used to calculate the rate of enzymic hydrolysis. Since hydrolytic enzymes other than AChE will produce some hydrolysis the method cannot be considered highly specific. Subsequently Voorhorst (1971) suggested the use of the 8-butyroxy analogue to distinguish between AChE and BuChE. A gas chromatographic method in which 3,3-dimethylbutyl acetate is the substrate has been recently described (Cranmer and Peoples 1971). It is rapid and sensitive but attention must be paid to the question of specificity.

The thiocholines have been employed as substrates in methods other than the successful DTNB spectrophotometric technique of Ellman et al. (1961; see §3.2.2.5). In all these procedures it is the free —SH groups produced by hydrolysis which are measured (see Augustinsson 1957; Ho et al. 1965). According to Augustinsson these methods have no particular advantages and may, indeed, be less accurate than many others. Autoradiography lies somewhere between quantitative biochemistry and staining histochemistry. Ostrowski et al. (1963) showed how [^{3}H]-labelled DFP could be used in the autoradiographic demonstration of the distribution of cholinesterases in motor end-plates of mouse diaphragm. Subsequently the method has been quantitated (see Barnard and Rogers 1967) and it can be used in this way in electron microscopy (see Salpeter 1969; Salpeter et al. 1972).

3.2.2.9 Diagnostic work   With the widespread use of anticholinesterases as insecticides it became imperative to develop simple routine methods for measuring blood cholinesterases in large-scale screening tests of factory and farm workers. Augustinsson (1957, 1971a) indicates a number of methods that are suitable for this purpose (e.g. that of Davies and Nicholls 1955). Those such as the Hestrin method, which can be applied either to liquid blood or to blood spots on filter paper, are particularly useful. The fast potentiometric method described by Vitale and Blincoe (1971) may prove effective as a laboratory tool in certain specialized clinical work such as the diagnosis of liver disorders but since it involves precision instruments and the operating conditions are critical it may be of more limited value in routine work.

For diagnosis in the field in cases of suspected anticholinesterase poisoning, speed and simplicity rather than accuracy are the primary requirements. Limperos and Ranta (1953) developed a method with bromothymol blue specifically for this purpose, all the necessary solutions and glassware can be carried in the pocket and the result is available within 20 min. Subsequently test-papers impregnated with ACh and bromothymol blue were introduced for field work. These papers can be used on whole blood, unlike the commercially available Acholest type which are suitable only for serum and plasma (Augustinsson 1971a).

Occasionally, cholinesterase levels may be measured for forensic rather than diagnostic purposes in cases of unexplained death. In the baffling Bogle-Chandler case in Australia in 1963, for instance, anticholinesterase poisoning was among the possible causes of death that were considered. Two general points should be noted here about measurements made on tissues which may have been exposed to inhibitors. First, there is always the possibility that inhibitor trapped in the tissue may be released during processing, with the result that the cholinesterase activity of the homogenate is very much lower than that obtaining in the intact tissue (see for example Hoskin et al. 1969). Second, if tissues are not removed from the body until some time after death – and this is very likely in forensic cases – it is especially important that adjustments are made to the analytical method to take account of any acid liberated from the tissue (see Jensen-Holm et al. 1959). If this factor is ignored the rate of hydrolysis given by most methods will be too high and a very low level of cholinesterase activity indicative of poisoning could be masked. Using histochemical methods Bergner and Durlacher (1951) showed that 'normal' levels of ChE activity persisted in human intercostal end-plates for 26 hr postmortem in unrefrigerated corpses and up to 180 hr with refrigeration. In a later paper Moore and Petty (1958) reported histochemically demonstrable AChE in muscles from bodies which had been refrigerated for 157 days. They also detected activity in muscles from a corpse which had been immersed in a river for 4 months during the winter.

3.2.2.10 Detection of isozymes   The isozymes of AChE and pseudoChE (ch. 2 §3.2, ch. 10 §5.1) can be separated by conventional starch, agar or polyacrylamide gel electrophoresis. A comprehensive list of references is given by Bajgar and Žižkovský (1971). The specific substrates AcThCh and BuThCh can be used with most techniques (see Bajgar and Žižkovský 1971) but Guilbault et al. (1970a) recommend the use of N-methylindoxyl acetate for the rapid detection of low activity in polyacrylamide gels. N-methylindoxyl acetate is not specific for cholinesterases, being hydrolysed by lipase and cellulase as well, but under the conditions used these enzymes showed no appreciable activity. This is relevant to studies of preparations from cellulase-containing invertebrates such as *Helix* (see Baldwin 1967). Augustinsson (1971a) gives references to the use of α-naphthylacetate in screening tests for pseudoChE variants. Other methods for these tests are given in ch. 10.

3.3 Histochemical methods

3.3.1 General considerations

Cholinesterases can be demonstrated histochemically at the optical and at the electron microscopic level. The principles of the technique and, to a large extent, the technical requirements and associated problems are common to both fields. In brief, the enzyme in the tissue hydrolyses a substrate present in the incubation medium and the hydrolysis product then reacts with some other component of the medium to form an insoluble precipitate at the site of enzyme activity. The precipitate may be converted to a coloured compound by a simultaneous or subsequent reaction or, in some methods, it may be observed directly.

The risk of artifacts is far greater in histochemical than in biochemical procedures. Apart from the problem of establishing the specificity of the enzyme (see §3.1.1) dangers inherent in the method include the inaccurate localization of end-product due to diffusion, and misleading negative results caused either by non-penetration of the substrate or by inhibition of the enzyme during processing. A bewildering number of variants have been introduced into the basic histochemical methods in attempts to improve their accuracy. A few of these variants will be discussed below (§§3.3.2, 3.3.3) and others are indicated in table 3.4; some practical details are given in the appendix at the end of the chapter.

3.3.2 Optical microscopy

3.3.2.1 History The first method for demonstrating cholinesterases histochemically was introduced by Gomori in 1948. The substrates he used were higher fatty-acid esters of choline, such as lauroyl- or palmitoylcholine. The acid released on hydrolysis by esterases formed a salt with the cobalt which was present in the medium; subsequently the section was immersed in alcoholic ammonium sulphide and the salt converted to black cobalt sulphide. Gomori himself had reservations about the specificity of the method for cholinesterases but it remains useful for demonstrating esterases in general. It is important not to confuse this method with Gomori's modification of the thiocholine technique (Gomori 1952, see table 3.4, no. 4).

In 1949 Koelle and Friedenwald showed that thiocholine esters were suitable histochemical substrates for cholinesterases. The method as originally published had a number of drawbacks and has subsequently been extensively modified by Koelle himself and by other workers. The modified methods are often referred to by the names of those who adapted them and although this prevents confusion it tends to obscure the immense debt that histochemists owe to Koelle and Friedenwald for the initial development of the method.

Of the other substrates which have been introduced subsequently, thiolacetic acid (§3.3.2.5) has proved the most popular. It is less specific than the thiocholines but gives an end-product which is more precisely localized. The sudden spate, in the last decade, of new candidates for histochemical substrates, has probably been triggered-off by the demands of the electronmicroscopists for something which yields a really fine end-product. The thiocholine methods do fall short of the ideal in this respect but it is of overriding importance that any substrate which supplants them should be at least as specific. There seems a certain danger that the question of specificity may be considered only secondarily and this is a retrograde step.

3.3.2.2 Principles of the thiocholine technique   Basically, Koelle's thiocholine medium contains acetylthiocholine or butyrylthiocholine as substrate, copper (in the form of copper sulphate) as the capture agent, and glycine which chelates the copper and so minimizes its inhibitory effect on the enzyme. The reactions which occur in a tissue section during incubation in the medium are not completely understood. Bergner and Bayliss (1952) suggested the sequence which is given below and which yields copper thiocholine as the end-product. This sequence takes no account of the sulphate radical and, according to Malmgren and Sylvén (1955), the end-product may be copper thiocholine sulphate rather than copper thiocholine. Thiocholine substrates are most commonly obtained as the iodide. The fate of the iodide radical will be considered in the appendix (§3.4.4).

The thiocholine reaction proposed by Bergner and Baylisss (1952) is:

$$2 \text{ CH}_3\text{COSCH}_2\text{CH}_2\text{N(CH}_3)_3\text{OH} + (\text{NH}_2\text{CH}_2\text{COO})_2\text{Cu} + \text{H}_2\text{O} \xrightarrow{\text{cholinesterase}}$$
$$\underset{\text{acetylthiocholine}}{} \qquad \underset{\text{copper glycinate}}{}$$

$$\text{Cu[SCH}_2\text{CH}_2\text{N(CH}_3)_3\text{OH]}_2 + 2 \text{ NH}_2\text{CH}_2\text{COOH} + 2 \text{ CH}_3\text{COOH}$$
$$\underset{\text{copper thiocholine}}{} \qquad \underset{\text{glycine}}{} \qquad \underset{\text{acetic acid}}{}$$

The copper thiocholine, which is precipitated in the region of the enzyme, is white and the unconverted salt can be examined by phase-contrast or conventional microscopy (Zajicek et al. 1954; Giacobini 1956a; Holmstedt 1957b). Most workers, however, prefer to 'develop' the sections in a sulphide-containing solution in which the crystals of copper thiocholine are replaced by copper sulphide. Areas which contain enzyme thus appear dark brown. The reaction occurring when yellow ammonium sulphide is used as the developer is thought to be:

$$\text{Cu[SCH}_2\text{CH}_2\text{N(CH}_3)_3\text{OH]}_2 + (\text{NH}_4)_2\text{S} \longrightarrow \text{CuS} + 2 \text{ NH}_4\text{SCH}_2\text{CH}_2\text{N(CH}_3)_3\text{OH}$$
$$\underset{\text{copper thiocholine}}{} \quad \underset{\substack{\text{ammonium} \\ \text{sulphide}}}{} \quad \underset{\substack{\text{copper} \\ \text{sulphide}}}{} \quad \underset{\substack{\text{ammonium} \\ \text{thiocholine}}}{}$$

Malmgren and Sylvén (1955) found that there may not be a complete replacement of crystalline copper thiocholine by amorphous copper sulphide; under some conditions a mere coating of the crystals with copper sulphide occurs. In the so-called 'direct coloring' modification introduced by Karnovsky and Roots (1964) (see §3.3.2.4 below), the copper thiocholine is converted not to copper sulphide but to copper ferrocyanide (Hatchett's brown).

Before turning to the practicalities of the various methods which have stemmed from the original thiocholine technique it is perhaps worth considering certain theoretical requirements in histochemistry (see also Holt 1958; Holt and O'Sullivan 1958; Pearse 1968). If histochemistry is to have any significance at all in terms of function, it is vital that the distribution of the end-product should mirror the distribution of the enzyme; furthermore, the activity of the enzyme should not be seriously affected by the processes to which the tissue is subjected. To achieve an accurate deposition of end-product at the enzyme sites, there must be no translocation during processing of the enzyme itself nor of the final end-product. The problem of enzyme diffusion and the complications resulting from the existence of isozymes of AChE with differing solubilities is discussed in §3.3.2.3.

To prevent the thiocholine diffusing away from the site at which it is released it must be immediately precipitated as an insoluble salt by reacting with the capture agent (Cu^{2+}). Rapid precipitation requires that the concentration of copper ions in the immediate vicinity of the enzyme be high: according to Holt and O'Sullivan (1958) the theoretical ideal for accurate deposition of end-product is that the enzyme reaction should obey zero-order kinetics and the capture reaction first-order kinetics (see glossary). This requirement for a high concentration of copper ions cannot be fully met without causing inhibition of the enzyme, thereby defeating the object of the experiment. This means that conditions in the incubation medium have to be a compromise between those which give maximum enzyme activity and those which give the best localization. It is the continuing search for the most satisfactory compromise that has produced so many variations in the histochemical medium and in the technical details of the procedure.

In his papers of 1950 and 1951 Koelle showed how, by the use of the separate substrates, AcThCh and BuThCh, together with the appropriate inhibitors as discussed in §3.1.1.3, the method could be made sufficiently specific to distinguish AChE from pseudocholinesterase. He also made a number of modifications aimed at minimizing diffusion artifacts. He recommended that sections should be preincubated in the histochemical medium, complete except for the substrate (see appendix) so that the copper had time to penetrate into the tissue before the enzyme reaction occurred. Second, he added 24% sodium sulphate to the medium because biochemical tests showed that although somewhat inhibitory it would precipitate the enzyme and so prevent any diffusion; $MgCl_2$ was also added since Nachmansohn (1940a) had found that it activated cholinesterases. Thirdly, Koelle suggested that diffusion of copper thiocholine in the tissues would be reduced if all solutions used (up to the developer stage) were saturated with copper thiocholine; fourthly he added buffer to the incubation medium and the pH was reduced from the original value of 8.06 to 6.4 (Koelle 1950) or 6.0 (Koelle 1951). Subsequently Couteaux (1951), Couteaux and Taxi (1952) and Coërs (1953) found that a further lowering of pH to 5.0 or below, and the use of fixed tissue, reduced the artifacts still further. Coërs also showed that saturating the solutions with copper thiocholine did not improve localization but merely complicated the procedure. This observation, that the method works well in the absence of added copper thiocholine, is interesting in view of a report by Zajicek et al. (1954): they found the presence of copper thiocholine in the medium to be essential for the deposition of crystals of copper thiocholine sulphate by the action of purified AChE obtained from *Torpedo*.

Some of the modifications to the method have been the result of trial and error but Lewis (1961) designed experiments so that he could analyse systematically the consequence of any particular change. When the pH is lowered enzyme activity is reduced. This reduction seems to result from two factors; first, at an acid pH the enzyme is well below its optimum of 8–8.5 and second, glycine chelates the copper less efficiently when the pH is low, so inhibition of the enzyme by free copper ions could be increased. The reduction of activity at a low pH is accompanied by a more precise localization of end-product and Lewis pointed out the probable interrelation of the two effects. The kinetic criteria (Holt and O'Sullivan 1958) for accurate deposition of end-product, which were discussed earlier, are more nearly met under the conditions of the slow hydrolysis which occurs at a low pH than with the faster hydrolysis which occurs when the pH is high. If hydrolysis is very rapid the capture agent may be depleted and diffusion of unprecipitated thiocholine could result. Another advantage of lowering the pH (but which Lewis did not discuss) is that it will reduce the risk of artifactual binding of the highly charged nitrogen atom in the thiocholine to any acidic tissue constituents (such as cell nuclei) which have free negative charges (see Malmgren and Sylvén 1955).

Taking account of these theoretical considerations, Lewis (1961) advocated that incubation should be at room temperature and at a pH below 6.0. When the pH is reduced, the period of incubation has to be increased to produce a given depth of staining. Microphotometric measurements of the density of end-product showed that for a fall in pH of 0.3–0.5 the incubation time had to be approximately doubled. The medium Lewis recommended (see appendix) has a higher substrate and a lower glycine concentration than Koelle's (1951) medium but within certain limits the concentrations do not appear critical. The pros and cons of fixation, which improves localization while decreasing enzyme activity, are discussed in §3.3.2.3 but Lewis concluded that with fixed sections incubated under the conditions listed above, diffusion artifacts are minimized.

A point which emerges all too clearly from Lewis's careful analysis is the one made earlier: conditions which reduce diffusion and improve localization of the reaction-product may, at the same time, decrease enzyme activity. Although this does not matter in tissues in which the activity is high, it becomes increasingly important where activity is low. This is one of the reasons (see also §3.3.2.3 re isozymes) why conditions derived from studies on one tissue may not be directly applicable to a different tissue or species. In particular, methods which give encouraging results when tested on the motor end-plate may prove disappointing when applied to ganglia or the CNS (see Koelle 1970). As a preliminary to any study of an unfamiliar tissue it is a good practice to test various combinations of pH and incubation times, and to compare fresh and fixed material. The most difficult tissues to stain satisfactorily are those such as the brain stem in which regions of high and low activity may be present in the same section: conditions necessary to show up the enzyme in the less active areas may cause diffusion in those in which the activity is very high. Foetal brain poses special problems since the reaction-product tends to form very large crystals. I have experienced the same difficulty with the carotid body (see Biscoe and Silver 1966) and with some invertebrate nervous tissue.

3.3.2.3 The use of fixed tissue   Enzymologists are often astounded that histochemists are prepared to study cholinesterases in fixed tissue and since fixation has become a generally accepted stage in some of the methods to be discussed below it seems important to give the justification for this practice. First, cholinesterases are relatively resistant to fixation especially if this is done in the cold (4°C) and for only a short period, say 3–5 hr (see Taxi 1952; Fukuda and Koelle 1959; Austin and Phillis 1965). Hardwick and Palmer (1961) found that in sheep brain 13% of activity persisted even after 65 hr in 10% formaldehyde solution. Second, the whole point of histochemistry is to demonstrate the enzyme in relation to structure. The object is not to show that it is present – biochemistry can do that – but to show where it is present. If the morphology is poor, as it may be in unfixed sections of certain tissues such as brain, the value of the histochemical approach is lost. Third, even with fresh tissue, histochemical results cannot be interpreted quantitatively because, as already discussed, conditions in the incubation medium imposed by the needs for accurate localization are themselves inhibitory. In view of this, there seems no real reason why the inhibitory effect of fixation should be singled out for special condemnation. As indicated above, a comparison of fixed and fresh tissue can always be made to ensure that fixation is not giving false results.

Fixation not only preserves the structural integrity but it also improves the localization of end-product. This stems partly from the fact that it helps to prevent diffusion of enzyme during processing. Since AChE is considered to be a tissue-bound enzyme the problem of diffusion might seem relevant only to pseudocholinesterase – but this is not so. It has been shown (see Eränkö et al. 1964; Koelle et al. 1970) that certain of the isozymic forms of AChE are easily solubilized and, as an added complication, the proportions in which the different types occur vary from tissue to tissue. This is a further reason (see §3.3.2.2) why histochemical techniques which give good results at the end-plate, where most of the AChE is in the insoluble (desmo-) form, may be unsuitable for demonstrating the more soluble (lyo-) form present, for example, in ganglia. Koelle et al. (1970) examined the histochemical reaction for AChE in unfixed sections of cat ganglia which had been exposed to saline or Triton X-100 before they were stained. They found a marked diffusion of enzyme and considerable loss of activity compared with control sections. If sections were fixed in formaldehyde prior to treatment with saline or Triton X-100, diffusion was minimized and staining preserved. Muscle behaved differently. Fresh sections treated with saline or Triton X-100 showed no evidence of diffusion and the staining reaction was intensified compared with that in untreated controls. These results support the idea that the isozymic form of AChE in the ganglion is more easily solubilized than that at the end-plate.

In addition to stabilizing the enzyme, fixation may improve localization in another way. Brzin et al. (1966) have shown by electron microscopy that reagents penetrate fixed membranes more readily than unfixed ones and it is possible that, as a result, the concentration of capture agent in the region of the enzyme is increased in fixed sections.

For work in the CNS, fixation by perfusion (see appendix) is the method of choice but fixation by immersion is adequate for many other tissues so long as the specimens are small. In my own experiments fixation does not exceed 4 hr but some workers use much longer periods. Table 3.3 lists some of the fixatives which have been used for optical and electron microscopical histochemistry. An extra precaution which can be taken to minimize inhibition in the case of formaldehyde fixation is to prepare the fixative from paraformaldehyde, immediately before use, instead of using the commercial analytical (AR) reagent.

3.3.2.4 The thiocholine technique – practical applications Faced with the numerous methods indicated in table 3.4 (and these represent only some of the proposed variants), the novice embarking on cholinesterase histochemistry may well be at a loss to know which technique to choose. Because no single method is ideal for all requirements it may be necessary to test several to find the one best suited to a particular problem. It is not feasible to describe all the methods listed in table 3.4 so the original papers must be consulted. Three of the methods (nos. 8, 9 and 10) will, however, be discussed here and technical details of these and of Koelle's (1951) method (no. 3) are given in the appendix.

Lewis's method (table 3.4, no. 9, see §3.4.5.3) which I use routinely, was developed partly for class use and is simple and very versatile. Tissues studied in my laboratory have included central nervous tissue from monkeys, cats (adult, young and foetal), guinea-pigs, foetal and adult sheep, goldfish, an immature wallaby, leeches, earthworms and cockroaches. Peripheral tissues have included mammalian and avian muscles, cat carotid bodies, duck salt-glands and the urinogenital tract of the pig.

The technique can be used on sections of fresh-frozen or fixed-frozen tissue including fragile specimens which have been embedded in alginate (Lewis and Shute 1963; see § 3.4.1.3). Tissues such as muscle which have an open type of structure need not be sectioned but can be stained in bulk (see §3.4.3 for technical details). The experimental design in terms of the combination of substrate and inhibitors obviously depends on the purpose of the experiment. If no distinction is to be made between AChE and pseudocholinesterase, AcThCh can be used without an inhibitor. When the enzymes are to be distinguished the procedure (summarized in table 3.1) is as follows. To demonstrate AChE, sections are preincubated at room temperature for 30 min in 0.2 M sodium acetate : acetic acid buffer (pH 5.5) containing 10^{-4} M ethopropazine (EP) to inhibit pseudocholinesterase. They are then incubated in the AcThCh-containing medium in which EP is also present.

TABLE 3.3

Fixatives used in optical and electron microscopic histochemistry.

Fixative	Buffer/pH	Microscopy	Reference
1 10% formaldehyde in 0.12 M (1.67%) Na$_2$SO$_4$	pH 3.4	Optical	Atherton (1963); Krnjević and Silver (1965, 1966)
2 10% formaldehyde in 0.44 M (15%) sucrose	pH adjusted to 7.4 with 0.880 ammonia	Optical	Bell (1966)
3 10% formaldehyde with 1% CaCl$_2$	pH 6.5	Optical	Karnovsky and Roots (1964)
4 10% formaldehyde, 15% sucrose, 1% 0.880 ammonia	pH 6.7	Optical	Pearson (1963)
5 5% formaldehyde in 0.9% saline		Optical	Coupland and Holmes (1957)
6 5% formaldehyde in 0.9% saline	'Neutralized'	Optical	Phillis (1968b)
7 4% formaldehyde with 1% CaCl$_2$	pH 9.0	Electron	Eränkö et al. (1967a)
8 4% formaldehyde with 7.5% sucrose	0.056 M Maleate; pH 7.2 0.028 M Maleate; pH 7.4 0.028 M Maleate; pH 7.4	Optical Optical Electron	Koelle and Horn (1968) Koelle and Gromadzki (1966) Davis and Koelle (1967)
9 4% formaldehyde with 0.3 M (10.2%) sucrose	Phosphate; pH 7.4	Electron	Eränkö et al. (1967a)
10 4% formaldehyde, 2% glutaraldehyde and 10% sucrose	Cacodylate; pH 7.5	Electron	Kása and Silver (1969)
11 4% formaldehyde, 1% glutaraldehyde	Phosphate; pH 7.3	Optical and electron	Reale et al. (1971)
12 3.5–4% formaldehyde in Krebs–Ringer calcium	Phosphate; pH 6.9	Optical and electron	Eränkö et al. (1967b)
13 2% formaldehyde, 1% glutaraldehyde	Phosphate; pH 7.2	Optical and electron	Eränkö et al. (1970)
14 4% glutaraldehyde	Cacodylate; pH 7.4	Electron	Kása (1968a)
15 2.5% glutaraldehyde	Cacodylate; pH 7.2	Electron	Eränkö et al. (1967a)
16 2.0% glutaraldehyde	Calcium acetate and cacodylate; pH 7.0	Electron	Lewis and Shute (1964)
17 2.0% glutaraldehyde for perfusion and immersion followed by 4% formaldehyde in the same buffer	Calcium acetate and cacodylate; pH 7.4	Electron	Shute and Lewis (1966a)

Sections to demonstrate BuChE are similarly preincubated in buffer but with 5×10^{-4} M BW284C51 to inhibit AChE and the incubation medium contains BuThCh and BW284C51. The duration of incubation, at room temperature, depends on the tissue. Usually in this laboratory it is for 2 hr but 30 min proved adequate for certain experiments on spinal cord while up to 4 hr was needed to demonstrate activity in foetal brain. At the end of the incubation period the sections are washed 3 times in distilled water and then immersed for 2 min (4 min for foetal tissue) in sodium sulphide developing solution (see §3.4.5.3). After two further washes the sections are laid out in order on trays under a light. When dry they are taken through 3 changes of absolute alcohol to xylene and mounted in Dammar-xylene. DPX and Canada balsam should not be used since they may cause oxidation and consequent solubilization of the end-product. With Dammar-xylene, on the other hand, the staining seems to be unaltered after ten years or more.

Coupland and Holmes's (1957) method (table 3.4, no. 8) was developed from Koelle's technique specifically for work on peripheral tissue and in my experience the end-product in peripheral structures such as the salt-gland is finer and less crystalline than with Lewis's medium (see plate 3.1). The medium and its preparation are described in the appendix (§3.4.5.2). A technical disadvantage of the method is its requirement for sodium sulphate in a concentration of 40 g/100 ml. It is very difficult to get this amount into solution and rather more difficult to keep it in solution. Even if the stock solution is kept in an incubator at $38°C$ it may crystallize out when added to the rest of the medium; because of this, the method is less straightforward than Lewis's.

The fixative recommended by Coupland and Holmes is 10% formol-saline (14–24 hr at $4°C$) but I use 10% formaldehyde:isotonic Na_2SO_4 as for Lewis's method. According to the original procedure 10^{-6} or 10^{-7} M DFP in saline can be used for selective inhibition but I found that when sections were preincubated for 20 min in saline, even without an inhibitor, the subsequent staining was considerably less than that in sections preincubated in buffer. For this reason I prefer preincubation in 0.2 M sodium acetate: acetic acid buffer containing either EP or BW284C51 as described earlier. Both these inhibitors mix satisfactorily with the incubation medium itself (cf. below, re Karnovsky and Roots's method). The pH of the medium and the duration of incubation (at room temperature) can be varied to suit different species. Coupland and Holmes suggest a pH of 4.6–5.0 for man, dog and cat, 5.0 for rat and 5.6 for rabbit. The developer given in the original paper is 'dilute ammonium sulphide' but the sodium sulphide solution described in the appendix is very satisfactory.

Karnovsky and Roots's (1964) direct-colouring method (table 3.4, no. 10, see §3.4.5.4) differs in principle from the other thiocholine techniques. It is called the direct-colouring method because the sites of enzyme activity become stained (a reddish brown) during the incubation itself and not as a result of subsequent development. The advantage of this is that the reaction can be watched throughout

TABLE 3.4
Thiocholine methods

Author	Substrate		Buffer, pH
	Compound	Concentration etc	
1. Koelle and Friedenwald (1949)	AcThCh	≃ 4.0 mM. Iodine precipitated with $CuSO_4$	None; pH 8.06
2. Koelle (1950)	AcThCh or BuThCh	as above	Phosphate; pH 6.4
*3. Koelle (1951)	”	as above	Maleate; pH 6.0
4. Gomori (1952)	AcThCh	≃ 6.9 mM. Iodine not precipitated	” ”
5. Coërs (1953)	”	≃ 4.0 or 2.0 mM. Iodine precipitated with $CuSO_4$	Acetate; pH 5.0
6. Gerebtzoff (1953, 1959)	AcThCh or BuThCh	≃ 4.0 mM. Iodine precipitated with $CuSO_4$	Acetate; pH 5.0, 6.2, 6.8
7. Zajicek et al. (1954)	”	as above	Maleate; pH 6.4
*8. Coupland and Holmes (1957)	”	as above	Acetate; pH 4.6−5.6 depending on species
*9. Lewis (1961)	”	≃ 6.0 mM. Iodine precipitated with $CuSO_4$	Acetate; pH 5.0−6.0
*10. Karnovsky and Roots (1964)	”	≃ 1.7 mM. Iodine ⟩ not precipitated	Maleate; pH 6.0
11. Eränkö et al. (1967b)	”	≃ 1.7 mM. Iodine precipitated with lead acetate	Tris-acetate; pH 6.0
12. Koelle and Gromadzki (1966)	”	≃ 4.0 mM. Iodine precipitated with $AgNO_3$	Phosphate; pH 5.6

Methods marked with an asterisk are considered further in the text and/or appendix.

Chelating agent	Capture agent	Other additions	Inhibitor
Glycine with KOH	Cu (sulphate)	Traces of CuThCh	DFP in saline
Glycine	" "	" "	" "
"	" "	$MgCl_2$, Na_2SO_4 and traces of CuThCh	DFP in Na_2SO_4
"	" "	$MgCl_2$, Na_2SO_4	" " "
"	" "	None	None
"	Cu (acetate)	"	DFP in H_2O
"	Cu (sulphate)	NaCl, CuThCh	Eserine in incubate
"	" "	$MgCl_2$, Na_2SO_4	DFP in saline
"	" "	None	Ethopropazine or 62C47
Na citrate	" "	Potassium ferri-cyanide	Eserine sulphate
Tris-acetate	Lead (acetate)	" "	Eserine
None	$AuNa_3(S_2O_3)_2$	AuThCh phosphate	Eserine salicylate; BW284C51 or DFP in saline, or DFP i/v

TABLE 3.4 (continued)

	Conditions			Developer	Comments
	Time	Temperature	Fixation		
1.	10–60'	?	Fresh tissue	Yellow ammonium sulphide	This is the original thiocholine technique
2.	15–60'	38°C	" "	" "	All solutions up to developer saturated with CuThCh. Developer and rinse saturated with CuS. Sections fixed in formaldehyde after development
3.	5–60'	"	" "	" "	All solutions up to developer saturated with CuThCh. Developer, rinse, alcohols and xylol saturated with CuS. Post-fixation as above
4.	10–60'	37°C	" "	" "	Ingredients as in Koelle (1951) but no CuThCh and some concentrations slightly different. Preparation simpler since iodine not precipitated
5.	60'	"	10% formaldehyde	" "	No CuThCh. Fixed tissue. Sections 50 μ, cf. Koelle, 20–35 μ
6.	5–120' (18 hr for embryos)	37°C (warm-blooded); 22°C (cold-blooded)	10% neutral formaldehyde	" "	Acetate thought to be the rational buffer since acetyl radical is liberated
7.	Up to 100'	?	Unfixed blood	Yellow ammonium sulphide or phase contrast examination	Method used on blood cells, similar method used on nerve cells by Giacobini (1956a)
8.	20'–16 hr	?	10% formal saline	Yellow ammonium sulphide	Method specifically designed for peripheral tissue – see text
9.	1–4 hr varies with pH and temp.	Room temp. or 37°C	10% formaldehyde in Na_2SO_4	Na_2S in acetic acid	See text
10.	?	?	10% formaldehyde, 1% $CaCl_2$	Colour appears without development	See text
11.	1–60'	0°C	≃ 4% formaldehyde in Krebs–Ringer-Ca solution	Yellowish ppt. of $Pb_2Fe(CN)_6$ visible without development but can be converted with yellow ammonium sulphide	
12.	1.5'–3 hr	Room temperature	4% formaldehyde, sucrose and maleate pH 7.4	Alcoholic yellow ammonium sulphide	AuThCh phosphate in rinse

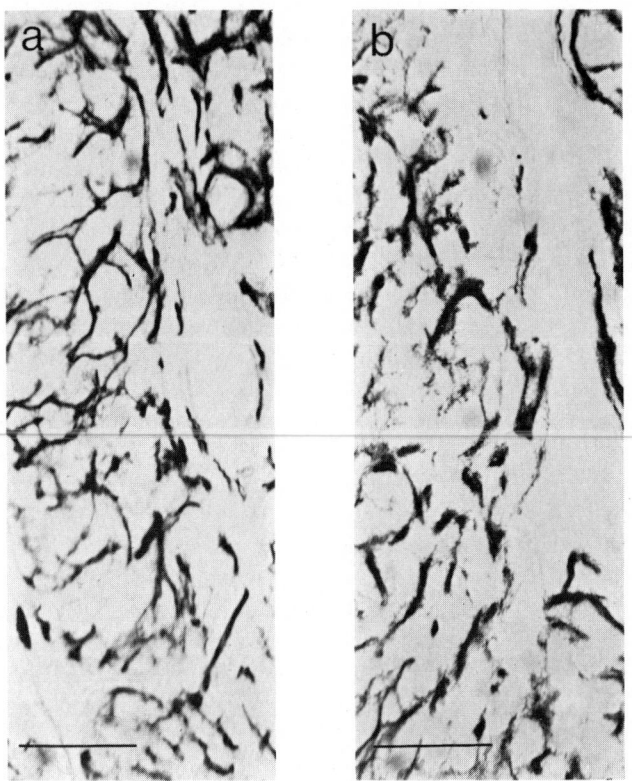

Plate 3.1. Sections of salt-gland of duck stained for AChE. a) Coupland and Holmes's method; b) Lewis's method. Note that the staining in a) is crisper. Scale bar: 100 μ.

the experiment and the incubation can be terminated as seems appropriate rather than at an arbitrarily predetermined time. Another advantage is that the end-product is finer and more precisely localized than that given by the conventional methods. The basis of the method is thought to be as follows: enzymically released thiocholine preferentially reduces ferricyanide in the medium to ferrocyanide and this in turn reacts with copper to give insoluble copper ferrocyanide which is reddish brown. A direct reaction between copper and ferricyanide is prevented because the copper ions are chelated with citrate.

Karnovsky and Roots mention only eserine sulphate as an inhibitor, 1.0 ml of a 1 mM solution being added to 9.0 ml of incubate in place of 1.0 ml of water. BW284C51 is satisfactory for inhibiting AChE and iso-OMPA can be used to inhibit pseudocholinesterase. Ethopropazine hydrochloride produces cloudiness in the medium but, nevertheless, it has been used satisfactorily (Fillenz and Wood 1966).

Phillis (1965a) found certain differences in the staining produced in cat cerebellum by the direct-colouring method and that given by Gerebtzoff's (1959) modification of Koelle's medium. With the former technique, the molecular layer and fibres in the white matter reacted more strongly but granule cells, which showed marked intracellular activity with Gerebtzoff's method, were often unstained. Phillis showed that such unstained cells could be 'developed' if the sections were subsequently immersed in sulphide solution. His conclusion, that the ferricyanide ion had been unable to cross the membranes of the negative cells to react with the thiocholine, has been confirmed by Bell (1966) who showed that intracellular staining occurred only in sections which had been subjected to some treatment, such as freeezing and thawing or osmotic shock, which would disrupt cell membranes. If sections were transferred from a sucrose-containing solution to one from which sucrose was absent, intracellular enzyme activity was demonstrated; if, on the other hand, sucrose was present at all stages or absent at all stages, there was no intracellular reaction.

Eränkö et al. (1967b) modified the direct-colouring method by using lead rather than copper as the capture agent; the very fine end-product, which is yellowish white, can be viewed by conventional or phase contrast microscopy or it can be converted to lead sulphide. In a preliminary comparison they used copper, lead, cobalt, gold and silver and also tested a number of different chelating agents and a range of pH. They found that conditions have to be critically adjusted because chelation of the heavy metal must be adequate to prevent it reacting with thiocholine and yet the binding must be sufficiently loose to allow the metal to precipitate with ferro- but not with ferricyanide. Tris-acetate was recommended in place of citrate but in another modification Kokko et al. (1969) replaced citrate with tartrate. Recently Norvell et al. (1971) reported that dark-field microscopy was of value in showing up the reaction product, particularly in areas of low activity.

3.3.2.5 Thiolacetic acid and other non-thiocholine substrates   Crevier and Bélanger (1955) showed that thiolacetic acid (ThAc) could be used as a histochemical substrate for cholinesterases. The enzymes release acetic acid and hydrogen sulphide and the latter is captured by lead, present in the medium as the acetate or nitrate, to give an insoluble precipitate of lead sulphide. The method is simple, the substrate is cheap and the end-product finely localized but its major disadvantage is that it is less specific than the thiocholine technique. ThAc is hydrolysed not only by AChE and BuChE but also by various other esterases and certain acetyltransferases (for references see Koelle and Gromadzki 1966) and most of these cannot be differentially inhibited. Despite this disadvantage the method has been developed by the electron microscopists (see §3.3.3) and some of their findings apply also at the optical level.

Particular care is needed in the selection of inhibitors for use with the thiolacetic acid method since this substrate combines with cholinesterase only at the esteratic site. DFP also reacts with the esteratic site of the enzyme and can, therefore, effectively prevent the histochemical reaction. BW284C51, on the other hand, has virtually no effect on the staining because it attacks only the anionic site (ch. 11 §2.2). Koelle and Gromadzki (1966) found that 10^{-6} M ambenonium, $[N,N'$-bis-(2-chlorobenzyldiethylammoniummethyl) oxamide dichloride], seemed suitable as a selective inhibitor of AChE in the AuThAc method (see below).

Koelle's group (for numerous references see Koelle and Gromadzki 1966; Davis and Koelle 1967) showed that when gold replaced lead as the capture agent in the ThAc method (and also in the thiocholine method), the reaction product was much finer. To allow the capture reagent to penetrate, the sections were preincubated in a solution which contained aurous sodium dithiosulphate, $AuNa_3(S_2O_3)_2$, and an amount of thiolacetic acid which was just sufficient to combine stoichiometrically with the gold salt. The resulting bis-(thioacetoxy) aurate (I) complex, $Au(CH_3COS)_2$ or $Au(TA)_2$, penetrates more readily than does the aurous ion. In the incubation medium itself, the thiolacetic acid was present in a higher concentration and it acted as the substrate. The primary end-product formed by the reaction of the gold with sulphide released during hydrolysis of the substrate could be viewed directly or could be intensified by immersion of the section in ammonium sulphide solution. In the original method, $Au(TA)_2$ was acting merely as a vehicle for the capture agent but later Koelle et al. (1968) found that this complex could also act as a substrate for cholinesterase. The method they developed for electron microscopy has recently been refined (Koelle et al. 1974, see §3.3).

In 1968 Koelle and Horn showed that when commercial thiolacetic acid is redistilled it is practically useless as a histochemical substrate and they concluded that the component responsible for the staining obtained with impure ThAc is acetyl disulphide. Koelle et al. (1968) studied the behaviour of acetyl disulphide as a substrate and discussed the reasons why, with lead as the capture agent, the type of reaction which occurred depended on whether or not ThAc was present. They found that in the absence of added thiolacetic acid the staining was reddish brown but in its presence, the stain was black (due to the formation of black lead sulphide) and its intensity was greatly increased. Subsequent work led Koelle's group (see Koelle et al. 1974) to reject the acetyl disulphide method because of the very rapid spontaneous hydrolysis of the substrate. With the brief incubation period needed to demonstrate AChE at the motor end-plate this is not a serious problem but for tissues requiring a longer period of incubation the heavy background deposit which is formed is a major snag. This provides one more example of the problem that methods which are useful at the end-plate may be useless elsewhere.

The use of the same compound as both substrate and capture agent was mentioned above in connection with the bis-(thioacetoxy) aurate (I) complex. Another method employing this principle has been introduced by Mednick et al. (1971). The compounds they used were benzene and toluene diazonium salts of thiolesters. The products resulting from enzymatic hydrolysis of the thiolacetate

group are the p-mercapto-benzenediazonium ion and α-mercapto-*m*-toluenediazo-
nium ion respectively. These subsequently form pale yellow polymeric diazoethers
which are osmiophilic, thus osmication of the stained sections produces a dense
black deposit at the site of enzyme activity. The method can be used for optical
and electron microscopy. Disadvantages would seem to be the complexity — at least
for the non-chemist — of the synthesis of the substrates and their relative instability;
this is particularly true of acetyl thiolbenzene diazonium. The substrates are
thought to be highly specific for cholinesterases by virtue of the positive charge on
the diazonium group. The authors showed they were not hydrolysed by A- and
B-esterases in rat kidney but the only inhibitor used was eserine in the rather high
concentration of 10^{-3} M. In a subsequent study with this type of substrate Davis et
al. (1972) found that 10^{-5} M eserine and BW284C51 were 90—100% effective in
inhibiting the reaction at the end-plate.

Booth and Metcalf (1970) have made out a case for using phenylthioacetate (PT)
as a substrate and they list several apparent advantages over AcThCh: it is cheap,
easily synthesised, more stable and being lipid-soluble is able to penetrate membranes more easily. There is evidence, however, that the two compounds may not
be interchangeable, at least for some species. McEnroe (1971) has found that in the
spider mite, *Tetranychus urticae,* PT stained the midgut and was insensitive to 1 X
10^{-7} M Paraoxon while AcThCh stained the synaptic areas in the brain, and also
the surfaces of the nerves.

Apart from the thiocholine and thiolacetic acid type of compound, other
substrates tested have included indoxyl compounds, which eventually yield blue
indigo (Barrnett and Seligman 1951; Holt and Withers 1952), and β-naphthyl esters
used in a diazo-coupling method (Nachlas and Seligman 1949a, b; Ravin et al.
1953). Both methods are, however, unspecific and this is especially unfortunate in
the case of indoxyls because in other respects they have certain advantages over
thiocholine. Holt and Withers pointed out that because they are chemically and
electrically neutral the risk of artifacts due to non-specific binding of quaternary
cations by acidic components in the tissue is avoided. Unlike other methods, the
indoxyl technique is capable of demonstrating AChE in the erythrocyte. Holt and
Withers attributed this to the ease with which indoxyls can penetrate membranes,
aided by the fact that with the absence of copper from the medium, the RBC
retains its special permeability characteristics. More recent work suggests that the
absence of copper from the indoxyl medium can be important in another way.
Skaer (1973) found that the particular isozymic form of AChE present in red cells
is especially sensitive to inhibition by copper and also by fixatives and acidity. In
contrast, the form of enzyme in red cell precursors is more resistant to inhibition
and can be demonstrated by the thiocholine technique at the electron microscope
level.

3.3.2.6 Immunofluorescence   An immunofluorescence technique for demonstrating AChE in electric tissue has been described by Benda et al. (1970). In this method antibodies to AChE are produced by injecting purified AChE into rabbits. Sections of *Electrophorus* electroplax are exposed to a dilution of the rabbit immune serum which has been absorbed with eel serum. Subsequently the sections are washed, treated with fluorescein-conjugated anti-rabbit globulins and viewed under fluorescence microscopy. The method has recently been used to demonstrate nerve cells in brain and cord of *Electrophorus* (Tsuji et al. 1972).

3.3.3 Electron microscopy

3.3.3.1 Introduction The electron microscope histochemist, like the histochemist working at the optical level, needs a method which localizes maximum enzyme activity with maximum accuracy but in electron microscopy there is the additional requirement that the integrity of the fine structures must be preserved throughout all stages of processing. This latter condition requires that the tissue be really well fixed and, at once, the histochemist is in difficulties because, as mentioned earlier, the longer the period for which a tissue is fixed the less active will its enzymes be. Barrnett and Palade (1958) discussed the general applicability of histochemistry to electron microscopy and said that at first sight the needs of the two techniques seemed 'mutually antagonistic' but they went on to say that compromises could be achieved. That such compromises have been successful in the case of cholinesterases is clear from the impressive photomicrographs to be found in the literature (see also ch. 4 §2.3). Nevertheless, as with optical microscopic histochemistry the techniques still require further refinement. Koelle (1970) has summarized some of the criteria to be met. His first and very pertinent point is that it is useless to increase the resolution of the microscopy unless diffusion artifacts are reduced in proportion. Similarly, he pointed out that it is no use revealing the subcellular elements unless these have been fixed sufficiently to be identifiable. Thirdly, since tissues for electron microscopy may be stained as small pieces rather than as sections, the samples are often relatively thick compared with the already sectioned tissue which is stained for light microscopy; as a result the problems of penetration of substrate and capture agent assume greater importance. The final point Koelle makes is that an end-product which seems suitably amorphous and opaque under the optical microscope may be too coarse and insufficiently electron-dense to give precise localization at the subcellular level. Other characters required of the end-product are that it should be insoluble in the reagents used during the processing and embedding of the tissue and that it should be stable in the electron beam.

3.3.3.2 Practical points Although the details of the timing and the composition of the fixative, incubation and rinsing solutions may vary from method to method most procedures have a common design. The fixed tissues are well washed and chopped into small pieces or are sectioned, preferably without freezing. Preincubation with inhibitors in the absence of substrate is followed by incubation with substrate; the samples are then rinsed and, if appropriate, developed. After further washing, and either before or after post-fixation, minute fragments are dissected from the stained tissues. These are dehydrated, embedded in resin and sectioned on an ultramicrotome.

Because of the need to preserve the ultrastructure as well as the enzyme activity, the composition, pH and molarity of the subsidiary solutions such as rinsing solutions are more critical than in optical microscopy. It is equally important, however, that precautions aimed at maintaining structural integrity should not hinder substrate penetration. Bell's (1966) finding (§3.3.2.4) that with Karnovsky and Roots's method, intracellular staining was prevented if sucrose was present in all solutions illustrates this point. Care is also needed in the selection of any stain to improve the contrast of the ultrathin sections; Lewis and Shute (1966) found, for instance, that uranylacetate caused some loss of the histochemical end-product.

3.3.3.3 Fixation for electron microscopy   Most electron microscopists are agreed that fixation is necessary for satisfactory results. As with optical microscopy, fixation by perfusion is recommended, especially for the nervous system, but immersion in fixative may be adequate for some other tissues.

Fixatives can change the permeability of membranes and may aid the penetration of reagents to intracellular sites (see ch. 4 re Tennyson et al. 1968) but because various fixatives can affect penetration to a different degree the precise distribution of end-product may depend on the one used. Eränkö et al. (1970) found, for example, a difference in the distribution of end-product in pineal glands fixed in glutaraldehyde and those fixed in formaldehyde. Whatever the fixative chosen, it is imperative that tissues should be washed very thoroughly between fixation and incubation.

Much of the early electron microscope histochemistry was done on motor end-plates where AChE activity is high and where considerable inhibition does not necessarily vitiate the result. Fixatives used for muscle have included cold buffered (pH 7.4) 10% formaldehyde (Lehrer and Ornstein 1959), cold buffered (pH 7.5) 1% osmic acid (Birks and Brown 1960) and cold buffered (pH 7.4) 0.25% osmic acid (Barrnett 1962). Torack and Barrnett (1962) also used the latter on brain tissue but found it inhibited the enzyme almost completely if fixation exceeded 3 min. Glutaraldehyde (Sabatini et al. 1962) was criticized by Torack and Barrnett because, while preserving the morphology, it inhibited cholinesterase but subsequently it has been used by many workers at various concentrations and in various buffers, either alone or in combination with formaldehyde. Some of the fixatives which have been recommended during the last decade are shown in table 3.3.

3.3.3.4 Staining methods   Lehrer and Ornstein (1959), using a diazocoupling method, seem to have been the first to demonstrate cholinesterases at the ultrastructural level. Shortly afterwards, modifications of the thiocholine method (Birks and Brown 1960; Brown 1961) and the thiolacetic acid technique (Zacks and Blumberg 1961; Barrnett 1962) were described and, as with optical microscopy, it is these two types of method which are currently in general use. Some of these techniques will be outlined here but the original papers should be read for details. For a review of ultrastructural studies of AChE at the motor end-plate see Friedenberg and Seligman (1972).

Lewis and Shute (1966, see also 1969) used a medium which is basically their buffered copper glycinate-thiocholine incubate of optical microscopy but with some adjustment of concentrations and with the addition of isotonic sodium sulphate. The pH can be altered according to the tissue; the more active the tissue the lower the pH used. Skaer (1973) acting on a suggestion of P.R. Lewis showed that localization of end-product was improved if, following incubation, samples were rinsed for 30 min in an acetylcholine wash made by substituting ACh for AcThCh in the reaction medium.

In another modification of Koelle's medium, Kása and Csillik (1966) introduced lead (as lead nitrate) in addition to copper, as the capture agent. They suggested that this combination increased the electron density of the end-product. In this method, as in some of those for optical microscopy (table 3.4), iodine is not precipitated out. As mentioned earlier (§3.3.2.5) gold has been used as a capture agent for thiocholine at both the optical and EM levels (see Koelle and Gromadski 1966). The advantage of gold, like lead, is that its mercaptide is more electron-dense and is finer than that of copper. A quaternary carbon analogue of acetyl-thiocholine, 3,3-dimethylbutylthioacetate, has been tested as an alternative substrate in the Koelle medium by Nyberg-Hansen et al. (1969). Since it is uncharged it should, in theory, penetrate membranes more efficiently and it has produced good results at the motor end-plate.

Karnovsky and Roots's (1964) direct-colouring method (§3.3.2.4, table 3.4 no. 10) has formed the basis for several EM variants. Eränkö et al. (1967a) used the original medium but in some instances sodium citrate has been replaced by tartrate (Kokko et al. 1969; Eränkö et al. 1970). Acetyl-β-methylthiocholine (MeThCh) and acetylselenocholine (AcSeCh) were tested as substrates for the direct-colouring method by Kokko et al. (1969) and both gave a well localized fine end-product. The reaction was fastest with AcSeCh but histochemical tests with inhibitors, and biochemical studies with purified enzymes, showed MeThCh to be the more specific. It should be noted that this thioanalogue is not, however, as specific for AChE as is acetyl-β-methylcholine itself (Augustinsson 1957).

I mentioned earlier that the thiolacetic acid methods are less specific than the thiocholine type but their advantage, the precision with which the end-product is localized, gains an added importance at the EM level. So long as their questionable specificity is remembered they will remain useful. Koelle and his co-workers (see Davis and Koelle 1967) have described an EM version of their thiolacetic acid method with gold as the capture agent. This method, and also an adaptation of the one outlined in §3.3.2.5 in which $Au(CH_3COS)_2^-$ [for brevity termed $Au(TA)_2$] acts both as substrate and capture agent, gave well-localized staining at motor end-plates but were not satisfactory for ganglia and central nervous tissue (see Koelle 1970). Recently Koelle et al. (1974) have described a modification of the $Au(TA)_2$ technique which gives reliable results for AChE and pseudocholinesterase in autonomic ganglia as well as at the end-plates. Among the factors necessary for success are the purity of the reagents, a ratio of thiolacetic acid to gold of exactly 2 : 1, and the replacement of the medium every hour.

The osmication method, with benzene or toluene diazonium salts of thiolesters acting again as both substrate and capture agent (§3.3.2.5), has been used for electron microscopy at the end-plate (Mednick et al. 1971). Davis et al. (1972) have recently reported the synthesis of an indoxyl substrate, 3-acetoxy-5-indole diazonium ion, which works in the same way. The reaction product formed at the end-plate was finer than that produced by the thiol type of substrate.

3.4 Appendix

3.4.1 Preparation of tissue for sectioning

The procedures described here are those normally used in this laboratory on tissue destined for staining in Lewis's medium but they are, in the main, applicable to other methods as well.

3.4.1.1 Fresh frozen tissue   Robust specimens such as liver are easy to handle and the fresh tissue can be cut without difficulty on a freezing microtome, cooled either by CO_2 or electrically. Very small objects, for example the cat carotid body, or very delicate tissue, for example foetal brain, are more conveniently cut in the cryostat. Various methods may be used to freeze samples for the cryostat; these include exposure to gaseous and solid CO_2 or immersion in liquid nitrogen, propane or Arcton (ICI). The choice of the freezing agent depends to some extent on the tissue. The rapid freezing produced by liquid quenching agents may cause cracks in some specimens, particularly brain, but Winckler (1970a) has found these can be avoided if the tissue is first wrapped in plastic film; he has also shown that addition of talcum powder to nitrogen-propane minimizes ice crystal artifacts. Carbon dioxide gas freezes tissues more slowly and will not crack them, but if the tissue block is thicker than 1 cm cooling may be too slow and ice crystals will then form, producing holes in the sections (see plate 3.2). Freezing can be speeded up if the stage is fitted with a simply-made collar which directs the CO_2 onto the tissue (plate 3.3). Recently Winckler (1970b) has introduced the use of freeze-dried cryostat sections (as opposed to freeze-dried blocks) for histochemistry in general and has obtained good results with AChE techniques.

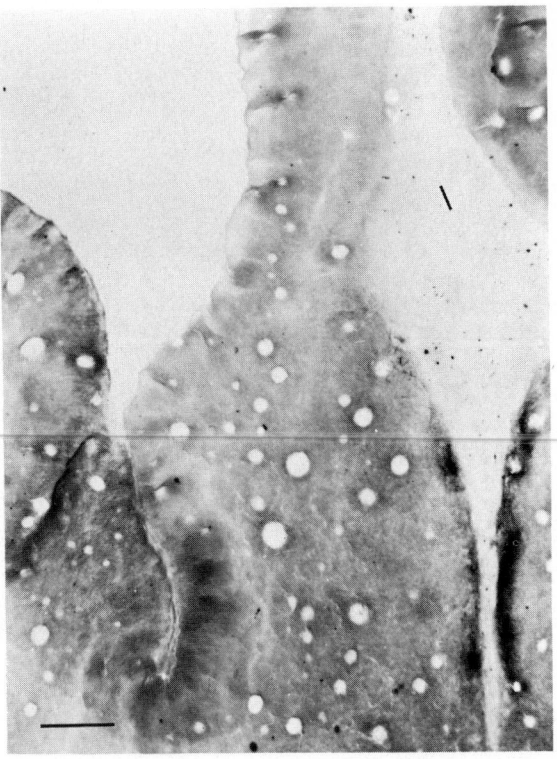

Plate 3.2. Section of brain from foetal cat showing the ice holes which result when the tissue is frozen too slowly. Scale bar: 250 μ.

3.4.1.2 Fixation Unless samples are very small, those fixed by perfusion are generally more satisfactory than those fixed by immersion. This is particularly true of the CNS and the routine in this laboratory for perfusion of the brain is as follows: animals are anaesthetized with an intraperitoneal injection of sodium pentobarbitone (Nembutal), the chest is opened and the pulmonary vessels tied off. The right ventricle is pierced and immediately a cannula is inserted either into the descending aorta (directed headwards) or into the innominate artery, near the heart; in the latter case, the descending aorta is subsequently tied off. The initial perfusion, to flush out the blood, is with isotonic (16.76 g/l) sodium sulphate solution at 37°C. When the effluent from the right ventricle is relatively clear, the perfusate is changed to isotonic sodium sulphate containing 10% (v/v) formaldehyde solution (AR, 36% w/v); the total volume circulated is usually about 100 ml/kg body weight. The brain is then dissected out and the whole of it, or the relevant portion, put into 10% formaldehyde:Na_2SO_4 (at 4°C) in the refrigerator.

Plate 3.3. Reichert freezing microtome fitted with a collar (arrowed) which directs CO_2 on to the specimen stage. (Collar designed by Mr D.V. Barker in this Institute.)

After a total of 4 hr, timed from the start of the perfusion with formaldehyde, it is washed several times in water at $4°C$. It is then transferred to 20% ethanol (at $4°C$) in which it can be kept under refrigeration ($4°C$), until sectioned. Under these conditions AChE activity survives apparently unaltered for well over a week but it is preferable to section the tissue within 1–3 days of removal. Storage in alcohol overnight is routine in this laboratory because this improves the quality of the sections (Marshall 1940). Karnovsky and Roots (1964) advocate 0.88 M sucrose with 1% gum arabic as a storage solution.

3.4.1.3 Alginate embedding   Lewis and Shute (1963) developed a method suitable for cholinesterase histochemistry in which fixed tissues are embedded in sodium alginate – this is a salt of polymannuronic acid which is extracted from seaweed. The processing involved does not harm the enzyme and the blocks do not crack on freezing. In both these respects alginate is superior to gelatin as an embedding medium. The method is useful for tissues which tend to fragment and is specially good for maintaining samples in a particular orientation. For example, it proved very satisfactory for embedding the goldfish optic system so as to give sections showing the continuity of the eyes, optic nerves and tectum.

3.4.2 Handling of sections

The thickness at which blocks are cut depends on the object of the experiment. To a certain extent detail is clearer in thinner sections and this is particularly so if counterstains are used. On the other hand, with thicker sections (say up to $70\,\mu$) nerve fibres can be traced over longer distances and interconnections are more obvious. Where enzyme activity is low the incubation time must be increased for thinner sections.

Sections of fresh tissue, whether cut in the cryostat or on the freezing microtome, are generally mounted directly on to coverslips. So long as these have been thoroughly cleaned in absolute alcohol and ether, and then polished with 'Velin' tissue (General Paper and Box Co.) sections will normally stick adequately. Once it has been established that fixation does not produce artifacts, mounted sections of fresh tissue can be lightly fixed by exposing them, when dry, to formaldehyde vapour for 1–2 min. This improves the morphology and enzyme localization and also increases the attachment of the section to the coverslip. The apparatus used in this laboratory for controlled formaldehyde treatment is shown in plate 3.4.

Sections of fixed tissue are best cut on a freezing microtome rather than in a cryostat. If they are to be attached to coverslips for processing, they can be floated out on the coverslip on a drop of 20% alcohol. To ensure that a fixed section will stick, it is necessary to coat the coverslip very lightly with glycerin albumin. Adhesion is improved if the sections, when dry, are exposed to formaldehyde vapour as described above. As an alternative to the coverslip method, fixed sections, if sufficiently tough, can be floated in the histochemical media and mounted on slides after processing is complete. An advantage of the free-floating technique is that sectioning is quicker since no time is lost in arranging each section on a coverslip as it is cut. On the other hand, sections are more liable to damage during transfer from solution to solution and some, such as those of cerebellum, may become irreparably tangled. An important factor to be considered in deciding which method to use is the cost. The thiocholine substrates are expensive and the method which requires the smaller volume of medium should be used if other factors are equal. Eight coverslips, 22×22 mm, can be immersed in about 8 ml medium in a columbia jar, and with care, more than one small section can be put on each coverslip without the serial order being lost. Sections of a similar range of size can be floated in medium contained in the wells of a haemagglutination tray; individual wells need only 0.5 ml but if the serial order is important only one section can be put in each well.

Special methods have been developed in this laboratory for the economical handling of sections that are too large to be incubated in either columbia jars or haemagglutination trays. Free-floating sections can be processed in shallow watch-glasses; to maintain serial order and for ease of transport these may be arranged in groups of 10–12 on a thin piece of board or metal to which they are stuck with Chatterton's compound. As an alternative, sections of monkey and sheep brain, up

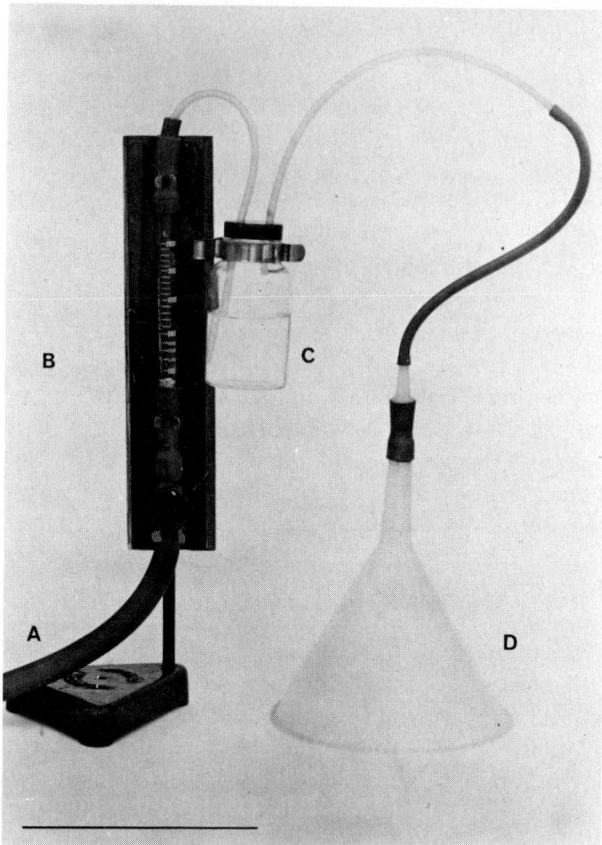

Plate 3.4. Apparatus for delivering a controlled flow of formaldehyde gas during fixation of tissue sections. A) tube from CO_2/O_2 cylinder. B) rotameter. C) screw-capped bottle containing $\simeq$ 75 ml formaldehyde solution (AR, 37% w/v). D) 18 cm plastic funnel which is placed over the petri dish on which the section-bearing cover-slips are arranged. Scale: 20 cm. (Apparatus designed by Mr G.R. Marshall in this Institute.)

to 7 X 4 cm in size, have been mounted on coverslips and incubated face downwards on a puddle of incubation medium pipetted on to a layer of paraffin wax in the bottom of an enamel dish.

3.4.3 Bulk staining

Certain kinds of tissue which have an open type of structure (e.g. muscle), can be stained without being sectioned (Lewis 1961). This is a useful way of demonstrat-

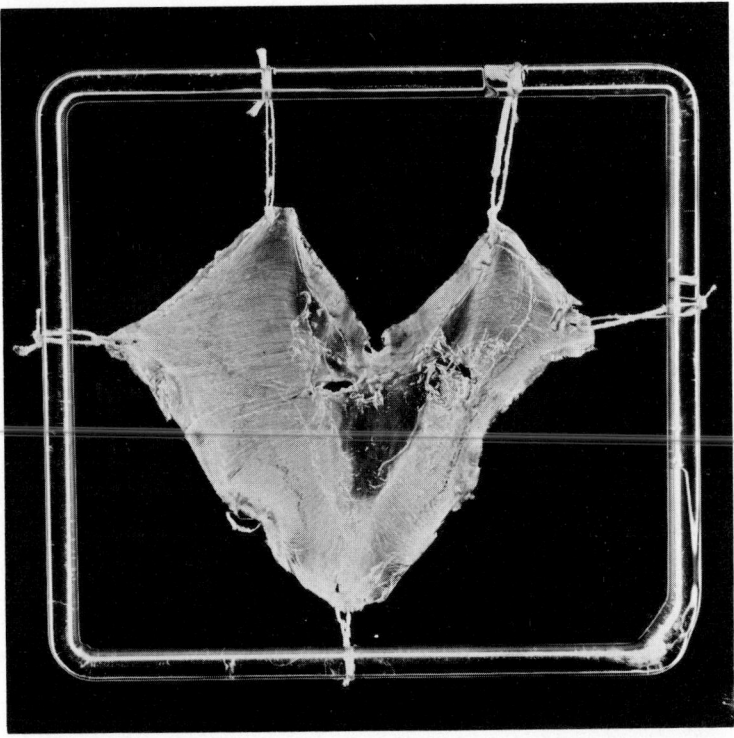

Plate 3.5. To show use of glass frame for maintaining the shape of tissue (guinea-pig diaphragm) during fixation and processing.

ing the distribution of elements such as motor end-plates (Silver 1963) or peripheral nerve fibres (Hebb and Linzell 1970) over a wide area. Pieces of tissue should be mounted on some sort of support to prevent distortion. Plate 3.5 shows the type of glass frame which I use for holding diaphragms; Hebb and Linzell (1970) used similar ones for rabbit mammary glands. The tissues should be fixed by immersion in 10% formaldehyde: Na_2SO_4 for up to 4 hr and then washed in water. Bulky muscles must be dissected into strips not greater than 2 mm thick but whole mounts of diaphragm and mammary glands of the diffuse type can be incubated intact.

Lewis's concentrated medium (see §3.4.5.3) should be used and the length of the incubation period determined by trial and error. The longer the incubation period, the more even is the staining, hence the pH should be adjusted until a period of at least 6 hr can be used without over-staining. Following incubation the tissue is washed in water, kept at room temperature in 10% formaldehyde: Na_2SO_4 containing 0.5% glacial acetic acid for 6–18 hr, rewashed in water and then

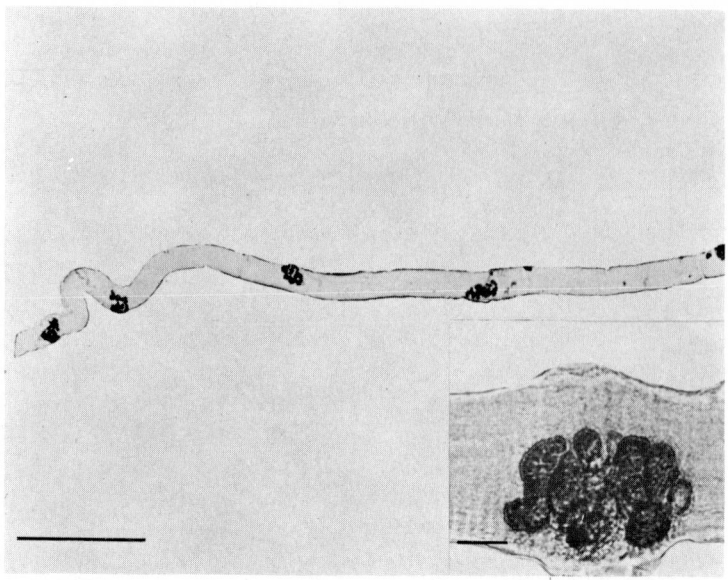

Plate 3.6. Demonstration of AChE in whole mounts. Fibre teased from superior rectus oculi of goat after the muscle had been stained to show the AChE at the end-plates. Inset, high-power view of a single end-plate. Scale bar: 250 μ (inset 10 μ).

developed in sodium sulphide developer. Lewis recommends development for 6–8 hr but I have found much shorter periods (½–1 hr) are adequate for showing up motor end-plates in diaphragm. Whole mounts should be dehydrated in alcohol and cleared in xylene; they can be stored in methyl benzoate for several years without loss of stain. Alternatively, pieces of muscle can be soaked in glycerol and the fibres teased and mounted in glycogel so that individual motor end-plates can be examined (plate 3.6). If sections are required, the specimens are dehydrated in alcohol after development, cleared in xylene and then vacuum-embedded in paraffin.

3.4.4 Note on thiocholines as substrates

As mentioned already, the thiocholines used as histochemical substrates are usually in the form of iodides. In the original Koelle method, and in many of its variants, the iodide is precipitated with copper sulphate (or acetate). The reaction which occurs is: $2\,Cu^{2+} + 4\,I^- \rightarrow 2CuI + I_2$. According to Malmgren and Sylvén (1955) most of the free iodine formed is adsorbed onto the precipitate of cuprous iodide. In a number of methods including those of Gomori (1952) and Karnovsky and

Roots (1964) prior precipitation of the iodide is omitted. That precipitation does not occur when the substrate is mixed with the other ingredients is due to the conditions existing in the medium and it is therefore important to add the constituents only in the order prescribed. When, as a control (see §3.1.1.3), substrate is omitted from any medium in which the iodide is normally removed by the reaction with copper, the concentration of the copper sulphate should, in theory, be reduced. If this is not done the copper concentration of the so-called control will be higher than that normally present. The literature suggests, however, that little attention is given to this point.

Berry and Rutland (1971) have suggested that the iodide radical may in some way be essential to the histochemical reaction since neither acetylthiocholine chloride nor acetylthiocholine methane sulphonate produced any staining of the rat motor end-plate under conditions in which acetylthiocholine iodide was an effective substrate. In this laboratory, however, acetylthiocholine perchlorate has been used successfully as a histochemical substrate so some factor other than the lack of iodide must have been responsible for Berry and Rutland's negative findings. Possibly those substrates which did not give staining had undergone hydrolysis during storage. In my experience deterioration of the substrate is the commonest cause of staining failures – any substrate with a marked fishy smell is suspect. Because different samples of substrate, even from the same manufacturer, may vary in performance it is worth comparing the staining given by each new batch with that given in the same experiment by the previous batch. Methods for preparing thiocholine esters in the laboratory are described by Hansen (1957).

3.4.5 Preparation and use of thiocholine incubation media

3.4.5.1 Koelle's (1951) medium (table 3.4, no. 3)   In this, as in all methods, glass distilled water must be used throughout. The method was developed for fresh tissue but can be used on fixed material.

   i) Stock solutions

   Copper sulphate: 0.1 M

   Copper glycinate: 3.75 g glycine, 2.5 g $CuSO_4 \cdot 5H_2O$, make up to 100 ml with H_2O

   Maleate buffer: 9.60 g sodium hydrogen maleate, 52.2 ml N NaOH, make up to 100 ml with H_2O

   Sodium sulphate: 40 g Na_2SO_4 (anhydrous), make up to 100 ml with H_2O, adjust pH to 6.0, store at 38°C

   Magnesium chloride: 9.52 g $MgCl_2 \cdot 6H_2O$, make up to 100 ml with H_2O

   Copper thiocholine (CuThCh): this can be specially made from AcThCh (Koelle 1950) but the more economical method is to retrieve it from used incubation media as follows: filter the medium after use and store at 38°C for 2–4 days to allow spontaneous hydrolysis of the substrate. Spin down the precipitate and wash it

ii) Substrates: Prepare just before use

AcThCh: 23 mg AcThChI, dissolve in 1.2 ml H_2O, add 0.4 ml 0.1M $CuSO_4$, centrifuge, retain supernatant

BuThCh: 43 mg BuThChI, dissolve in 1.8 ml H_2O, add 0.6 ml 0.1M $CuSO_4$, centrifuge, retain supernatant

For use, add the specified volume in the order given

Ingredient	Volume (ml)	
	For AChE	For BuChE
Copper glycinate	0.6	0.6
H_2O	2.1	0.6
Maleate buffer	1.5	1.5
Sodium sulphate	9.0	10.5
Magnesium chloride	0.6	0.6
CuThCh	trace	trace
AcThCh substrate solution	1.2	–
BuThCh substrate solution	–	1.2

At the end of the incubation sections are rinsed sequentially in the following solutions

20% Na_2SO_4 with CuThCh

10% Na_2SO_4 with CuThCh

H_2O with CuThCh

They are then developed with CuS-saturated ammonium sulphide. To prepare this, dilute concentrated (.880) ammonia 50:50 with distilled water and saturate the solution with H_2S from a Kipp's apparatus; store in the refrigerator. Immediately before use, dilute 1:25 with H_2O and saturate with CuS by the dropwise addition of 0.1 M $CuSO_4$. After development (20 sec) wash sections in water, fix in CuS-saturated 10% formalin and dehydrate and clear in CuS-saturated alcohols and xylene.

3.4.5.2 Coupland and Holmes's (1957) method (table 3.4, no.8) For original fixative see table 3.3, no.5, and also comment §3.3.2.4.

i) Stock solutions

Copper sulphate (as for Koelle 1951)

Copper glycinate (as for Koelle 1951)

Magnesium chloride (as for Koelle 1951)

0.2 M acetic acid: sodium acetate buffer at required pH

Sodium sulphate (as for Koelle 1951)

ii) Substrates: Prepare just before use. Both AcThCh and BuThCh are made up as for AcThCh in Koelle's method.

To demonstrate AChE or BuChE add, in order

0.6 ml copper glycinate

0.6 ml magnesium chloride

5.0 ml acetate buffer

7.6 ml sodium sulphate

1.2 ml AcThCh or BuThCh as appropriate

At the end of the incubation wash sections in water, develop as for Koelle's method, rewash in water, dehydrate and mount.

3.4.5.3 Lewis's (1961) method (table 3.4, no. 9)   See table 3.4, no. 1 and §3.4.1.2 re fixative. The medium can be prepared at least one week before use if stored in the refrigerator.

Stock solutions

0.1 M copper sulphate

1.0 M sodium acetate

Dissolve 100 mg AcThChI or 110 mg BuThChI in 4.0 ml H_2O in a centrifuge tube. Add 7.0 ml 0.1 M $CuSO_4$; to ensure complete removal of iodide (see §3.4.4) the initial 2–3 ml should be added drop by drop. Leave the tube to stand for 10 min and then centrifuge for 15–20 min at 2000 r.p.m. Pipette off 10 ml supernatant and in this dissolve 62 mg glycine. Adjust pH to the required value (usually 5.5 in this laboratory) with 1.0 M sodium acetate. With water, make up to a final volume of either 20 ml for bulk staining (see §3.4.3) or 50 ml for sections, and then filter.

At the end of the incubation period rinse sections three times in water and treat with developer (for 2 or 4 min, §3.3.2.4) prepared as follows. Dissolve 1 g sodium sulphide (analytical grade) in 45 ml 0.2 N acetic acid and reduce the pH to between 5 and 6 with 1 N acetic acid. Filter through Whatman's No. 1 paper folded fan-wise. The developer should be freshly prepared for each experiment and should be kept in the refrigerator until immediately before use. This developer, like yellow ammonium sulphide, gives off H_2S and should be used only in a fume cupboard. Particular care should be taken to avoid contamination of the incubation medium. Following development sections are dehydrated, cleared and mounted as described in §3.3.2.4.

3.4.5.4 Karnovsky and Roots's (1964) method (table 3.4, no.10)   See table 3.3, no.3 for fixative

Stock solutions

0.1 M sodium hydrogen maleate buffer (pH 6.0)

0.1 M sodium citrate

30 mM copper sulphate

5.0 mM potassium ferricyanide

Dissolve 5 mg AcThChI or BuThChI in 6.5 ml sodium hydrogen maleate buffer and add the following in order, stirring between each addition

0.5 ml sodium citrate
1.0 ml copper sulphate
1.0 ml H_2O
1.0 ml potassium ferricyanide

Although the stock solutions can be stored for some weeks in the refrigerator the final medium is stable for only a few hours. As explained earlier (§3.3.2.4) no developer is necessary. At the end of the incubation the sections are rinsed in water, dehydrated, cleared and mounted as usual.

3.4.6 The combination of histochemistry with other techniques

3.4.6.1 Silver staining   A number of methods are available for the silver staining of nerve fibres in preparations already treated to show AChE activity. Thiocholine staining of fixed tissue can be followed by Hortega's silver method (Wolter 1964) or by Bielschowsky's method (Gwyn and Heardman 1965). Namba et al. (1967) stained fresh tissue by Koelle's method followed by a modification of Honjin's (1956) version of Cajal's technique while Ip (1967) has used Coupland and Holmes's method followed by the silver method of Barker and Ip (1963). The thiolacetic acid method for cholinesterases has been combined with a protargol technique (Csillik and Sávay 1958) and with the Bielschowsky method (Nakata and Nishijima 1971).

3.4.6.2 Catecholamine techniques   Jacobowitz and Koelle (1963, 1965) modified the Falck (1962) fluorescence method for catecholamines so that it did not totally inactivate AChE. With their method both components could be demonstrated in the same section but not simultaneously. Cryostat sections are exposed to para-formaldehyde vapour at $37°C$ for 2 hr, photographed and then stained for AChE by Koelle's technique, and rephotographed. The distribution of fluorescence and of staining can thus be compared. In the method of El-Badawi and Schenk (1967b) the sections are treated in the reverse order, first for AChE by a modification of Karnovsky and Roots's procedure and subsequently for noradrenaline (NA) by an adaptation of Eränkö and Räisänen's (1966) cold formaldehyde version of the Falck method. Under combined fluorescence and phase-contrast microscopy, the adrenergic nerves show yellow-green fluorescence and the cholinesterase-containing fibres are brownish-black. Ellison and Olander (1972) combined the Koelle and the Falck techniques and used a dual illumination system which shows AChE-positive structures as red and adrenergic elements as green against a black background. As well as these methods for use on the same section there are a number of others (Eränkö 1967) in which AChE and NA are separately demonstrated on adjacent cryostat sections. In the method of Hebb et al. (1966) the two are shown on consecutive sections from paraffin-embedded freeze-dried tissues; with this technique the cytology is particularly well preserved.

Methods for the simultaneous demonstration of catecholamines and cholinesterases at the ultrastructural level have been developed by a number of workers who exploit the convenient fact that NA-containing granules in glutaraldehyde-fixed tissue are extremely osmiophilic, and this property is retained after treatment with the thiocholine technique. Palkama (1967) gives details of his method in which he stained glutaradehyde-fixed tissue for cholinesterase and then post-fixed in Dalton's (1955) chrome-osmium fluid.

Graham et al. (1968) developed a technique whereby the tissue is perfused with [³H]-NA, fixed in glutaraldehyde, stained for AChE by the method of Lewis and Shute (1966) and finally processed for autoradiography at the ultrastructural level. The silver grains and AChE-reaction product can be visualized simultaneously under the electron microscope.

3.4.6.3 Application of cholinesterase techniques to lesioned tissue, combined with measurement of choline acetyltransferase When a cholinesterase-containing fibre or tract is sectioned, enzyme accumulates on the cell body side of the lesion and is lost distal to it. This behaviour is a useful guide to the polarity of fibres and where the object of an experiment is mainly anatomical (Shute and Lewis 1961; Gwyn and Wolstencroft 1966) the accumulation of enzyme can be demonstrated histochemically. More information about the pharmacology of the tract, that is whether or not it is likely to be cholinergic, can be gained if fibres showing this type of reaction are also analysed for ChAc activity. In the method used by Lewis et al. (1964, 1967), on rat hippocampus following unilateral lesions in the fimbria, unfixed frozen sections were cut at alternately 400 μ and 100 μ. The latter were stained for AChE while the others were stored overnight at −70°C. With the stained sections as a guide, areas corresponding to those showing an accumulation or loss of enzyme on the side with the lesion were dissected from the thick sections with the aid of a low-power microscope. Equivalent samples were taken from the control side. The tissues were smeared on pieces of 'Visking' (Hude Manufacturing Co.) dialysis tubing and analysed for ChAc by the method of Bull et al. (1963). A special tareing method (Silver 1969) can be used to weigh very small samples, such as cross-sections of individual spinal tracts dissected from cryostat sections (see Gwyn et al. 1969, 1972).

3.4.6.4 Blood vessel markers In some experiments it is important to demonstrate the exact relation of nerve fibres to blood vessels. Biscoe and Silver (1966) obtained satisfactory cholinesterase staining (Lewis's method) in cat carotid bodies which had been perfused with carmine gelatin (Moore 1929). Ballantyne (1970) added 2% ferric potassium ferrocyanide (soluble Berlin Blue) to the fixative with which he perfused rat brain. Electron microscopy showed that the dye did not pass into or across the epithelium of the blood vessels and experiments on homogenized tissue proved that in any case it had no effect on the ChE activity.

3.4.6.5 Decalcification etc   Cholinesterases are sufficiently robust to survive certain processes which have practical advantages but which might well inhibit more sensitive enzymes. Nomura and Schukneckt (1965) decalcified the cat cochlea by immersion in cold 10% EDTA (pH 7.4) for 40 days prior to cutting cryostat sections which were then stained for cholinesterases by Gomori's (1952) method. Similarly, Winkelmann et al. (1967) have shown that the staining and visualization of dermal nerves in whole-mount preparations is greatly improved if the skin specimen is first soaked for 1–4 hr in aqueous 2 M sodium bromide; this loosens the epidermis which can be teased off and discarded.

CHAPTER 4

Subcellular localization, synthesis and transport of acetylcholinesterase

4.1 Introduction

This must necessarily be one of the more speculative chapters in the book. The three topics it covers are all controversial: different laboratories have produced different answers, and similar results have been subjected to a variety of interpretations. To give a universally acceptable outline of current concepts of AChE synthesis, subcellular distribution and transport is virtually impossible and, in any case, an authoritarian précis is inappropriate to such a dynamic field. If a sample of data and ideas is given instead, the reader will at least have a chance to form his own, probably different, theory and onto this he can graft fresh information as it becomes available.

It might seem logical to begin by discussing the synthesis of AChE, but since this involves some consideration of ultrastructural organization the latter is proabably the better starting point.

4.2 Subcellular localization

Ultrastructural localization has been examined biochemically, by subcellular fractionation techniques, and by electron microscopy. Despite continuing work on the subject a number of doubts persist and for some time at least these may increase rather than decrease. The development of progessively more refined techniques has shown up apparent errors in earlier results, and conclusions need frequent revision.

4.2.1 Neuronal morphology

Fig. 4.1 shows the major components of the neurone and although it takes no account of the finer details it may be a guide to ultrastructural terminology and provide a framework on which to fit the results discussed below. The cell body or perikaryon is surrounded by a unit membrane called the plasmalemma or cell membrane; this is continuous with the axolemmal membrane enclosing the axon. In addition to the nucleus which is bounded by a nuclear membrane (or envelope) the

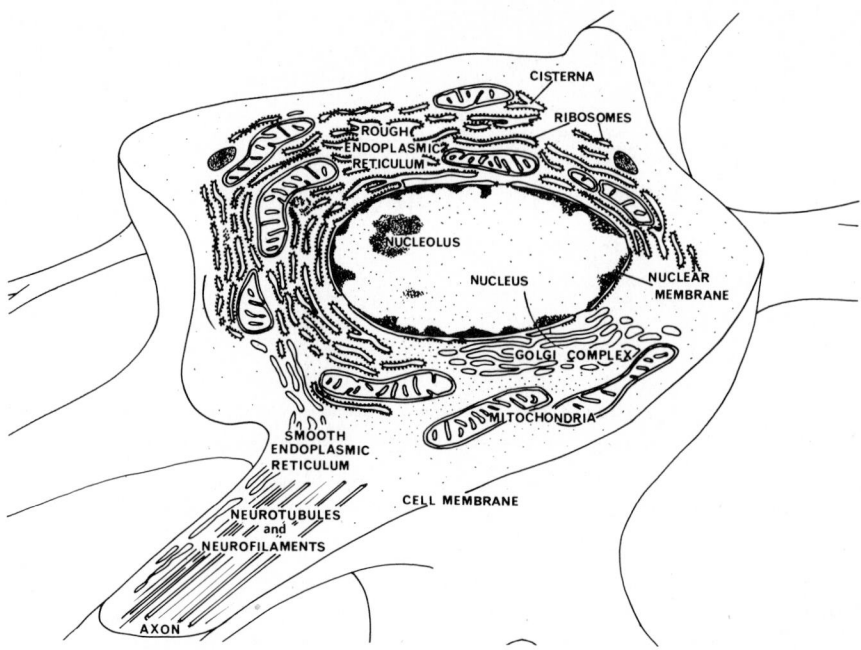

Fig. 4.1. Generalized diagram of the ultrastructure of a neurone. (Drawn by Dr. E.A. Munn.)

perikaryon is crowded with organelles – this crowding is a special feature of neurones and secretory cells. The major cytoplasmic component is the granular or rough endoplasmic reticulum (RER), an extensive membrane system composed of piled sheets of flattened vesicles known as cisternae. The cisternae are interconnected by tubules and are often perforate. Attached to the membranes are ribosomes and it is these which react with Nissl stains in histological preparations; in other words, the RER of the electron microscopist is the Nissl substance of the histologist. The Golgi complex (GC) is also membranous but the membranes are smooth and form large vacuoles and smaller vesicles. Other structures seen in the cell body include mitochondria, lysosomes and multivesicular bodies. Extending out of the cell body, from the axon hillock, is the agranular or smooth endoplasmic reticulum (SER) which is probably continuous with the RER but is devoid of granules. The axon contains various other structures amongst which are nonmembranous neurotubules composed of protein molecules, and neurofilaments which, with the neurotubules, run longitudinally from the cell body towards the terminals. According to some authors (see Grafstein 1969) the tubules do not extend into the terminals themselves.

Some confusion is evident in the literature about the recognition of these latter structures in the axoplasm, one reason being that their appearance may be variably altered by different fixatives. Martinez and Friede (1970) suggest that the conflicting reports about the precise localization of AChE in the axon (§4.2.3) stem mainly from differences in interpretation.

At the nerve terminals mitochondria are numerous and at sensory endings they are the major feature. Motor fibres contain a high concentration of vesicles, the type predominating depending on the type of nerve. Small synaptic vesicles containing ACh are characteristic of cholinergic terminals, while dense-cored vesicles are indicative of catecholaminergic fibres. Some terminals contain vesicles of both types and, in addition, various other vesicles, including some which appear empty, can often be found (see Bodian 1970; Burnstock and Iwayama 1971). The pre-synaptic membrane of the terminal is separated from the plasmalemma of the post-synaptic cell body by the synaptic cleft which is about 200 Å wide. The cleft itself contains organized structures (De Robertis et al. 1961; Gray 1966) and the post-synaptic membrane (i.e. that part of the plasmalemma of the post-synaptic cell which lies immediately beneath the terminal) shows some specialization known as the post-synaptic thickening or sub-synaptic web. The structures, known as synaptosomes, which can be separated from homogenates of neural tissue (§4.2.2) were first described by Gray and Whittaker (1962) who identified them as pinched-off nerve

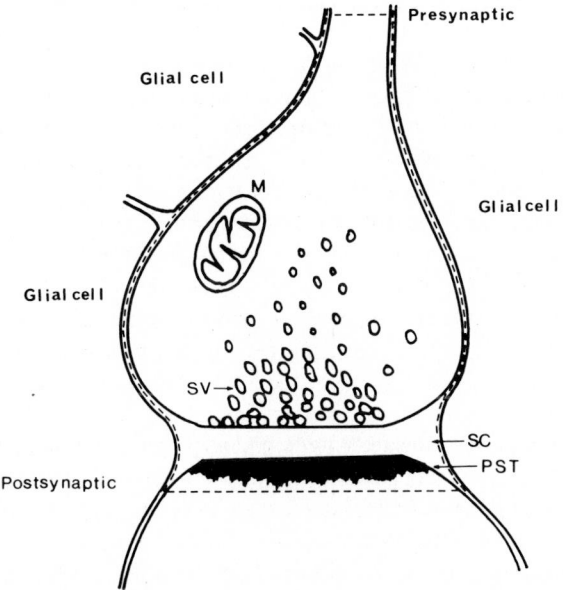

Fig. 4.2. Diagram to illustrate the pre- and post-synaptic components of a synaptosome. M, mitochondrion; SC, synaptic cleft; PST, post-synaptic thickening; SV, synaptic vesicles. Broken lines indicate sites of cleavage. (After E.G. Gray 1959.)

endings. They consist of the pre-synaptic axolemmal membrane which has sealed itself off to form a bag enclosing the synaptic vesicles and mitochondria. Attached to this pre-synaptic component is the post-synaptic thickening of the plasmalemma of the cell with which the synapse is made (fig. 4.2).

4.2.2 Subcellular fractionation

The association of AChE in the central nervous system with particulate material has been recognized since the early attempts to purify the enzyme (see Ord and Thompson 1951). Nathan and Aprison (1955) showed by subcellular fractionation of a homogenate that 90% of the AChE of rabbit caudate nucleus was present in the cytoplasmic particulate, the greatest part being in the mitochondrial and microsomal fractions, the latter having the highest specific activity. These findings were broadly confirmed in the rat by Aldridge and Johnson (1959), Holmstedt and Toschi (1959) and by Toschi (1959); results essentially the same as those for rat were later obtained from frog brain by Clouet and Waelsch (1961). Electron microscopic examination of microsomal fractions in parallel with biochemical analyses (Hanzon and Toschi 1959) showed that the cholinesterase was confined to the membrane component, in contrast to RNA which was associated with vesicular structures.

After the initial agreement that most of the AChE is firmly bound to membranes (thus differing from pseudocholinesterase which is soluble) controversy arose over the finer details of the localization. These controversies may be easier to follow if a few general remarks about fractionation are made first. The designation of various fractions, and more particularly subfractions, is different in different laboratories and the terms can be confusing. Special care is necessary over the crude mitochondrial fraction, generally called P_2. This contains a good deal besides mitochondria and it is in this fraction that the bulk of the synaptosomes sediment. Synaptosomes are not the same size in all regions of the brain and Whittaker (1965) found that some of those in the guinea-pig cerebellar cortex are so big that they appear in the nuclear fraction, P_1; others are small and appear in P_3, the microsomal fraction. Although, as Whittaker (1965) emphasizes, the division between the fractions is arbitrary rather than strictly morphological, the bulk of the microsomal fraction is probably derived from the endoplasmic reticulum (Robertson 1959). Since perikaryal AChE seems to be associated predominantly with the reticulum, and terminal enzyme with synaptic membranes (§4.2.3), the relative distribution of enzyme between the mitochondrial and microsomal samples may be influenced by the proportions of nerve endings to nerve cells in the area examined.

De Robertis et al. (1962) fractionated whole brain of rats and reported that most of the AChE was in the synaptosomes. When these were ruptured and refractionated the coarse particulate matter possessed the greatest total activity but the highest specific activity was associated with the fine particulate fraction which was taken to represent synaptic vesicles and post-synaptic membranes (De Robertis et al. 1963). Subsequently, Whittaker et al. (1964) showed that the various fractions examined by De Robertis and his colleagues probably suffered severe microsomal contamination, hence the distribution they observed was misleading. Reviewing the problem in 1965, Whittaker concluded that the methods in current use gave fractions which were not sufficiently homogeneous to show the subcellular localization of AChE with absolute precision but, clearly, it was not associated with synaptic vesicles.

Rodríguez de Lores Arnaiz et al. (1967) osmotically shocked their crude mitochondrial fraction from rat cortex and subfractionated it into 3 components. One of these, M, which was composed of myelin, mitochondria and synaptic membranes was layered on a discontinuous sucrose gradient and recentrifuged. Three of the 6 resulting fractions contained synaptic membranes. Membranes remaining above the 1.0 M concentration layer were rich in AChE but those in the 1.2 M layer had very little AChE activity and were thought to be derived from non-cholinergic terminals. This study provides clear evidence of the association of AChE with synaptic elements but since the structures isolated by the procedure include both pre- and post-synaptic membranes the localization is still somewhat imprecise. Application of the microscale fractionation technique (Giacobini et al. 1971) which was mentioned in ch. 3 may well help to clarify the problem.

It is understandable that so much of the work on AChE should have been directed towards the enzyme at the synapse because its functional importance at that site is generally recognized. In contrast, the role of AChE in the axolemma is uncertain. One possibility, that it has no role but is merely in transit to the nerve endings, is discussed in §4.3.2. Another possibility, strongly advocated by Nachmansohn and his colleagues, is that it is as essential for axonal conduction as for synaptic transmission. Although not widely accepted (see ch. 7 §2.4), this belief has stimulated some very elegant quantitative work on AChE activity in axolemmal membranes of invertebrate nerves (Brzin et al. 1965). The measurements were made with Cartesian divers on sonically disintegrated squid and lobster axons from which the axoplasm had been removed. Further details are given in the following chapter but, in brief, the axolemma of both motor and sensory nerves showed AChE activity; values when expressed per unit area were greater in motor than sensory axons but on a weight basis the order was reversed. Subsequent fractionation experiments on whole bundles of lobster walking-leg nerves (Welsch and Dettbarn 1970) have shown that the membrane fraction contains 30% of the total AChE activity of the homogenate.

4.2.2.1 Relation of AChE to membrane   The possible identity of AChE with the cholinergic receptor protein has been considered in ch. 2. Evidence against the idea includes the demonstration by De Robertis and his colleagues (De Robertis et al. 1967; De Robertis 1971) that synaptic membranes still retain the ability to bind d-tubocurarine after much of the AChE has been removed. Solubilization of the enzyme can be achieved in a number of ways (for references see chs. 2 and 3) and it is clear that the final state of the product depends on the method of extraction. As indicated in ch. 3, Ho and Ellman (1969) compared the AChE extracted from rat brain by bacterial protease with that extracted by the non-ionic detergent Triton X-100 and they also used a combination of both extractants. They concluded that the enzyme solubilized by the detergent is a 'precursor' of the more water-soluble form which results from protease treatment. Crone (1971) suggested that since the detergent treatment is a mild one, the released AChE might exist in a state which is intermediate between the lipoprotein complex that is bound to the membranes in vivo and the purified form. His experiments involved solubilization of AChE in rat cerebral cortex with Triton X-100 and the subsequent removal of contaminating protein and phospholipids by EDTA and exclusion chromatography on Sepharose and Sephadex columns. His failure to improve the specific activity by a separation based on particle size indicated that when the membrane is solubilized the particles formed are all much the same size but only some, a rather small proportion, carry AChE. Since the solubilized particles tended to re-aggregate when the detergent was removed, Crone postulated that they might well represent actual subunits of the membrane (Korn 1966), and he commented on the homogeneous way in which his enzyme preparation behaved. This finding suggests that all the enzyme, whether derived from cell bodies or terminals, exists in a similar form but, in the light of evidence for isozymes (see ch. 2 §3.2), this may be too simple a view.

The high concentration of gangliosides in subcellular fractions rich in AChE has been reported by many workers (see Dekirmenjian et al. 1969). Whether there is any functional significance in the association is not clear although a relation of ganglioside to receptor protein has been mooted (see Lapetina et al. 1967; De Robertis 1971). Morgan et al. (1973) have shown that in contrast to synaptosomal plasma membranes, the membranes of synaptic vesicles are almost totally devoid of gangliosides. Trotter and Burton (1969) list evidence from the literature which indicates that gangliosides may in some way activate AChE but in their own experiments on rat brain, low concentrations of gangliosides inhibited rather than enhanced enzyme activity.

4.2.2.2 Relation of AChE to ACh   In concentrating on AChE it is important not to ignore ACh and ChAc both of which are of paramount importance in the context of the cholinergic synapse. It is now generally accepted that synaptic vesicles contain ACh but not AChE (Whittaker et al. 1964; for conflicting view see Matsuda et al. 1971). At first sight this seems the 'logical' arrangement because one would expect ACh to be separated from AChE in a way which prevents hydrolysis. The situation is complicated, however, by the fact that ChAc is also apparently excluded from the vesicles (Whittaker et al. 1964; Fonnum 1966, 1967, 1973; Tuček 1966, but cf. Matsuda et al. 1971) hence the actual production of ACh must occur extravesicularly. In other words, newly synthesised ACh could be at risk if AChE were present in the vicinity of the synaptic vesicles. Surprisingly little direct information is available about the presence or absence, in the CNS, of AChE inside the terminals as opposed to its localization outside, on the synaptic membranes (see §4.2.3 for electron microscopic findings). The problem has been examined in autonomic ganglia but even here the approach has been mainly indirect and has depended on the different effects of lipid-soluble anticholinesterases (see ch. 11) which can enter the axon (Nachmansohn 1950) and lipid-insoluble ones which cannot (Koelle and Steiner 1956). Koelle and his co-workers (Koelle and Koelle 1959; McIsaac and Koelle 1959) studied a number of different ganglia and concluded that it is the 'external' AChE (i.e. that accessible in vivo to bisquaternary anticholinesterases of low lipid solubility) which is responsible for hydrolysing ACh released during transmission. Consequently they designated the external enzyme 'functional' to distinguish it from the internal or 'reserve' enzyme which apparently plays no part in hydrolysis of transmitter. In a parasympathetic ganglion such as the ciliary, the internal, reserve enzyme might be attributable to the post-ganglionic cells and not to the pre-ganglionic fibres. On the other hand, in sympathetic ganglia the post-ganglionic cells are, with some exceptions, adrenergic and probably contain only a little AChE, hence it is reasonable to conclude that most reserve enzyme indicated by lipid-soluble anticholinesterases would be located inside the pre-synaptic fibres (but see ch. 7 §2.2 for further discussion). Birks and MacIntosh (1961) independently invoked the existence of intra-axonal enzyme in the interpretation of their finding that when the cat superior cervical ganglion was perfused with the appropriate eserine-containing medium, ACh accumulated above the resting level. Because this 'surplus' ACh could not be released by nerve stimulation (a finding confirmed and extended by Collier and Katz 1971) the authors suggested that it represented ACh which had leaked out of synaptic vesicles and which would, under normal conditions, have been destroyed by AChE in the vicinity. In the light of the subsequent findings, mentioned earlier in this section, that ChAc apparently occurs only outside the vesicles, an equally good explanation might be that it represents not ACh which has escaped from vesicles but ACh which has yet to reach them. If this is indeed the case the next question to ask is why there should be this seemingly wasteful hydrolysis of newly synthesised transmitter. At this stage there seems no simple answer although possible explanations suggest themselves: maybe it represents some sort of safety mechanism to stop ACh accumulating should the synthetic process outstrip the storage capacity of the vesicles. Admittedly such an accumulation is unlikely since excess ACh would probably inhibit ChAc (Kaita and Goldberg 1969; Morris et al. 1971; but cf. Fonnum 1973). Furthermore, surplus ACh is not apparently harmful to the pre-synaptic membrane since Collier and Katz (1971) were able to release more or less normal amounts of transmitter ACh from superior cervical ganglia in which surplus ACh was present.

One of the difficulties in speculating about the purpose of the 'reserve' AChE is

TABLE 4.1

Variations in reports of ultrastructural localization of AChE: central and peripheral sites. As far as possible the most generally accepted result for each site (i.e. + ve or − ve) has been given first (but see text).

Site	Species and Tissue	Result	Fixation	Method (see ch. 3)	Reference
Cell membrane (plasmalemma)	Rat: trigeminal ganglion	+ ve	Glutaraldehyde	Karnovsky & Roots	Kalina & Bubis (1969)
	Rat and guinea-pig: neurones of facial nucleus	+ ve	Glut. & form.	Modified Koelle	Kreutzberg et al. (1973)
	Frog: sympathetic ganglion	+ ve	Glut. or form.	Modified Koelle	Brzin et al. (1966)
	Rat: hippocampus	− ve even though ER + ve	Glut. followed by form.	Modified Koelle	Shute & Lewis (1966a)
Rough endoplasmic reticulum	In virtually all cells which show any activity except	+ ve	Various	Various	Various
	Frog: sympathetic ganglion	− ve	Unfixed (fixed tissue + ve)	Modified Koelle	Brzin et al. (1966)
Ribosomes (free)	Rat: ventral horn cells, hypoglossal nucleus, dorsal motor nucleus of vagus	− ve	Glut. followed by form.	Modified Koelle	Lewis & Shute (1966)
	Rat: ventral horn cells	− ve	Formaldehyde	Karnovsky & Roots	Eränkö et al. (1967a)
	Rat: locus coeruleus; dorsal motor nucleus of vagus	+ ve (?)	Glutaraldehyde	Karnovsky & Roots	Shimizu & Ishii (1966)
	Kitten, postnatal: Purkinje cells	+ ve (around ribosome clumps)	Glut. & form.	Pb-Cu Thiocholine	Kása & Csillik (1968)
Golgi complex	Rabbit: retinal ganglion cells	+ ve (inner saccule)	Glut. & form.	Karnovsky (1964)	Reale et al. (1971)
	Rat: ventral horn cells	+ ve (some saccules)	Formaldehyde	Modified Karnovsky & Roots	Eränkö et al. (1967a)
	Kitten, postnatal: Purkinje cells	+ ve (some saccules)	Glut. & form.	Pb-Cu thiocholine	Kása & Csillik (1968)
	Rat and guinea-pig: neurones of facial nucleus	+ ve (some saccules)	Glut. & form.	Modified Koelle (Lewis & Shute 1969)	Kreutzberg et al. (1973)
	Cat: Purkinje cells	− ve	Glut. & form.	Pb-Cu thiocholine	Kása & Csillik (1968)

TABLE 4.1 (continued)

Site	Species and Tissue	Result	Fixation	Method (see ch. 3)	Reference
	Guinea-pig: cerebellar Golgi cells	–ve	Glut. & form.	Pb-Cu thiocholine	Kása & Csillik (1968)
	Rat: hippocampal Golgi cells (stratum oriens)	–ve	Glut. followed by form.	Modified Koelle	Shute & Lewis (1966a)
Nuclear membrane (nuclear envelope)	Rat: ventral horn cells	+ve	Formaldehyde	Modified Karnovsky & Roots	Eränkö et al. (1967a)
	Rat: dorsal root ganglion	+ve	Glutaraldehyde	Karnovsky (1964), & Karnovsky & Roots	Novikoff et al. (1966)
	Rat: ganglion cells in adrenal medulla	+ve	Glutaraldehyde	Modified Koelle	Lewis & Shute (1969)
Axonal membrane (axolemma)	Virtually all axons with AChE +ve synaptic membranes except	+ve but discontinuous	Various	Various	Various
	Rat: fibres in medulla oblongata	–ve (synaptic membranes +ve)	OsO_4	Thiolacetic acid	Torack & Barnett (1962)
Neurotubules/filaments or other intra-axonal structures	Rabbit: sciatic nerve	+ve	Glut. & form.	Pb-Cu thiocholine	Kása (1968b)
	Rat: sciatic nerve	+ve	Glut. & form.	Pb-Cu thiocholine	Kása (1968b, 1971); Kása et al. (1973)
	Rat: medulla oblongata	+ve	OsO_4	Thiolacetic acid	Torack & Barnett (1962)
	Man, foetal: spinal cord	+ve (SER)	Glutaraldehyde	Modified Koelle	Tennyson et al. (1968)
	Lobster: walking-leg nerve	+ve	Glutaraldehyde	Modified Koelle	De Lorenzo et al. (1969)
	Rat: sciatic nerve	+ve only occasionally	Various, including formaldehyde	Karnovsky & Roots	Schlaepfer & Torack (1966)
	Rat: spinal cord	+ve only occasionally	Glut. & form.	Modified Karnovsky & Roots with AcSeCh	Kokko et al. (1969)
	Rat: hypoglossal nerve	–ve (except for? axolemmal invaginations)	Not stated, probably glut.	Modified Koelle	Lewis & Shute (1965a)

TABLE 4.1 (continued)

Site	Species and Tissue	Result	Fixation	Method (see ch. 3)	Reference
Mitochondria	Rat: diaphragm neuro-muscular junction, ventral horn cells, hypoglossal nucleus, dorsal motor nucleus of vagus	− ve	Glut. followed by form.	Modified Koelle	Lewis & Shute (1966)
	Rat: medulla oblongata	+ ve (in small neurones only)	OsO₄	Thiolacetic acid	Torack & Barrnett (1962)
	Rat and guinea-pig: axon and dendrites of motoneurones, facial nucleus	+ ve occasionally	Glut. & form.	Modified Koelle (Lewis & Shute 1969)	Kreutzberg & Tóth (personal communication)
Synaptic vesicles	Rat: diaphragm neuromuscular junction	− ve	Glut. followed by form.	Modified Koelle	Lewis & Shute (1966)
	Rat: medulla oblongata	+ ve	OsO₄	Thiolacetic acid	Torack & Barrnett (1962)
Other structures within terminals	Rat, adult: diaphragm neuromuscular junction	− ve	Glut. & form.	Pb-Cu thiocholine	Kása & Csillik (1968)
	Rat, 2 months old: diaphragm neuromuscular junction	+ ve	Glut. & form.	Pb-Cu thiocholine	Kása & Csillik (1968)
	Hen: kalyform ending in ciliary ganglion	+ ve	Formaldehyde	α-naphthylacetate-hexazonium pararosaniline	Lehrer (1966)
Synaptic membranes	Rat: caudate nucleus	+ ve pre-synaptic + ve post-synaptic + ve cleft	Glutaraldehyde	Modified Koelle	Lewis & Shute (1964)
	Rat: caudate nucleus, cerebral cortex, spinal cord	− ve pre-synaptic − ve post-synaptic + ve cleft	Glut. & form.	Modified Karnovsky & Roots with AcSeCh	Kokko et al. (1969)
	Rat: pineal gland	− ve pre-synaptic − ve post-synaptic + ve cleft	Glutaraldehyde	Karnovsky & Roots	Eränkö et al. (1970)
	Rat: pineal gland	+ ve pre-synaptic + ve post-synaptic − ve cleft	Formaldehyde (brief)	Karnovsky & Roots	Eränkö et al. (1970)

TABLE 4.1 (continued)

Site	Species and Tissue	Result	Fixation	Method (see ch. 3)	Reference
Intrasynaptic organelles	Rat: Guinea-pig: } cerebellum Cat:	+ ve	Glut. followed by OsO_4	Pb-Cu thiocholine	Kása (1968a)
	Rat: spinal cord, cerebral cortex, caudate nucleus }	– ve	Glut. & form.	Modified Karnovsky & Roots with AcSeCh	Kokko et al. (1969)
Schwann cell: membrane	Rat: sciatic nerve	– ve	Various, including formaldehyde	Karnovsky & Roots	Schlaepfer & Torack (1966)
	Mouse: intercostal muscle neuromuscular junction	+ ve	Formaldehyde	Au-thiolacetic acid	Davis & Koelle (1967)
	Guinea-pig: uterine artery neuromuscular junction	+ ve + ve	Formaldehyde	Karnovsky & Roots	Bell (1969)
cytoplasm	Usually – ve but cf. Frog: sympathetic ganglion + ve 'wrappings'		Formaldehyde	Modified Koelle	Brzin et al. (1966)

that its exact morphological position is by no means as clearly defined as some workers would seem to believe. Koelle and Koelle (1959) concluded on the basis of optical microscopy that in the superior cervical ganglion the enzyme at the terminals was predominantly external, but along the course of the pre-synaptic fibres it was both external and internal – a localization which, as will be discussed below (§4.2.3), has been partly confirmed by electron microscopy. Collier and Katz (1971) showed by denervation experiments that in the superior cervical ganglion their surplus ACh arose in 'preganglionic terminals' as opposed to adrenergic cell bodies but these experiments would not differentiate between the terminals proper and the intraganglionic portion of the preganglionic fibres. Maybe the surplus ACh is axonal rather than terminal, a possibility which might explain why it is not apparently releasable by nerve stimulation (Birks and MacIntosh 1961; Collier and Katz 1971).

In their work on slices of cat cerebal cortex, Szerb et al. (1970) used eserine to preserve released ACh and they therefore assumed that surplus ACh would be accumulating within the tissue. Subsequently, Bourdois and Szerb (1972) found that so long as the K^+ concentration was not elevated, the ACh content of the slices was the same whether or not eserine was present. They took this as an indication that the situation in the CNS may not be analogous to that in the ganglion. In contrast, Collier et al. (1972) found that rat and mouse brain slices did accumulate non-vesicular ACh and they considered this accumulation to be equivalent to the synthesis of surplus ACh by peripheral cholinergic systems. Parallel studies on ganglia and cerebral tissue could be useful in pin-pointing the source of surplus ACh.

4.2.3 Electron microscopy

The various methods available for the demonstration of AChE under the electron microscope have been outlined in ch. 3. Table 4.1 indicates the discrepancies in the reported presence or absence of histochemical end-product at some ultrastructural sites in central and peripheral nervous tissue. The variations may be attributable to a number of factors amongst which species, age and type of tissue could well be significant. Where the variations seem due to differences in technique it is difficult to know which, if any, of the methods accurately demonstrates the in vivo situation. In certain instances where the specificity of the reaction is in doubt (e.g. with the thiolacetic acid method) or where diffusion is suspected, the results could be misleading. On the other hand, discrepancies between fixed and unfixed tissue may be reflecting the behaviour of permeability barriers or the relative susceptibility of enzyme at different sites to inhibition by fixatives and could, if properly interpreted, be quite informative. Brzin et al. (1966; see also Tennyson et al. 1968) have dealt in some detail with the effect of fixation on the location of demonstrable AChE activity in frog dorsal root and sympathetic ganglia. Although synaptic stain-

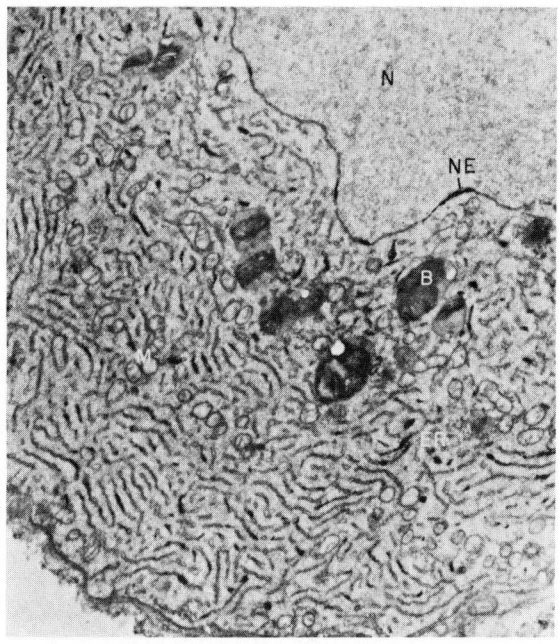

Plate 4.1. To show spotty appearance of the AChE-reaction product in the endoplasmic reticulum (ER) of a dorsal root ganglion cell of frog. B, heterogeneous bodies; M, mitochondrion; N, nucleus; NE, nuclear envelope. (X 19,500) (From Brzin et al. 1966.)

ing was much more obvious in unfixed than in fixed sympathetic ganglia the reverse was true of staining in the endoplasmic reticulum. Some of the tissues were subjected to microgasometric enzyme analysis and it was found that the cells with the highest measurable AChE activity were those which were shown by subsequent electron microscopy to have damaged cell membranes and disrupted sheath cells. The authors concluded that the unfixed, intact membranes probably form an appreciable permeability barrier and that fixation, although inhibiting extracellular enzyme, might break these down thus exposing at least some of the intracellular enzyme. They suggested that the spotty nature of the stain in the endoplasmic reticulum (plate 4.1) could represent fixative-resistant enzyme; in other words the discontinuity of the distribution could be an artifact. This is a point to be kept in mind in any consideration of histochemistry but it is of particular importance when the absence of stain appears to present problems in interpretation. Despite the discrepancies in the results, which may or may not be attributable to artifacts, there is sufficient agreement to allow some conclusions to be drawn with a fair degree of certainty. In particular, the association of AChE with membrane structures, deduced from subcellular fractionation techniques, has been confirmed.

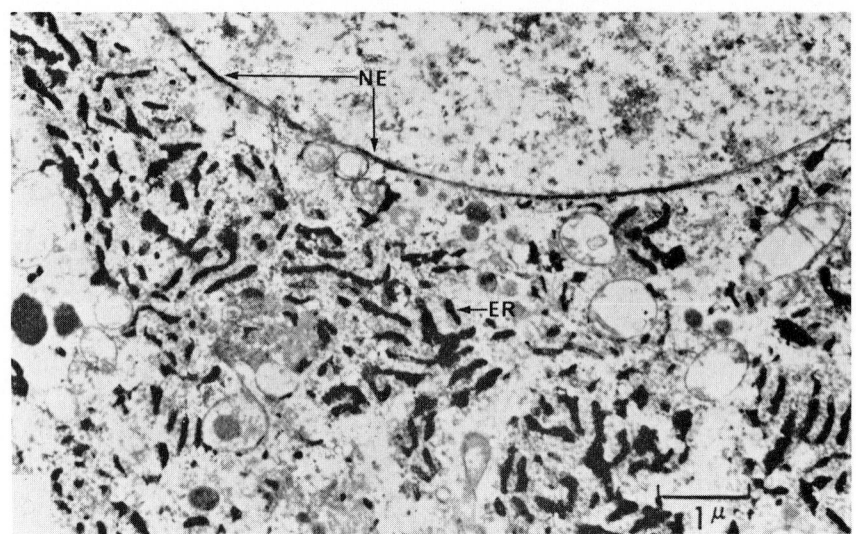

Plate 4.2. 'Cholinergic' ganglion cell in adrenal medulla of rat showing a marked reaction for AChE in the endoplasmic reticulum (ER) and discontinuous staining of the nuclear envelope (NE) (From Lewis and Shute 1969.)

In a cell which by other criteria could be classified as cholinergic, the endoplasmic reticulum is always positive (plates 4.2 and 4.3) although the distribution of the reaction end-product may be patchy. The enzyme appears to be associated with the internal surface of the membranes rather than with the ribosomes (Eränkö et al. 1967) but some ribosomal staining has been reported (see table 4.1). Staining of the nuclear envelope (plate 4.2) and parts of the Golgi apparatus (see plate 4.3) seem a general but not an invariable feature. Reports about the activity of the plasmalemmal membrane show some variation. Thus Lewis and Shute (1966) found that in motor neurones of the rat spinal cord and medulla, the plasma membrane was generally inactive except where there was a synapse; Shute and Lewis (1966a) reported a similar lack of activity in the plasmalemma of AChE-containing Golgi cells in the rat hippocampus. On the other hand, Brzin et al. (1966) demonstrated activity in the plasmalemma of presumably non-cholinergic cells in frog sympathetic ganglia and Kreutzberg et al. (1973) showed strong but patchy plasmalemmal staining, more particularly in the dendrites (plate 4.4) of rat and guinea-pig motor neurones. Cholinergic neurones show a reaction on the outer surface of the axolemmal membrane and Lewis and Shute (1966) suggest that any cell without appreciable axolemmal activity is almost certainly non-cholinergic. Like the enzyme in the RER, axolemmal AChE may have a patchy distribution, some axons may be almost completely outlined by end-product (plate 4.5a) while others show spotty,

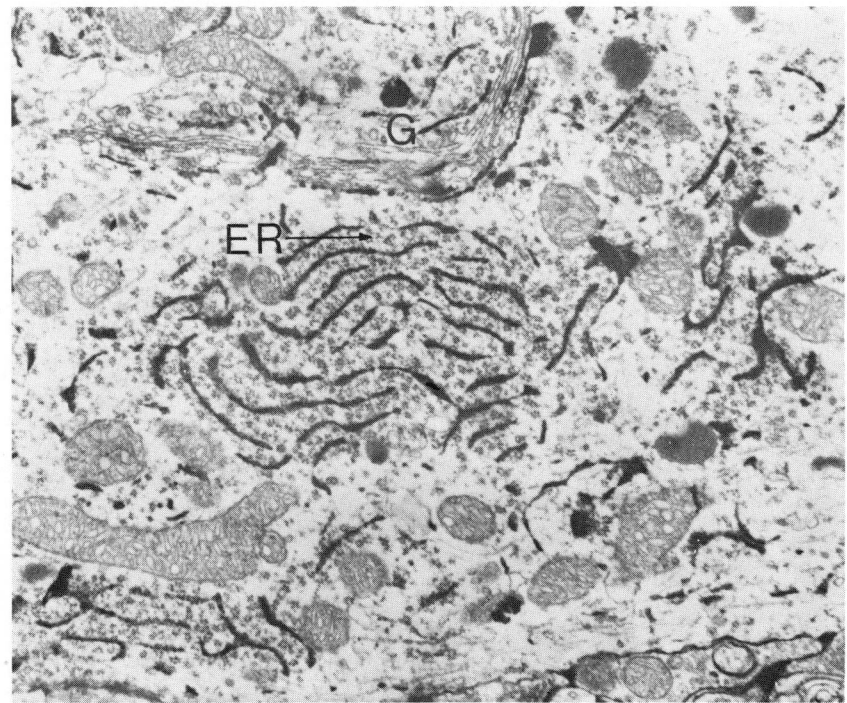

Plate 4.3. AChE activity in a motor neurone in the facial nucleus of rat. Most of the cisternae of the endoplasmic reticulum (ER) are stained but the reaction in the Golgi complex (G) is patchy. (X 18,000) (G.W. Kreutzberg and L. Tóth, unpublished.)

discontinuous staining (plate 4.5b). Histochemical end-product is frequently found between the axolemma and Schwann cells and in some, but by no means all, cases the membrane of a Schwann cell associated with a stained axon may also be positive (Brzin et al. 1966; cf. Reale et al. 1971; see also chs. 6, 7 and 8). Reports of activity in the Schwann cell cytoplasm are much less common. Brzin et al. (1966) describe end-product 'at the wrappings of Schwann cell cytoplasm' and in satellite cell cytoplasm in the sympathetic ganglia of the frog. On the other hand, Lewis and Shute (1969) noted a complete absence of reaction in Schwann cells associated with the intensely stained unmyelinated axons in the rat adrenal medulla.

It seems fairly safe to say that at the synapse the pre-synaptic enzyme is localized on the external surface of the terminals and, likewise, on the external surface of the cholinoceptive post-synaptic cell. What is still uncertain is whether AChE, as opposed to the histochemical reaction product, is also present within the synaptic cleft (see for instance Eränkö et al. 1970). There are doubts, too, about the precise relation of AChE to intra-axonal structures. Some of the uncertainties arise because

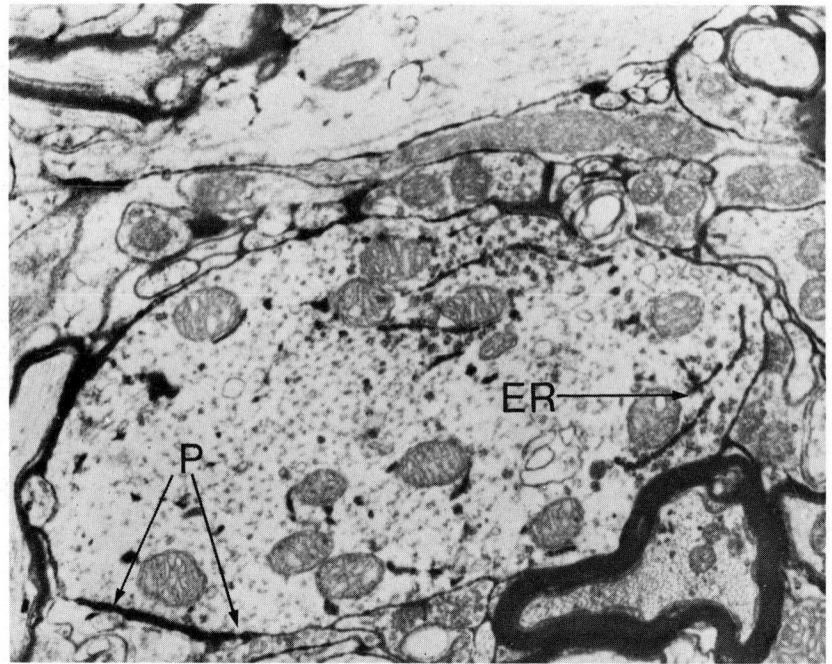

Plate 4.4. To show the distribution of AChE reaction-product in a dendrite of a motor neurone in the facial nucleus of the rat. Staining occurs in the endoplasmic reticulum (ER) and in the plasmalemmal membrane (P). (X 18,000) (From Kreutzberg et al. 1973.)

different methods give different results. Controversies also stem from discrepancies in the identification and terminological specification of different organelles (see Martinez and Friede 1970). For instance, are the 'invaginations of axolemma' seen in rat hypoglossal nerve by Lewis and Shute (1965a) and the 'subsurface cisternae' described by Brzin et al. (1966) in frog sympathetic axons the same as the structures which Kása (1968b) identified in rat sciatic nerve as neurotubules fused to the axolemma? What can be said is that the evidence for the association of enzyme with some sort of linear structures, be they neurotubules or neurofilaments, is becoming quite strong (see table 4.1).

Early work with the thiolacetic acid method suggested that AChE might be associated with mitochondria and synaptic vesicles (see Zacks and Blumberg 1961; Torack and Barrnett 1962). When more specific substrates were used no reaction was seen on these organelles and it became generally accepted that the mitochondria and vesicles, as well as lysosomes and dense bodies lacked AChE activity. Recently, however, G.W. Kreutzberg and L. Tóth (personal communication), using a thiocholine method have found occasional AChE-containing mitochondria in

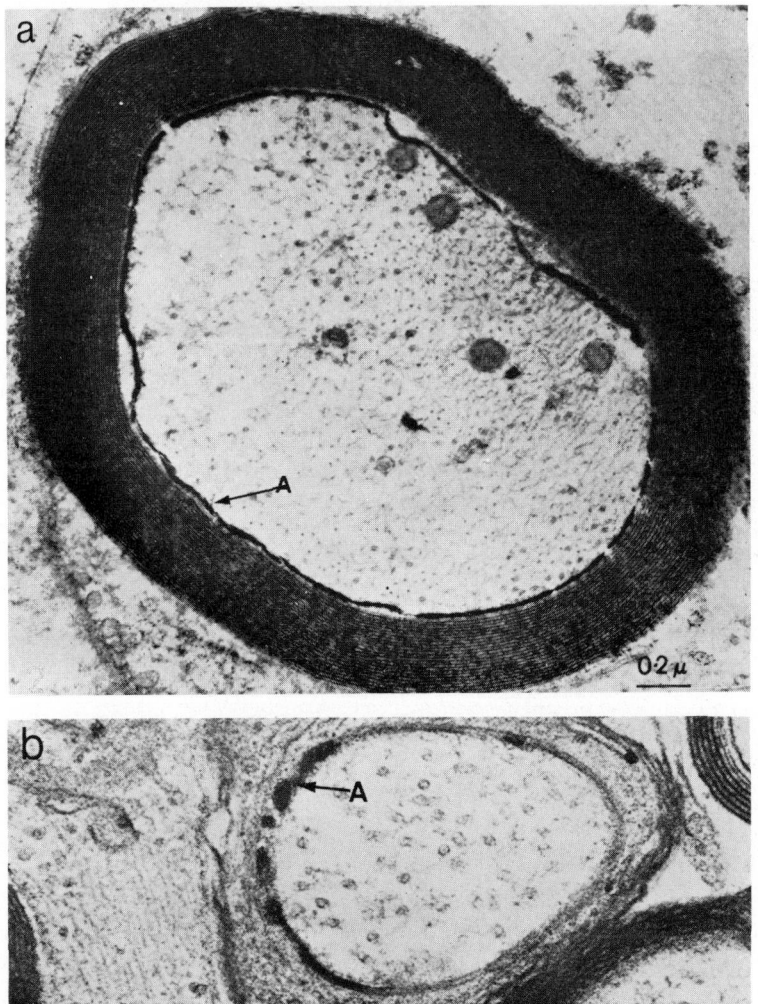

Plate 4.5. AChE activity in the axolemma. a) Myelinated axon in the adrenal medulla of rat showing intense and relatively uniform staining of axolemma (A) (from Lewis and Shute 1969); b) unmyelinated axon in optic nerve of rabbit showing discontinuous staining of the axolemma (A). (X 57,000) (From Reale et al. 1971.)

axon terminals and dendrites and, much more rarely, in the perikaryon. The end-product is always localized in a characteristic way. This is illustrated in plate 4.6 which shows a reactive mitochondrion in a dendrite (a) and an axon terminal (b) in the facial nucleus of the rat. In both, the staining is confined to a small stretch of the mitochondrial membrane and to a few of the cristae connected with this region.

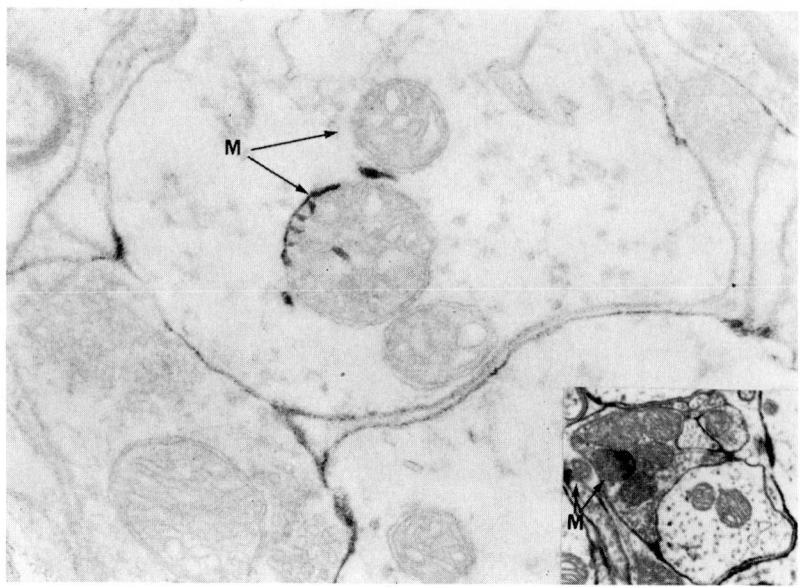

Plate 4.6. AChE activity in mitochondria in the facial nucleus of rat. a) In a dendrite (X 54,000); b) in a nerve terminal (X 18,000). (G.W. Kreutzberg and L. Tóth, unpublished.)

Mazza et al. (1973) also using a thiocholine technique have found AChE activity in mitochondria of neurones in the rat trigeminal ganglion but the end-product was not localized as discretely as in Kreutzberg and Tóth's experiments. With BuThCh as substrate, a reaction occurred in mitochondria of both the neurones and satellite cells.

Apart from Kreutzberg and Tóth's observation of rare reactive mitochondria in nerve endings there have been few reports of AChE-positive structures within the axon terminals themselves. A search of the literature produced only two firm accounts of such a localization. The first was Lehrer's (1966) report of a deposition of end-product in the cytoplasm of the kalyform ending on the ciliary ganglion of young adult chickens. The electron micrograph he shows is not very convincing and although this could be due to poor reproduction the reaction is unequivocal in pictures of other structures stained by the same method which employed d-naphthylacetatehexazonium pararosaniline as substrate. The second report of positive structures inside the terminals was that of Kása and Csillik (1968) who found AChE activity on what they identified as neurofilaments in phrenic nerve endings of rats of 2 months old. What is particularly interesting is that no such stained structures were seen in the adult rat. The staining could be of some special significance in the young animal in connection with Koenig and Koelle's (1961) idea (§ 4.3.3) that new enzyme is laid down during growth and thereafter is turned over in situ.

Alternatively, the lack of similar staining in the adult might be a warning that methods suitable for revealing certain structures in the immature animal are no longer reliable in the adult, in which diffusion barriers may be more efficient (see ch. 3).

The comments above apply mainly to reputed cholinergic cells. In their study of rat hippocampus, Shute and Lewis (1966a) were able to identify various classes of cells showing different degrees of activity. The Golgi II cells, which may or may not be cholinergic, had appreciable activity in the RER but their axons could not be identified. The authors mention that some non-cholinergic cholinoceptive cells (e.g. some cells in the dentate nucleus of the rat cerebellum) show appreciable RER activity; this agrees with the findings of Brzin and his co-workers, referred to earlier, in ganglion cells. On the other hand, in the non-cholinergic hippocampal pyramidal cells staining of the RER was only faint while in the dentate granule cells, on which there were AChE-positive synapses, the RER was completely un-stained. In some cells the staining of the RER seems particularly marked beneath an AChE-reactive synapse. Plate 4.7a shows a rat motor horn cell in which the RER is more positive beneath an axon which is stained than beneath one which is not (plate 4.7b).

The morphological localization of the enzyme raises a number of questions. The most puzzling is how the enzyme which is apparently synthesised inside the neu-rone gets to the outer surface of the axon. This will be considered again in § 4.3.2 in the context of enzyme transport but one or two general points can be made here. First, it is difficult to be certain that the methods used are showing up all the enzyme that is present and second, the position of the reaction-product may not ac-curately reflect the position of the enzyme in vivo. Kokko et al. (1969) have com-mented on the artifactual deposition of dense aggregates of end-product in transverse bands (plate 4.8) across neuronal processes in rat spinal cord and they suggest these reflect some sort of penetration barrier. Where these bands occurred, the neurotubules on one side of the aggregation contained finely dispersed end-product but on the other side of the band they were unstained. Only a few processes showed this type of reaction, in the majority the neurotubules did not stain at all. These results suggest that at any one moment enzyme is present in neurotubules of a very small proportion of cholinergic axons – but is this the true situation? Other alternatives readily suggest themselves: enzyme may, in fact, be present in tubules in all the axons but, because of some selectivity of the method, is demonstrable in only a few. On the other hand, it may be present in none and the fine deposits might, like the heavy linear aggregates, be some sort of artifact as observed under certain conditions by Schlaepfer and Torack (1966). Kása (1968b, 1971; see also Kása et al. 1973) showed that the staining of neurotubules was greatly enhanced in lesioned nerves (plate 4.9a, b) and he took this as evidence of an increase in enzyme concentration, produced by the damming up of AChE in transit (§ 4.3.2.2). Again though, this result is open to other interpretations. Morphological changes resulting from the lesion might reduce penetration barriers so that previously invisible en-

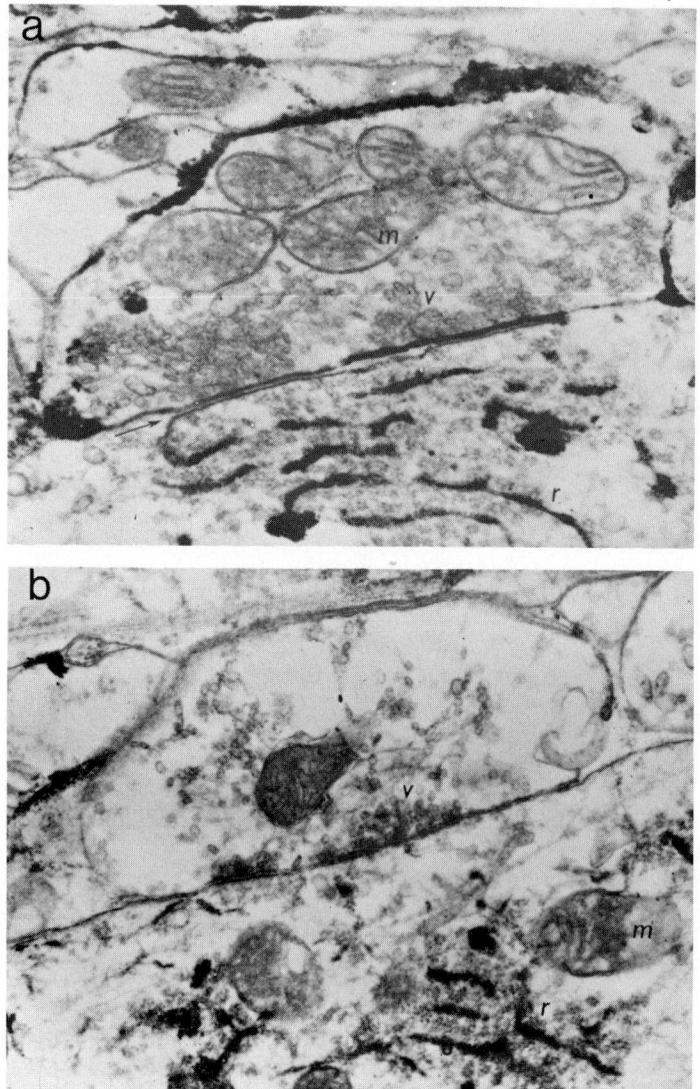

Plate 4.7. Rat spinal cord. Note that the reaction for AChE in the endoplasmic reticulum (r) of a motor horn cell is stronger beneath an AChE-positive ending (a) than beneath a negative ending (b). Note also absence of stain within the AChE-positive terminal. M, mitochondria; v, synaptic vesicles. The arrow marks a tongue of RER. (X 40,000) (From Lewis and Shute 1966.)

zyme becomes demonstrable: conversely, the same changes may allow some non-specific deposition of end-product.

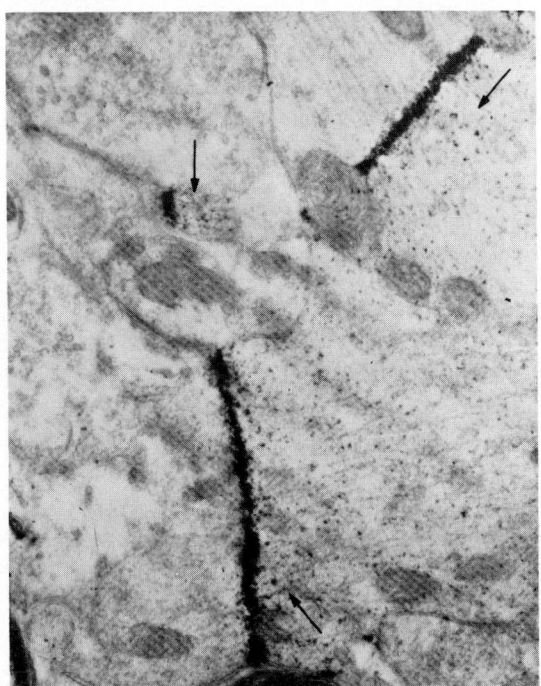

Plate 4.8. To show artifactual deposition of dense bands of AChE reaction-product in neuronal processes of rat spinal cord. Note finely dispersed end-product in neurotubules on one side (arrowed), but not on the other, of the dense bands. The substrate was AcSeCh. (X 26,000) (From Kokko et al. 1969.)

By using subcellular fractionation techniques and electron microscopy in the same experiment it might be possible to determine whether the latter method is sufficiently sensitive to detect low concentrations of enzyme. Since in an animal treated with a lipid-insoluble anticholinesterase, all the external enzyme, i.e. axo-lemmal and synaptic, should be inhibited, any residual activity could be fairly safely attributed to intracellular sites. If one hypoglossal nerve were removed for subfractionation and biochemical analysis of AChE, and the other were examined by electron microscopic histochemistry the results could be useful. Any AChE in the microsomal fraction should represent intra-axonal enzyme; if the electron micrographs failed to reveal activity there, this would be evidence that the histo-chemical method lacks the required sensitivity. Elliott (1968) has suggested that AChE in transit may be bound in such a way that it is non-reactive histochemically.

The other site at which the apparent absence of enzyme is to some degree surprising is inside the terminals, and the functional implications have been indicated earlier (§ 4.2.2.2). The point to cover here is again the reliability of the method. Lewis and Shute (1966) describing the massive cholinergic innervation of the rat caudate nucleus mention specifically that the general cytoplasm of the synaptic processes is unstained. That this negativity could be due to penetration barriers seems unlikely because in tissues such as the spinal cord where cholinergic cells as well as cholinergic fibres are present, strong staining of the intracellular RER provides a sharp contrast to the lack of intraterminal stain in the apparently cholinergic fibres which synapse with them; fig. 4.7a makes this point well.

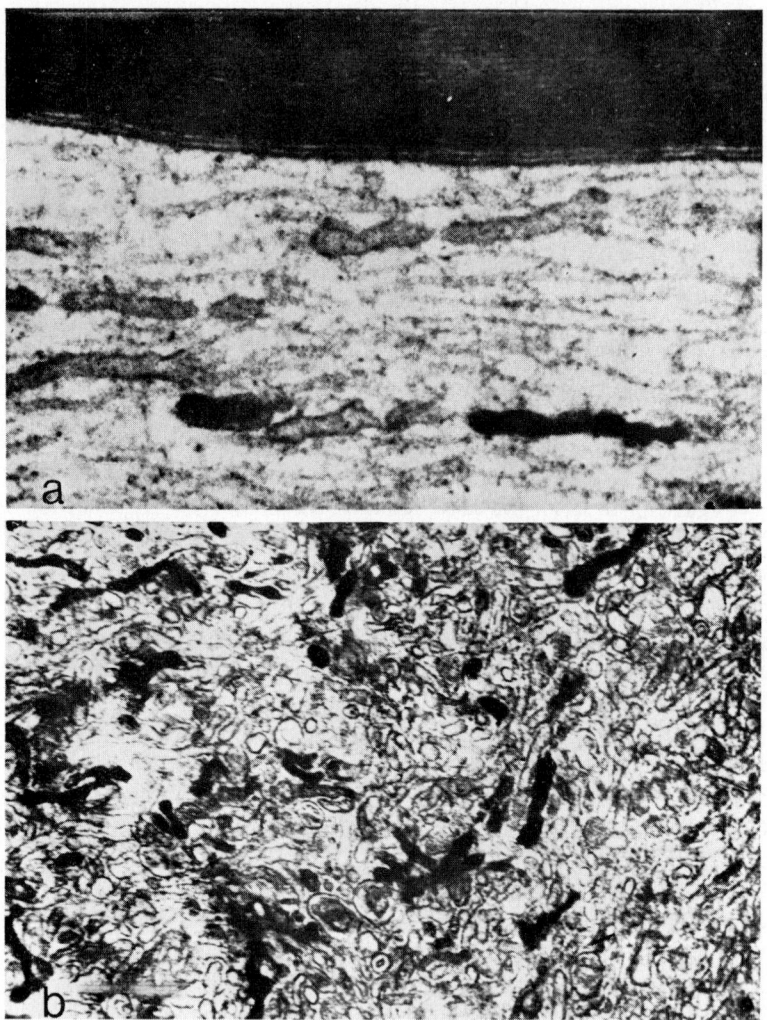

Plate 4.9. AChE reaction product in neurotubules of rat sciatic nerve. a) Normal nerve (X 50,400); b) following a crush (X 29,400). (From Kása 1968b.)

4.3 Synthesis and transport of AChE

4.3.1 Synthesis of neuronal AChE

It would appear from the literature that little work has been done to determine the specific steps involved in the synthesis of AChE, but it seems safe to assume that, in the neurone, it is elaborated on the endoplasmic reticulum under the influence of the ribosomal RNA. Fukuda and Koelle (1959) demonstrated in cat ciliary ganglion cells that following inhibition of AChE with DFP, the pattern of distribution of newly synthesised enzyme is equivalent to the distribution of Nissl substance, that is, the RER (see § 4.2.1).

This brief statement leaves unanswered a number of questions which stem from the ultrastructural localization described in the preceding section. The first is whether the enzyme associated with the nuclear envelope reflects a stage of synthesis or whether it is there to perform some function, perhaps one connected with permeability. Most workers describe the distribution of enzyme on the nuclear membrane as patchy but whether this is functionally significant or is an artifact is again unknown. The next question concerns the Golgi complex. According to Droz and Koenig (1970), proteins destined for neurofilaments and neurotubules may bypass the Golgi complex while those involved in the elaboration of lysosomes, smooth ER and various vesicles migrate from the RER and accumulate within the Golgi membranes. As table 4.1 shows (see also fig. 4.3), AChE is either absent from the Golgi complex or, if present, is confined to occasional saccules. From this rather meagre evidence it is not possible to tell whether or not AChE is 'packaged' by the Golgi membranes. The third point to consider is the source of the axolemmal enzyme but this, together with the question of axonal synthesis, can be tackled more conveniently in the next section. As will become clear, the bulk of evidence favours the idea that the greater part of the enzyme associated with the axon and its terminals originates in the cell. That being so, any discussion of axonal enzyme must include the question of enzyme transport.

4.3.2 Transport

4.3.2.1 Development of the concept of axonal transport   As with so many theories, it is difficult to pin-point exactly who originated the idea that material arising in the cell body can travel down the axon to the terminals. Waller (1850) seems to have been the first to state formally that the cell body is essential for the maintenance of the axon, but this may well have been implicit in earlier theories about the influence of one part of the body on another. In 1905 Scott proposed that a material, which was formed in the nucleus and in the Nissl substance of a neurone, could pass into the axon and he subsequently concluded (Scott 1906) that recovery in a fatigued nerve was due to the arrival of more of this material at the terminals.

The experimental demonstration of the passage of substances along the axon did not occur, however, for a further 40 odd years (Weiss 1947; Weiss and Hiscoe 1948). After that, the belief that proteins manufactured in the perikaryon travel to the terminals via the axon gained ready acceptance and was soon applied to the cholinergic system (Feldberg and Vogt 1948; Dale 1955; Hebb and Waites 1956).

The idea that AChE is exported from the perikaryon rapidly became an integral part of the thinking about the structure and function of cholinergic elements (see Feldberg 1957; Lewis and Hughes 1957; Fukuda and Koelle 1959; Lubińska et al. 1961). However, doubts about the exclusiveness of the role of the perikaryon in AChE synthesis developed equally quickly, and the possibility of local synthesis in the axon was mooted (Koenig and Koelle 1960, 1961; Clouet and Waelsch 1961). This idea has not found universal favour but the evidence on which it is based will be re-examined later in this section.

4.3.2.2 The transport of AChE   Axonal transport of proteins has been studied by a number of techniques. The commonest methods are those in which autoradiography or subfractionation are used to follow the fate of radiochemically labelled amino acids injected into nerve cells or terminals (for reviews see Grafstein 1969; Lasek 1970; Lubińska 1974). In the case of enzymes, activity can be measured in samples of tissue taken from above and below a lesion or ligature involving a central or peripheral nervous pathway (see ch. 3). Biochemical and histochemical experiments (Sawyer 1946; Lubińska et al. 1961; Shute and Lewis 1961; Lewis et al. 1964, 1967; J.L. Johnson 1970) have shown that when a nerve is cut or ligated AChE activity increases immediately proximal to the lesion (i.e. in the stump of nerve still in continuity with the cell body) while below the lesion, enzyme activity decreases. The most generally accepted interpretation of the increased activity is that it represents a damming up of the AChE which is in transit from the cell body to the periphery. With this as a working hypothesis the next point to consider is the state in which the enzyme is travelling. The alternatives seem to be passive diffusion, some sort of active mechanism, or the translocation of the enzyme by the forward growth of the axonal structure to which it is bound.

Lubińska (1964) reviewed the evidence for axoplasmic streaming and suggested that particles containing AChE are carried along in the stream of axoplasm. From the rate at which AChE accumulated at the cut end of the peroneal nerve of dog she and her co-workers estimated that the rate of streaming was roughly 30–60 mm/day. Since that time, the whole subject of protein transport within the CNS has been under constant study with radiochemical methods and it has become clear that transport can occur by 'rapid' and by 'slow' mechanisms but the actual range of speeds encompassed by these terms varies in different species and tissues. McEwen and Grafstein (1968) injected [³H]-leucine into the goldfish eye and found evidence, from autoradiography and tissue analysis, of a fast-moving protein component travelling at 40 mm/day and a slower one moving at 0.4 mm/day. In a later series of experiments on goldfish, Grafstein et al. (1972) obtained a figure for the fast component of 60 mm/day at 20.5°C. The Q_{10} for this component was at least 2.6 but the slow component seemed relatively insensitive to temperature indicating that the underlying mechanisms were probably different. Cuénod and Schonbach (1971) examined the passage of [^{3}H]-leucine in the optic system of the pigeon and observed the 'rapid' wave to have two phases: the radioactivity in subfractions of tissue was associated partly with components migrating at 100–200 mm/day and partly with those travelling at 400 mm/day. The 'slow' wave had a speed of 1–2 mm/day. In parallel experiments involving autoradiography, Schonbach and Cuénod (1971) detected waves within the range of 100–500 mm/day and in addition to these, and to the very slow phase, they distinguished another travelling at 20–60 mm/day. Four different velocities were recorded in rabbit optic nerve by Karlsson and Sjöstrand (1971), these were 150, 40, 6–12 and 2 mm/day. Evidence so far available suggests that the fast component is predominantly membrane-bound, being associated with the particulate fraction, and the slow component seems to represent soluble protein (McEwen and Grafstein 1968; Karlsson and Sjöstrand 1971; Schonbach et al. 1971). This working hypothesis may prove inadequate, more particularly in view of the findings mentioned above (see also Bradley et al. 1971) that transport mechanisms may be operating over a whole range of velocities and not merely at two speeds, 'fast' and 'slow'.

The experiments with labelled amino acids do not differentiate between the various proteins and to obtain information specifically related to AChE it is necessary to use the techniques, outlined earlier, of measuring the rate of accumulation of enzyme in a cut or ligated nerve. The disadvantage of this method is that although the results give information about the behaviour of AChE in a lesioned nerve, they may bear little real relation to the situation in an intact nerve (see Lubińska and Niemierko 1971). Not only may the lesion disrupt the mechanism of transport, but changes in enzyme synthesis accompanying chromatolysis of the cell body (see ch. 9) could introduce confusing factors. The migration rate for AChE, quoted earlier, of 30–60 mm/day was obtained in the dog peroneal nerve. More recent experiments by Lubińska and Niemierko (1971) have revealed that 15% of the AChE in this nerve may be transported much faster, at up to 260 mm/day. A value of this magnitude, lying as it does well within the reported range for the rapid phase obtained with radioisotopes seems to indicate that a sectioned nerve is still capable of transporting protein at a high rate. Support for this assumption has come from the work of Ranish and Ochs (1972). Their analysis of the ligated sciatic nerve of cat showed an even higher rate for the movement of AChE, the mean figure being 431 mm/day. This value correlates well with the figure of 407 mm/day obtained by Ochs and Smith (1971; see also Ochs 1972b) for the migration of [^{3}H]-leucine in the sensory fibres of the non-ligated sciatic nerve of the cat. Ranish and Ochs suggested that the difference between their results and those of Lubińska and Niemierko stemmed from differences in experimental design. Like Lubińska and

Niemierko, they concluded that only a small proportion of the total AChE was moved by the fast process and suggested that of the rest, some was moved by slow transport and some was attached to membranous structures.

In contrast to the evidence for rapid transport of AChE, Frizell et al. (1970) reported that in the ligated vagus and hypoglossal nerves of the rabbit, both AChE and ChAc moved slowly, the rate being 15 mm/day in the vagus and only 5 mm/day in the hypoglossal. It seems probable that a more rapidly moving component could have escaped detection, despite the fact that the first measurements were made 2 hr after the lesion, which is the same interval as that used by Lubińska and Niemierko (1971). That rapid transport of protein does occur in rabbit nerves was shown by Sjöstrand (1970; see also Ochs 1972b). Using [^{3}H]-leucine Sjöstrand obtained values of 400 and 300 mm/day for the vagus and hypoglossal nerves respectively. He also observed a slow phase travelling at 5 mm/day in the hypoglossal nerve; this is identical with the value obtained both for AChE and for ChAc by Frizell et al. In the vagus nerve, the figure for leucine was somewhat higher than the value for the enzymes, being 26 mm/day.

The rates of transport do not pin-point the underlying mechanisms but they can provide some clues, albeit partly negative ones. Hebb and Silver (1961) had evidence that ChAc was bound to some axonal structure and moved as it moved; they tentatively suggested that transport could depend on the growth of the axon as a whole. Since regenerating axons may grow at rates of about 3–5 mm/day (for references see Lubińska 1964), this process might be fast enough to account for the lowest values reported. Lubińska and Niemierko (1971) concluded that except for the 15% of AChE which was moving very rapidly, axonal AChE was stationary or moving only very slowly — as they point out, their short-term experiments would not detect low rates of movement.

A possible passenger for a very slow transport system — be it propelled by growth or by some other mechanism — is the axolemmal enzyme. If 'growth' of mature axons does indeed occur, it is not difficult to imagine AChE, attached as it is to the external surface of the axon, being borne along by this process. What is not so easy to imagine is how the enzyme gets to the outside of the axon in the first place — how, as it were, it takes its seat. If, as seems probable (§ 4.3.2.1) axolemmal enzyme originates in the synthetic system of the perikaryon, it must migrate in one or more stages from the inside of the cell to the outside of the axon. Kása (1968b) using electron microscopic histochemistry observed AChE activity within a fairly small proportion of neurotubules in rat and rabbit sciatic nerves. Some of the enzyme-positive tubules appeared to have fused with the internal surface of the axolemma, and the histochemical reaction product within the neurotubules was in continuity with that on the outer surface of the axolemma. Kreutzberg et al. (1973) reported a similar apposition of AChE-positive structures to the plasma membrane of dendrites, particularly at post-synaptic sites. It is dangerous to attribute functional significance to purely morphological data and it is especially difficult to decide how far the static situation shown by tissue processed for electron microscopy is likely to reflect the dynamic events occurring in vivo. Further work is needed to establish whether or not the enzyme found on external surfaces has been transferred from intracellular structures which fused with the membrane. If this is the mechanism of transfer it might account for the spotty appearance of the end-product seen in electron micrographs of the plasmalemma and axolemma; but the possibility remains that the patchiness is artifactual.

As mentioned earlier, Kása (1968b, 1971; see also Kása et al. 1973) found that the histochemical reaction for AChE in neurotubules was much more pronounced in ligated nerves than in control nerves and the appearance suggested the damming up of enzyme in transit. He postulated that the neurotubules might be involved in the rapid transport of unbound enzyme. Schmitt (1968) has also suggested that neurotubules have a role in rapid transport but according to his scheme the migrating material is vesicle-bound. The vesicles move along the neurotubules in a series of jumps, the necessary power being provided from the splitting, by vesicular ATPase or GTPase, of the ATP or GTP of the neurotubule itself. In the scheme proposed by Ochs (1971, 1972a) the neurotubules act as the stationary member over which a specially synthesised 'transport filament' can slide. The filament is visualized as binding a variety of components including both particulate and soluble proteins. Ochs suggested that a unified system of this type could account for his finding that different materials all travelled at the same fast rate of 410 mm/day. In contrast to these ideas linking the neurotubules with fast transport, McEwen and Grafstein (1968) suggested that they might be responsible for the slow component. They pointed out that over 50% of the protein associated with the neurotubules is soluble and that soluble protein seems to migrate only slowly (see also McEwen et al. 1971). Because of these conflicting data no firm conclusion can, as yet, be drawn about the role of neurotubules in AChE transport (see also Kreutzberg 1969). All that can be said is that the AChE present in some neurotubules is probably undergoing translocation but whether to the axolemma or to the terminals and whether by fast or slow processes is not established.

Tennyson et al. (1968) followed the time-course of development of AChE in the dorsal root ganglia and spinal cord of rabbits by electron microscopy. Observable AChE activity appeared first in the RER in the ganglion cell and then on the axolemma of the distal part of the axon, only later did it appear in the SER of the intervening dorsal root. If it is assumed that the distally situated enzyme has arrived there from the cell — as opposed to being synthesised locally (see § 4.3.3) or being formed in situ from inactive subunits of perikaryal origin (see Koenig and Koelle 1961; Tennyson and Brzin 1970) — then it follows that at some stage it was in the axon but in an undetectable form. The same may well be true in adult cholinergic nerves — the nature of the bond between the enzyme and the particulate may render the active site unavailable to the histochemical substrate. If this is the case, then the demonstration of rapid transport mechanisms by histochemical methods may not be feasible.

Once the rapidly transported AChE has arrived at the terminals, where does it go? The dearth of ultrastructural data about its location within the terminals in adult nerves has been discussed above as has Kása and Csillik's (1968) observation of AChE-containing tubules in terminals of the phrenic nerve in rats of 2 months-old but not in adults. Unless the apparent lack of enzyme is a histochemical artifact, one must conclude that any enzyme reaching the terminal is rapidly translocated to the external surface of the synaptic membrane but, as with axolemmal enzyme, the process by which it crosses from the inside to the outside is obscure. An added complication is that on the theory discussed earlier, the slow moving axolemmal enzyme may be destined for the terminal too and could also contribute to the synaptic AChE.

Although the details of the mechanism are still unknown the basic concept of a proximodistal movement of AChE is firmly established. The idea that flow also occurs in the opposite direction, that is from the terminal to the cell body, has been accepted less readily. Lubińska and her co-workers (Zelená and Lubińska 1962; Jankowska et al. 1969) have shown in short-term experiments that AChE accumulates distal to a lesion, as well as proximally, and they have recently demonstrated (Lubińska and Niemierko 1971) that the rate of accumulation in dog peroneal nerve is equivalent to a distoproximal flow rate of 134 mm/day. This is in contrast to a rate of 260 mm/day for the proximodistal movement measured in the same series of experiments. Ranish and Ochs (1972) found in their experiments on cat nerves, mentioned above, that the rate of retrograde flow was 220 mm/day compared with a rate of 431 mm/day for anterograde movement. The relevance of these results to the situation in intact nerves has been questioned on the grounds that injury currents or general trauma might cause an abnormal redistribution of axonal contents. However, results obtained with $[^{14}C]$-glutamate from the brain-nerve-muscle preparation of snail and the spinal cord-nerve-muscle preparation of frog (Kerkut et al. 1967) give support to the belief that bidirectional axonal movements do occur in intact nerves. Watson (1968) working on the rat hypoglossal nerve labelled via the geniohyoid muscle with $[^3H]$-lysine came to much the same conclusion. Further evidence of retrograde movement has been obtained by Kristensson et al. (1971) from rabbits, rats, mice and guinea-pigs. In these experiments horseradish peroxidase (for the rationale of this method see Krishnan and Singer 1973) or a fluorescent complex of Evans blue and albumin were used as markers in place of radioisotopes. Kirkpatrick et al. (1972) who viewed hen sciatic nerves by Normarski differential interference microscopy were able to see particulate matter moving distoproximally as well as proximodistally.

Although this evidence supports the possibility of a passage of AChE from terminal to cell body, the purpose of the process is obscure. Inevitably the idea of a feed-back system controlling enzyme synthesis suggests itself but this is the facile answer – the true reason may be much more complex (see Lubińska 1974).

4.3.3 Synthesis of AChE within the axon

Koenig and Koelle (1960) were the first to question whether the concept of export of AChE from the perikaryon down the axon was entirely valid. They pointed out that all the evidence of proximodistal movement was based on experiments on ligated or growing nerves and these might not be equivalent to the normal adult nervous system in the steady state. They postulated that if the 'somatoaxonal convection' hypothesis applied to AChE then the restoration of enzyme activity following irreversible inhibition should show a proximodistal gradient. Such a gradient was not, however, observed in nerves (for example, the hypoglossal) from cats injected with DFP. Instead there was a simultaneous reapppearance of AChE activity along the whole length of the nerve. This, they argued, was evidence of the relative independence of axonal enzyme from perikaryal enzyme. The following year they pursued the problem further (Koenig and Koelle 1961) and carried out several series of experiments in which they followed the rate of return of enzyme activity after the differential inhibition of neuronal and/or axonal AChE. When DFP was injected intravenously, both axonal and neuronal enzyme was inhibited; if it were given into the ventricle the AChE in the soma of the hypoglossal cells was inhibited but that in the trunk was not appreciably affected. Conversely, intravenous injections of di*iso*propylphosphostigmine (DPS), which does not cross the blood–brain barrier resulted in inhibition of axonal but not neuronal enzyme. In some experiments cats were given a single intravenous injection of DFP to inhibit axonal AChE and this was followed by a daily injection of DFP intraventricularly. The rate of return of axonal enzyme was the same in these cats as in those which had received only the intravenous injection. The level of AChE in the hypoglossal nucleus after intraventricular injection was never more than 34% of that in intravenously treated animals. When axonal enzyme was inhibited by intravenous DPS – this treatment should not affect neuronal enzyme – the rate of regeneration was no greater than that following suppression of neuronal as well as axonal activity by DFP.

In analysing their results, Koenig and Koelle believed that the possibility of the reactivation of phosphorylated enzyme could be eliminated and they preferred the interpretation that axonal AChE was independent of neuronal synthesis. The possibility of de novo synthesis of AChE in the axon was thought to be rather remote because of the apparent lack of axonal RNA, but formation of AChE within the axon from precursors exported by the cell body seemed possible. The other idea they suggested was that AChE, initially exported from the perikaryon during growth, might be continually broken down and reconstituted in the axon. The reappearance of activity after inhibition could represent the replacement, via this process, of the phosphorylated sites by functionally active sites. Koenig and Koelle's criticism of experiments on lesioned nerves, that the situation is abnormal, has some relevance to their own work: mechanisms called into play following total, unphysiological enzyme inhibition may be a poor mirror of what occurs under physiological conditions.

Clouet and Waelsch (1961) were also of the opinion that axonal AChE might be formed locally. In their experiments on frog sciatic nerves in which AChE had been inhibited by DFP, they found not a uniform return, as Koenig and Koelle had done, but a distoproximal gradient. This they felt to be incompatible with the theory of axonal flow and tentatively raised the question of local synthesis but, like Koenig and Koelle, they recognized that the lack of axonal ribosomes could prove a stumbling block.

By the time Koenig (1965a) published his extended studies this objection had to some extent, been met. Edström (1964) had found RNA in the axons of Mauthner cells in goldfish and Koenig (1965b) had himself detected very small amounts in myelin-free axons from the glossopharyngeal nerve of the cat. Koenig (1965b) proposed that axons might contain permanent messenger-RNA which would serve as a template for AChE synthesis or, alternatively, that the perikaryon might export an enzyme precursor-RNA complex. Axonal RNA has subsequently been detected by a number of investigators in both mammalian and non-mammalian tissues (for references, see Bondy 1972; Jarlstedt and Karlsson 1973) and, furthermore, there is now an indication that ribosomes may be present in at least some nerves (Zelená 1970, 1972). In keeping with these findings, studies with tritiated amino acids are providing results which suggest that protein synthesis can take place in the axon and, particularly, in synaptosomes (Koenig 1967; Gordon and Deanin 1968; Morgan and Austin 1968; Bosmann and Hemsworth 1970; Cotman and Taylor 1971; Goldberg 1971, 1972). These results do, however, refer to protein in general and not to a specific enzyme. The mechanisms of local synthesis may not be related to enzyme production at all; they may well be elaborating membranes or the intravesicular proteins (Whittaker et al. 1971).

Evidence against local production of AChE comes from two types of experiment. First, Lubińska et al. (1964) isolated segments of dog peroneal nerve by sectioning them proximally and distally. These segments were left in situ for periods of 2–22 hr after operation and then analysed for AChE, their content being compared with that in a segment of control nerve equivalent in length and position. No difference was found between the control and experimental segments and the absolute values depended on their lengths. Clearly, under the conditions of the experiments no local synthesis had occurred. It could be argued that the lesions might have introduced a totally abnormal situation; however, if that were so, values in the isolated segment might well be higher or lower than those in the control, but they are unlikely to be the same.

The second piece of evidence against local synthesis comes from work on non-lesioned nerves. Austin and James (1970) injected DFP into rat brains and killed the rats at intervals up to 28 days later. They subfractionated the cortex and analysed the microsomal and synaptosomal fractions for AChE. The enzyme was found to regenerate more slowly in the synaptosomal than in the microsomal fraction, the synaptosomal fraction taking 24 hr longer to reach any given value. Values were expressed as specific activity, that is units AChE/mg protein, where one unit hydrolysed 1 μmol substrate/hr at 37°C. The regeneration ratios were calculated as the specific activity of a subfraction at a given time after DFP, divided by the specific activity of a similar fraction from a control animal. The authors interpreted their results as evidence against local synthesis: the lag between restoration in the microsomal and the synaptosomal fractions suggests that the synaptosomal enzyme had been derived from the perikaryon. Austin and James rightly point out, however, that these experiments do not entirely rule out the possibility that some axonal synthesis can occur as well.

If the idea of a significant local synthesis is to be rejected, what is the explanation for Koenig's findings and those of Clouet and Waelsch? This is a difficult question to answer but it may hinge on two points. The first is that Koelle (1957a) and Fukuda and Koelle (1959) showed that a return of AChE activity was detectable in cat ganglion cells well within 80 min of inhibition with DFP; the second is the finding in the cat (Ranish and Ochs 1972) that 10% of the axonal AChE is travelling at over 400 mm/day. With rates of this sort, the time-course of Koenig's experiments may not have been sufficiently precise to detect any gradient of restoration. Furthermore, DPS being lipid-insoluble will have inhibited only the extracellular part of the axonal enzyme and this could complicate the interpretation.

Clouet and Waelsch worked on the frog in which the rate of synthesis is probably slower than in the mammal; if perikaryal enzyme is exported to the periphery as fast as it is produced, proximal levels could remain below those in the distal areas to which the enzyme is being dispatched. Tennyson and Brzin (1970) suggested that this type of situation might explain the similar distribution of enzyme activity in the developing rabbit neuroblast. This purely speculative attempt to interpret the results of Koenig and of Clouet and Waelsch in terms other than local synthesis can be faulted on various grounds; in particular it makes no attempt to accommodate the complications introduced by the existence of slow as well as fast export mechanisms. A combination of electron microscopy and subfractionation techniques together with the use of lipid-soluble and lipid-insoluble inhibitors might help to settle the question. The conflicting evidence for and against the synthesis of AChE in the axon probably arises from the insensitivity of the methods used so far.

When a cholinesterase-containing nerve is cut, the AChE does not disappear from the isolated stump in parallel with the degeneration of the axons. Sawyer (1946) found that in guinea-pig tibial and peroneal nerves in which the axons had degenerated, the AChE activity was still 40% of that in intact nerves; Lubińska et al. (1963) give a figure of 25% for dog nerves. To explain the persistence of enzyme, Sawyer suggested it might be present in the Schwann cells or fibroblasts. Later work on hen (Cavanagh et al. 1954) and rat (Tewari and Bourne 1960) indicated that in some species, however, Schwann cells might lack AChE. As mentioned in § 4.2.3, electron microscopic evidence of AChE in Schwann cell cytoplasm is sparse although the enzyme may be present on the cell membrane and between this and the axonal membrane. Where this enzyme originates is unknown. After a nerve is sectioned the number of Schwann cells increases greatly and various metabolic changes occur (Lubińska 1961 a,b). The possibility that these changes include the acquisition (see ch. 9) of the ability to synthesise AChE was first raised by Lubińska et al. (1963). On the other hand, Eränkö and Teräväinen (1967) showed that the cholinesterase activity appearing in Schwann cells of the sectioned sciatic nerve of rat was due entirely to BuChE.

4.3.4 Synthesis of AChE at the motor end plate

Before leaving the question of synthesis, brief consideration must be given to the motor end-plate. At the neuromuscular junction AChE is localized mainly post-synaptically (Couteaux and Nachmansohn 1938). Davis and Koelle (1967) and Salpeter (1967), among others, have investigated the ultrastructural localization of the enzyme and have shown that it is restricted to the postjunctional sarcoplasmic membrane; Teräväinen (1967, 1969a) has, however, reported reaction product in the sarcoplasm as well under certain experimental conditions (see also Tennyson et al. 1973 re-developing muscle). Padykula and Gauthier (1970) have demonstrated the presence of granular endoplasmic reticulum and free ribosomes in the junctional sarcoplasm in rat diaphragm and they discussed the possibility that AChE and receptor protein might be among the products of protein synthesis occurring in the junctional area. Once again the problem of translocation arises. How is the enzyme reaching the external surface of the membrane? The sarcoplasmic localization reported by Teräväinen (1967, 1969a) may perhaps represent recently synthesised enzyme which has not reached the membrane, but the ever-present possibility of a histochemical artifact cannot be ignored. The sort of experiment which Fukuda and Koelle (1959) did to determine the area – in the case of the ciliary ganglion it was the RER – in which enzyme activity first reappears after inhibition, could perhaps pick out the site of synthesis.

Kupfer (1951) and subsequently many others (for references see Eränkö and Teräväinen 1967) have shown that following denervation an appreciable amount of the enzyme persists in the post-synaptic membrane long after the nerve terminals have degenerated. This persistence, together with the presence in the sarcoplasm of the apparatus for protein synthesis, supports the generally accepted view that post-junctional enzyme is not supplied by the neurone. Nevertheless, the innervating nerve appears to exert a considerable influence on the enzyme levels in the junction. This will be discussed again in ch. 9 but, in brief, the evidence for such an influence is this: the fall in AChE produced in a junction by denervation is reversed by re-innervation but, in addition, a nerve can make its presence felt even though it has no structural connection with an end-plate. Guth et al. (1966) found that when the hypoglossal nerve was implanted into the denervated sternomastoid muscle of rat, in such a position that it did not re-innervate the original end-plates, AChE in these end-plates was restored to 60% of the control value in 8 weeks whereas, in muscles without an implant, activity at that time was only 20% of normal. Implantation of the hypoglossal nerve into the normally innervated muscle had no effect on end-plate AChE. Guth et al. (1967) tried to find out how the effect on AChE was mediated. Although they eliminated a number of factors such as muscle contraction they could not identify the mechanism. Their results suggested that ACh was not involved but subsequent work (see ch. 9) indicates that it probably is.

4.4 Concluding remarks

As yet there are still far too many uncertainties and far too few data to allow a definitive account of the synthesis, location and translocation of AChE. But the interest in the general subject of axonal transport is currently enormous and it may not be too long before the answers to at least some of the questions are forthcoming. For a start we need to know what controls AChE synthesis in the perikaryon; we need to know whether or not significant synthesis occurs in the axon, and if so, for what purpose; and we need to know how AChE moves and if, as available data suggest, it moves both fast and slowly why this should be. Finally we need to know the extent and significance of the distoproximal movement of enzyme. To answer these questions the paramount requirement is for methods capable of demonstrating, as closely as possible, the situation as it exists in the intact living system. Only then can we be rid of the fear that the observations we seek to explain are mere artifacts of our own creating.

CHAPTER 5

Cholinesterases in invertebrates

5.1 Introduction

Rothschild (1961) prefaced his book, 'A Classification of Living Animals', by giving the approximate number of species in each phylum. When the figures for the invertebrates are added up they give the staggering total of well over 900,000 species. In the light of this sort of figure it is obvious that any investigation of the invertebrates must be highly selective. Studies of cholinesterase activity have been made on what amounts to a tiny proportion of the total invertebrate population but even so, the literature on those species that have been examined is daunting and the references alone would more than fill a chapter.

The problem is to decide what should be included and what omitted in a chapter that must in any case be incomplete. Considering that this book is primarily on neuronal cholinesterases, and that some invertebrates do not possess a nervous system and some apparently do not contain ChE, the easiest solution might be to ignore the invertebrates altogether. Such a course, however tempting, would be quite unjustifiable: no matter how incomplete the treatment must be in relation to the invertebrates themselves, consideration of the lower animals is important in the much broader context of the evolution of the nervous system and of mechanisms of nervous transmission. Furthermore, the widespread use of many pesticides which apparently depend for their action on their anticholinesterase activity has transferred the subject from an academic one to a matter of practical importance (see ch. 11). Because of the economic aspects of insecticides a disproportionate amount of research on invertebrate cholinesterases has been devoted to the insects and, inevitably, this specialization will be reflected here. It should, though, be pointed out that physiologists and pharmacologists owe a debt to a number of invertebrate species of other phyla including the leech, *Hirudo medicinalis*, the sea cucumber, *Holothuria forscali*, the gaper, *Mya arenaria*, and the clam, *Venus mercenaria*, which have provided test objects for the bioassay of ACh.

5.2 Occurrence and functional significance

It is probably safe to make the generalization that cholinesterases are present in at least some members of almost every invertebrate phylum but it is much less safe to state that in some species cholinesterases are absent. First, various factors which will be discussed below may interfere with the experimental detection of the enzymes; second, papers reporting work on cholinesterases are scattered through the scientific literature in journals which are almost as diverse as the invertebrates themselves, and positive findings are all too easily overlooked. Thus, although Bayer and Wense had reported the hydrolysis of ACh by cultures of *Paramecium* in 1936, statements that ChE is absent from protozoans appear in papers published as late as 1948.

Table 5.1 shows some of the species in which cholinesterases have been found. All the animals listed contain cholinesterase, but it is not necessarily acetylcholinesterase. As will become apparent, invertebrate cholinesterases do not fit neatly into the classification used to distinguish acetylcholinesterase from other cholinesterases in mammals. In many cases, whole animals or non-nervous tissues have been analysed but those species in which cholinesterases have been demonstrated in nervous tissues are marked in the table with an asterisk. Quantitative values have not been given even where biochemical, rather than histochemical, methods have been used because, as will be shown later, experimental conditions can affect the results to an extent which renders meaningless the comparison of values obtained in different studies.

Nachmansohn (1939) suggested that the presence of cholinesterase in a primitive animal was evidence for the existence of nervous structures. He and Bullock (Bullock and Nachmansohn 1942) took the view that the evolution of a differentiated nervous system was accompanied by the development of 'the mechanism connected with acetylcholine'. This idea, that when the need for a nervous system arose the chemical components necessary for its function were developed simultaneously, could be termed the ad hoc theory. Augustinsson and Gustafson (1949) also subscribed to the idea that enzymes are synthesised to fulfil a need but Pantin (1956) proposed a rather different scheme. He suggested that as the nervous system evolved it appropriated for its function chemicals already present in the cells. In view of the evidence that cholinesterase is found in the sponges (Lentz 1966) in which the presence of a nervous system is controversial, and in the protozoans (Bayer and Wense 1936; Seaman and Houlihan 1951; Nakajima and Hatano 1962) in which it is non-existent, Pantin's idea is the more tenable. Pantin suggested that the electrical and chemical properties which became available for evolutionary selection were 'accidental' features of the primitive cell. By extrapolation, one might conclude that the protozoans are endowed with a collection of substances which need not be serving any specific purpose. Certain evidence suggests that this is not the full story. Seaman and Houlihan (1951) found that the cholinesterase present in the protozoan *Tetrahymena geleii* seemed to be involved in the coordination of ciliary movement. Addition of the cholinesterase inhibitors, eserine or DFP, in concentrations of 3.85×10^{-3} M reversibly immobilized the animals but was without effect on the rate of glycolysis or oxidative phosphorylation. It must be noted,

TABLE 3.1

Selected papers describing cholinesterases in invertebrates. An asterisk indicates that the presence of ChE in nervous tissue has been established.

Phylum/Class	Genus/Species	Method	Comment	Reference
Protozoans				
	Paramecium	Biochemical	Cultured at 5°C	Bayer & Wense (1936)
	Tetrahymena geleii	Biochemical	Whole animals	Seaman & Houlihan (1951)
	Tetrahymena pyriformis	Electron microscopy & autoradiography	Whole animals	Schuster & Herschenov (1969)
	Physarum polycephalum	Biochemical	Homogenate of plasmodium	Nakajima & Hatano (1962)
Porifera (Sponges)				
	Sycon ciliatum	Histochemical	Presence of nervous tissue controversial	Lentz (1966)
Coelenterates (Hydra, sea anemones etc)				
	Hydra fusca	Biochemical	Whole animals	Mitropolitanskaya (1941)
	*Hydra littoralis	Histochemical	Enzyme in sensory & ganglion-type cells	Lentz & Barnett (1961)
	Tubularia crocea	Biochemical	Tissue analysed included nerve	Bullock & Nachmansohn (1942)
	Tubularia	Biochemical	Heads	Bullock et al. (1947a)
	Metridium marginatum	Biochemical	Part of animal	Bullock & Nachmansohn (1942)
	Sagartia luciae	Biochemical	Whole animals	Bullock & Nachmansohn (1942)
	Sagartia parasitica	Biochemical	Tissue analysed included nerve net	Augustinsson (1948)
	Actinia equina	Biochemical	Whole animals (?)	Mitropolitanskaya (1941)
	Actinia	Biochemical (electrophoresis)	See text	Haites et al. (1972)
Platyhelminths (Flat worms) N.B. Additional references in text				
Turbellarians (Non-parasitic flatworms)	*Procotyla fluviatilis*	Biochemical	Whole animals	Bullock & Nachmansohn (1942)
	*Procotyla fluviatilis	Histochemical	See text	Lentz (1968)
	Planaria maculata	Biochemical	Whole animals	Bullock & Nachmansohn (1942)
	Planaria dorotocephala	Biochemical	Whole animals	Hawkins & Mendel (1946)
Cestodes (Tapeworms)	*Hymenolepis diminuta*	Biochemical	Whole animals	Graff & Read (1967)
	*Taenia taeniaformis	Biochemical & histochemical	See text	Eränkö et al. (1968)
	Dicrocoelium	Biochemical	Whole animals	Polyakova (1967)
Trematodes (Flukes)	*Fasciola hepatica	Histochemical	See text	Krvavica et al. (1967)
	*Fasciola hepatica	Histochemical	See text	Halton (1967)
	Fasciola hepatica	Biochemical	Whole animals	Frady & Knapp (1967)

135

Phylum/Class	Genus/Species	Method	Comment	Reference
Trematodes (continued)	*Schistosoma mansoni	Histochemical	Whole animals	Fripp (1967a)
	*Schistosoma rodhani			
	*Schistosoma haematobium			
Nemertines (Ribbon worms)	Cerebratulus lacteus	Biochemical	Whole animals	Smith et al. (1940)
	Prostoma rubrum	Biochemical	Whole animals	Kamemoto (1957)
Aschelminths	N.B. Additional references in text			
Nematodes (Roundworms)	Dictyocaulus filaria	Biochemical	Whole animals	Polyakova (1967)
	Haemonchus contortus	Biochemical	Whole animals	Lee & Hodsden (1963)
	Nippostrongylus brasiliensis	Biochemical	See text	Sanderson (1969)
	Ascaris suum	Biochemical (electrophoresis)	See text	Haites et al. (1972)
Brachiopods	Terebratulina caput serpentis	Biochemical	Whole animals	Augustinsson (1946a)
Molluscs	N.B. Additional references in text			
Polyplacophora	Tonicella marmorata	Biochemical	Whole animals	Augustinsson (1946a)
Gastropods (Slugs, snails etc)	Patella vulgata	Biochemical	Whole animals	Augustinsson (1946a)
	Purpura lapillus	Biochemical	Whole animals	Augustinsson (1946a)
	Aplysia depilans	Pharmacological	See text re indirect evidence	Tauc & Gerschenfeld (1962)
	*Aplysia californica	Biochemical	Abdominal ganglion	Giller & Schwartz (1971b)
	*Aplysia californica	Biochemical	Abdominal ganglion	McCaman & Dewhurst (1971)
	*Lymnaea stagnalis	Biochemical	Ganglia	Varanka (1968b)
	*Lymnaea stagnalis	Biochemical & histochemical	Ganglia	Jurchenko et al. (1973a, b)
	Helix pomatia	Biochemical	Whole animals	Augustinsson (1946b)
	*Helix pomatia	Histochemical	See text	Zsoltan-Nagy & Salánki (1965)
	*Helix aspera	Biochemical	Brain analysed	Korn (1969)
	*Helix aspera	Histochemical	Electron microscopy	Newman et al. (1968)
	Dentalium entalis	Biochemical	Whole animals	Augustinsson (1946a)
Lamellibranchs (Mussels etc)	Mytilus edulis	Biochemical	Muscle	Twarog (1954)
	Modiolus demissus	Biochemical	Heart	Smith & Glick (1939)
	Venus mercenaria	Biochemical	Heart	Smith & Glick (1939)
	Astarte sulcate	Biochemical	Whole animals	Augustinsson (1946a)
	Mya arenaria	Biochemical	Whole animals	Augustinsson (1946a)
	*Anodonta cygnea	Biochemical	Ganglia	Salánki et al. (1966)
Cephalopods (Squid, octopus etc)	*(Eu) sepia officinalis	Biochemical	Brain-ganglion	Bacq & Nachmansohn (1937)
	*Sepia officinalis	Histochemical	Optic lobes	Drukker & Schadé (1965a)
	*Loligo pealeii	Biochemical	Axons	Boell & Nachmansohn (1940)

TABLE 5.1 (continued)

Phylum/Class	Genus/Species	Method	Comment	Reference
Cephalopods (continued)	*Loligo pealeii	Biochemical	Axons	Bullock et al. (1947b)
	*Loligo pealeii	Biochemical & histochemical	Axons (see text)	Brzin et al. (1965)
	*Loligo pealeii	Biochemical & histochemical	Axons	Bryant & Brzin (1966)
	*Loligo pealeii	Biochemical (subcellular)	Optic ganglia	Welsch & Dettbarn (1972a)
	*Ommatostrephes sloanei-pacificus	Biochemical & histochemical	Optic ganglia (see text)	Turpaev et al. (1968)
	*Octopus dofleini	Biochemical	Nervous tissue	Loe & Florey (1966)
Sipunculoids	Phascolosoma	Biochemical	Retractor muscle	Nachmansohn (cited by Prosser 1950)
	Sipunculus nudus	Biochemical	Blood	Bacq (1935)
Annelids	N.B. Additional references in text			
Polychaets (Lugworms etc)	Nereis	Biochemical	Body wall	Nachmansohn (cited by Prosser 1950)
	Spirographis spallanzani	Biochemical	Blood (low activity)	Augustinsson (1948)
Oligochaets (Earthworms etc)	Eisenia foetida	Histochemical	See text	Vigh-Teichmann & Goslar (1968)
	Lumbricus terrestris	Biochemical	Muscle	Bacq & Oury (1937)
	Lumbricus terrestris	Histochemical	See text	Silver (unpublished)
Hirudineans (Leeches)	Hirudo (medicinalis ?)	Biochemical	Muscle (species not given)	Ammon (1935)
	*Hirudo medicinalis	Histochemical	Nervous tissue	Gerebtzoff (1970)
	*Hirudo medicinalis	Histochemical	See text	Silver (1972)
	*Haemopsis sanguisuga	Biochemical	Abdominal nerve cord	Schwab (1949)
Arthropods Insects	See text for numerous references, also:			
	*Locusta migratoria	Histochemical	CNS	Mandel'shtam (1967)
	*Rothschildia orizaba	Biochemical	Increases during development	Mansingh (1967)
	*Oncopeltus fasciatus	Histochemical	Increases during development	Salkeld (1961)
	*Periplaneta americana	Histochemical	See text	Hess (1972)
Crustaceans (Lobsters, crabs, barnacles etc)	*Homarus vulgaris	Biochemical	Abdominal nerve cord	Marnay & Nachmansohn (1937c)
	*Homarus americanus	Biochemical	Abdominal nerve cord	Bullock et al. (1947b)
	*Homarus americanus	Biochemical	Walking-leg nerve (see text)	Dettbarn (1963)
	*Homarus sp.	Biochemical	Walking-leg nerve (see text)	Brzin et al. (1965)
	{ *Homarus americanus *Panulirus argus *Panulirus guttatus	Biochemical (electrophoresis)	Proportions of isozymes differ in central and peripheral tissue	Maynard (1964)
	Libinia emarginata	Biochemical	Walking-leg nerve (crab)	Dettbarn (1963)

TABLE 5.1 (continued)

Phylum/Class	Genus/Species	Method	Comment	References
Arachnids (Spiders, mites)	See text for references, also: *Boophilus microplus*	Biochemical	Homogenized larvae	Roulston et al. (1968)
Echinoderms				
Crinoids (Feather stars)	*Antedon petasus*	Biochemical	Whole animals	Augustinsson (1946a)
Holothurians (Sea cucumbers)	*Holothuria nigra*	Biochemical	Muscle	Bacq & Nachmansohn (1937)
	Mesothuria intestinalis	Biochemical	Whole animals	Augustinsson (1946a)
	Cucumaria lactea	Biochemical	Whole animals	Augustinsson (1946a)
	Thyone briareus	Biochemical	Muscle	Bullock & Nachmansohn (1942)
Echinoids (Sea-urchins)	*Echinus esculentus*	Biochemical	Whole animals	Augustinsson (1946a)
Asteroids (Starfish)	*Asterias forbesi*	Biochemical	Radial nerve cord	Bullock & Nachmansohn (1942)
	Asterias rubens	Biochemical	Gut & podia mainly	Augustinsson (1946a)
	Asterias rubens	Histochemical	Radial nerve & podia	Pentreath & Cottrell (1968)

however, that the concentration of anticholinesterase which prevented movement was far greater than that needed (4×10^{-7} M) to inhibit the cholinesterase activity in homogenates. Earlier evidence of a possible relation between cholinesterase activity and ciliary action came from experiments of Augustinsson and Gustafson (1949) who found that in developing sea urchin's eggs, ChE activity increased greatly at the time when the ciliated tufts appeared. If formation of the tufts was prevented by treating the embryos with lithium, the increase in ChE was also prevented. Willmer (1960) cited several examples of motile structures in which acetylcholine seemed to be effective in the absence of a nervous system. Ciliary activity in the gill plates of the mussel, *Mytilus edulis,* is stimulated by ACh (Bülbring et al. 1953a) and ACh increases the motility of pig spermatozoa (Sekine 1951). Similarly, Bülbring et al. (1949) found ACh (and also hydrolysis of ACh) in *Trypanosoma rhodesiense* which is extremely motile, but not in the non-motile, blood-borne stage of the malarial parasite, *Plasmodium gallinaceum.* More recently Gustafson and Toneby (1970) have shown that ACh (and, at an earlier stage, 5-hydroxytryptamine) may initiate various types of cell movement essential to morphogenesis in the sea urchin.

All these findings indicate the possible importance of the ACh-ChE system in cellular movements in some species, but it should not be assumed that it has a universal role. Willmer (1960) could find no good evidence that ACh, eserine or DFP had any effect on the amoeba, *Naegleria gruberi*, which exists in a flagellated as well as an amoeboid form. Presumably, in these animals, movement depends on a different chemical constituent. Similarly, Manukhin and Buznikov (1965) reported that in embryos of opisthobranch molluscs (these are marine gastropods) ChE was absent from the motor cells of all but one of the species examined and ACh was without effect on the movement of cilia; serotonin, on the other hand, accelerated ciliary beating. According to Burnasheva and Efremenko (1962) the motility of *Tetrahymena pyriformis* can be directly correlated with the ATP content. How this dependence of ATP is related to the apparent dependence on cholinesterase is not clear but in a general discussion of cellular excitation, Duncan (1967) has postulated that ATPase and pseudocholinesterase may form a complex which can affect membrane permeability. Shuster and Hershenov (1969) found what could be pseudoChE in the ciliary shafts of *Tetrahymena pyriformis* and they raised the possibility that it might complex with ATPase as Duncan suggested and, by changing permeability, could indirectly control the ciliary beat. They point out that the localization of ATPase (Burnasheva and Jurzina 1968) in the cilia parallels the distribution of cholinesterase. It should, perhaps, be emphasized that the ATPase-cholinesterase complex would seem to involve *pseudocholinesterase*. In a recent paper Abdel-Latif et al. (1970) have shown that *AChE* is independent of ATPase, at least in rat brain. Working with AChE from ox caudate nucleus, Maheshwari et al. (1971) obtained some evidence that there might be an association between ATPase and AChE activity but other results suggested that the two enzymes worked independently.

The probability that as the nervous system developed materials present in the primitive cells were incorporated as functional components has already been mentioned and ACh is a good example of a substance which has been appropriated as a neurotransmitter. Accumulating evidence suggests that it may be a transmitter at certain invertebrate synapses but interesting points emerge when a comparison is made between the use to which it and other putative transmitters have been put by different animals. In mammalian pharmacology there are still questions to settle about the role of ACh in central nervous transmission (see ch. 7) but its role as the transmitter at the neuromuscular junction is firmly established. In the insects and crustacea, on the other hand, it is almost equally certain that ACh is not the neuromuscular transmitter although it may act in the CNS (see §5.3.1.4). Faeder et al. (1970) have re-examined evidence for and against cholinergic transmission at the insect neuromuscular junction and have reconfirmed the belief that the junction is non-cholinergic. They concluded that certain anomalous results reported in the literature and interpreted as evidence for cholinergic transmission have other explanations.

One of the substances proposed as a transmitter at the crayfish neuromuscular junction is glutamate (Takeuchi and Takeuchi 1964) and this has also been found to have pre-synaptic effects on motor nerves of crab (Florey and Woodcock 1968). In contrast, in the cat, Galindo et al. (1967, 1968) found evidence that glutamate might be associated with afferent fibres. Conversely, ACh which is virtually absent from mammalian sensory nerves is present (though not necessarily as a transmitter) in the peripheral sensory fibres of the crab, *Cancer magister* (Florey and Biederman 1960), and there is now strong evidence that it functions as the transmitter in lobster sensory neurones (Barker et al. 1972). These differences open up a whole field of speculation about the chemical evolution of the nervous system. Ross (1965) concluded, somewhat despondently, that there was no clear evolutionary pattern in the distribution of chemical transmitters in the animal kingdom, but this lack of pattern, if genuine and not due to the lack of necessary data, is itself of interest. How and why did animals evolving along different lines happen to select particular transmitters from among the chemical constituents on offer? Even more intriguing, how did some creatures, for example the snake, *Bungarus bungarus,* develop toxins which are specific blockers of transmitters employed by their prey?

The indication that the ACh–ChE system may be put to different uses by different species raises the question of the extent to which the properties of the choline ester-splitting enzymes may also vary from species to species. Species variations among vertebrates are mainly of theoretical interest but the differences in the cholinesterases of the various invertebrates and, more particularly, the differences between invertebrates and vertebrates have a practical implication. The use of anti-cholinesterases as pesticides is widespread but, because of the occurrence of cholinesterases throughout the animal kingdom, it is not easy to confine their effects to the particular pest under attack. The need to develop pesticides of highly selective toxicity has stimulated an enormous volume of research both into the mechanism of action of anticholinesterase agents and into the properties of cholinesterases. As mentioned earlier, much of this work has been directed towards the insects and for this reason insect cholinesterases will be considered in most detail.

5.3 Cholinesterases of the arthropods

5.3.1 Insects

The first point that must be made is that the behaviour of insect cholinesterases can be considerably influenced by a variety of factors including pH, salt concentration and enzyme concentration. Vertebrate cholinesterases also are affected by experimental conditions (ch. 2) but the point about the insects is that the class includes members of widely differing structure and habits and since, with many of the smaller species, it is necessary to homogenize whole heads or even whole insects, preparations from different insects may be quite disparate with respect to pH, electrolytes, activating or inhibitory materials and enzyme concentration. This last point is illustrated in the very thorough paper by Wolfe and Smallman (1956) who found that the optimal concentration of ACh as a substrate for ChE in homogenized bee heads was lower than that for the ChE in homogenized fly heads. While this might be taken as evidence of a fundamental difference between the two enzymes the authors mention the possibility that the difference could be due to a difference in enzyme concentration. When the fly head preparation was diluted its optimal substrate concentration fell to a value comparable to that for the bee. Obviously it is important to recognize whether discrepancies between enzyme preparations from various species are attributable to genuine differences in the enzymes themselves or to differences in other components which, in turn, affect enzyme function. Similarly, variations in the experimental method could account for some of the variation in values reported for 'optimum' pH and 'optimum' substrate concentration for the ChE of a particular species. As an extreme example of this, Babers and Pratt (1950) found 5.75 to be the optimum pH for fly head cholinesterase but Chadwick et al. (1954) and also Wolfe and Smallman (1956) give values of between 8.0 and 9.0.

Some factors may have more than one effect; for instance, addition of NaCl to fly head ChE increases the rate of hydrolysis but reduces the affinity of the enzyme for the substrate (see ch. 2 §2.2 re qualitatively similar effect on mouse brain enzyme). Wolfe and Smallman (1956) make the point that in view of this dual action the effects of salts should be studied over a range of substrate concentrations; the same probably holds true for other variables. Other salts which increase the rate of hydrolysis by fly head ChE include $CaCl_2$, KCl, $MgCl_2$ and $NaNO_3$; however, above about 0.5 N these salts reduce the enzyme activity. Lord and Potter

TABLE 5.2
Some properties of insect cholinesterases.

Property	Enzyme source	Substrate	Value	Reference
pH optimum	Housefly head	ACh	8.0–9.0	Chadwick et al. (1954)
				Wolfe & Smallman (1956)
	Cockroach CNS	ACh	7.4	Stegwee (1951)
	Cockroach gut	ACh	6.5–7.0	Kooistra (1950)
pK	Housefly head	ACh	7.6	Wolfe & Smallman (1956)
K_M	Housefly head	ACh	1×10^{-5} M	van Asperen (1962)
	Housefly head	ACh	6.3×10^{-4} M	Wolfe & Smallman (1956)
	Housefly head	MeCh	2×10^{-2} M	Wolfe & Smallman (1956)
	Housefly head	α-naphthylacetate	1×10^{-4} M	van Asperen (1962)
	Housefly head	β-naphthylacetate	2.3×10^{-4} M	van Asperen (1962)
	Bee head	ACh	6.5×10^{-4} M	Wolfe & Smallman (1956)
	Bee head	MeCh	8.5×10^{-3} M	Wolfe & Smallman (1956)
Temperature optimum	Housefly head (soluble fraction)	ACh	30° C	Wolfe & Smallman (1956)
	Housefly head (particulate fraction)	ACh	35° C	Wolfe & Smallman (1956)
ΔE^* for denaturation (see glossary)	Housefly head (soluble fraction)	ACh	79 Kcal/mole	Wolfe & Smallman (1956)
Turnover number	Housefly head (particulate fraction)	ACh	37 Kcal/mole	Wolfe & Smallman (1956)
	Housefly head (partially pure)	ACh	100,000	Dauterman et al. (1962)

(1953) found a material in extracts of the flour beetle, *Tribolium casteneum,* which inhibited hydrolysis of ACh; similar anticholinesterase-like substances have since been described by J.F. Thomas and T.L. Hopkins (unpublished) and Menn and McBain (1968) in the cockroaches, *Blaberus craniifer* and *Blatella germanica.* Inhibitory compounds may well occur in other species and their presence in a tissue homogenate could result in low ChE activity which might be mistakenly attributed to some feature of the enzyme.

Insects are, of course, cold blooded while mammals are warm blooded but this does not seem to be reflected in any difference in the Q_{10} of their cholinesterases. However, although ChE from insects like that from mammals (ch. 2 §2.3) is rapidly inactivated above 55°C (see Chadwick 1963), the process of denaturation in insects starts at a lower temperature (approx. 35°C; see Wolfe and Smallman 1956). As part of their purification process Dauterman et al. (1962) heated partially purified preparations of fly head ChE to 54—55°C in the presence of ACh and in this case denaturation did not occur. Wolfe and Smallman (1956) had found NaCl to be similarly protective. Protection of this type, by some material present in the preparation, might account for the rather different rate-constants for heat denaturation which have been obtained by various workers.

The message of this section has been that preparations of insect ChE may behave variably depending on the experimental conditions. For this reason, properties have so far been considered more qualitatively than quantitatively. The figures now given in table 5.2 have been obtained under a variety of conditions and must be of limited significance unless the original papers are consulted for experimental details.

5.3.1.1 Substrate specificity and active sites   The characters which distinguish cholinesterases from other esterases were discussed in ch. 2 and it was shown that in the mammals a further distinction could be made between acetylcholinesterases and other cholinesterases. In the invertebrates this division is less clear-cut. Casida (1955a, b) examined substrate preferences of acetylesterases, including cholinesterases, in a variety of insects. He found an enzyme with properties resembling the acetylcholinesterase of mammals in whole leaf-hoppers, in the heads of adult house-flies and wax moths, and in the nerve cord of cockroaches; in pea aphids and in the larvae of the carpet beetle, the enzyme resembled mammalian pseudoChE. In addition, esterases with properties intermediate between the two types were present in some of the insects. In view of this and many other similar reports Chadwick (1963) pointed out that the classification used for mammalian cholinesterases is not applicable to the invertebrates without qualification. Nevertheless, some authors do use the term 'acetylcholinesterase' regardless of species. What is particularly misleading is that some writers designate anything that splits ACh, 'acetylcholinesterase', and, on this basis, reviewers may list an animal as possessing AChE activity when the original report indicated nothing more than an ability to hydrolyse ACh.

Wolfe and Smallman (1956) compared the hydrolysis of various substrates by

TABLE 5.3

Hydrolysis of esters by the soluble and particulate fractions of fly and bee head cholinesterase[*]. Activity expressed in μl CO_2/mg head tissue/hr. (From Wolfe and Smallman 1956.)

Substrate	Substrate concentration (M)	House-fly		Bee	
		Soluble	Particulate	Soluble	Particulate
Acetylcholine	0.1	11.0	3.0	2.8	1.0
Acetylcholine	0.01	26.6	8.4	6.4	2.0
Acetyl-β-methylcholine	0.1	4.2	1.5	9.6	3.2
Acetyl-β-methylcholine	0.01	7.8	2.0	5.2	1.7
Butyrylcholine	0.1	10.0	3.0	1.8	0.4
Butyrylcholine	0.01	14.0	5.0	2.9	1.1
Benzoylcholine	0.1	1	0	0	0
Benzoylcholine	0.01	1	0	0	0
Triacetin	0.1	19	6	–	–
Triacetin	0.01	13	4	–	–
Tributyrin	0.1	3	1	–	–
Tributyrin	0.01	1	0	–	–

[*]Manometric method, pH 7.4, 25°C. Augustinsson's bicarbonate–Ringer's solution. Determinations in triplicate

the cholinesterases in the soluble and particulate fractions of house-fly head and bee head homogenates. Table 5.3 shows that fly head cholinesterase suffered substrate inhibition with 0.1 M ACh, MeCh and BuCh; BzCh was hydrolysed only very slightly but triacetin was split appreciably, especially by the soluble enzyme. The bee head enzyme differed in that at a concentration of 0.1 M, MeCh was hydrolysed faster than ACh and did not cause substrate inhibition; BzCh was not hydrolysed at all. With substrate concentrations below 0.01 M both the fly and the bee preparations hydrolysed ACh faster than any other ester tested. It was concluded that fly head ChE resembled the acetylcholinesterase of mammals and that no significant amount of pseudoChE was present in either fly or bee preparations. Implicit in this conclusion is that bee enzyme is considered a 'true' cholinesterase despite the lack of substrate inhibition with MeCh.

Dauterman et al. (1962) developed a technique for the partial purification of fly head cholinesterase and measured the activity in preparations at successive stages of purification. They compared the values with those for bovine erythrocyte cholinesterase and with those for a homogenate of fly heads (termed AD in table 5.4) in which the aliesterases had been denatured. Unlike Wolfe and Smallman (1956), they considered that fly head cholinesterase was significantly different from the erythrocyte enzyme. As table 5.4 shows, the specificity pattern, which changed only slightly during purification, was certainly not the same as that for erythrocyte ChE, the activity towards BuCh, triacetin and phenylbutyrate being much greater. Dauterman et al. were in agreement with Wolfe and Smallman that only one en-

TABLE 5.4

Substrate specificity of enzyme preparations. (From Dauterman et al. 1962.)

Nature of enzyme preparation	Purifi- cation factor	Substrates						
		ACh 0.015 M	BuCh 0.015 M	TA 0.036 M	MB 0.048 M	AAc emulsion	PA emulsion	PB emulsion
Flyhead homogenate	1	100	57	116	38	65	162	107
* AD-treated homogenate	1	100	55	114	2	51	140	73
I. Freeze-dried	1	100	53	109	4	52	129	84
II. Phosphate-extract	3	100	55	110	2	50	130	74
IV. AS-fraction	28	100	52	101	0	54	139	76
VII. Enzyme powder III	94	100	57	84	3	54	90	65
I. Freeze-dried	1	100	59	118	–	65	128	81
II. Phosphate-extract	3	100	53	105	0	53	134	79
IV. AS-fraction	23	100	56	81	0	46	125	75
VIII. Acetone-fraction	157	100	52	99	–3	50	115	62
Bovine erythr. ChE		100	2	32	0	32	104	2

* AD-treatment consists of keeping the homogenate for 1 hr at 37°C in a 0.025 M-bicarbonate-solution (pH ca. 8.3). This causes denaturation of ali-esterases but leaves ChE-activity unaffected. Not included in the table are the figures for propionylcholine and acetyl-β-methylcholine both at 0.015 M conc. They are 71 and 25 respectively, whereas those for bovine ChE were found to be 79 and 43 respectively. AS, ammonium sulphate; for other abbreviations see p. 491. Technique: Total contents of Warburg flasks 2.2 or 2.4 ml; substrate (at indicated concentration) and enzyme dissolved in 0.025 M $NaHCO_3$ and 0.5 M NaCl. Gas phase: 95% N_2 + 5% CO_2; pH 7.5; temp. 37°C. Enzyme activities to the different substrates are expressed as percentages of the activity to acetylcholine (taken as 100%). Roman figures refer to the last purification step completed.

zyme was responsible for the hydrolysis of all the substrates but they questioned whether it should be termed 'cholinesterase' since its turnover number to some aliphatic and aromatic acetates and possibly to phenylbutyrate might be as high as that for ACh. However, because of its high affinity for ACh and from physiological considerations, they felt that it could be designated a 'true' cholinesterase if any classification were required.

Casida (1955b) provided useful comparative data on substrate specificity in a variety of adult insects and in their developing eggs and larvae; he also included some arachnid mites in his study. His values, shown in table 5.5, can be compared with those for bovine erythrocyte enzyme expressed in the same way. By using mixed substrates, as shown in the last two columns of the table, Casida determined whether the same or different enzymes were responsible for the hydrolysis of ACh, TA and O-nitrophenylacetate (NPA) in the various insects. Where hydrolysis of ACh, NPA and TA was apparently due to a single enzyme (e.g. with house-fly heads), inhibition with eserine, choline or TEPP was generally marked with all 3 substrates. On the other hand, in the carpet beetle where ACh and NPA appeared to be hydrolysed by separate enzymes, 10^{-6} M eserine inhibited hydrolysis of ACh only. An interesting point about the house-fly is that in the egg and early larval stages two different enzymes are responsible for splitting ACh and NPA but by adulthood one enzyme is attacking both substrates (see ch. 8 §4 for other references to papers concerned with developmental aspects).

Casida examined the ChE in the pea aphid, *Macrosiphum pisi,* in some detail and showed it did not hydrolyse MeCh and was inhibited by excess BzCh but not by excess ACh. In other words, its properties were more like those of mammalian serum ChE than erythrocyte ChE. The enzyme hydrolysing ACh was, however, distinct from those hydrolysing NPA and TA. More recent work has indicated that the ChE in the pear aphid, *Toxoptera piricola,* behaves in much the same way, hydrolysing BuCh faster than ACh (Sakai 1967). In *Tribolium confusum* (the confused flour beetle) the enzyme prefers ACh to BuCh and is inhibited by excess of ACh; electrophoresis indicates that only one cholinesterase is present (Chaudhary et al. 1966). The activity is high in the early larval stage but then falls off and is very low in the adult (see also O'Brien 1953). A possibility to be kept in mind is that this low activity could be due to an increase in an inhibitory material such as that described by Lord and Potter (1953) in *Tribolium casteneum.*

The arachnid mites are often included in studies on insects and they are considered here for convenience. Voss and Matsumura (1965) and Motoyama and Saito (1968) showed that in 3 species of mite propionylcholine was the preferred substrate and that in excess it caused inhibition. In the two-spotted spider mite, *Tetranychus urticae,* ACh was hydrolysed faster than BuCh but the reverse was true for *Tetranychus kanzawai* (see Sakai 1967). Neither substrate caused inhibition in these species, nor in the citrus red mite. Voss and Matsumura (1965) suggested that the preference for PrCh indicated that the esteratic site in the mites is larger than that

TABLE 5.5

Substrate specificity of certain esterases from various insects. (From Casida 1955b.) Tissues from various species were homogenized in bicarbonate buffer and a 1.0-ml sample was added to each flask to yield the indicated amount of tissue per ml. Substrates (0.4 ml) were tipped in from the side arm to give a final molar concentration of 0.03 for ACh, BzCh and MeCh, 0.01 for TA and 0.003 for NPA. Mixed substrates were prepared by adding 0.2 ml of a double concentration of each substrate to the side arm of the flask. The enzyme was incubated with the flask contents (1.6 ml) for 30 min at 28°C and 30 min at 38°C before tipping in the substrate. The CO_2 evolution in 95% N_2 and 5% CO_2 was followed manometrically and corrected for tissue and substrate blanks. A known sample of each insect preparation was analysed for total N by nesslerization.

Order and common name	Tissue (mg/ml)	Total N (μg/ml)	Substrate specificity (μl CO₂/30 min)					% activity of mixed substrates	
			ACh	BzCh	MeCh	TA	NPA	ACh+ TA	ACh+ NPA
Orthoptera									
German cockroach	Adult heads 1.88	140	40	17	105	175	147	98	114
Homoptera									
Leafhopper	Whole adults 13.7	120	38	3	46	24	144	54	53
Aphid	Whole adults 122	2200	37	17	3	309	145	102	105
Whitefly	Whole adults 1.36	43	41	6	7	8	69	94	65
Scale insect	Whole adult females 50	980	27	3	21	69	132	100	131
Hemiptera									
Milkweed bug	Adult heads 1.88	100	17	5	36	26	122	93	95
Lepidoptera									
Armyworm	Larval heads 1.98	280	10	4	12	71	112	107	120
Waxmoth	Larval heads 2.50	260	9	4	8	14	104	91	109
Waxmoth	Adult heads 2.50	98	18	6	52	12	141	96	55
Coleoptera									
Carpet beetle	Whole larvae 25	580	26	38	32	69	137	80	129
Bean beetle	Adult heads 2.1	120	17	10	21	15	59	71	89
Granary weevil	Whole adults 25	490	15	10	14	37	160	94	118

TABLE 5.5 (continued)

Order and common name	Tissue (mg/ml)	Total N (µg/ml)	Substrate specificity ($\mu l CO_2$/30 min)					% activity of mixed substrates	
			ACh	BzCh	MeCh	TA	NPA	ACh+ TA	ACh+ NPA
Diptera									
Housefly	Eggs 25	980	10	6	2	16	121	77	129
Housefly	Larvae 32	600	32	5	9	45	128	87	114
Housefly	Pupae 29	880	93	8	37	76	164	69	84
Housefly	Adult heads 1.25	57	130	13	48	78	178	61	48
Hymenoptera									
Honey-bee	Adult heads 1.5	150	67	15	69	33	131	39	53
Acarina									
Mite	Whole adults 7.5	220	13	6	8	28	151	115	116
Purified bovine AChE*	0.150 µg	18	113	10	51	11	165	94	47

* Purified bovine cholinesterase plus stabilizer. For abbreviations see p. 491.

TABLE 5.6

The hydrolytic activity of the cholinesterases in the Leverkussen strains of spider mite against different cholinesters *. (From Voss and Matsumura 1965.)

Strains (No. replicates)	Acetyl-choline (2)	Propionyl-choline (2)	Butyryl-choline (2)	Benzoyl-choline ** (2)	Acetyl-β-methyl-choline (2)
Leverkussen normal	6.82	9.79	1.50	0.33	1.42
Leverkussen resistant	4.31	5.61	1.71	0.33	0.85

* The results are expressed in μmoles cholinester hydrolysed by 100 mg mites per hr at 30°C. The enzyme extract was 5.4 mg mites in 0.3 ml.

** Not accurate because of the low activity towards this substrate. Enzyme prepared by homogenizing mites in 0.067 M phosphate buffer, pH 7.4. 1 min ultrasonic disintegration, 10 min centrifugation, 3,000 r.p.m. Supernatant used.

in insect or mammalian ChE. An interesting feature of *Tetranychus urticae* is that some strains have arisen which are resistant to poisoning by Paraoxon and Malaoxon, and Smissaert (1964) showed that the ChE in the resistant strains was insensitive to these compounds. Voss and Matsumura compared the properties of the enzyme in the two strains and found that in resistant mites the affinity of the enzyme for the substrate was lower than in the normal ones. From the data in table 5.6 they calculated a K_M of 1.14×10^{-3} M for the resistant strain and 0.25×10^{-3} M for the normal strain. They attributed the difference to a 'weakness' of the esteratic site in the former. Voss and Matsumura also compared the effect of pH on the activity towards ACh in the two strains; the results are shown in table 5.7. Under the condition of their experiments the pH optimum for both strains of mite was about 7.5, somewhat lower than Wolfe and Smallman's figure for the house-fly head enzyme. The difference between the two strains with respect to pH decreased with increasing pH but Voss and Matsumura were not prepared to put a firm interpretation on this observation because pH changes can influence many factors.

Underlying much of the research on insect ChE has been the hope that it will unearth some property which can be exploited in the design of highly selective pesticides. Casida (1955b) summed up his comparative study of insects and mites by saying that their cholinesterases appear to be a group of related enzymes with widely varying properties among themselves and in comparison with mammalian cholinesterases. He felt that the differences in specificity were sufficiently great to be used as a basis for selectivity; Guibault et al. (1970b) reached a similar conclusion when they examined the properties of ChE in honey bees and boll weevils. Selectivity is not, however, simply a question of differentiating mammals from insects nor even of distinguishing insects which are obviously beneficial to man from those which apparently are not. The problem is very much broader and embraces the whole question of ecological interaction. In destroying insects and other creatures which man finds 'undesirable' he may be upsetting food chains which in the long-term are essential to him. It may be poetic justice that pharmacologists who, as a group, have contributed to the development of insecticides, are among those inconvenienced by the almost universal shortage of frogs caused by their indiscriminate use.

All too easily, control and pollution could become synonymous and if this is to be prevented we require far more precise knowledge about the insects under attack. It is not enough to be aware of variations in the properties of ChE in different species. The real question to ask is what, exactly, confers these differences, and the most likely explanation is that the differences reside in the configuration of the active sites (see ch. 2). O'Brien (1963), looking for differences between the active sites in house-fly head ChE and the human erythrocyte enzyme, compared the effects on the two enzymes of alkyl ammonium salts. These are presumed to bind to the anionic site and can interfere with the phosphorylation of the esteratic site by organophosphorus compounds. With ChE from red blood cells, tetraethylammonium (TEA) and tetrapropylammonium (TPA) were equally effective in reducing the degree of inhibition caused by the organophosphorus compounds; this suggests that when their N^+ is attached to the anionic site, both TEA and TPA extend far enough to mask the esteratic site. However, TPA was 32 times more effective than TEA in

TABLE 5.7

The influence of pH changes on the activity of the cholinesterases of the Leverkussen strains * of spider mites. (From Voss and Matsumura 1965.)

Strains (No. replicates)	pH								
	6.0 (2)	6.5 (2)	7.0 (3)	7.5 (5)	8.0 (5)	8.5 (5)	9.0 (2)	9.5 (2)	10.0 (2)
Leverkussen normal	1.87	2.39	2.89	2.96	2.44	2.28	1.98	1.48	1.24
Leverkussen resistant	0.85	0.96	1.70	2.40	2.28	2.02	1.79	1.15	1.02
Absolute difference	1.02	1.43	1.19	0.56	0.16	0.26	0.21	0.33	0.22
Relative difference	2.20	2.48	1.69	1.23	1.07	1.13	1.10	1.29	1.22

* The results are expressed in μmoles ACh (2×10^{-3} M final concentration) hydrolysed by 100 mg mites per hr at 30°C. Phosphate buffer (0.067 M) and Tris buffer (ionic strength = 0.05) were used in these experiments. The enzyme extract was 5.4 mg in 0.3 ml.

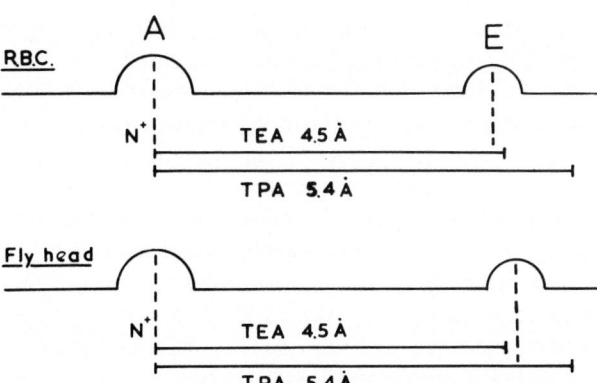

Fig. 5.1. Diagram to illustrate a possible difference in the dimensions of the active sites in mammalian erythrocyte (R.B.C.) and fly head cholinesterases, as suggested by the differing effectiveness of tetraethylammonium (TEA) and tetrapropylammonium (TPA) in preventing inhibition. A, anionic site; E, esteratic site.

protecting fly head ChE from phosphorylation which indicates that it masks the esteratic site much more efficiently. O'Brien concluded that in the erythrocyte ChE, the distance between the anionic and esteratic site must be less than the diameter of TEA, that is, less than 4.5 Å but in the ChE from fly heads, the distance must be greater than 4.5 Å but not more than 5.4 Å, the diameter of TPA. This situation is illustrated diagrammatically in fig. 5.1.

Dauterman and Mehrotra (1963) found differences between the N-alkyl group specificity of cholinesterase in the house-fly and in the two-spotted spider mite, *Tetranychus telarius*; in another paper (Mehrotra and Dauterman 1963), they reported similar investigations on rat brain AChE. Dimethyl- or diethyl-alkyl analogues of ACh were equally good substrates both for the rat and the fly enzyme and di-*n*-propyl- and di-*n*-butyl-alkyl esters were equally poor for both enzymes; in contrast, the spider mite ChE was equally active towards all these types of substrate. The results could indicate that for rat and fly enzyme the anionic site fits best with groups which do not exceed 8.5 Å in diameter, this being the maximum distance of the diethyl-alkyl analogues. However, as Mehrotra and Dauterman have pointed out, the poor performance with the long-chain compounds does not necessarily mean that their fit is bad. If, on the contrary, the fit at the anionic site were particularly good, removal of this part of the substrate from the enzyme after hydrolysis might be slow, thus reducing the overall speed at which the enzyme could function. Since the anionic site in the mite seems able to fit all the substrates equally well it appears to be 'broader' than that of rat and fly and probably less demanding in the degree of fit required. Voss and Matsumura's (1965) suggestion that the esteratic site could also be rather large in the mite has been mentioned earlier. This was based on the observation that PrCh was a better substrate than

ACh. In addition, as table 5.6 shows, there was a large difference in the rate of hydrolysis of PrCh by enzyme from normal strains and from organophosphorus-resistant strains but virtually no difference in the rate of hydrolysis of BuCh. This suggests that when the substrate molecule is increased by one carbon atom in the ester, neither enzyme can make a good fit. In turn, this implies that if the development of resistance to inhibitors depends on an alteration in the size of the esteratic site, the difference in size in the two strains must be less than one carbon atom. Similar reasoning suggests that in _Tetranychus kanzawai_ the esteratic site is different from that in either strain of _Tetranychus urticae_ since, as mentioned earlier, it splits BuCh appreciably faster than ACh (Sakai 1967). However, since the experiments on the two species of mite were done by different groups of workers, the possibility that technical differences could account for the discrepancy must be considered. It was mentioned in ch. 2 §5.3 that Hellenbrand and Krupka (1970) demonstrated that the mechanism responsible for substrate inhibition of AChE in the insects is apparently different from that in the mammals. They postulated that fly head AChE contained not one, but two, anionic sites and that it was this feature which conferred the difference in behaviour. This hypothesis merits further exploration.

5.3.1.2 Subcellular distribution   Babers and Pratt (1951) found that when homogenates of fly heads prepared in 30% glycerol were subfractionated, most of the ChE activity was in the supernatant but with bee head homogenates the activity was mainly in the particulate fraction. On the other hand, Smallman and Wolfe (1956) reported that with aqueous homogenates the distribution was the same for bee and fly heads. In both, there was slightly more enzyme activity in the supernatant than in the particulate and some, but not all, of the activity in the particulate could be removed by further washing. The effect of sucrose was much the same as water with both bee and fly head preparations but the proportion of activity in the supernatant was slightly increased. With 0.5 M KCl there was a species difference: fly head preparations possessed considerably more activity in the particulate than in the supernatant but for bee heads the activity was still greatest in the soluble fraction. When bee brains, as opposed to whole heads, were examined, activity was highest in the particulate fraction in all three media. With cockroach nerve cord, the soluble fraction was slightly more active than the particulate in both water and sucrose, but in KCl over 60% of activity was in the particulate fraction.

Smallman and Wolfe (1956) showed that not only salts but also pH affected the distribution of enzyme between the particulate and soluble fraction, the higher the pH, the greater the activity in the supernatant. They suggested that the difference between their results and those of Babers and Pratt (1951) could be due to the fact that bee head preparations are acid whereas fly head preparations are almost neutral; only when bee head preparations are neutralized do they behave like fly head homogenates.

A question considered by Smallman and Wolfe (1956) was whether the enzyme in the supernatant was completely solubilized or whether it was still attached to particulate matter which could be removed by adequate centrifugation. Since considerable activity remained in the supernatant after it had been centrifuged for 1 hr at 50,000 *g* they concluded that a soluble form of ChE exists in vitro but they emphasized the danger of extrapolation from these results to the situation in vivo. Knowledge of the state of the enzyme in vivo could have certain practical implications in the quest for anticholinesterases which are species specific. Over 20 years ago Peters (1951), in his now classic Croonian Lecture on 'Lethal Synthesis', suggested that soluble and mitochondrial aconitase might behave differently towards citrate inhibition. His idea, that bound and free enzymes may differ in behaviour, is now well established. Some of these differences depend on the presence or absence of co-enzymes and as such are not applicable to bound and free ChE but nevertheless it might be worth investigating whether it is feasible to incorporate into pesticides certain ingredients capable of altering the soluble: particulate distribution. This might render the cholinesterase of certain species more vulnerable to inhibition while in others, in which the binding behaved differently, the enzyme might be less affected. Wolfe and Smallman (1956) showed that the particulate ChE of fly heads required less energy for heat inactivation than did the soluble form. They pointed out that when adsorbed onto surfaces, proteins are less folded and more easily disorganized; it may well be that in the unfolded state the active sites of the enzyme are more accessible to inhibitors.

5.3.1.3 Summary of biochemical properties   The salient points about insect and arachnid cholinesterases may be summarized as follows: the enzymes capable of splitting esters of choline in different species show considerable variation with respect to substrate specificity, including their ability to hydrolyse non-choline esters. Their activity in vitro is influenced by salts, pH, and enzyme concentration and, again, the way in which they are affected varies between the species. It seems very probable that some of these differences are a manifestation of greater or lesser variation in the dimensions and spatial arrangement of the anionic and esteratic sites in the active centre.

5.3.1.4 Function   In concentrating on the biochemistry of ChE it is easy to lose sight of the all-important question of function. Histochemistry is valuable in countering this tendency because in relating enzyme to structure it provokes the query 'What does it do there' rather than 'How does it do it'? In the case of insects, histochemical studies are not very numerous probably for two reasons. First, the very small size of so many insects introduces technical difficulties but with the development of histochemistry at the ultrastructural level this problem is disappearing. Second, in many insects some sort of diffusion barrier (Hoyle 1953) is present outside the nervous tissue and penetration of histochemical substrates may be prevented. Winton et al. (1958) using acetylthiocholine as substrate produced staining in the ganglia of the American cockroach, *Periplaneta americana,* only where the surface was cut; this suggests that the substrate could not penetrate elsewhere. The use of radioactive compounds and the comparison of the effects on ChE of ionized and non-ionized inhibitors has shown that there is, indeed, some impedance to the penetration of ionized substances. O'Brien (1967), in summarizing the situation in the cockroach, wrote that ionized compounds penetrate as if there were a barrier slowing ionic penetration 10 to 15 times, but the barrier seems to be attributable to some property of the ganglion as a whole and not to a specific cell layer. A useful description of the morphology of insect ganglia is given by Maynard (1967).

Smith and Treherne (1965) examined the ultrastructural localization of ChE in the abdominal ganglion and nerves of the cockroach, *Periplaneta americana,* using Barrnett's (1962) thiolacetic acid method. Activity was observed in glia around the axons in the cercal nerves and around the cell bodies in the ganglia. Activity was also demonstrable in the neuropil of the ganglia, the end-product being localized to some areas of the axonal membranes. In this site it was often, but not invariably, flanked by synaptic vesicles. Smith and Treherne considered their results to provide good evidence for the existence of cholinergic synapses in the neuropil and suggested that the enzyme in the glia round the perikarya and nerves might be protective, preventing a build-up of 'extraneous' ACh in the limited extracellular spaces. Hess (1972), using optical microscopy, has mapped the localization of AChE in the constituent ganglia of the cockroach brain. The neuropil, nerves and commissures were stained in most areas but the degree of activity varied and in the alpha and beta lobes the neuropil did not react at all. Some large neurones in the pons intercerebralis reacted either weakly or strongly but none of the other neurones showed any staining. In no area was there any reaction towards BuThCh. Hess pointed out that in several regions there was an inverse relation between the distribution of AChE and that of catecholamines, as reported by Frontali (1968).

Table 5.8 exemplifies the long-established fact that ACh and ChAc, as well as ChE, are present in appreciable quantities in the insect central nervous system (see also Smallman and Pal 1957; Mehrotra and Dauterman 1963) but a major difficulty in accepting the idea of ACh as a central transmitter in the CNS had been that externally applied ACh had no effect on ganglionic function. The demonstration of the apparent diffusion barriers mentioned earlier seemed to afford a possible explanation for this. However, Treherne, in a series of papers (for references see Treherne and Smith 1965), showed that certain ions and molecules could pass this 'barrier' quite easily and finally he and Smith (Treherne and Smith 1965) demonstrated that radioactive ACh did, in fact, enter the ganglia rapidly. The ineffectiveness of applied ACh seemed to be due to its rapid destruction by the high concentration of ChE present in the nervous system and not to any failure of penetration. This idea has been borne out by the later work of Kerkut's group (Kerkut et al. 1969; Pitman

TABLE 5.8

The distribution of acetylcholine, cholinesterase, and choline acetylase in conductive tissue of the cockroach. (From Colhoun 1959.)

Tissue	ACh (μg/g)	QChAc(mg/g/hr)	QChE(mg/g/hr)
Brain	143.6 ± 4.0*	50.6	137.7 ± 15.4
Brain + suboesophageal ganglion	135.2 ± 2.0	53.0	153.4 ± 21.2
Ventral cord	63.2 ± 4.5	10.6	270.4 ± 16.3
Thoracic cord	79.0 ± 6.1	11.4	221.3 ± 15.9
Thoracic ganglia	95.4 ± 4.3	20.8	331.8 ± 12.4
Thoracic connectives	31.3 ± 2.4	2.6	238.7 ± 18.5
Abdominal cord	65.2 ± 1.6	6.2	187.5 ± 10.1
6th abdominal ganglia	63.0 ± 1.5	18.0	314.9 ± 20.6
5th leg nerve	1.21 μg per 60 nerves	2.0	176.5 ± 26.8
Cercal nerves	1.43 μg per 60 nerves	3.6	150.8 ± 30.3
Coxal muscle	0	0	0
Flight muscle	0	0.08	?
Heart	0	0	0
Blood serum	0	0	?
Blood cells	0	0	?

* Standard deviation.

and Kerkut 1970). They applied ACh and GABA to neurones in the 6th abdominal ganglion of cockroach and the values they obtained for the reversal potentials suggested that ACh was likely to be the excitatory transmitter and GABA the inhibitory one. Subsequently Kerkut and his co-workers (Kerkut et al. 1970a, b; Oliver et al. 1971; Kerkut et al. 1971) have proposed that cholinergic mechanisms may be involved in the process of 'shock-avoidance learning' by intact or headless cockroaches or by preparations composed only of the metathoracic segment. In a preparation trained to avoid an electric shock by keeping its leg out of water, the ChE level fell rapidly. When training ceased the preparation 'forgot' the lesson and ChE rose again. Beesley et al. (1972) found that although the V_{max} did not alter the K_M of ChE in trained animals was 1.33 × 10^{-4} M compared with a value of 5.88 × 10^{-5} M in controls. The conclusion, that the changes observed represent a biochemical correlate of learning, has been questioned by Woodson et al. (1972). These authors could detect no change in the level of ChE in trained cockroaches showing clear-cut evidence of learning.

Evidence for the existence of cholinergic synapses in the ant has come from the very elegant work of Steiner and Pieri (1969). They compared the reaction to iontophoretically applied drugs of neurones in rat, cat and ant. Neurones in the corpora pedunculata and β-lobe of the protocerebrum of the ant (anaesthetized with CO_2) responded to ACh in much the same way as those in various regions of the mammalian brain although proportionally fewer ACh-sensitive cells were found in the ant. Except in one instance, where it was inhibitory, ACh was excitatory; it increased the rate of firing of spontaneously active cells and fired the silent ones after a latency of 3 to 10 sec and, as with mammals, the firing continued after the application ended. Steiner and Pieri questioned whether ant receptors had a relatively high threshold for ACh (Smith and Treherne 1965 raised the same possibility for cockroach cord) or whether the paucity of sensitive neurones actually detected was, in fact, due to the high concentration of ChE. The effect of eserine was less pronounced in the ant than in vertebrates — in view of the high level of ChE in the ant it seems possible that the percentage inhibition produced by the eserine might have been less than in rat and cat. Ultrastructural studies on the ant (Landolt and Sandri 1966) showed that the enzyme is most concentrated in the cytoplasm adjoining the post-synaptic membranes and is rarely found within the synaptic cleft.

In the ant a few cells responded to glutamate with a small increase in spontaneous activity and one cell was inhibited; this is strikingly different from the universal strongly excitatory effect of glutamate in cats. If glutamate is indeed one of the physiological transmitters (for a review of evidence see Kerkut 1967) this could be an example of the phenomenon discussed earlier, of different animals employing transmitters in different ways. Steiner and Pieri do, however, point out that the vast stores of GABA in the ant are maintained by decarboxylation of glutamic acid, so applied glutamate may possibly be destroyed by this mechanism. Furthermore, the apparent differences in sensitivity of ant and vertebrate neurones to ACh and glutamate under experimental conditions may not accurately reflect what happens under conditions of normal physiological function when the transmitters may well be released so close to the receptors that they do not have to run the same gamut of destructive enzymes. It is of interest that dopamine, like GABA, was inhibitory in the ants as well as in the vertebrates; earlier work had shown it to be excitatory on the abdominal ganglia of the cockroach (Gahery and Boistel 1965).

Although positive evidence for central cholinergic transmission in insects is still sparse compared with that for vertebrates, there is no overwhelming reason for rejecting the idea now that the problem of the lack of sensitivity to applied ACh has been resolved. In the peripheral nervous system the weight of evidence, including the absence of ChE from motor end-plates of species examined (Wigglesworth 1958), indicates that transmission at the neuromuscular junction is, however, non-cholinergic (see Faeder et al. 1970).

Venkatachari and Naidu (1969) assumed the existence of cholinergic transmission in the arachnid, *Heterometrus fulvipes*, an Indian scorpion, because of the high levels of ChE in nervous tissue, heart and muscle. This is a dangerous extrapolation but the earlier observation by Venkatachari and Dass (1968), that the level of ChE in the ventral nerve cord showed a circadian rhythm which closely followed the pattern of electrical activity, is suggestive of some functional connection. Here again though, extreme caution is needed: enzyme and electrical activity may be totally unrelated except that they are subject to a common variable (see ch. 9).

5.3.2 Crustaceans

The crustaceans, like the insects and arachnids, belong to the Phylum Arthropoda. On the whole they impinge very little on the average man. Although certain barnacles can cause serious trouble for shipowners and the racing yachtsman, most people are more concerned with the edible members of the class, lobsters, crabs, prawns and shrimps. Because of this, crustaceans have not been a major target for pesticides and there has been little urgency to examine their cholinesterases. However, many species possess large axons which have been extensively used in studies of various aspects of electrical activity in neuronal membranes. Nachmansohn and his co-workers have long been interested in the contentious question of whether ACh plays any part in axonal, as opposed to synaptic, transmission (see Nachmansohn 1962; Dettbarn 1967) and much of their work has been done on the lobster, *Homarus*. Their claim that the ACh-AChE system is vital to axonal transmission is partly based on observations that applications of ACh depolarize axonal membranes and that high concentrations of anticholinesterases block conduction. Nachmansohn's ideas have not received wide acceptance (see ch. 7 §2.4) and his results are open to more than one interpretation (Dettbarn and Bartels 1968; Bartels et al. 1969; Hoekman and Dettbarn 1971; Hoskin 1971). Notwithstanding the doubts about the significance of the results, the work has yielded useful data on the ACh and ChE in the axons.

Dettbarn (1963) concluded that the cholinesterase in axons of the lobster and spider crab is of the specific type despite some atypical features. ACh was preferred to MeCh or BuCh and it alone caused substrate inhibition. Eserine was inhibitory but other anticholinesterases were not tested. More recently, Haites et al. (1972) have examined the esterases in the tissues of the king prawn, *Penaeus plebjus*, by polyacrylamide gel electrophoresis. In muscle, two ChE isozymes were present. Both had molecular weights of 87,000 and showed a preference for acetyl substrates. In most other tissues, including the spinal cord, only one isozyme was found. It had a molecular weight of 180,000 and was equally active towards acetyl and butyryl substrates.

In experiments with the microgasometric-magnetic diver technique, Brzin et al. (1965) obtained values for the ChE associated with the axonal membranes of lobster walking-leg nerve. For sensory and motor axons respectively, the activities per mm^2 of axon surface were 1.2×10^{-3} and 1.93×10^{-3} μmole ACh hydrolysed/hr. On a weight basis, the activity in the sensory nerves was much higher than that in the motor nerves, the respective values being 741 and 116 μmole ACh/g/hr. So far there seems to be no satisfactory explanation for the presence of the high enzyme activity in the motor axons since peripheral transmission in crustaceans, as in insects, is apparently non-cholinergic (see Florey 1962; Treherne 1966). At one time there appeared to be some evidence for a cholinergic action by the cardio-accelerator nerves of crab and lobster, various workers (see Treherne 1966) having observed that application of ACh to the heart produced acceleration. Florey (1963) showed, however, that neither the cardiac ganglion nor the cardiac accelerator nerves contained detectable ACh, and the ChE content of the ganglion and heart muscle was low. Moreover, although eserine increased the accelerating effect of applied ACh, it did not cause any change in the response to stimulation of the cardio-accelerator nerves. Atropine, hexamethonium, decamethonium and benzo-quinone were similarly ineffective. In contrast to cardiac muscle, skeletal muscle, although containing ChE, is insensitive to applied ACh and also to curare (see Treherne 1966).

Keyl et al. (1957) found ACh in whole limb nerves of the lobster, *Homarus americanus*. Subsequent work by Barker et al. (1972) indicates that this is likely to be associated with sensory rather than with motor fibres. Florey and Biederman (1960) had earlier reached a similar conclusion in the case of the crab, *Cancer magister*. The possibility that ACh-induced permeability changes are involved in a transducer mechanism at sensory endings (cf. ch. 7 §3.7) was mooted by Nadol et al. (1970) who examined the ultrastructural localization of AChE in the stretch receptor (abdominal muscle receptor organ) of *Homarus*. They found end-product on the cell membranes, including the axonal and dendritic membranes, of sensory neurones. It was also present on the external surface of the motor axons but was lacking from the neuromusclular endings. Other histochemical studies on the locali-zation of ChE in *Homarus* include the optical microscopic investigation of the stretch receptor by Maynard and Maynard (1960) and the ultrastructural examina-tion of axons by De Lorenzo et al. (1969).

5.4 Cholinesterases in other invertebrates

Data for cholinesterase in invertebrates other than arthropods are not abundant although a good deal is known about the enzyme in the platyhelminths and nema-todes (aschelminths) — as with insects, these have been studied in relation to the development of the anticholinesterase type of pesticide. A certain amount of sys-tematic work has been done on molluscs and annelids and, in addition, observations have been made on a few animals scattered through other phyla. These last will be considered together, irrespective of their correct phyletic position. The probability, touched on already, that ChE plays a part in morphogenesis of some invertebrates belongs to the general topic of ChE in relation to development and will be discussed in ch. 8.

5.4.1 Platyhelminths and nematodes

The liver fluke, *Fasciola hepatica*, is parasitic in sheep and cattle and as such is of agricultural significance. Frady and Knapp (1967) used radiometric and manometric techniques to examine the enzyme in whole homogenates of the fluke from sheep and, on the basis of activity towards ACh and because of evidence of substrate inhibition, concluded that a single specific cholinesterase with several properties common to mammalian AChE was responsible for the hydrolysis of both ACh and BuCh. What they do not comment on, and what is more striking, is the strange behaviour with inhibitors. DFP, even at 10^{-4} M did not cause any inhibition but 10^{-4} M atropine sulphate, which Augustinsson (1948) described as inhibiting ChE only slightly, produced 45% inhibition. Eserine at 10^{-4} M inhibited 84% of the activity but for total inhibition a concentration of 2×10^{-3} M was needed. Such a pattern of inhibition is very unlike that for mammalian AChE but the possibility that the enzyme is an aliesterase capable of hydrolysing choline esters (as described in cockroach nerve cord by Metcalf et al. 1956) seems ruled out by its apparent inability to hydrolyse tributyrin. Frady and Knapp showed that the AChE in the sheep erythrocytes present in the gut of the fluke was still active. Obviously in a study of this type it is essential to be sure that the enzyme under examination belongs to the parasite and not to the host.

Halton (1967) stained *Fasciola hepatica* histochemically, using acetyl- and butyrylthiocholine as substrates for cholinesterases, and a variety of other carboxylic esters to demonstrate non-specific esterases. Like Frady and Knapp, he concluded that the ChE activity was due to a single enzyme hydrolysing AcThCh more actively than BuThCh and should be classed as 'acetylcholinesterase'. He did, however, make the point that this ChE was 'unaffected' not only by iso-OMPA (10^{-6} M) but also by 62C47 and that eserine at 10^{-5} M was not totally inhibitory. Using polyacrylamide gel electrophoresis, Haites et al. (1972) found 5 cholinesterase isozymes in homogenates of *Fasciola hepatica* from sheep; all showed a preference for acetyl rather than butyryl substrates. Two of the isozymes had molecular weights of 300,000, the figures for the other forms were 182,000, 160,000 and 40,000.

In contrast to the conclusion of Frady and Knapp, and of Halton that the cholinesterase activity in the sheep fluke is due to only one enzyme, Krvavica et al. (1967) found evidence of a separate enzyme hydrolysing BuThCh in *Fasciola hepatica* from cattle. This hydrolysis was inhibited by 10^{-7} M DFP but hydrolysis of AcThCh was unaffected at this concentration. Differential effects were also found with Paraoxon and methyl paraoxon, in this case activity towards AcThCh was inhibited more strongly than that towards BuThCh. Eserine inhibited hydrolysis of AcThCh and BuThCh at 10^{-5} and 10^{-4} M respectively. Halton did not describe his technique in such detail as Krvavica et al. but the limited comparison that is possible provides no obvious technical explanation of Halton's inability to demonstrate two cholinesterases. It seems unlikely that the difference is due to the use of fixed material by Halton and of fresh specimens by Krvavica et al.: some of the developing larvae studied by the latter had been fixed and these, like the unfixed adults, could be stained for both enzymes. The possibility that there may be a genuine strain difference between the flukes in sheep and cattle should be followed up.

A number of schistosomes, which are members of the same subclass as *Fasciola,* were examined histochemically by Fripp (1967a). These, too, apparently possess a pseudoChE in the nervous system which hydrolyses BuThCh and is inhibited by iso-OMPA. AcThCh plus iso-OMPA revealed what is presumably an acetylcholinesterase but it was inhibited only slightly by 62C47 although eserine in a concentration of 10^{-5} M produced total inhibition. The histochemical findings confirm, in essence, the earlier biochemical observations of Bueding (1952) who recognized the existence of two cholinesterases though he did not test 'selective' inhibitors.

Another atypical pattern of ChE inhibition has been found in the cat tapeworm, *Taenia taeniaformis* (Eränkö et al. 1968). In this case, low concentrations of eserine (10^{-5} M or less) inhibited activity towards ACh, MeCh and BuCh in biochemical experiments and towards AcThCh and BuThCh in histochemical experiments, but the enzyme was relatively resistant to BW284C51 and 62C47 as well as to iso-OMPA. For example, BW284C51 totally inhibited rat brain enzyme at 10^{-5} M but a concentration of 10^{-2} M was necessary for 100% inhibition of tapeworm enzyme. The substrate preference was in the order MeCh, ACh, BuCh but all produced substrate inhibition. The histochemical distribution of enzyme in ganglia, nerve trunks and fibres was identical whether AcThCh plus iso-OMPA or BuThCh plus 62C47 were used but, in keeping with the biochemical findings, the staining was weaker with BuThCh than with AcThCh. In another tapeworm, *Hymenolepis diminuta,* Graff and Read (1967) demonstrated a high rate of hydrolysis of ACh but, as in the sheep fluke, *Fasciola hepatica,* it was not inhibited by 10^{-4} M DFP. In contrast, atropine sulphate (10^{-4} M) and eserine (10^{-4} M) produced 51% and 90% inhibition respectively. Neither the rationale nor the significance of the use of atropine sulphate in the experiments on platyhelminths is obvious; as mentioned earlier it is generally regarded as a relatively weak anticholinesterase (see also Long 1963). The apparent lack of inhibition by DFP and other organophosphorus compounds in some platyhelminths raises the question of enzyme classification: should an enzyme which is resistant to DFP be considered as a cholinesterase at all, when one of the criteria on which these are identified (see ch. 2) is their susceptibility to inhibition by organophosphorus compounds and carbamates?

Bueding (1952) made a passing reference to the presence of relatively low AChE activity in the nematodes, *Ascaris lumbricoides* and *Litomosoides carinii.* More recently, Haites et al. (1972) found that the cholinesterase present in *Ascaris suum* had a preference for butyryl rather than acetyl substrates. Electrophoresis on polyacrylamide gel showed three cholinesterase isozymes; two had molecular weights of 180,000 and the other had a weight of 40,000. The ChE of another nematode, *Haemonchus contortus,* which is parasitic in sheep was investigated by Lee and Hodsden (1963). They obtained data for pH and temperature dependence and for subcellular distribution as well as for other properties. ACh was the preferred substrate and 10^{-4} M eserine caused total inhibition but, as with the platyhelminths, the enzyme was relatively resistant to 62C47. Paraoxon and Haloxon were better inhibitors, the latter being 89% effective at 2×10^{-7} M. With both, the hydrolysis of BuCh was inhibited slightly less than the hydrolysis of ACh but Lee and Hodsden did not accept this difference as sufficient evidence for the existence of two separate cholinesterases. They concluded that a single enzyme was responsible for the hydrolysis of both substrates and this may well prove correct but, in the light of the inhibitor patterns, more exhaustive tests are needed before the existence of a second ChE can be eliminated with certainty.

The rat nematode, *Nippostrongylus brasiliensis*, has been the subject of a number of studies. Sanderson (1969) concluded on the basis of substrate and inhibitor patterns that the enzyme present is AChE. Histochemical experiments have shown that this is distributed not only in the nervous system but also in secretory glands opening into the oesophageal lumen and in the anterior excretory gland opening to the outer surface (Lee 1970). Sanderson and Ogilvie (1971) found that the ChE activity, expressed on a weight basis, was relatively low in the eggs and infective larvae but increased markedly during the parasitic phase to reach an adult level which was 15 times that in the egg. It appears that AChE secreted by the worm stimulates antibody production by the host. Worms which are damaged by the immune reaction of the host show an increased level of AChE (Sanderson et al. 1972) and also a change in the isozyme pattern (Edwards et al. 1971). The isozymes, B and C, increase and isozyme A decreases. Subsequently Jones and Ogilvie (1972) found that in worms that had adapted to immune hosts, there was a marked increase in A as well as some increase in B and C. When such worms were transferred to non-immune rats, the level of isozyme A fell rapidly. Jones and Ogilvie's results to some extent supported the idea, proposed earlier by Edwards et al., that AChE was likely to be involved in stimulating host immunity but they concluded it was not the only antigen responsible for the reaction.

The value of histochemical studies in assessing the possible function of enzymes has been discussed earlier in this chapter. Fripp (1967a) and Krvavica et al. (1967) considered the likely role of the ChE in schistosomes and in *Fasciola hepatica* respectively. According to histochemical findings, the enzyme is present not only in nervous tissue but also in the suckers, pharynx and cirrus; the latter is a muscular structure through which the male reproductive tract opens. In the flukes it is the pharynx rather than the sucker which is responsible for extracting the host's blood and Krvavica et al. found the degree of staining in the pharynx was appreciably higher than in the sucker. In support of the idea that ChE may be important in pharyngeal sucking movements, they cited their unpublished observation that the pharyngeal muscle was only poorly stained in *Ascaris lumbricoides* which lies free in the intestinal contents of its host (man) and has no need to apply powerful suction. They also suggested that the cirrus may perform an ejaculatory function in reproduction and, here again, ChE could have a role in muscle contraction. Eränkö et al. (1968) showed that in the tapeworm, *Taenia taeniaformis*, ChE activity in the cranial segments was largely confined to the nervous system but in the more mature caudal segments a weak reaction was seen in the sperm ducts, testes and vagina. The cuticle of these segments gave an intense reaction and the underlying skin was moderately stained. Strong staining of the cuticle and of parts of the reproductive system was also observed by Lee et al. (1963) in a number of other tapeworms including *Hymenolepis diminuta* and *Hydatigera taeniaformis*.

In his histochemical study of the planarian, *Procotyla fluviatilis,* Lentz (1968) using the thiolacetic-lead method (Crevier and Bélanger 1955) found ChE activity to be localized almost exclusively in nervous tissue. Enzyme was present in both the cells and fibres of the submuscular and subepidermal nerve plexuses. The cephalic ganglia showed less reaction while the large nerve bundles and the sensory cells in the epithelium were negative. Lentz concluded that this distribution of enzyme was indicative of a cholinergic mechanism in the nervous system and, in support, he cited reports that structures resembling ACh-containing synaptic vesicles have been observed in planarian neurones and nerve endings (Lentz 1967). Despite these suggestive findings, the question of whether or not the cholinesterase in the nervous tissue of flukes and worms is involved in transmission remains uncertain. As yet there is no unequivocal evidence of cholinergic transmission in any of the platyhelminths or nematodes but Smith et al. (1940) found that eserinized strips of muscle from *Cerebratulus* (a member of the closely related phylum **Nemertina**) were sensitive to ACh in the rather high concentration of 5×10^{-5} M. It seems quite likely that the enzyme in the muscle of the flukes and worms is involved in processes, as yet undefined, akin to those in the motile structures mentioned earlier. Rybicka (1967) showed that the appearance of AChE in developing embryos of *Hymenolepis diminuta* seemed to be associated with muscle formation. Several other suggested roles for ChE in various parasites have been listed by Fripp (1967a). These include the control of glucose transport across the wall of the hydatid cyst of the tapeworm *Echinococcus* (Schwabe et al. 1961), control of glucose uptake by schistosomes (Fripp 1967b) and involvement in sodium transport in *Hymenolepis* (Lee et al. 1963).

In summary, the details of substrate specificity vary between different parasitic flukes and worms and in some species it is questionable whether one enzyme or two are present. The property which seems to emerge most clearly as a characteristic of the cholinesterases of the two groups is that their behaviour with inhibitors tends to be rather different from that of their hosts. Unfortunately from the practical point of view, this difference is generally one of greater resistance rather than one of greater susceptibility. The case of Haloxon in *Haemonchus contortus* (Lee and Hodsden 1963) is, however, a hopeful exception: the worm's ChE is inhibited irreversibly, that in the erythrocytes of the sheep (the host) is inhibited less and the inhibition is rapidly reversible. The physiological role of the enzymes both in nervous and non-nervous structures has yet to be finally established as has the question of whether AChE secreted by the parasitic nematodes plays a major role as an antigen.

5.4.2. Molluscs

Some of the earliest investigations of invertebrate ChE and ACh were done on members of the phylum Mollusca by Bacq and his colleagues (see Bacq 1947). It is clear from these and later studies that levels of ACh, ChAc and ChE may be extremely high in parts of the molluscan nervous system (see Feldberg et al. 1951; Florey 1963; Bryant and Brzin 1966; Loe and Florey 1966; Florey and Winesdorfer 1968; Heilbronn et al. 1971; Welsch and Dettbarn 1972a). Like the lobster axons discussed already, axons from squid (*Loligo*) have been extensively studied by Nachmansohn's group in connection with mechanisms of axonal conduction. Brzin et al. (1965) found axonal membranes of the small stellar nerves of squid to hydrolyse 3.2×10^{-4} μmole ACh/mm^2/hr; the figure for the giant axons was rather lower, 9.5×10^{-5} μmole ACh/mm^2/hr.

The properties of molluscan cholinesterases have not been investigated in great detail. Augustinsson (1946b) examined the enzymes in the land snail, *Helix pomatia,* and found a difference between that in blood and that in the dart sac. The ChE in blood was inhibited by an excess of both ACh and MeCh and was capable of the rapid hydrolysis of acetylaneurin; the dart sac enzyme behaved somewhat like a pseudoChE since it was not inhibited by excess substrate but, on the other hand, the order of substrate preference was ACh, MeCh, BzCh. Neither enzyme split ethyl acetate although dart sac enzyme showed appreciable activity towards tributyrin. In the garden snail, *Helix aspera*, Korn (1969) found high ChE activity in the heart and this he attributed to an acetylcholinesterase possibly mixed with some pseudoChE. Subsequently Haites et al. (1972) detected 3 ChE isozymes in non-nervous tissue of this snail. All showed a preference for acetyl substrates; two had molecular weights of 40,000, the other gave a value of 180,000. The ChE in brain tissue of *Helix aspera* has been examined by Emson and Kerkut (1971). It appears to be an acetylcholinesterase with a pH optimum of 8.4 and it shows substrate inhibition by ACh at a concentration of 10^{-3} M. The four constituent polypeptides have a total molecular weight of 240,000. Working on the pond snail, *Lymnaea stagnalis* Varanka (1968b) reported that the enzyme in the CNS hydrolysed propionylcholine somewhat faster than ACh and was inhibited by excess ACh and BuCh; PrCh was not tested for substrate inhibition. Activity towards MeCh was 10%, and towards BzCh 2%, of that towards ACh. Recently Jurchenko et al. (1973a, b) have presented evidence of considerable differences in the nature of the cholinesterases in *Lymnaea stagnalis* and another gastropod, *Planorbarius corneus*, especially with respect to inhibitors. A point of special interest is their observation that eserine in concentrations up to 1×10^{-4} M did not potentiate the depolarizing effect of ACh on the ganglion cells of *Lymnaea* nor did it inhibit the histochemical reaction towards AcThCh. Eserine did, however, produce some potentiation of the depolarization caused by BuCh and at 10^{-4} M it abolished the staining reaction with BuThCh. In histochemical experiments on *Planorbarius*, eserine at a concentration of 10^{-5} M abolished the reaction to both AcThCh and BuThCh.

According to Twarog (1954) the cholinesterase in the muscle of *Mytilus edulis* splits ACh but not MeCh or BzCh. Dettbarn (1963) showed that in the axon of the squid, *Loligo*, the substrate inhibition produced by excess of ACh was less marked than in crustacean nerves and this, together with differences in K_M values, led him to suggest that the 'enzyme protein' might show some species variation. He did not enlarge on this except to indicate that the difference was unlikely to be in the active sites — in view of the situation in insects this may, however, be an over-simplification. Turpaev et al. (1968) examined another squid, *Ommatostrephes sloanei-pacificus* and found that the optic ganglia contained one cholinesterase which hydrolysed optimal concentrations of ACh, MeCh and BuCh at the same rate. The K_M values were different, however, that for ACh being the lowest. Benzo-ylcholine was not hydrolysed. The enzyme was particularly sensitive to a number of organophosphorus compounds. What seems to emerge from the biochemical experiments is that there is probably no single cholinesterase typical of the molluscs. The substrate patterns seem to vary to a greater or lesser extent in different members, and the same may be true with different inhibitors although these have not been systematically examined.

In a histochemical study on formalin-fixed material, Zsoltan-Nagy and Salánki (1965) compared the cholinesterase present in gastropods and lamellibranchs. They showed appreciable ChE activity in the CNS of the gastropods (including *Helix pomatia*) but their assertion that this was due to BuChE is based on an unusual interpretation of inhibitor effects. In contrast, they were unable to demonstrate any reaction at all in the CNS of the lamellibranchs which they tested. Subsequent biochemical studies on homogenates (Salánki et al. 1966) have shown that ChE is, in fact, present in the nervous tissue of the lamellibranch, *Anodonta cygnea* (the swan mussel), and in some respects but not in all it resembles mammalian AChE. The discrepancy between the histochemical and biochemical results may be explicable on the grounds of inhibition by formalin or of non-penetration of histochemical substrates, or it may be related to the evidence that in the ganglia (but not in the muscle) part of the active site of the ChE in *Anodonta cygnea* is apparently masked by lipid material (Varanka 1968a). Whatever the explanation, it underlines the importance of applying more than one method before accepting negative results. In the experiments of Jurchenko et al. (1973b), referred to earlier, it was shown that in *Lymnaea stagnalis* all the ganglia gave a strong reaction towards AcThCh and BuThCh. With AcThCh the reaction occurred round the cell bodies and fibres but the cells themselves and the ganglion sheath were not stained. The neuropil of the pedal and cerebral ganglia was strongly positive but in the visceral and parietal ganglia the reaction was very much weaker. With BuThCh the staining, though less intense, was similar in distribution with, in addition, a strong reaction in the ganglionic sheath. In *Planorbarius corneus* the reaction to AcThCh was largely confined to the neuropil. No staining occurred in the neurones and that round nerves was weak. With BuThCh a weak reaction occurred in the neuropil.

Histochemical experiments have also been done on optic tissue from the squid, *Ommatostrephes* (Turpaev et al. 1968) and the cuttlefish, *Sepia* (Drukker and Schadé 1965); in both, most of the ChE activity is confined to parts of the neuropil. Barlow (1971) examined the AChE in the vertical lobe of the brain of *Octopus vulgaris* and found very strong staining in the efferent fibres in the centre of the lobe. The large efferent cells also reacted but the amacrine cells and their processes did not.

As always, the question is whether the occurrence of the components of the ACh-ChE system is adequate evidence for the existence of cholinergic synapses. Bacq (1947) pointed out the danger of the unqualified acceptance of such evidence and cited the existence of ACh in some plants, mushrooms and bacteria. Applying his criteria to a number of molluscs including *Octopus, Helix* and *Aplysia* he found acceptable evidence of peripheral cholinergic transmission only in the inhibitory nerve to the heart of the clam, *Venus mercenaria*. Pilgrim (1954) was equally cautious in assessing the effects of ACh on a number of molluscan hearts but thought the possibility of peripheral cholinergic transmission 'by no means unlikely'. In agreement with this, Ten Cate (1955) presented evidence suggesting that in *Anodonta cygnea* the heart receives an inhibitory cholinergic innervation from the visceral ganglion. Whether or not ACh is the physiological transmitter, the sensitivity of the muscle to ACh makes these hearts very valuable for bioassay (see Cottrell and Laverack 1968; Cottrell et al. 1970).

Evidence for cholinergic activity in the central nervous system of molluscs is becoming quite strong. In 1962 Kerkut and Walker showed that in the brain of *Helix aspera* 12 out of 18 cells tested responded to ACh applied by micropipettes, and the following year Kerkut and Cottrell (1963) established that ACh was a normal constituent of *Helix* nervous tissue. Later Newman et al. (1968) speculated whether the ACh was contained in the two types of vesicle that they showed, by electron microscopy, to be associated with synapses which stained for ChE. Both nicotinic and muscarinic receptors are apparently present in *Helix* nervous tissue (Walker et al. 1968); for further work on the snail, see Stefani and Gerschenfeld (1969).

Another mollusc which has been studied particularly thoroughly is the sea slug (or sea hare), *Aplysia*. Tauc and Gerschenfeld (1962) found that the ganglion cells were sensitive to ACh in the bathing fluid. Cells which receive both an excitatory and an inhibitory input were hyperpolarized and inhibited by ACh while cells which have only an excitatory input were depolarized and excited. Strumwasser (1962) postulated that, since ACh can cause depolarization as well as hyperpolarization in *Aplysia,* a single interneurone might be able to excite one cell and inhibit another depending on the properties of the respective post-synaptic membranes. Wachtel and Kandel (1967, 1971) extended this idea, suggesting that ACh might be both excitatory and inhibitory on a single cell where the two types of receptor had different thresholds. A similar proposal has recently been made by Levitan and Tauc (1972) in the case of another mollusc, *Navanax*.

Individual cells can easily be characterized in the *Aplysia* ganglion and it has been shown by Giller and Schwartz (for references see Giller and Schwartz 1971a) and by McCaman and Dewhurst (1970) that only 3 cells in the abdominal ganglion and one in the pleural ganglion contain significant amounts of ChAc (see McCaman et al. 1973 re ACh). In contrast, AChE seems to be present in all the identified cells and the level of activity does not depend on whether or not the cells are cholinergic (Giller and Schwartz 1971b). Some of the cells lacking ChAc had AChE activities equal to those in ChAc-containing cells while some had higher, and some lower, values. McCaman and Dewhurst (1971) have suggested that the AChE is not necessarily confined to neurones and could be present in closely associated glia. On the basis of inhibitor and substrate concentration studies all the cholinesterase in the nervous system seems to be AChE but the haemolymph may contain BuChE in addition to AChE, since it hydrolyses BuThCh at an appreciable rate. As in many other invertebrates (see Augustinsson 1948) the total activity in blood is high, about twice that in nervous tissue, and it would be interesting to know the function of the blood-borne enzyme. The possibility that the particularly high levels of ACh in molluscs reflect some metabolic role has been mooted by Florey (1963) and the circulating AChE could be involved in this. The role of the unusual choiine esters, urocanylcholine and β,β-dimethylacrylylcholine which occur in the hypobranchial glands of some molluscan whelks (Keyl et al. 1957) has not yet been elucidated.

5.4.3 Annelids

Augustinsson (1946a) lists some 20 marine invertebrates which he tested for ChE and only 3 were totally inactive towards ACh, MeCh and BzCh. Of these, two were the polychaets, *Aphrodite aculeata* and *Nereis virens* (the third was a crustacean). In contrast to this negative result, Nachmansohn (cited by Prosser 1950) reported considerable hydrolysis of ACh by the body wall of *Nereis* (the species is not given). Since Augustinsson analysed whole animals his samples might have contained some inhibitory material. Bacq (1947) reviewed the evidence for cholinergic transmission in the annelids and this included the presence of appreciable ACh, the sensitivity of the muscles to ACh and its potentiation by eserine, and the blocking of neuromuscular transmission by curare and atropine.

Little further basic information has been added subsequently but Vigh-Teichmann and Goslar (1968) examined histochemically the distribution of hydrolases including ChE in an earthworm, *Eisenia foetida*. Substrate and inhibitor patterns suggested the presence in ganglia of two cholinesterases, one resembling mammalian AChE and the other, mammalian BuChE. Glial cells contained both AChE and BuChE, and strong AChE activity was seen in structures at the border of the neuronal layer and in the neuropil. In this laboratory, I have stained a few formaldehyde-fixed sections of the earthworm, *Lumbricus terrestris*, using Lewis's (1961) modification of the Koelle method. With acetylthiocholine as the substrate and

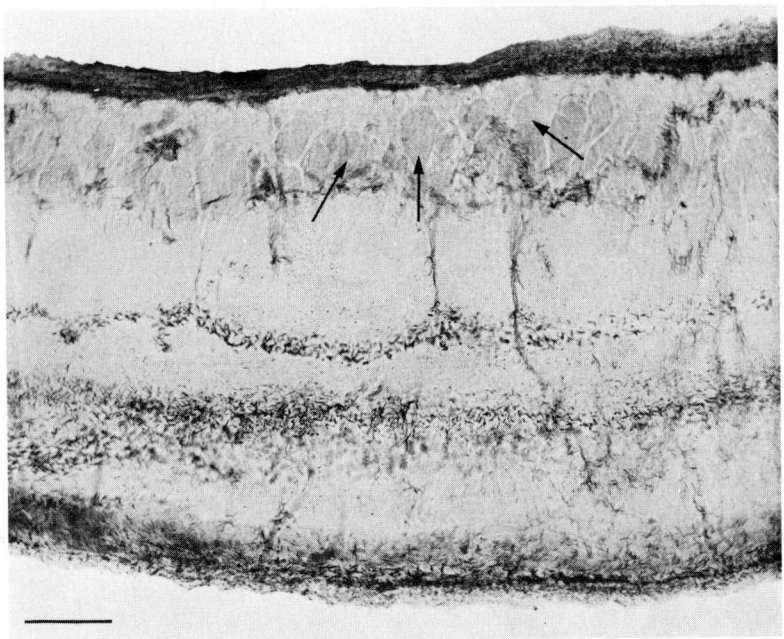

Plate 5.1. *Lumbricus.* AcThCh + 10^{-4} M EP. Sagittal section of ventral nerve cord. Note absence of stain from cells (arrowed). Scale bar: 50 μ.

10^{-4} M ethopropazine as the pseudocholinesterase inhibitor, staining was present along the border of the ventral nerve cord and in elements in the middle, but the cell bodies were totally unstained (plate 5.1). Cells in the supraoesophageal ganglia were similarly negative. As plate 5.2 shows, there is considerable activity in what appear to be nerve fibres on the muscle but in view of the lack of any obvious AChE-containing neurones, there is a possibility that the staining is associated with satellite cells. In contrast to Vigh-Teichmann and Goslar's finding in *Eisenia foetida,* little BuChE activity was seen in *Lumbricus* nervous tissue; in sections incubated with butyrylthiocholine and 5 × 10^{-4} M BW284C51 the only distinct staining was in blood vessel walls. Eserine (10^{-5} M) totally inhibited activity towards both substrates. Teräväinen (1969b) using the methods of Gomori (1952) and of Karnovsky and Roots (1964), examined the cholinesterases in the ventral nerve cord of *Lumbricus* at both the optical and electron microscopic level. His results are in general agreement with the reports that the enzyme is mainly in the glia. He showed that whereas the supportive glia round the neurones were strongly reactive, the so-called migratory glial cells (Coggeshall 1965) were unstained. This finding prompted the suggestion that the glial enzyme could be involved in the maintenance of the extra-

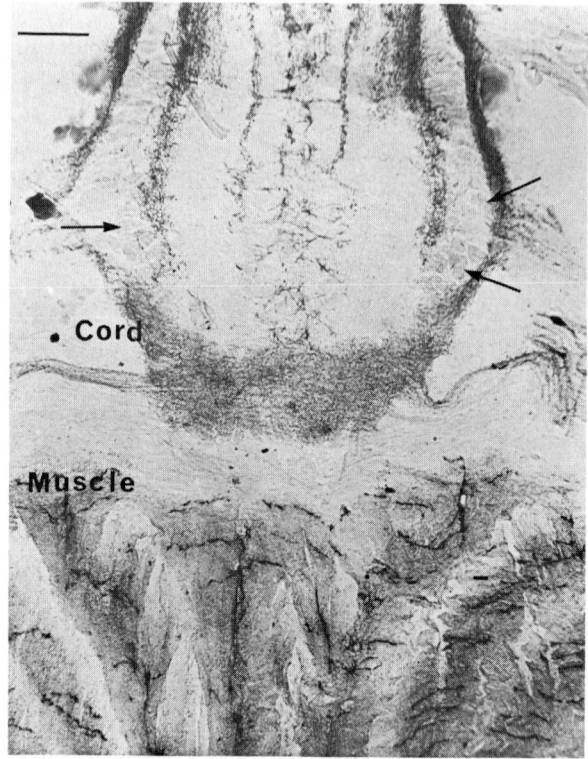

Plate 5.2. *Lumbricus*. AcThCh + 10^{-4} M EP. Horizontal section cut at an angle, so as to include both ventral nerve cord and muscle. Note some stain in the neuropil of cord but absence of cell stain (arrowed). Stain is strong in nerve fibres on the muscle. Scale bar: 100 μ.

neuronal ion balance (see ch. 8 for general discussion). In keeping with the findings from other studies, Teräväinen saw no reaction in the neuropil nor in the majority of nerve cells. Electron microscopy did, however, reveal a relatively weak reaction for AChE in the RER, nuclear membrane and lamellated bodies of the large median neurones. BuChE activity was demonstrable in glia but the reaction was very weak compared with that for AChE. This suggests that in my own experiments the conditions were unsuitable for showing up this low activity.

Haites et al. (1972) found two bands of ChE when they ran homogenates of *Lumbricus* tissue (excluding nerve) on polyacrylamide gel. Both showed a preference for acetyl substrates and had a molecular weight of 220,000.

The leech, *Hirudo medicinalis*, is well known for its extreme sensitivity to ACh and the eserinized dorsal muscle is a valuable preparation for bioassay – the need for eserine being proof of the efficiency of its cholinesterases (see Ammon 1935;

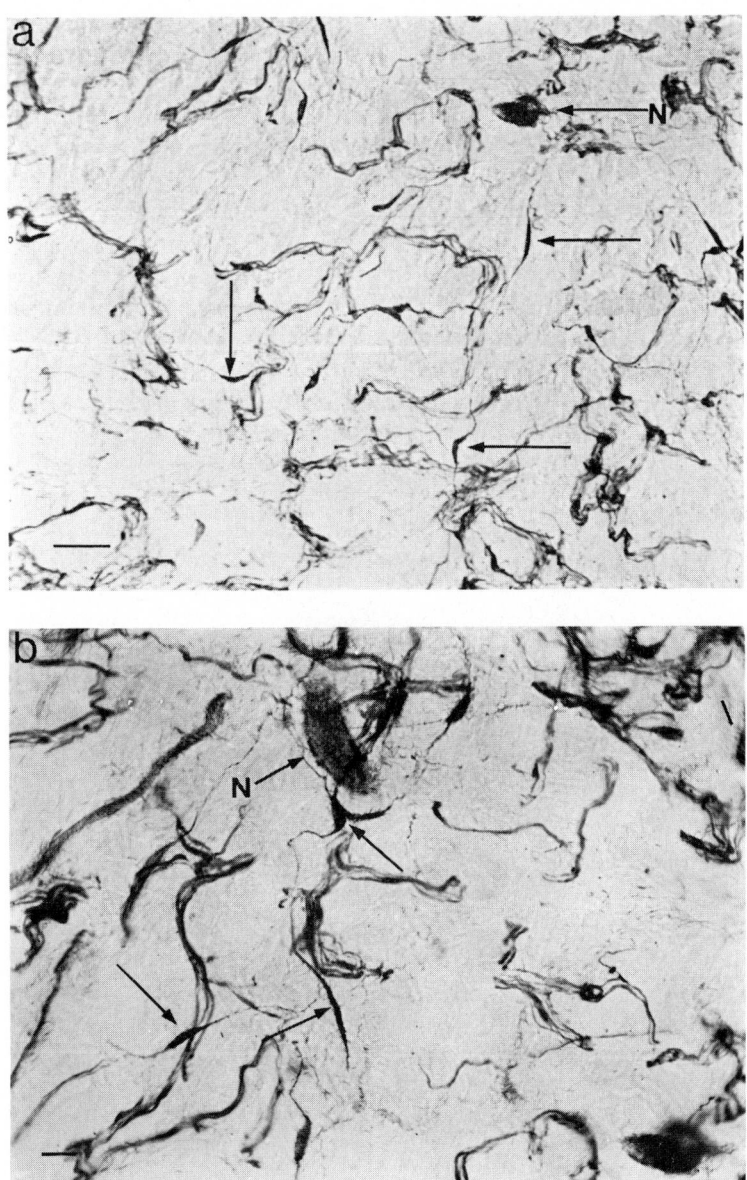

Plate 5.3a, b. *Hirudo.* AcThCh + 10^{-4} M EP. Horizontal section of muscle. Note network of stained bipolar cells. To prevent confusion with pigment, some of these are arrowed. N, nerve trunk. Scale bars. a) 50 μ; b) 25 μ.

Bacq and Coppée 1937). Using Lewis's (1961) method I have looked at the distribution of cholinesterases in nerve and muscle in formaldehyde-fixed sections from seven leeches. The muscle is pigmented but, nevertheless, as plate 5.3a,b shows it is possible to recognize a network of AChE-containing bipolar cells in sections incubated with AcThCh and 10^{-4} M ethopropazine. Large stained nerve trunks crossing the muscle can also be distinguished from the strands of pigment. In addition, there is marked activity, apparently associated with the muscle fibres themselves, in the anterior sucker (plate 5.4). Staining in the ventral nerve cord is pronounced, particularly in Faivre's nerve (see Gaskell 1914) but the perikarya in the segmental and oesophageal ganglia (plates 5.5 and 5.6) give no reaction. The ganglionic neuropil which is where the synapses are located is, however, stained. The stomatogastric nerve ring (Mann 1962) is particularly strongly stained (plate 5.7) but there again cell bodies, recognizable in counter-stained sections, are apparently non-reactive. Because the cell bodies are unstained it seems unlikely that the staining in the ganglionic neuropil and fibre trunks is attributable to neuronal processes and this

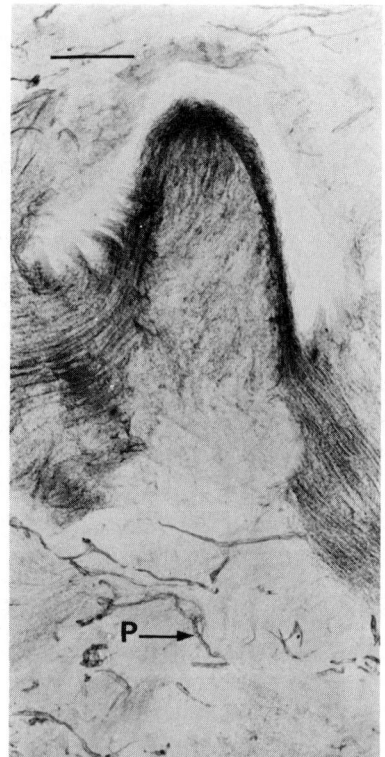

Plate 5.4. *Hirudo*. AcThCh + 10^{-4} M EP. Horizontal section of anterior sucker. P, pigment. Scale bar: 200 μ.

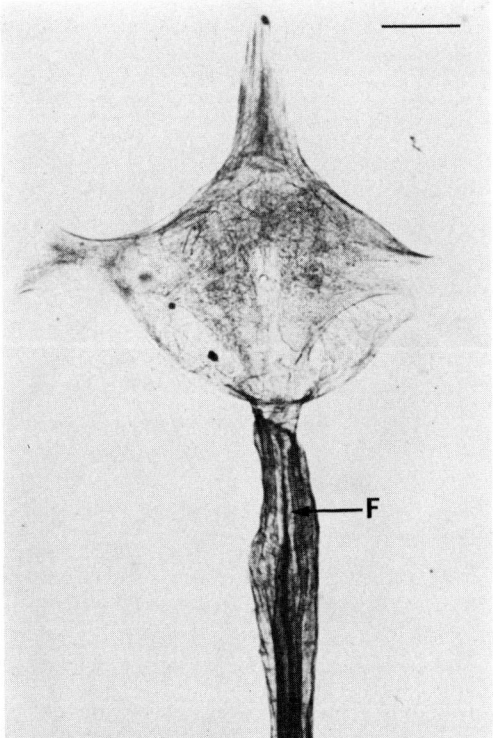

Plate 5.5. *Hirudo*. AcThCh + 10^{-4} M EP. Horizontal section of ventral nerve cord with segmental ganglion. Note weak staining of ganglionic neuropil but more marked activity in cord, especially in Faivre's nerve (F). Scale bar: 200 μ.

would suggest that, as in the earthworm, the enzyme is associated with glial cells (for morphology see Coggeshall and Fawcett 1964). Gerebtzoff (1970) took the view, however, that the enzyme was localized not in glia but at synapses. This idea raises a number of questions, not least the site of enzyme synthesis. Since the cell bodies are unstained it would seem that any synaptic AChE would have to be synthesised in situ at nerve endings.

With BuThCh as substrate I found that the staining in the ventral nerve cord and ganglion was similar in distribution to that seen with AcThCh. The reaction was slightly less intense but Faivre's nerve showed up very clearly (plate 5.8). The staining was not inhibited by 10^{-4} M ethopropazine but it was eliminated by 5×10^{-4} M BW284C51. The results suggested that 'AChE' in the leech may be capable of appreciable hydrolysis of BuThCh. Further specificity tests on muscle, which included the elimination of the possibility that BW284C51 inhibited pseudoChE as well as AChE, strengthened this view. As in *Lumbricus*, 10^{-5} M eserine totally inhibited the reaction towards AcThCh and BuThCh.

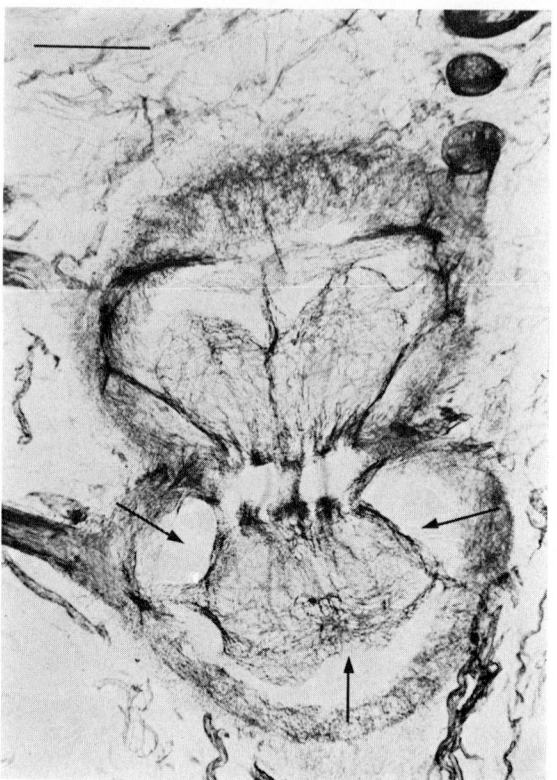

Plate 5.6. *Hirudo.* AcThCh + 10^{-4} M EP. Horizontal section of suboesophageal ganglion showing activity in neuropil and nerve trunks but not in cell bodies (arrowed). Scale bar: 200 μ.

My experiments produced one other finding of interest. As sometimes happens
(see ch. 7) the cholinesterase technique showed up some hitherto unrecognized
structures which, as yet, I have been unable to identify. These, as plate 5.9 shows,
are two strongly-stained bipolar cells orientated between the peripheral nerves leaving the segmental ganglion.

Recently, at a time when the development of radiochemical techniques for
measuring ACh is supplanting the bioassay, the leech has found a new role: it is
being used in studies on the behaviour and nature of ACh receptors (Flacke and
Yeoh 1968a, b; Rang and Ritter 1969, 1970). In leech, the muscle receptors react
only slightly, if at all, with neostigmine and DFP — but they do react with eserine —
so studies of the effects of enzyme inhibition are not complicated, as they are in
other species, by simultaneous effects on the sensitivity of the receptors (chs. 7 and
11); they seem to be nicotinic rather than muscarinic.

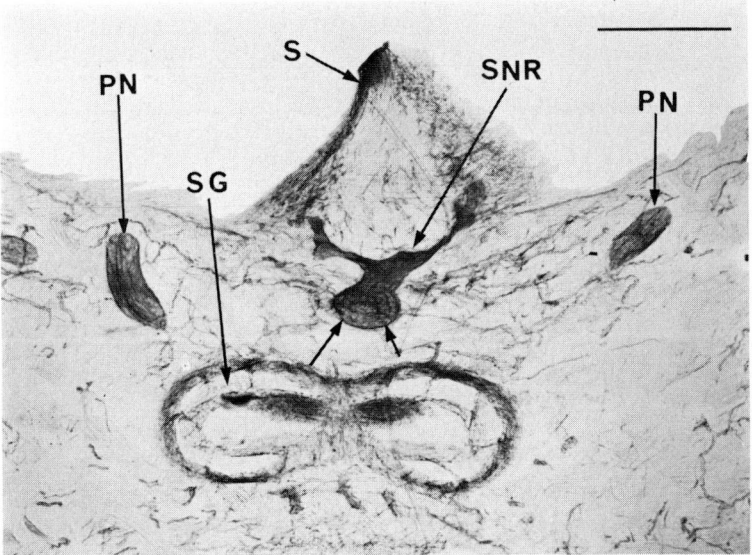

Plate 5.7. *Hirudo.* AcThCh + 10^{-4} M EP. Horizontal section at level of supraoesophageal ganglion (SG). The light areas (arrows) in the stomatogastric nerve ring (SNR) are unstained cells. PN, peripheral nerve; S, sucker. Scale bar: 500 μ.

Kerkut and Walker (1967) found good evidence for nicotinic receptors in the cells of Retzius in the leech ganglia when they applied ACh iontophoretically and confirmatory results were obtained by Newton et al. (1970). Results with curare (see Sakharov 1970) do suggest, however, that the ganglionic receptors, although nicotinic, are not identical with those at the periphery. Furthermore, it is not safe to assume that nicotinic receptors are characteristic of the annelids as a whole and there is evidence that in the polychaet, *Aphrodite,* neuromuscular transmission may be muscarinic since it is blocked by atropine (Bacq 1947). In the light of this evidence suggesting the presence of cholinergic mechanisms in the leech, the apparent lack of neuronal cholinesterase is surprising but it could be argued that, if hydrolysis of ACh at synapses can be adequately accomplished by glial enzymes, a contribution from the neurone would be unnecessary (see Smith and Treherne's ideas re cockroach § 5.3.1.4).

5.4.4 Echinoderms

The sensitivity of echinoderms to applied ACh has long been recognized (see Bacq 1947) and the holothurians (sea cucumbers) have been used in bioassay of ACh. Bullock and Nachmansohn (1942) found ChE in the radial nerve cord of the starfish, *Asterias forbesi,* and the value for ACh hydrolysis by the feather star, *Antedon*

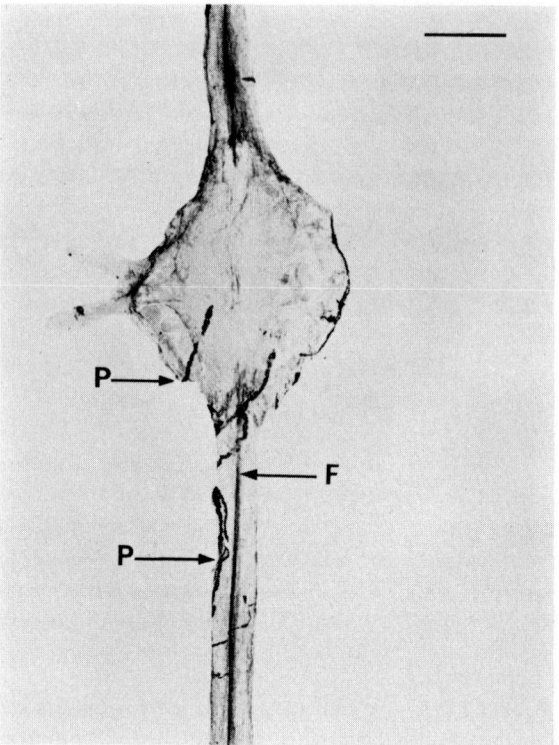

Plate 5.8. *Hirudo*. BuThCh (no inhibitor). Horizontal section of ventral nerve cord with segmental ganglion (cf. plate 5.5). Note activity in Faivre's nerve (F). P, pigment. Scale bar: 200 μ.

petasus was the highest obtained in any of the invertebrates which Augustinsson (1946a) studied. MeCh was hydrolysed to some extent by all the echinoderms in Augustinsson's experiments, but only two were active towards BzCh, the starfish, *Asterias rubens* and the sea urchin, *Echinus esculentus*. Generally whole animals were used and in none was the nervous tissue examined separately. In 1964, Roberts and also Cottrell (see Welsh 1966) described an ACh-like material in the radial nerve cord of starfish. This was subsequently identified as ACh by Pentreath and Cottrell (1968) who also concluded that the cholinesterase which they demonstrated histochemically in the nerves of *Asterias rubens* was AChE. To obtain staining they had to incubate the tissue for very long periods (15 hr with AcThCh and 40 hr with BuThCh) which suggests there may be penetration barriers. Staining was strongest in the ectoneural layer of the nerve cord but the types of cell which stained were not identified; the activity in the podia (tube feet) seemed to be associated with the ectoneural plexus.

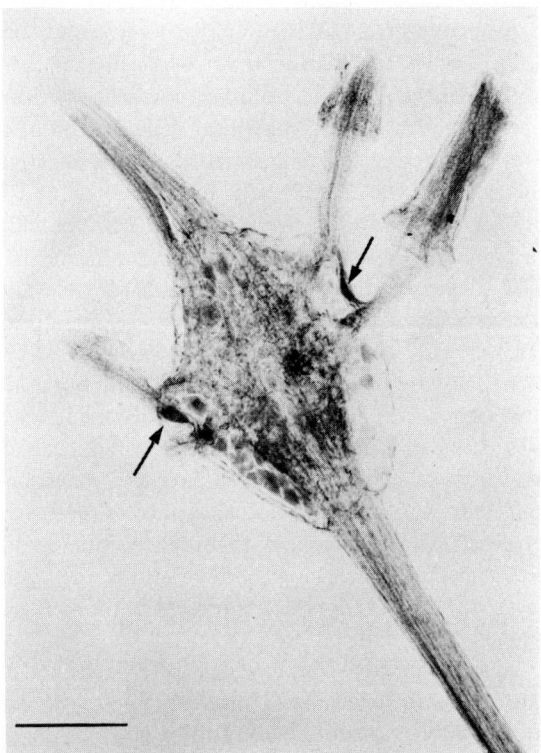

Plate 5.9. *Hirudo.* AcThCh + 10^{-4} M EP. Horizontal section of segmental ganglion. Two cells (arrowed) at the periphery stain strongly for ChE but the neurones in the body of the ganglion, counterstained with methylene blue, do not show enzyme activity. Scale bar: 250 μ.

5.4.5 Miscellaneous phyla

The order in which the phyla have been discussed so far has been determined by the amount of available data and not by their position within the invertebrate sub-kingdom. An inevitable consequence is that this final section has to accommodate a variety of creatures (see table 5.1) which have little in common beyond the paucity of data about their cholinesterases.

The evidence of ChE activity in *Paramecium, Tetrahymena* and *Physarum* (see table 5.1 and §5.2) belies the sweeping generalization, all too often repeated, that protozoa do not contain ChE and, very probably, ChE will be demonstrated in other protozoa when the appropriate experiments are done. 'Appropriate' is used advisedly to re-emphasize the point that negative results may reflect a deficiency in technique rather than a deficiency of enzyme. Bayer and Wense's (1936) experiments illustrate this point: they added ACh to cultures of *Paramecium* at 5 and at 37°C and made the unexpected observation that hydrolysis occurred only in the cold cultures. As a possible explanation they suggested that the warmed animals produced some interfering compound. Another possibility is that at 37°C the enzyme may have been inactivated. Although heat denaturation of vertebrate ChE is not apparent below 50°C, insect enzyme starts to become inactivated between 30 and 35°C (Chadwick and Lovell 1958). It is not unreasonable to suggest an even lower temperature for inactivation in *Paramecium*, a creature which, in its natural environment of a pond, will be existing at temperatures nearer to 5 than to 37°C. The figure of 37°C, so dear to the enzymologist, is hardly relevant to these studies and even 'room temperature' scarcely equates to pond temperature.

The sponges are variously classed as Porifera or Parazoa (see Rothschild 1961). Like the protozoans they are often listed as lacking ChE, but again this generalization may well be wrong since Lentz (1966) has demonstrated what appeared to be ChE in some cells of the sponge *Sycon ciliatum*. The failure to show the enzyme biochemically (Mitropolitanskya 1941; Bullock and Nachmansohn 1942) could be due to the low concentration in the tissue analysed or to species differences. Lentz raised the further possibility that the histochemical reaction could be non-specific and pointed out that electron microscopy shows nothing resembling a nerve ending or synapse in the sponge. Any ChE which is present in the sponges is thus likely to be involved in non-nervous activity.

Coelenterata was the lowest phylum in which Bullock and Nachmansohn (1942) could demonstrate ChE biochemically. Using the thiolacetic acid histochemical method with various inhibitors, Lentz and Barrnett (1961) showed the widespread distribution of ChE in *Hydra*. Staining was associated partly with ganglion-like cells and with a fibrillar network which resembled the reported arrangement of the nervous system. Some of the species which Bullock and Nachmansohn examined, notably various jelly-fish, were apparently inactive but the possibility that this was due to some interfering substance in the jelly seems strengthened by the similar lack of activity in the comb-jellies (phylum Ctenophora). Haites et al. (1972) found evidence of 5 ChE isozymes, all with a preference for acetyl substrates, in tissues of the sea anemone, *Actinia*. Four had molecular weights of 120,000 and one a molecular weight of 50,000.

The danger of accepting negative results too readily has been repeatedly emphasized but this does not mean all negative results are necessarily wrong: some members of some phyla may manage without cholinesterases but this is difficult to establish. What does seem clear is that there is no sharp phylogenetic cut-off below which cholinesterases are absent. This is well illustrated by the demonstration of the enzyme not only in the protozoa but also in the bacterium, *Pseudomonas fluorescens* (Laing et al. 1967) following culture in an ACh-containing medium. To conclude, it may be appropriate to quote an aphorism which Lwoff (1969) propounded in a totally different context 'If caution is always needed in the interpretation of what is seen more caution is necessary when one sees nothing'.

CHAPTER 6

The distribution of cholinesterases in vertebrates

6.1 Introduction

This chapter is for reference rather than reading. Its aim is to provide some sort of catalogue of those areas which have been examined histochemically and biochemically for cholinesterase activity. The main emphasis is on the nervous system (see also chs. 8 and 10). In most cases findings are indicated only briefly and the original papers must be consulted for details but results from a few key papers are given more fully in separate tables.

6.2 Arrangement of tables

In compiling the tables various schemes were considered. These included arranging them area by area regardless of species or area by area within a species. The final format, in which papers are listed in alphabetical order of author rather than area eliminates repetition when a paper deals with more than one area or structure. Species such as cat, rat, rabbit and man which have provided an enormous number of papers (those listed represent only a fraction of those published) have been tabulated separately but amphibia, birds, fish, primates (excluding man) and reptiles have been considered as groups. Table 6.31 accommodates those miscellaneous species which lack sufficient data to justify a table of their own, and tables 6.32.1 and 6.32.2 have been recompiled from the very comprehensive work on the retina published by Esila (1963). Where appropriate, histochemical papers have been split into those dealing with the CNS and those dealing with peripheral structures; this division has not been made for the biochemical papers which are less numerous.

6.3 Significance of the data

The question of the functional significance of cholinesterases and also the problems associated with species differences and developmental changes are covered in the next 4 chapters but some general comments on the interpretation of data may be useful here.

6.3.1 Histochemical staining at cholinergic and cholinoceptive sites

In ch. 1 cells which release ACh as their transmitter were defined as cholinergic and those which respond to ACh, as cholinoceptive (Dale 1934, 1954). Known cholinergic cells, for example the somatic motor neurones, contain ACh, ChAc and AChE. These three components of the cholinergic mechanism are present in the axon as well as in the cell body (ch. 4) and the histochemical reaction for AChE is generally intense. In areas containing cholinoceptive cells the cholinergic neuropil normally gives a strong reaction for AChE but the staining of the neurones is variable. Some cholinoceptive cells are themselves cholinergic and in this case they give the characteristically intense reaction for AChE. If they are not cholinergic they lack ACh and ChAc but AChE present in the cell body and dendrites may give a weak or moderate histochemical reaction. In the CNS, axons of non-cholinergic cells do not show appreciable staining (Lewis and Shute 1959, 1966) but in the autonomic nervous system AChE activity has been observed in adrenergic as well as cholinergic axons (see ch. 7) and somatic sensory fibres may stain under certain conditions (Gruber and Zenker 1973).

6.3.2 Histochemical studies in peripheral tissue

The histochemical technique for cholinesterases is a useful anatomical tool and some workers who are interested in the distribution of peripheral nerves have used the method in preference to the more capricious silver stain (for a comment on the comparison of methods see El-Bermani 1973). There is some danger that such studies, which were not designed to reveal the pharmacology of the nerves and in which no particular attention was paid to substrate specificity and inhibitors, may subsequently be quoted as revealing a 'cholinergic' innervation. In reading any paper it is important to take note of the inhibitor concentration and the conditions of incubation. In the CNS fairly firm pronouncements can be made that certain neurones contain AChE, BuChE or both but results in the peripheral nervous system are sometimes less clearcut. For example, Novikoff et al. (1966) and also Adams et al. (1967) suggest that unmyelinated fibres in motor and mixed nerves contain 'non-specific' ChE whereas Schlaepfer and Torack (1966) described them as containing mainly AChE. Adams et al. suggest two possible explanations for the discrepancies, either that the unmyelinated fibres do contain both enzymes or that there are two separate populations of fibres, each containing one of the enzymes. The exact nature of the 'non-specific' enzyme was questioned by Adams et al. because it behaved like BuChE when BuThCh was used in combination with inhibitors but the hydrolysis of AcThCh was not prevented by iso-OMPA (nor by BW62C47). This inconsistency may have arisen because the tissues were not preincubated with the inhibitors before addition of substrate (see ch. 3). Grant et al. (1967) have demonstrated that the perineurium and/or the myelin sheath of peripheral nerves may form an effective penetration barrier to thiocholine, ferricyanide or sulphide. The degree to which this barrier is overcome will vary with different techniques and this, too, may account for some contradictory findings.

6.3.3 Biochemical data

The absolute values obtained for cholinesterase activity are greatly influenced by the method of assay (chs. 2 and 3). A valid and meaningful comparison between levels in tissue from different areas or different species can be made only when analyses are done under identical conditions. For this reason no attempt has been made to convert values given by different workers into comparable units.

6.4 Notes about tables

6.4.1 Biochemical analyses of avian tissue

The avian cholinesterases are not identical with those of mammals (ch. 2). The main difference is that the pseudocholinesterase is able to hydrolyse MeCh which in mammals is considered 'specific' for AChE. Some of the papers which deal with the characterization of bird cholinesterases are listed in table 6.3b. Papers dealing with enzyme levels in particular regions are given separately in table 6.3a.

The chick, long favoured by the embryologists, has been adopted by biochemists interested in enzyme development. As a result it figures in a tremendous number of those papers which deal with the onset and specificity of cholinesterase activity in tissues or tissue culture. Results of this sort of work are not easily tabulated but the references in table 6.3c are intended to provide an entry into the literature. The significance of cholinesterases in developing tissue will be discussed in ch. 8.

6.4.2 Arrangement of data from the rat

For obvious reasons the rat has been used in an overwhelming number of histo-chemical and biochemical investigations and the respective tables can do little more than provide an entrance into the literature. In the histochemical table the alpha-betical arrangement of papers has not been strictly adhered to in the case of those by Lewis and Shute and their collaborators. As an aid to clarity many of the related papers by this group have been arranged chronologically under the subheading of Lewis, Shute and co-workers.

The rat has been a favourite subject for experiments on the effect of training on enzyme levels in the brain; references to measurements of cholinesterases in such studies are given in ch. 9.

Species	Reference	Area	Method	Result	Comment
URODELA					
1) Amblystoma	Boell and Shen (1950) (*Amblystoma punctatum*)	Spinal cord, hind, mid- and fore-brain (developing animals)	Biochemical: Cartesian diver, ACh as substrate	Embryonic enzyme of same type as in adult. Develops in caudal-rostral sequence. Final level in cord > hind brain > midbrain > forebrain	See also Youngstrom (1938), Sawyer (1943a,b) and Boell and Shen (1944) for other studies on amblystoma embryos, particularly in relation to development. Sawyer (1955) showed affinity for MeCh to be as great or greater than that for ACh
2) Newt	Dickson et al. (1971) (*Triturus viridescens*)	Retina	EM histochemistry: Lewis and Shute (1966)	Localization of AChE in outer plexiform layer in processes related to bipolar, horizontal and amacrine cells. No pseudoChE found	Results discussed in relation to those reported for other species
ANURA					
1) Frogs	Boell et al. (1955) (*Rana pipiens*)	Optic lobes	Histochemical: Koelle (1951)	Cholinesterase activity particularly marked in synaptic layers. Activity reduced by extirpation of contralateral eye. ChE appears in blood vessels late in development	Developmental study; histochemical and biochemical
			Biochemical: Cartesian diver and Warburg. Substrate not stated (?ACh)	ChE shows 3 phases of development. Extirpation of contralateral eye, while reducing level of activity, did not alter shape of curve relating activity to development	

180

Species	Reference	Area	Method	Result	Comment
Frogs (continued)	Brzin et al. (1966) (*Rana pipiens*)	Sympathetic and dorsal root ganglia	EM histochemistry: Modified Koelle technique; details given in paper	Reaction product on ER, nuclear envelope, subsurface cisternae and agranular reticulum of perikaryon and axon	Effects of fixation on penetration of substrate and localization of enzyme discussed (see also Tennyson et al. 1968)
			Biochemical: Micro-gasometric	High levels of enzyme activity were measurable only in cells with damaged membranes. Results reflect limited substrate penetration	
	Burt and Dettbarn (1972) (*Rana catesbeiana*)	Dorsal root ganglia; dorsal and ventral roots	Histochemistry: Karnovsky and Roots (1964)	In ganglion, AChE in many neurones but not in Schwann cells and connective tissue cells. With prolonged incubation AChE demonstrable in all ventral root fibres and, less strongly, in 15% of dorsal root fibres. No BuChE in any neurones or processes. Some activity in supportive cells in ganglion and dorsal roots but not in ventral roots	Choline acetyltransferase also examined histochemically
	Chacko and Cerf (1960) (*Rana catesbeiana*)	Spinal cord	Histochemical: Koelle (1951)	AChE + ve cells mainly confined to motor nuclei. Neuropil + ve especially in dorsal horn. Capillaries + ve. Ventral root section reduced reaction in motor cells. No reaction towards BuThCh	Toad (*Bufo boreas halophilus*) gave similar results. cf. Hardwick and Hebb (1956) re BuChE
	Giacobini (1956a)	Isolated neurones from spinal cord and spinal, sympathetic and parasympathetic ganglia	Histochemical: Koelle method as modified by Zajicek et al. (1954)	AChE in cytoplasm of cells. Some reaction towards BuThCh	Rat and cat also studied. See also Novikoff et al. (1966) re dorsal root ganglion and peripheral nerves (some EM). cf Burt and Dettbarn (1972) re BuChE
	Hardwick and Hebb (1956)	Brain, spinal cord and sciatic nerve	Biochemical (?Warburg) with ACh and BuCh	AChE highest in cord. Low in brain. In brain and cord BuCh is split to small but measurable extent. All activity inhibited by 10^{-6} M DFP	Results with BuCh, instead of BzCh, showed early conclusions re lack of pseudo-ChE to be wrong (see Shen et al. 1955). See also Hebb (1954)

TABLE 6.1 (continued)

Species	Reference	Area	Method	Result	Comment
Frogs (continued)					
	Hebb (1957)	Retina	Histochemical: Koelle (1950)	AChE present. No details given	Other species discussed
	Koblick (1958)	Skin	Biochemical: Warburg; modified Hestrin; ACh, MeCh, BuCh and BzCh	Enzyme predominantly pseudo-cholinesterase and is localized mainly in tela subcutanea	
	Marnay and Nachmansohn (1938) (Rana esculenta)	Sciatic nerve and sartorius muscle (nerve-free and innervated parts analysed separately)	Biochemical: Warburg; ACh	Rate of hydrolysis of ACh by muscle varied according to proportion of nerve endings included in sample	See also Marnay and Nachmansohn (1937b)
	Miledi (1964)	Sartorius muscle	EM histochemistry: Modified Koelle method, no details given	End-product localized on basement membrane in region of nerve ending, also in ER of muscle fibres. Some particles associated with synaptic vesicles	Author points out likelihood of artifacts due to diffusion of end-product
	Pecot-Dechavassine (1962) (Rana esculenta)	Rectus abdominis and gastrocnemius muscle, neuromuscular and myotendinous junctions	Histochemical: Couteaux and Taxi (1952)	AChE at end-plates and tendinous junctions in fast and slow fibres. BuChE activity at both types of junction in rectus. Distribution not so clear in gastrocnemius	Author finds enzyme very sensitive to formaldehyde. Comparison with several other species. See also Landmesser (1972) re end-plate in reinnervated sartorius muscle
		Rectus abdominis muscle	Biochemical: Warburg, with ACh, PrCh and BuCh plus various inhibitors	For general results with ACh and BuCh see table 6.35. Difference in activity between neural and tendinous areas confirms histochemical finding that BuChE more pronounced at tendinous junction that at end-plates. PrCh hydrolysed by both AChE and BuChE	
	Shen et al. (1955) (Rana pipiens)	Brain, with special emphasis on optic lobe	Histochemical: Modified Koelle (1951) on paraffin sections	ChE present in cells and neuropil at a number of sites, also in capillary walls. AChE and pseudoChE not distinguished	

182

TABLE 6.1 (continued)

Species	Reference	Area	Method	Result	Comment
	Shen et al. (1955) (continued)		Biochemical: Warburg or Cartesian diver with ACh, MeCh and BzCh	ChE activity highest in optic lobe, lowest in cerebral hemisphere	Authors' assumption that frog lacked pseudoChE arose from use of BzCh in biochemical studies, cf. **Hardwick and Hebb** (1956)
	Youngstrom (1938) (*Rana sphenocephala*)	Whole embryos	Biochemical: Bioassay of unreacted substrate (ACh)	ChE activity increased during stage of increasing motility	Toad and amblystoma examined too
2) Toads	Bell and Burnstock (1965) (*Bufo marinus*)	Bladder	Biochemical: Warburg; ACh, BuCh, BzCh, MeCh, SuccinylCh and tributyrin	90% ACh hydrolysis due to AChE. Activity towards tributyrin very high and towards BzCh and SuccCh very low. See Comment column re eserine resistance	Comparison with guinea-pig. AChE in toad bladder 100 times more resistant to eserine than that in guinea-pig bladder. With neostigmine difference is only 8-fold. (See also Hawkins and Mendel 1946 re eserine resistance in frog brain enzyme)
	Brightman and Albers (1959) (*Bufo marinus*)	Brain and spinal cord	Histochemical: Koelle (1951)	AChE activity strong in ventral horn cells and in blood vessels. No appreciable BuChE activity	Comparison with several other species
	Phillis and Tebēcis (1967) (*Bufo marinus*)	Spinal cord	Histochemical: Karnovsky and Roots (1964); Gerebtzoff (1959)	Strong AChE activity in motoneurones and in interneurones of dorsal horn. Substantia gelatinosa + ve, also much of the white matter. Blood vessels + ve	Report forms part of a more extensive study of response of cord to cholinomimetic drugs
	Robinson and Bell (1967) (*Bufo marinus*)	Bladder	EM histochemistry: Karnovsky (1964)	Majority of axons reacted strongly for AChE. Activity associated with axolemma and adjacent Schwann cell membrane. Many vesicles present in stained axons and a few vesicles react for AChE. Post-junctional muscle membrane stained only if incubation was prolonged	Tests showed enzyme liable to diffuse in unfixed, excised tissue
	Tsuji (1964) (*Bufo vulgaris*)	Heart	Histochemical: Koelle (1951)	Reaction for AChE in various sites including neurones in sinus and a–v bundle	Reproduction of pictures very poor

TABLE 6.2
BIRDS HISTOCHEMISTRY
CNS and peripheral tissue

Species and Author	Area	Method	Result	Comment
Canary				
Friede and Fleming (1964)	Cerebellum	Optical: Koelle (1954); Gerebtzoff (1959)	As for pigeon (q.v) except that Purkinje cells did not react for BuChE and capillaries did	Comparison with several species
Chicken (Domestic fowl)				
Akester and Mann (1969)	Renal portal valve	Optical: Lewis (1961) N.B. sections cut from freeze-dried tissue	AChE-containing fibres are distributed throughout valve. Localization identical to that of adrenergic fibres. AChE + ve cells in ganglia associated with arteries	Adrenergic innervation also examined. Choline acetylase measured
Brightman and Albers (1959)	Brain and spinal cord	Optical: Koelle (1951)	AChE in neurones. BuChE in astrocytes of grey and white matter. No activity in blood vessels	Comparison with various species
Capurro et al. (1958)	Retina			See table 6.32
Cavanagh and Holland (1961a)	Brain and spinal cord	Optical: Gomori (1952)	AChE activity uniformly distributed in many neurones including those of dorsal root ganglia, cord, brain stem. Activity in, rather than on, these cells but there is also heavy staining in synaptic area e.g. cerebellar molecular layer	Brief report drawing attention to uniformity of neuronal activity in hen compared with some mammals. See also Cavanagh and Holland (1961b) for distribution of BuChE
Francis (1953)	Retina			See table 6.32
Ginsborg and Mackay (1961)	End-plates of various muscles in chick	Optical: Modified Koelle (1950)	AcThCh used without inhibitor. In some muscles fibres had a multiple innervation; in other muscles the fibres had a single end-plate and in yet others both types were found	Cholinesterase staining used as an anatomical tool and not to provide evidence about the enzyme itself. Latissimus dorsi of pigeon examined

TABLE 6.2 (continued)

Species and Author	Area	Method	Result	Comment
Shen et al. (1956)	Retina	Optical: Modified Koelle (1951); (see Shen et al. 1955)	AChE associated with ganglion and amacrine cells and, in the inner plexiform layer, with their processes. Horizontal cells are slightly + ve. Bipolar cells and outer plexiform layer-ve	Developmental study
Silver (1963)	End-plates in anterior and posterior latissimus dorsi and superior rectus muscles	Optical: Lewis (1961)	Activity towards AcThCh and PrThCh at both single (focal) and multiple (fine) endings. BuThCh hydrolysed to almost same extent as AcThCh in the two types of ending in rectus but no reaction to BuThCh in posterior (focal endings) nor anterior (fine endings) latissimus dorsi	Discussion of nature of enzyme in different types of end-plate. Lack of activity towards BuThCh in latissimus muscle is contrary to result of Hess (1961b). Comparison with rabbit, goat, monkey and guinea-pig
Tingari and Lake (1972b)	Reproductive tract (male)	Optical: Gomori (1952)	Marked BuChE activity in the epithelium of connecting ductules, ductus epididymidis and the cranial part of ductus deferens. AChE in the same structures but is confined to basal part of the epithelium and to nerve fibres. Neither enzyme found in seminiferous tubules, rete testis, ductuli efferentes and narrow connecting ductules	Pattern of staining does not vary with sexual activity. See Tingari and Lake (1972a) for description of AChE-containing nerve fibres
Duck				
Ash et al. (1964)	**Salt-gland and its innervation**	Optical: Coupland and Holmes (1957); Lewis (1961)	AChE mainly in nerve trunks and branches within gland. Strong AChE activity in ganglion cells of secretory nerve. Distribution of BuChE in gland is not quite the same. Level of BuChE activity unaffected by salt loading, cf. Fourman (1966b, 1969) and other references in ch. 10 § 6	Staining abolished when secretory nerve cut inside, but not outside, the orbit. Paper includes pharmacological studies on mechanisms of secretion
Fourman (1969)	**Salt-gland and its innervation**	Optical: Coupland and Holmes (1957)	AChE in nerves and, in some cases, in the region of capillaries. Localization of BuChE is variable. Activity present inside secretory cells of salt-loaded birds	cf. Ash et al. (1969). See also Ballantyne and Fourman (1967)

TABLE 6.2 (continued)

Species and Author	Area	Method	Result	Comment
Fourman and Ballantyne (1967)	Harderian gland	Optical: Coupland and Holmes (1957)	AChE and BuChE in nerves related to blood vessels. BuChE also in epithelial cells	Biochemical analysis confirms histochemical observation that salt loading does not alter enzyme, cf. Ballantyne and Fourman (1967) re salt-gland
Khera and Laham (1965)	End-plates in 'thigh' muscles	Optical: Koelle (1951) and myristoylcholine method (Gomori 1948)	End-plates demonstrable with myristoylcholine from 19th day of incubation but after hatching reaction decreased and activity towards AcThCh appeared. BuThCh hydrolysed from 19th day, reaction present in adult	Developmental study
Phillis (1965b)	Cerebellum	Optical: Gerebtzoff (1959)	AChE strong in molecular layer, weaker in granular layer. Purkinje cell membrane + ve	Comparison with various species. See Austin and Phillis (1965) for biochemical data (table 6.3.a)
Goldfinch, Greenfinch, Hawfinch, Italian sparrow, Parrot, Rock dove, Song thrush				
Capurro et al. (1958)	Retina			See table 6.32
Parakeet				
Friede and Fleming (1964)	Cerebellum	Optical: Koelle (1954); Gerebtzoff (1959)	As for pigeon	
Pigeon				
Friede and Fleming (1964)	Cerebellum	Optical: Koelle (1954); Gerebtzoff (1959)	Marked AChE in neuropil of molecular layer. Some activity in glia and blood vessels. Granular layer reacted very weakly. Purkinje cell membrane + ve. BuChE much stronger in molecular than granular layer and Purkinje cells + ve. Glia	Several species compared. Pattern in birds resembles that in man. See also Capurro et al. (1960) for essentially similar results on Italian sparrow and 2 finches

TABLE 6.2. (continued)

Species and Author	Area	Method	Result	Comment
Nichols and Koelle (1968)	Retina	Optical: Koelle (1951; 1955); Koelle and Gromadski (1966)	AChE activity in amacrine and ganglion cells. Bipolar cells – ve. 2 heavy bands of stain in plexiform layer. BuChE in some of the horizontal cells of inner nuclear layer	Comparison with cat, rat, rabbit and ground squirrel. Pigeon resembles latter which also has a pure cone retina. Only pigeon and ground squirrel reacted for AChE in clearly defined cells
White-crowned sparrow (Zonotrichia leucophrys gambelii)				
Kobayashi and Farner (1964)	Hypothalamus and pituitary gland	Optical: Coupland and Holmes (1957)	In hypothalamus AChE activity strong in cells and proximal part of their axons in supraoptic, paraventricular and infundibular nucleus: weaker in median eminence. No reaction for BuChE except in blood vessels and occasional slight activity in median eminence. In pituitary, weak but definite AChE activity in posterior lobe. BuChE in other lobes, most marked in pars tuberalis and cephalic of anterior lobe	See also Follett et al. (1966) for AChE in relation to MAO. For biochemistry see Russell (1968), table 6.3.a

TABLE 6.3a
BIRDS BIOCHEMISTRY
General

Reference	Species and area	Method/substrate	Result	Comment
Aprison et al. (1964)	Pigeon: major divisions of brain analysed separately. Also liver	Potentiometric: Aprison et al. (1954) ACh. No inhibitor	All activity attributed to AChE which may not be entirely accurate (see Blaber and Cuthbert 1962). Activity in midbrain > cerebellum > telencephalon ≡ pons-medulla. Activity in liver low compared with brain. Absolute values tabulated	ChAc, 5-HTPD and MAO also analysed. Very little relation between ChE and ChAc values in different areas. Ratio in cerebellum > 10,000
Austin and Phillis (1965)	Duck: cerebellum	Manometric: Warburg, ACh, MeCh, BuCh. No inhibitors	Activity towards BuCh greater than that in other species tested. High rate of hydrolysis of ACh, exceeded only by that in cat	Comparison with several species
Goldberg and McCaman	Pigeon: cerebellum	Spectrophotometric: Ellman et al. (1961); AcThCh	Activity greatest in molecular layer. Level higher than in other species tested. Granular layer also very active but white matter and nuclei showed less activity than in other species. See table 6.33	ChAc also measured. Level in molecular layer highest. Comparison with cat, rat, guinea-pig and rabbit
Russell (1968)	White crowned sparrow: hypothalamus and pituitary gland	Spectrophotometric: Ellman et al. (1961); AcThCh, BuThCh and various inhibitors	Appreciable AChE activity in median eminence and anterior pituitary gland, low in posterior pituitary; AChE activity in anterior lobe 50 times greater in *photosensitive than photorefractory birds. On some light/dark regimes AChE activity in median eminence showed daily cycle in photosensitive birds	* Photosensitivity and photo-refractoriness defined in terms of growth of gonads in response to fixed periods of exposure to light
McCaman et al. (1965)	Pigeon: cerebral hemispheres	Spectrophotometric: Ellman et al. (1961); AcThCh. Subcellular fractions	Subcellular distribution shown in table 6.34	ChAc, 5-HTPD and MAO also analysed. Comparison with rat, rabbit and guinea-pig
Namba (1971)	Hen: muscle fibres and end-plates in red and white muscles	Colorimetric: Hestrin (1949); ACh and BuCh. Subcellular fractions	In total homogenate and all subfractions, the ChE activity of red muscle exceeded that of white. 10% of activity of isolated muscle membrane was due to BuChE	Comparison with man, rat, mouse, dog and guinea-pig
Ruckebusch and Ruckebusch (1959)	Hen: blood serum and erythrocytes	Potentiometric: Michel (1949) with ACh	No distinction between AChE and pseudoChE. Activity in erythrocytes lowest of any species examined	Paper in french. Comparison with dog, horse, sheep, cow, goat, pig and rabbit

TABLE 6.3b
BIRDS BIOCHEMISTRY
Specificity studies

Reference	Species and tissue	Method/substrate	Result	Comment
Augustinsson (1948)	Domestic cock: blood and liver	Manometric: Warburg; substrates included ACh, MeCh, BzCh and TB but not BuCh	No ChE detected in erythrocytes. ChE level in serum low but activity towards MeCh exceeded that towards BzCh. High level of 'specific' activity in liver	
Blaber and Cuthbert (1962)	Hen: blood, brain, gut, muscle and amniotic membranes	Manometric: ACh, MeCh, BuCh, PrCh, TB and various inhibitors	No ChE in erythrocytes. Serum enzyme hydrolysed all substrates including MeCh. Greatest activity towards PrCh. Brain enzyme resembled mammalian AChE but another ChE probably present in addition. Enzyme in muscle similar to that in brain, enzyme in gut and amnion similar to that in serum	Specificity of brain enzyme different from that reported by Myers (1953), see ch.10. N.B. Effects of inhibitors differ from those in mammals
Earl and Thompson (1952a)	Hen: brain, spinal cord, peripheral nerve and serum	Manometric: Warburg; ACh, MeCh, BuCh and BzCh with inhibitors	Confirmation of reports that serum enzyme hydrolyses MeCh. Brain had small amount of this enzyme in addition to AChE	Paper predominantly concerned with effects of TOCP on activity of 'pseudoChE'. Some tissues from man, rabbit and rat also examined
Mendel et al. (1943)	Duck, pigeon: blood	Manometric: Warburg; ACh, MeCh and BzCh	No AChE in erythrocytes of pigeon and a trace in duck. MeCh hydrolysed faster than BzCh by duck serum (pigeon not tested)	A number of other species also examined (see ch.10)
Myers (1953)	Duck, pigeon, hen: serum			See ch. 10 re pseudocholinesterase
Whittaker (1953)	Pigeon: brain, also some data for chick brain	Manometric: Warburg; MeCh, TB, amylbutyrate and amylacetate	Specificity pattern suggests brain enzyme to be an acetocholinesterase. In pigeon and chick, activity towards MeCh was in the order: optic lobes > hemispheres > cerebellum > brain stem	Paper includes data on cat, mouse and rabbit. In these mammalian species the level of activity towards MeCh much lower than in avian species

The biology of cholinesterases

TABLE 6.3c
Developmental studies on cholinesterases in birds

Reference	CNS		Peripheral tissue	
	Histochemical	Biochemical	Histochemical	Biochemical
Atherton (1963)	+			
Atsumi (1971)			+	
Bonichon (1957, 1958, 1961, 1962)	+	+		
Burdick and Strittmatter (1965)		+		
Burkhalter et al. (1957a, b)				+
Filogamo (1960)	+			
Giacobini et al. (1970)				+
Iqbal and Talwar (1971)		+		
Jones et al. (1956)				+
Mumenthaler and Engel (1961)			+	
Shen et al. (1956)	+			
Sperti et al. (1960b)		+		
Strumia and Baima-Bollone (1964)	+		+	
Turbow and Burkhalter (1968)		+		
Werner et al. (1971)		+		
Zacks (1954)	+			

TABLE 6.4
CAT HISTOCHEMISTRY CNS

Reference	Area studied	Method	Result	Comment
Abrahams (1963a,b); Abrahams and Edery (1964)	Hippocampus; brain stem including thalamus and hypothalamus	Optical: Holmstedt (1957b)	Many cell groups and some glia AChE + ve. BuChE in fibre tracts and also in some neurones	Some biochemical tests of substrate and inhibitors
Austin et al. (1964)	Cerebellar cortex	Optical: Holmstedt (1957b)	AChE in granular layer and some fibres. Little, if any, BuChE	Comparison with several species
Brightman and Albers (1959)	Brain and spinal cord	Optical: Koelle (1951)	AChE + ve neurones in various regions. BuChE in neuroglia (fibrous astrocytes) of grey and white matter. Endothelium of blood vessels – ve	
Esila (1963)	Retina			See tables 6.32 and 6.32.1 for Esila's own results and other references
Friede (1967)	Brain and spinal cord	Optical: modified Gerebtzoff (1959)	Deals mainly with BuChE. BuChE + ve glia identified as oligodendroglia. In some white matter tracts glia were – ve and axons + ve. Marked neuronal BuChE activity in some nuclei	Comparison with man, monkey and rat. Identification of stained elements is in some respects unorthodox (see ch. 10). See also Roessmann and Friede (1966) re BuChE in oligodendroglia and astrocytes following lesions
Friede and Fleming (1964)	Cerebellum	Optical: Gerebtzoff (1959)	AChE in granular layer and Purkinje cell membrane. No obvious variations in reaction of different folia. BuChE gives diffuse reaction in granular layer. Glia in subcortical white matter more + ve than those situated more deeply	Comparison of distribution in many species
Gwyn and Wolstencroft (1967, 1968)	Medulla oblongata	Optical: Lewis (1961)	AChE demonstrated in some cell groups but paper mainly concerned with AChE and BuChE in capillaries of 'area postrema'	See ch. 8 re blood vessels

TABLE 6.4 (continued)

Reference	Area studied	Method	Result	Comment
Holmes and Wolstencroft (1964)	Medulla oblongata and pons	Optical: Coupland and Holmes (1957)	AChE + ve structures * listed, also − ve areas. No data on BuChE	* See table 6.4.1.
Kása and Csillik (1966)	Cerebellar cortex (also diaphragm)	EM: thiocholine method described in paper	AChE on membranes in glomeruli of granular layer and on a few axons in molecular layer. Capillary endothelium reacted with AcThCh (see Kása et al. 1965 re specificity). BuChE in white matter	Primarily a methodological paper illustrated also by tissues from guinea-pig and rat. See also Kása (1968a)
Kása et al. (1965)	Cerebellum	Optical: Gerebtzoff (1959)	AChE activity confined to the glomeruli of granular layer. Dentate nucleus also + ve. BuChE in fibrous astrocytes (bodies in granular layer, processes in molecular layer)	See ch. 8 re blood vessels. The enzyme in vessels was of unusual specificity
Koelle (1951)	Spinal cord (also many peripheral tissues)	Optical: thiocholine method described in the paper	AChE in lateral and anterior horn cells and in some axons. No BuChE activity detected	This paper supersedes Koelle (1950) in which localization was marred by diffusion artifacts
Koelle and Geesey (1961)	Pituitary	Optical: Koelle (1951)	Moderate AChE activity in fibres in posterior lobe, weak staining of fibres in stalk. Anterior and intermediate lobes − ve. BuChE activity in poorly defined structures at edge of posterior lobe and occasionally at border of anterior lobe	
Krnjević and Silver (1965)	Telencephalon and diencephalon	Optical: Lewis (1961)	In cortex, AChE + ve cells confined mainly to layers V and VI but extensive network of AChE + ve fibres present. BuChE absent from cortex but present in parts of diencephalon	Lesions made to determine polarity of fibres. See table 6.4.2 for other AChE + ve areas. Many pictures
Krnjević and Silver (1966)	Telencephalon and diencephalon	Optical: Lewis (1961)	Paper describes acquisition (and in some cases loss) of AChE activity during foetal and early postnatal development. 'PseudoChE' of unusual specificity present at an early stage	Brief description of monkey cortex. Many pictures

TABLE 6.4 (continued)

Reference	Area studied	Method	Result	Comment
McLennan (1969)	Red nucleus	Optical: Karnovsky and Roots (1964)	Some cells, both large and small, show strong AChE activity but others react weakly or not at all. Neuropil virtually unstained	
Nichols and Koelle (1968)	Retina	Optical: Koelle (1950, 1955) or Koelle and Gromadski (1966)	Diffuse AChE activity throughout plexiform layer. No BuChE activity	Stained amacrine and ganglion cells identifiable during recovery from inhibition by in vivo DFP. Comparison with pigeon, ground squirrel, rabbit and rat
Olivier et al. (1970b)	Diencephalon predominantly	Optical: Gomori (1952)	Striatopallidal and striatonigral fibres strongly stained for AChE. Selective inhibitors not used so significance of activity towards BuThCh uncertain	Lesions made to determine polarity of fibres. Cat compared with monkey (see table 6.22)
Osen and Roth (1969)	Cochlear nucleus	Optical: method essentially that of Couteaux (1951)	An apparently continuous AChE +ve system was traced from neurones in the superior olive, via olivo-cochlear bundle to molecular and granular cell layers of cochlear nuclei. BuChE in glia	Comprehensive paper. Some disagreement with Rasmussen (1967) over distribution of fibres within cochlear nucleus
Papp and Bozsik (1966)	Brain stem	Optical: Gerebtzoff (1959)	AChE in many cell groups (see table 6.4.3 below). Neuropil +ve in a few areas. BuChE in neurones of several nuclear groups. Neuropil +ve in many regions and was particularly strongly stained in some reticular nuclei	Comparison with rabbit. In general, AChE activity was less than in rabbit
Phillis (1968b; see also Phillis 1965a, b)	Cerebellum	Optical: Gerebtzoff (1959); Karnovsky and Roots (1964)	Some reaction for AChE in molecular layer particularly with Karnovsky and Roots method. Strong stain in granular layer, probably in mossy fibre endings. Weak stain in occasional Purkinje cells, majority non-reactive. Fibres in white matter and in peduncles +ve. Large cells in cerebellar nuclei +ve. BuChE not studied	Stain in molecular layer could not be related to structure. Lesions of peduncles used to determine polarity of fibres. Peduncle section or undercutting of cortex resulted in strong reaction for AChE in Purkinje cells

TABLE 6.4. (continued)

Reference	Area studied	Method	Result	Comment
Ramon-Moliner (1972)	Brain stem	Optical: Gomori (1952)	Comprehensive description of brain stem structures reacting towards AcThCh in unfixed tissue. Inhibitors not used. Note: the high pH (6.4) of the medium may have allowed some diffusion	Literature widely reviewed
Rasmussen (1967)	Cochlear nucleus	Optical: Koelle (not specified)	AChE + ve fibres, believed to come from superior olive, were distributed diffusely within the nucleus	cf. Osen and Roth (1969, who found a discrete, rather than diffuse, distribution of + ve fibres in the nucleus
Roessmann and Friede (1967)	Spinal cord	Optical: modified Gerebtzoff (1959)	Distribution of AChE in cells and neuropil of grey matter described for different levels of cord. Authors suggest all neurones react (cf. Silver and Wolstencroft 1971). BuChE scarcely considered	
Shute and Lewis (1966b)	Hypothalamus and subthalamus	Optical: thiocholine (exact method not specified)	Distribution of AChE in cells and neuropil of various nuclei listed below (table 6.4.4)	Results discussed in relation to distribution of mono-aminergic pathways. See also ch. 8
Silver and Wolstencroft (1971)	Spinal cord	Optical: Lewis (1961)	For distribution of AChE in grey matter, see table 6.4.5. BuChE present in white matter, neuropil of ventral horns and in a few scattered neurones	Activity at different levels discussed in relation to Rexed's laminae. Results not entirely in agreement with those of Roessmann and Friede (1967)
Snell (1961)	Brain, spinal cord and dorsal root ganglia	Optical: thiocholine (modified Koelle)	AcThCh used without selective inhibitors. Differentiation between areas of high and low activity clearer for brain stem than cord	Useful for gross distribution of ChE but detail obscured in many cases by over-long incubation

194

TABLE 6.4 (continued)

Reference	Area studied	Method	Result	Comment
		PERIPHERAL NERVES ETC.		
Ballard and Jones (1971; see also Jones and Ballard 1971)	Carotid body	EM; Karnovsky (1964)	No reaction product in cytoplasm of Type 1 or Type II cells. Reaction product on Type II cell membranes behaves as if due to BuChE, and that in axons between glomus cells as if due to AChE	Authors conclude that presence of BuChE round Type II cells indicates cholinergic transmission; cf. Biscoe and Silver (1966)
Biscoe and Silver (1966)	Carotid body	Optical: Lewis (1961)	BuChE activity predominated. AChE and BuChE present in walls of blood vessels, in plexuses associated with glomus cells and vessels, and in occasional ganglion cells. A few cells and fibres within sinus nerve + ve. BuChE in many cells of ganglia associated with IXth nerve; AChE in a few cells and fibres	Section of postganglionic branch of SCG abolished staining except in the ganglion cells in the carotid body
Cauna and Naik (1963)	Various sensory ganglia including dorsal root ganglia	Optical: Koelle (1951)	AChE in cytoplasm of all nerve cells but activity varied from very high to low. BuChE in satellite cells and nerve fibres	Comparison with man, guinea-pig, rat and mole. Prolonged anaesthesia did not influence staining
Cheng (1964a)	End-plates and musculotendinous junctions of extra-ocular and some skeletal muscles	Optical: Koelle (1951); Crevier and Bélanger (1955)	No distinction made between AChE and BuChE at end-plates and tendinous junctions. In spindles both compact endings and endings scattered along the length of intrafusal fibres were present	Paper in Japanese. English summary and figure legends. Comparison with other species
Coupland (1958)	Innervation of pancreas	Optical: Coupland and Holmes (1957), full details in paper	Nerves + ve for AChE and BuChE. Acinar and islet cells − ve. Muscle of arteries and large ducts contained BuChE, as did Pacinian corpuscles	See also Hebb and Hill (1955b) who found BuChE only in Pacinian corpuscles. Rabbit and rat also studied
D'Agostini and Rossatti (1959, 1961)	Lymphoid tissue	Optical: Gerebtzoff (1953)	AChE in blood vessel walls. The vascular network of the germinal centres especially noticeable in inflamed tissues	Comparison with man, rabbit and guinea-pig

TABLE 6.4 (continued)

Reference	Area studied	Method	Result	Comment
El-Badawi and Schenk (1966)	Innervation of the bladder	Optical: modified Karnovsky and Roots (1964)	AChE fibres present in various layers. Distribution described in detail. AChE-containing ganglion cells and also + ve interstitial cells at terminals present	Comparison with dog, rabbit and rat. Noradrenergic fibres also studied; considerable species differences in muscular innervation. No support for Burn and Rand hypothesis
El-Badawi and Schenk (1967a)	Innervation of the epididymis	Optical: modified Karnovsky and Roots (1964)	AChE + ve fibres to vessels and other structures. Distribution varies in different areas. AChE-rich ganglion cells in caput epididymidis	Comparison with dog, rabbit, and rat. Noradrenergic fibres examined too; innervate blood vessels etc. but run independently of AChE + ve fibres
Esterhuizen et al. (1968)	Nictitating membrane	EM: Lewis and Shute (1966)	AChE present in axolemma of a few fibres. AChE + ve and − ve fibres invested in same Schwann cell	Fluorescence studies and [³H]-noradrenaline auto-radiography indicate that adrenergic and AChE-containing fibres are separate. Burn and Rand hypothesis discussed
Fredricsson and Sjöqvist (1961)	Sympathetic, ciliary and nodose ganglia	Optical: Koelle (1951); Holmstedt (1957b)	AChE in preganglionic sympathetic fibres. In most sympathetic cells reaction very weak or moderate but a group of cells in stellate stains strongly. No stained fibres in nodose, cells very weakly + ve. In ciliary ganglion all cells moderately + ve. BuChE in elements (?glia) round cells and nerve fibres	Effects of pre- and post-ganglionic nerve section studied. Strongly reacting cells in stellate associated with cholinergic fibres to sweat glands. See also Sjöqvist (1962); paper mainly about 'cholinergic' sympathetic cells but gives many references to other papers by Swedish group
Garrett (1966a)	Salivary glands	Optical: Gomori (1952) EM: Barrnett and Palade (1958)	ChE associated with innervation. In parasympathetic fibres AChE activity exceeds BuChE; in sympathetic,	See Garrett (1966b) for effects of sympathetic and parasympathetic denervation

196

TABLE 5.4 (continued)

Reference	Area studied	Method	Result	Comment
Garrett et al. (1972)	Innervation of hind gut	Optical: method not specified	Neurones of intermyenteric ganglion gave weak to moderate reaction for AChE. AChE-positive nerves in both circular and longitudinal muscle but density of innervation varied in different regions. BuChE in smooth muscle of external anal sphincter but not in anococcygeus	Adrenergic innervation also studied
Graham et al. (1968)	Innervation of pancreas	EM: Lewis and Shute (1966); combined with autoradiography of [³H]-noradrenaline	Nerve bundles to arterioles may have exclusively adrenergic or exclusively AChE-containing fibres or may contain mixture of discrete 'cholinergic' and adrenergic axons which are juxtaposed	Results refute Burn and Rand hypothesis
Hamberger et al. (1965)	Various sympathetic ganglia; also ciliary and nodose	Optical: Holmstedt (1957b)	Sympathetic ganglia contained varying proportions of fluorescent and non-fluorescent cells. Non-fluorescent cells were identical with AChE-rich cells and were found mainly in the 'sweat secretory' ganglia. They were virtually absent from prevertebral ganglia and parts of the stellate	AChE and fluorescence demonstrated in consecutive sections
Hebb and Hill (1955b; see also Hebb and Hill 1955a)	Pancreas	Optical: Koelle (1950)	AChE activity confined to nerves. BuChE present in Pacinian corpuscles but not elsewhere	Comparison with several other species. Only cat possessed BuChE + ve Pacinian corpuscles; cf. Coupland (1958) who reported additional BuChE + ve sites
Jacobowitz and Koelle (1965)	Vas deferens, uterus, fallopian tubes and nictitating membrane	Optical: Koelle (1955)	AChE + ve fibres sparse in muscle of vas deferens but lamina propria well stained. Extensive network of weakly + ve fibres (corresponding to fluorescent fibres) in fallopian tube and uterus. Lamina propria well stained. Strong BuChE in muscles of fallopian tube. In nictitating membrane AChE + ve fibres sparse and do not correspond to fluorescent fibres. BuChE + ve fibres present	Comparison with rabbit and guinea-pig. Burn and Rand hypothesis discussed in light of species differences. After removal of SCG, AChE + ve fibres in nictitating membrane persist but fluorescence abolished

TABLE 6.4 (continued)

Reference	Area studied	Method	Result	Comment
Koelle (1951)	Liver, ganglia etc	Optical: thiocholine method described in paper	Sites of AChE and BuChE activity described. Presence of AChE in sensory cells and of BuChE in some cells of stellate noted	
Koelle et al. (1971)	Superior cervical ganglion	EM: modifications of disulphide (Koelle and Horn 1968) and bis-(thioacetoxy) aurate (Koelle et al. 1968) methods	AChE predominantly on post-synaptic membranes. Activity weaker or absent on presynaptic membranes. Presents evidence that ganglionic AChE is more soluble than that at end-plates	Results reverse of previous idea that enzyme mainly presynaptic (cf. Koelle 1955; Gromadski and Koelle 1965). Authors suggest that loss of activity after presynaptic section is due to loss of trophic effect of fibres on post-synaptic cells.
Nomura and Schuknecht (1965)	Cochlea	Optical: Gomori (1952)	Cholinesterase staining (AcThCh without inhibitors) used to trace efferent nerves	Man also studied
Novikoff et al. (1966)	Dorsal root ganglia and sciatic nerve	EM: Karnovsky (1964); Karnovsky and Roots (1964)	Reaction for 'AChE'* at axolemma of myelinated fibres. Unmyelinated fibres also reacted with AcThCh but not BuThCh; in other species the reaction on unmyelinated fibres appeared to represent non-specific ChE	Paper mainly on rat. Passing reference to rabbit, mouse, frog and toad. *Inhibitors not tested

TABLE 6.4.1

AChE-containing nuclei in cat medulla and pons. (From Holmes and Wolstencroft 1964.)

Nucleus	Intensity
Superior olive	Strong
Inferior olive	Strong
Gracile nucleus	Strong
Cuneate nucleus	Strong
Medial vestibular nucleus	Strong
Lateral vestibular nucleus	Strong
Nucleus of eminentia teres	Strong
Motor nucleus of Vth	Strong
Motor nucleus of VIth	Strong
Motor nucleus of VIIth	Strong
Motor nucleus of XIIth	Strong
Dorsal motor nucleus of Xth	Strong
Nucleus ambiguus	Strong
Nucleus of raphe	Moderate
Lateral and magnocellular reticular nuclei	Moderate

TABLE 6.4.2

AChE in telencephalon and diencephalon of cat. (Data from Krnjević and Silver 1965.)

Structure	Comment
Neocortex	
Betz cells, layer V	Moderate, superficial staining
Polymorph cells, layer VI	Strong intracellular staining
Extensive system of fibres mainly arising subcortically	Pronounced staining
Rhinencephalon	
Olfactory tubercle	Very intense staining
Septum	Strong staining
Basal ganglia	
Caudate nucleus	Very intense overall staining with some + ve cells
Nucleus accumbens	Very intense overall staining with some + ve cells
Putamen	Very intense overall staining with some + ve cells
Globus pallidus	Less intense overall staining but more cells stain than in striatum
Claustrum	Weak staining
Amygdala	
Lateral nucleus	Very weak staining
Central nucleus	Strong staining
Basal nucleus	Strong staining especially in magnocellular part
Diencephalon	
Habenula and habenulointerpeduncular tract	Very intense staining
Anteroventral thalamic nuclei	'Substantial' staining
Lateral geniculate body	'Substantial' staining
Thalamic reticular nucleus	'Substantial' staining
Ventrolateral thalamus	Weak staining
Medial geniculate body	Weak staining
Medial lemniscus	Weak staining
Subthalamic nucleus	Weak staining
Supraoptic nucleus	Strong staining
Paraventricular nucleus	Strong staining
Preoptic region of hypothalamus	Strong staining
Substantia nigra	Strong staining

TABLE 6.4.3

The intensity of AChE and pseudoChE activity in cell groups in the reticular formation of the lower brain stem of cat compared with rabbit. (From Papp and Bozsik 1966.)

	Cat			*Rabbit
	AChE	BuChE		AChE
	nerve cells	nerve cells	neuropil	nerve cells
N. medullae oblongatae centralis				
subnucleus ventralis	0–++	0	++	+–++
subnucleus dorsalis	0–+	0	+	+
N. parvocellularis	0–+	0	+	+
N. gigantocellularis	0–+++	0	+	+–+++
N. paragigantocellularis dorsalis	0–++	0	++	+–++
N. paragigantocellularis lateralis	0–+++	0	++	+–+++
N. pontis centralis caudalis	0–+++	0	+–++	++–++++
N. pontis centralis oralis	0–+++	0	+–++	+–++++
Griseum centrale pontis	+–+++	0–+	+	+–+++
N. raphe obscurus	++	0	0–+	++
N. raphe pallidus	++	0	0–+	++
N. raphe magnus	++	0	0–+	++
N. raphe pontis	++–+++	0	+–++	++–+++
N. raphe dorsalis	++–+++	0–+	+–++	++–+++
N. centralis superior	++–+++	0	+	++–+++
N. paramedium reticularis	+++	0	++	+++
N. lateralis reticularis	+++	0	+++	+++
N. reticularis tegmenti pontis	+++	0–+	+++	+++
N. interpeduncularis	++++	+	+	++++
N. cuneiformis	+–++	0–+	+	+–++
N. subcuneiformis	+–++	0–+	+	+–++
Griseum centrale mesencephali				
pars ventralis	++–+++	0	+–++	++–+++
pars lateralis	+	0	+	++
N. tegmenti pedunculopontinus	+–++	0	+	++–+++

The sections were of 60 μ thickness and were incubated for 2 hr at pH 6.5 for AChE or pH 5.0 for BuChE. Cell group terminology according to Taber (1961).

* Some pseudocholinesterase activity was present but was too weak to evaluate.

TABLE 6.4.4
AChE activity in the cat hypothalamus and subthalamus. (From Shute and Lewis 1966b.)

Areas with AChE-containing cells
        Dorsal hypothalamic area (stained cells form discrete islands)
        Interstitial nucleus of supraoptic decussation
        Lateral hypothalamic area
        Lateral preoptic area
        Posterior hypothalamic area
        Subthalamic area (stained cells form discrete islands)
        Supramamillary area
Areas with AChE-containing terminals or fibres of passage
        Anterior hypothalamic area
        Dorsal hypothalamic area
        Dorsomedial nucleus
        Lateral hypothalamic area
        Medial and lateral mamillary nuclei
        Medial preoptic nucleus
        Nucleus of H_1 field of Forel
        Paraventricular nucleus
        Paraventricular thalamic nucleus
        Premamillary nucleus
        Posterior hypothalamic nucleus
        Stria terminalis
        Subthalamic nucleus of Luys
        Supramamillary nucleus
        Supraoptic nucleus
        Ventromedial nucleus
Areas with AChE-containing tracts
        Fibres from paraventricular thalamic nucleus
        Fibres from subthalamic nucleus of Luys
        Fornix
        Supraoptic decussation

TABLE 6.4.5

AChE activity in the cat spinal cord. (From Silver and Wolstencroft 1971.)

Strong staining
    *Nuclei
        Lateral cervical
        Central dorsal
        Spinal accessory and other somatic motor nuclei
        Spinal reticular
        Commissural
        Intermediomedial
        Intermediolateral
    Scattered cells
        On surface of lamina I
        In laminae IV, V and VI
        In area of ventral horn from which arises ventral spinocerebellar tract
        In Clarke's columns among larger, weakly reacting cells
        In white matter
    Fibres
        Spinal accessory and other motor nerves
        Fibres connecting contralateral intermediomedial nuclei
        Fibres connecting intermediomedial and - lateral nuclei ipsilaterally
        Commissural fibres
        Some fibres in Lissauer's tract
        Some fibres in dorsal columns
    Neuropil
        Laminae III, IV, V and VI
Moderate staining
        Some cells in Clarke's columns
        Neuropil of laminae II, VII and VIII
Weak or negligible staining
        Central cervical nucleus
        Neuropil and majority of cells in Clarke's columns
        Neuropil of lamina I
        Neuropil of lamina X (no reaction)
        Most of white matter

* In most nuclei both cells and neuropil reacted strongly.

TABLE 6.5

CAT BIOCHEMISTRY

CNS and peripheral tissues

Reference	Area studied	Method/substrate	Result	Comment
Austin and Phillis (1965)	Cerebellum	Manometric: ACh, MeCh and BuCh. No inhibitors. Studies included subcellular fractionation	Activity towards ACh almost equal in molecular and granular layers. White matter, peduncles and nuclei also analysed. Subcellular fractionation: microsomal fraction had highest specific activity relative to total homogenate. Effect of peduncular section measured, also time-course of inhibition by formaldehyde	Note: values for peduncles and nuclei expressed per g wet wt, other values per mg protein. Comparison with several other species. See also Phillis (1965b)
Chu et al. (1971a)	Cerebral cortex (intact and undercut)	Spectrophotometric: Ellman et al. (1961); AcThCh. Measurements all made on subcellular fractions of intact or undercut cortex	Undercutting reduced AChE in fractions rich in synaptic membranes. Decrease prevented by electrical stimulation of slab	Study primarily directed at mechanism underlying supersensitivity of isolated cortical slabs. See also 1) Duncan et al. (1968) for similar studies but on non-fractionated tissues, 2) Green et al. (1970) who also measured ChAc
Goldberg and McCaman (1967)	Cerebellum	Spectrophotometric: Ellman et al. (1961); AcThCh	AChE in granular layer greater than that in molecular layer (cf. Austin and Phillis 1965). White matter and nuclei also measured. See table 6.33	Comparison with guinea-pig, rabbit, rat and pigeon. Conclude no correlation between AChE and ChAc activity
Holmstedt et al. (1963)	Sympathetic ganglia: superior cervical (SCG), stellate, paravertebral, coeliac and mesenteric	Titrimetric: details in paper. MeCh	AChE activity high in SCG, stellate, 7th lumbar and 1st sacral, lower in coeliac and mesenteric. After denervation 50% of AChE activity retained in L_7 and S_1 but only 20% in SCG	Paper includes some histochemistry on both intact and denervated ganglia
Kahlson and Renvall (1956)	Salivary glands, liver, gastric mucosa and small intestine	Manometric: Warburg; ACh, MeCh and BuCh. Homogenate and supernatant assayed	In salivary glands AChE predominates. In liver entirely, and in gastric mucosa, almost entirely, BuChE. Intestinal mucosa rich in both enzymes	Hypophysectomy and adrenalectomy increased AChE and BuChE in salivary glands. Liver and gastric mucosa not affected by either procedure, but enzymes in intestinal mucosa increased by adrenalectomy and de-

TABLE 6.5 (continued)

Reference	Area studied	Method/substrate	Result	Comment
Klingman et al. (1968)	Superior cervical ganglia, cervical sympathetic trunk, caudate nucleus	Spectrophotometric: Ellman et al. (1961); AcThCh plus BW284C51 and ethopropazine as inhibitors	Gives values for AChE and pseudoChE. In SCG and sympathetic trunk AChE levels similar to those in rat but pseudoChE lower. See table 6.27.2	Paper mainly on rat. Cat included for comparison
Pecot-Dechavassine (1962)	Anterior tibial, soleus and diaphragm muscle	Manometric: Warburg; ACh and BuCh	See table 6.35	Comparison with various species
Sawyer and Hollinshead (1945)	Sympathetic ganglia and fibres; also vagus and sciatic nerves, dorsal root ganglion and medulla	Microtitrimetric: see Sawyer (1943b); ACh, MeCh and BzCh	Values given for intact tissue and for ganglion and preganglionic fibres after section of cervical sympathetic trunk. Evidence that much of AChE, but not pseudoChE, in ganglion is in preganglionic fibres	Some observations on SCG and coeliac ganglion of guinea-pig
Strömblad (1957)	Salivary glands (intact and denervated)	Manometric: Warburg; ACh, MeCh and BuCh	Activity towards MeCh in submaxillary glands about twice that in parotid	Study mainly aimed at investigating mechanism of denervation supersensitivity. Includes effects of denervation and anticholinesterases
Utley (1966)	Medial geniculate body, retina and other parts of visual system; also parts of motor and somatic sensory system	Manometric: Warburg; MeCh and BuCh	See table 6.5.1	Retrograde degeneration technique demonstrates pre- and post-synaptic localization of enzyme in medial geniculate body

TABLE 6.5.1

AChE and BuChE in visual, somatic sensory and motor systems of the cat. (From Utley 1966.)

	AChE		BuChE			AChE		BuChE	
Dorsal	3.15		3.33		Retina	2.61		0.47	
columns	2.23	*2.52*	3.40	*2.93*		2.12	*1.92*	0.55	*0.34*
	1.70		2.05			1.04		0.01	
Cuneate	7.20		6.58		Optic	0.11		2.30	
nuclei	7.95	*6.79*	6.41	*5.83*	tract	0.57	*0.35*	1.84	*1.74*
	5.24		4.50			0.39		1.19	
Ventral	1.82		4.42		Superior	6.23		6.35	
posterior	1.23	*1.87*	3.93	*4.50*	colliculus	6.03	*6.17*	3.39	*5.11*
nucleus	2.79		3.80			6.27		5.60	
Somatic	1.06		3.80		Lateral	1.92		7.39	
radiation	0.22	*0.93*	3.64	*3.87*	geniculate	2.97	*2.87*	3.59	*5.13*
	1.81		4.19		body	3.74		4.43	
Somatic	1.68		1.11		Visual	0.25		4.16	
cortex	0.85	*1.16*	1.36	*1.28*	radiation	0.93	*0.53*	2.33	*3.09*
	1.16		1.38			0.41		2.79	
Motor	1.75		1.24		Visual	0.81		1.63	
cortex	1.40	*1.44*	1.27	*1.11*	cortex	1.01	*0.87*	0.54	*0.97*
	1.46		0.84			0.81		0.74	
Pyramidal	1.01		3.70						
tract	0.00	*0.42*	3.09	*3.50*					
	0.63		2.38						

Values are in μl CO_2/mg/hr. The italicized figures are the averages for the brain areas.

TABLE 6.6
COW HISTOCHEMISTRY
CNS

Reference	Area studied	Method	Result	Comment
Arvy (1961)	Pineal gland	Optical: method not specified	No AChE activity. BuChE very feeble	Paper in French. Comparison with pig and sheep. See also Arvy (1965)
Esila (1963)	Retina			See table 6.32 and 6.32.1 for Esila's own results and other references
Friede and Fleming (1964)	Cerebellum	Optical: Koelle (1954); Gerebtzoff (1959)	AChE weak in molecular layer; in granular layer activity varied from weak to very strong in different folia. Purkinje cell membrane + ve. BuChE in molecular layer of all folia. Bergmann glia stained, also perikarya of a few Purkinje cells	Comparison with several other species

PERIPHERAL NERVES ETC.

Reference	Area studied	Method	Result	Comment
Cheng (1964a)	Extraocular and retractor bulbi muscle	Optical: Koelle (1951); Crevier and Bélanger (1955)	No distinction between AChE and BuChE. ChE at end-plates, grape-like endings and musculotendinous junctions in extraocular muscles. Scattered and compact endings on intrafusal fibres. No grape-like endings on retractor bulbi fibres	In Japanese but with English summary and figure legends. Comparison with several other species
Klinge et al. (1970)	Innervation of re-tractor penis muscle	Optical: Karnovsky and Roots (1964)	AChE present in nerves. Some, but not complete, correspondence with fluorescent fibres	Fluorescence studies form main part of paper. See also Klinge and Pohto (1971). Results discussed in terms of Burn and Rand hypothesis

TABLE 6.6 (continued)

Reference	Area studied	Method	Result	Comment
Mann (1971)	Innervation of bronchial muscles	Optical: Lewis (1961; see Hebb and Silver 1970)	AChE + ve fibres in smooth muscle and peripheral bronchial ganglia. Some large nerve bundles and small cells + ve for AChE and noradrenaline. BuChE in smooth muscle and in ganglia	Young bull calves used. Comparison with rabbit, sheep, piglet and goat. Distribution of catecholamines studied too
Matsuura (1967)	Spinal ganglia	Optical: Koelle (no details)	AChE activity in nerve fibres 'slight', and negligible in capsular cells. In nerve cells depth and distribution of stain varied but all showed some activity	Other enzymes studied. Concentration of DFP used to inhibit BuChE was high (10^{-6} M) hence AChE could have been somewhat depressed

TABLE 6.7
COW BIOCHEMISTRY
CNS and peripheral tissue

Reference	Area studied	Method/substrate	Result	Comment
Anfinsen (1944)	Retina (pure rod)	Microtitrimetric: Glick (1938). Analysis of thin sections. ACh as substrate	Cholinesterase activity highest in ganglion cell layer and inner and outer plexiform layers. Others, including nerve fibre layer, had some activity	
Augustinsson (1948)	Blood, liver and kidney	Manometric: Warburg; substrates included ACh, MeCh, BzCh and TB but not BuCh. Various inhibitors	High level of AChE in erythrocytes but serum almost inactive towards the cholinesters and TB. ChE low in liver, activity attributed to AChE on grounds that MeCh was split faster than BzCh. In kidney, BzCh was split faster than ACh. High activity towards ASaCh and TB	Various species compared. Comprehensive data but note that BuCh was not tested. Cow blood used in a variety of tests on inhibitors and enzyme kinetics etc.
Banister et al. (1953)	Spleen	Manometric: Warburg; with ACh, PrCh, MeCh, BuCh, and BzCh	Activity greatest towards BuCh, least towards BzCh	Authors do not pursue question of whether specificity pattern is due to more than one enzyme
Esila (1963)	Retina	Starch gel electrophoresis		See table 6.32.2
La Bella and Shin (1968)	Anterior and posterior pituitary, pineal and whole brain	Colorimetric: McOsker and Daniel (1959); AcThCh, BuThCh	AChE levels, particularly in anterior pituitary, very low compared with whole brain. BuChE activity was low and of same order in all tissues	ChAc measured too. See also La Bella (1968)
Ruckebusch and Ruckebusch (1959)	Blood serum and erythrocytes	Potentiometric: Michel (1949) with ACh	No distinction between AChE and BuChE. ChE activity in erythrocytes high and in serum low	Paper in French. Comparison with dog, horse, sheep, goat, pig, rabbit and fowl
Williams and Cooper (1965)	Corneal epithelium and stroma	Spectrophotometric: Ellman et al. (1961); AcThCh. No inhibitors	In epithelium ChE activity twice that in stroma. Endothelium inactive	ACh level very high. Relatively low ChE activity surprising, see ch. 8 § 3.1.5

209

TABLE 6.8
DOG HISTOCHEMISTRY
CNS

Reference	Area studied	Method	Result	Comment
Abrahams et al. (1957)	Hypothalamus	Optical: Koelle (1951)	AChE confined to cells of supraoptic, paraventricular and suprachiasmatic nuclei. Neuropil − ve. BuChE in neuropil of the paraventricular and the supraoptic nuclei	Presence of AChE in suprachiasmatic cells but not neuropil cells is converse of situation in guinea-pig (see Cottle and Silver 1970b)
Esila (1963)	Retina			See table 6.32.1
Hard and Peterson (1950)	Spinal cord, brain stem (also sympathetic and dorsal root ganglia)	Optical: Gomori (1948); long-chain fatty acid method	In CNS, activity in neuropil rather than neurones. White matter − ve except for glia. In ganglia pronounced activity in capsular cells	Method not very specific for ChE. Ligation of sciatic nerve abolished staining in ipsilateral ventral horn; slight reduction contralaterally
Phillis (1965b)	Cerebellum	Optical: Gerebtzoff (1959)	Marked AChE activity in molecular and granule cell layer; also stain in middle peduncle	Comparison with several other species. Dog considered only briefly
			PERIPHERAL NERVES ETC.	
Aoki (1964)	Sweat glands in toe pad	Optical: Gomori (1952)	AChE and BuChE in nerves to glands. Abundant BuChE in secretory cells. End-product also seen in lumen of sweat ducts suggesting enzyme discharged into sweat. Myoepithelial cells − ve	Author points out that dog resembles marmoset and platypus in having ChE in secretory cells. In man and most other species studied the cells are − ve
Bell and McLean (1970)	Innervation of retractor penis muscle and vas deferens	Optical: Karnovsky and Roots (1964)	Distribution of AChE + ve fibres described. Reaction abolished by BW284C51 but BuThCh not tested	Pattern of distribution of adrenergic and AChE-containing fibres indicated two separate innervations
Cheng (1964a)	Extraocular and retractor bulbi muscles	Optical: Koelle (1951); Crevier and Bélanger (1955)	No distinction between AChE and BuChE. ChE at end-plates, grape-like endings and musculotendinous junctions of extraocular muscles. No grape-like endings in retractor bulbi muscles	In Japanese but with English summary and figure legends. Comparison with several other species

TABLE 6.8 (continued)

Reference	Area studied	Method	Result	Comment
El-Badawi and Schenk (1966)	Innervation of urinary bladder	Optical: modified Karnovsky and Roots (1964)	Profuse network of AChE-containing nerves and ganglia in muscle. Stained fibres in epithelium and subepithelium sparse compared with those of cat	Comparison with cat, rabbit and rat. Noradrenergic fibres studied too; muscular innervation shows considerable species differences. No support for Burn and Rand's hypothesis
El-Badawi and Schenk (1967a)	Innervation of epididymis	Optical: modified Karnovsky and Roots (1964)	AChE + ve fibres form rich plexuses round tunical, septal and interstitial blood vessels	Comparison with cat, rat and rabbit. Noradrenergic fibres examined too; innervate blood vessels but run independently of AChE + ve fibres
Hebb and Hill (1955b)	Pancreas and pancreatic juice	Optical: Koelle (1950). For juice the histochemical medium was used and production of a precipitate noted. Also, ACh hydrolysis (bioassay)	AChE confined to nerves. BuChE in acinar and Islet cells. Juice contained BuChE	Comparison with several species. Only in dog did glandular tissue and juice contain BuChE
Serafini-Fracassini and Frasson (1966)	Carotid body	Optical: Gerebtzoff (1959)	AChE in filamentous network on surface of cells. Without ethopropazine AcThCh gave a reaction in cytoplasm as well	Succinic dehydrogenase examined

TABLE 6.9
DOG BIOCHEMISTRY
CNS and peripheral tissue

Reference	Area studied	Method/substrate	Result	Comment
Augustinsson (1948)	Brain and muscle	Manometric: Warburg; ACh, MeCh, BzCh, TB etc	Activity towards substrates tabulated. ChE; low in striated muscle; relatively high activity towards BzCh attributed to a non-specific esterase	Various kinetic studies etc
Austin and Phillis (1965)	Cerebellum	Manometric: Warburg; ACh, MeCh, BuCh. No inhibitors	ChE activity is low compared with most other species examined	Comparison with several species
Burgen and Chipman (1951)	Brain and spinal cord	Manometric: Warburg; (also bioassay of residual substrate as a check) ACh, MeCh and BzCh	See table 6.9.1	Succinic dehydrogenase also measured
Esila (1963)	Retina	Starch gel electrophoresis		See table 6.32.2
Namba (1971)	Muscle fibres and end-plates in intercostal and other muscles	Colorimetric: Hestrin (1949); ACh and BuCh, subcellular fractions	ChE activity per end-plate lowest of the 6 species studied (approx.20% of that in rat). Maximum activity in microsomal fraction	Comparison with man, mouse, guinea-pig, rat and chicken
Ruckebusch and Ruckebusch (1959)	Whole blood serum and erythrocytes	Potentiometric: Michel (1949); ACh	Total ChE in plasma high, approximately twice that in erythrocytes	Paper in French. Comparison with horse, cow, sheep, goat, pig, rabbit and fowl
Vlk and Tuček (1962b)	Cardiac atria	Colorimetric: Hestrin (1949); ACh, MeCh and BzCh	During postnatal period of 1–80 days both AChE and BuChE decrease in all parts of the atria	Changes in ACh content also followed. This increases during the period when ChE is falling. For results on adult dog, also cat, rabbit and rat heart see Vlk and Tuček (1962a)

TABLE 6.9.1
Cholinesterase activity in different regions of dog brain and spinal cord.
(From Burgen and Chipman 1951.)

Area	(1) Q MeCh ± S.D.	(2) Q Benz	(3) Q ACh	(4) Q MeCh / Q Benz	(5) X Activity due to CHEII (%)
Proreal gyrus 8	171 ± 29 (5)*	—	331	—	—
Precentral gyrus 6α	237 ± 38 (4)	35	371	6.77	9
Postcentral gyrus 4α	178 ± 41 (6)	42	330	4.23	14
Primary sensory cortex 3	150 ± 25 (5)	—	282	—	—
Sylvian gyrus 52	230 ± 42 (6)	32	407	7.18	9
Postsplenial gyrus 17	107 ± 34 (5)	—	—	—	—
Suprasplenial gyrus 18, 19	114 ± 12 (5)	27	232	4.23	14
Posterior suprasylvian gyrus 20	162 ± 81 (4)	46	—	3.52	17
Cingulate gyrus 23, 24	203 ± 51 (5)	32	377	6.34	10
Hippocampal gyrus 28	238 ± 52 (3)	49	—	5.12	12
Uncinate gyrus 51	466 ± 148 (8)	91	1372	5.75	11
Cerebellar hemisphere	1075 ± 155 (7)	24	1864	44.8	2
Cerebellar vermis	1228 ± 489 (3)	—	—	—	—
Cerebellar flocculus	931 ± 350 (3)	—	—	—	—
Cerebellar anterior lobe	1756 ± 670 (3)	—	—	—	—
Cerebellar dentate nucleus	530 ± 300 (5)	—	—	—	—
Cerebellar superior peduncle	333 ± 97 (5)	90	654	3.70	16
Cerebellar middle peduncle	243 ± 68 (5)	50	582	4.85	13
Cerebellar inferior peduncle	294 ± 56 (5)	—	—	—	—
Caudate nucleus, head	3936 ± 396 (9)	360	7450	10.9	6
Thalamus, dorso-lateral nucleus	409 ± 97 (7)	161	808	2.54	21
Thalamus, massa intermedia	600 ± 94 (5)	194	1437	3.09	18
Lentiform nucleus	2606 ± 1200 (7)	318	4992	8.20	8
Hypothalamus	323 ± 37 (5)	358	866	0.90	43
Red nucleus	452 ± 162 (5)	—	1269	—	—
Periaquaductal grey matter	537 ± 205 (3)	—	—	—	—
Olfactory bulb	197 ± 27 (3)	—	—	—	—
Fornix	50 ± 22 (3)	69	—	0.73	49
Optic nerve	11 ± 7 (5)	222	283	0.05	93
Optic tract	86 ± 49 (6)	76	237	1.13	38
Lateral geniculate	230 ± 57 (4)	—	708	—	—
Medial geniculate	316 ± 86 (4)	—	938	—	—
Superior corpus quadrigeminum	932 ± 195 (6)	159	1619	5.89	11
Inferior corpus quadrigeminum	364 ± 73 (5)	184	811	1.98	26
** Corpus callosum	16 ± 13 (7)	27	62	0.59	54
* Subcortical white matter	10 ± 5 (6)	19	53	0.53	57
Pes pedunculi	84 ± 24 (4)	114	237	0.74	49
Medullary pyramids	82 ± 41 (6)	94	177	0.86	41
Lateral spinal columns	50 ± 20 (5)	36	—	1.39	33
Spinal grey matter	611 ± 132 (6)	218	—	2.80	20
Anterior spinal root	149 ± 57 (7)	20	212	7.45	8
Posterior spinal root	34 ± 21 (7)	20	82	1.70	29

TABLE 6.9.1 (continued)

Area	(1) Q MeCh ± S.D.	(2) Q Benz	(3) Q ACh	(4) $\dfrac{\text{Q MeCh}}{\text{Q Benz}}$	(5) [x] Activity due to CHEII (%)
Posterior columns	36 ± 7 (5)	39	–	0.92	43
Nucleus gracilis and cuneatus	477 ± 91 (4)	–	993	–	–
Pituitary posterior lobe	50 ± 34 (3)	–	252	–	–

Q MeCh, Q Benz and Q ACh = μl CO_2 evolved/g wet wt of tissue/10 min with respectively acetyl-β-methylcholine, benzoylcholine and acetylcholine as substrates.

 * No significant difference was found between frontal and occipital white matter
** No significant difference was found between different parts of the corpus callosum
 † Number of animals used for values in column 1
 x % of the total hydrolysis of ACh that is due to ChEII (pseudo ChE) is described thus:
   59 × Q Benz/0.85 Q MeCh + 0.59 Q Benz

TABLE 6.10
FISH HISTOCHEMISTRY
CNS

Reference	Species	Area	Method	Result	Comment
Brightman and Albers (1959)	Goldfish (*Carassius auratus*)	Brain and spinal cord	Optical: Koelle (1951)	AChE in some neurones. Walls of blood vessels contain both AChE and BuChE. Neuroglia non-reactive	Various species of animals compared
Esila (1963)	Angel fish Gambusi Golden bream Goldfish (*Carassius auratus*) Minnow Spotted herring	Retina			See table 6.32
Luppa et al. (1968); see also Luppa and Feustel (1966)	Carp (*Cyprinus carpio*)	Neurosecretory system in spinal cord	Optical: Gomori (1952); Karnovsky and Roots (1964)	Neurosecretory cells have variable AChE activity. Fibres to filum terminale also + ve	Papers in German with English summaries. Other enzymes studied; none affected by salt loading
Nichols et al. (1972)	Carp (*Cyprinus carpio*) Goldfish *Carassius auratus* Trout (*Salvelinus fontinalis*)	Retina	Optical: Koelle (1955); Nichols and Koelle (1968)	AChE in amacrine cells of all species, also in unidentified cells of inner nuclear layer and in ganglion cells but not their axons. No BuChE	Results discussed in relation to function
			PERIPHERAL TISSUES		
Couceiro et al. (1953)	Electric eel (*Electrophorus electricus*)	Electric organ	Optical: Koelle and Friedenwald (1949); Koelle (1951). **Naphthyl** acetate diazo-coupling method also used	Hydrolysis of AcThCh and naphthyl acetate demonstrated at innervated face of electroplaque. No reaction for ali-esterases	Paper in French. Histochemical reaction not prevented by inhibitors (including eserine), possibly due to poor penetration

TABLE 6.10 (continued)

Reference	Species	Area	Method	Result	Comment
Couteaux and Szabo (1959)	Gnathonemus	Electric organ	Optical: Couteaux and Taxi (1952)	AChE in aneural part only	cf. **Pecot-Dechavassine** (1962) below
Lundin (1962)	65 Swedish salt- and fresh-water fish	Muscles	Optical: Holmstedt (1957a, b)	AChE at end-plates. Salt-water, but not fresh-water, fish show diffuse activity (inhibited by BW284C51) on cell surface. In both types, reaction towards AcThCh of atypical specificity occurs at striations. BuThCh not tested	See table 6.11 re biochemical features
Lundin and Hellström (1968)	Plaice *(Pleuronectes platessa)*	Muscle	EM: Karnovsky (1964)	AcThCh and **BuThCh** hydrolysed to same extent on sarcolemmal surface. End-plates not described	Authors point out that localization is affected by pH etc. Danger of artifacts. For collected papers on plaice see Lundin (1968)
Pecot-Dechavassine (1962)	Catfish *(Ameiurus nebulosus)*	Lateral superficial and inferior muscle	Optical: Couteaux and Taxi (1952)	Neuromuscular and myotendinous junctions react for AChE. Very low activity towards BuThCh. Fast and slow muscle compared	Paper in French. Includes biochemical studies; also pharmacological study on catfish muscles. Author discusses possibility that hydrolysis of BuThCh is due to AChE and not BuChE. Comparative studies on other animals
	Electric catfish *(Malapterurus electricus)*	Electric organ		AChE activity greater on innervated than on aneural face. BuThCh gives diffuse stain	
	Gnathonemus *(Gnathonemus senegalensis elongattus)*	Electric organ		AChE activity strong on both innervated and aneural faces. Activity towards BuThCh less marked	

TABLE 6.11
FISH BIOCHEMISTRY
CNS and peripheral tissue

Reference	Species	Area	Method/substrate

Augustinsson (1948)	Cod (*Gadus callarias*)
	  Ballan wrasse (*Labrus berggylta*)
	  Dogfish shark (*Scylliorhinus canicula*)
	  Spiny dogfish shark (*Squalus acanthias*)
	  Common ray (*Raja radiata*)
	  Hagfish (*Myxine glutinosa*)
		Brain, also for some species, liver, muscle, blood, heart and air bladder	Manometric and titrimetric with ACh, ASaCh, MeCh and BzCh

Clos et al. (1957)	Tench (*Tinca tinca*)
	  Carp (*Cyprinus carpio*)
	  Eel (*Anguilla anguilla*)
	  Salmon (*Salmo irideus*)
		Brain, blood, spleen, liver, gut, kidney, ovary and testis	Manometric : Warburg; ACh, MeCh and BzCh

Clos and Serfaty (1958)	Perch (*Perca fluviatilis*) (results compared with those obtained in preceding paper)	Brain, muscle, liver, stomach and gut	As above
Dupé and Bockelée-Morvan (1968)	African lung fish (*Protopterus annectens*)	Brain (also whole blood, cells and serum)	Manometric : Warburg; ACh, MeCh and BuCh
Lüdtke and Ohnesorge (1966)	Tench (*Tinca vulgaris*)	Brain (also muscle and stomach)	Manometric : Warburg; ACh, PrCh, BuCh and BzCh with eserine and neostigmine
Lundin (1962)	65 Swedish salt- and fresh-water fish	Muscles	Electrometric: Tammelin (1953) with ACh, MeCh, BuCh and inhibitors
Nachmansohn et al. (1941)	Electric eel (*Electrophorus electricus*) and Goldfish (*Carassius auratus*)	Electric organ of eel; various areas of brain and spinal cord of eel and goldfish	Manometric : Warburg; ACh
Pecot-Dechavassine (1962)	Catfish (*Ameiurus nebulosus*); Electric catfish (*Malapterurus electricus; Gnathonemus senegalensis elongatus*)	Muscles electric organs	Manometric : Warburg; ACh, BuCh with various inhibitors
Tibbs (1960)	Trout (*Salmo trutto*) and Perch (*Perca fluviatilis*)	Spermatozoa	Colorimetric : Hestrin (1949); ACh, BuCh, MeCh, BzCh. Eserine

Result	Comment
Main finding is apparent absence of pseudo-ChE from all tissues. Cod has highest and hagfish lowest AChE activity in brain. Muscles more active than those of most mammals. Activity in liver varies widely with species. Erythrocytes lack AChE	Monograph contains many tables with comparisons of various animals, tissues and substrates. Also references to earlier work. See also Augustinsson (1959b, c) re probable presence of AChE in serum
No AChE in erythrocytes. Serum said to contain pseudoChE (see comment)*. Total ChE activity greatest in brain, least in gonads. Activity in liver higher in omnivores than carnivores	Paper in French. *Activity towards ACh and tributyrin in serum attributed to pseudoChE but possibility of nonspecific esterase not eliminated (no inhibitors used). See also Augustinsson (1959b, c) re serum (AChE)
Both AChE and pseudoChE reported in all tissues of perch. PseudoChE in perch higher than in the species in preceding paper	Paper in French. No inhibitors used
ACh hydrolysed very strongly by brain and erythrocytes. No measurable activity towards BuCh with any of the samples	Paper in French. Hydrolysis of ACh by erythrocytes contrasts with lack of activity of erythrocytes from teleosts and elasmobranchs. See also Dupé (1967) who found that total activity of telencephalon towards ACh shows seasonal changes. Low activity towards MeCh does not vary with season
ChE in all tissues more specific for ACh, but more resistant to inhibitors than that of rabbit. Overall levels lower than in rabbit. K_M higher. Temperature characteristics not altered by adapting fish to different temperatures	Paper in German, English summary. Comparison with rabbit
AChE activity per unit weight inversely related to length of fish. Muscles of salt-water, but not fresh-water, fish active towards BuCh	Author discusses nature of BuCh-splitting enzyme of salt-water fish; it is atypical in some respects. See table 6.10 for histochemistry
Absolute value for ChE in electric organ differed by a factor of 10 in the two eels tested but in both was high compared with that in CNS. CNS values comparable to those in goldfish	ChE activity parallels the number of electroplax per cm. See also Augustinsson (1955b); Karlin (1965)
See table 6.35	Author discusses possibility that hydrolysis of BuCh may be due to AChE. Various species compared. For histochemistry see table 6.10
AChE present in spermatozoa of both fish. Additional analyses from perch showed enzyme was mainly in the head. Seminal plasma also contained some AChE	See ch. 8 § 4 re functional significance

TABLE 6.12
GUINEA-PIG HISTOCHEMISTRY
CNS

Reference	Area studied	Method	Result	Comment
Cottle and Silver (1970a, b)	Hypothalamus	Optical: Lewis (1961)	AChE in cells and/or neuropil at sites given in table 6.12.1; also in some blood vessels. BuChE predominantly in white matter but also in unidentified globules in the median eminence and occasionally in neurones near IIIrd ventricle	Discussion of possible relation between AChE activity and reproductive state
Esila (1963)	Retina			See table 6.32 and 6.32.1 for Esila's own results and other references
Friede and Fleming (1964)	Cerebellum	Optical: Koelle (1954); Gerebtzoff (1959)	AChE activity strong in both molecular and granular layer. Purkinje cell membrane + ve. BuChE in granular layer and glia of white matter	Comparison with other species
Geneser-Jensen and Blackstad (1971)	Hippocampal region: Entorhinal area, parasubiculum and presubiculum	Optical: modified Koelle	This paper, together with the two below, provides a detailed map of the layered distribution of AChE in different parts of the hippocampus	These studies reveal that the distribution of AChE differs considerably from that in the rat hippocampus (see Storm-Mathisen and Blackstad 1964)
Geneser-Jensen (1972a)	Hippocampal region: Subiculum and hippocampus	Optical: modified Koelle	See above	
Geneser-Jensen (1972b)	Hippocampal region: Dentate area	Optical: modified Koelle	See above	

TABLE 6.12 (continued)

Reference	Area studied	Method	Result	Comment
Hall and Geneser-Jensen (1971)	Amygdala	Optical: modified Koelle	AChE activity varied considerably in the different nuclear subdivisions. The large-celled part of basal complex stained very strongly but the central and medial nuclei gave little reaction	Species differences discussed. MAΘ also studied
Ishii (1957a)	Brain and anterior horn cells	Optical: Koelle (1951)	AChE + ve and − ve structures listed very fully. Although BuThCh mentioned in method BuChE not given in results	Text in Japanese but summary, figure legends and table in English. Other rodents studied
Kása and Csillik (1965a)	Cerebellum	Optical: Gerebtzoff (1959)	AChE reported in climbing fibres and parallel fibres in molecular layer. Basket cells + ve, also Golgi cells, granule cell dendrites and mossy fibre endings in glomeruli of granular layer	Authors suggest that all synapses, both excitatory and inhibitory, are cholinergic
Kása and Csillik (1966)	Cerebellum (also diaphragm)	EM: method in paper	In granular layer, AChE on presynaptic endings in glomeruli and on granule cell dendrites and axons of Golgi cells. Golgi cell parenchyma occasionally + ve. Parallel fibres + ve in molecular layer. BuChE between myelin lamellae. No activity in capillaries	Primarily a methodological paper, illustrated also by tissues from cat and rat. See also Kása (1968a)
Leonardelli (1966)	Hypothalamus and pituitary	Optical: Gerebtzoff (1959)	AChE activity in magnocellular nuclei varies with oestrous cycle; in general maximal at proestrus. In anterior pituitary, capillaries + ve. In posterior pituitary no AChE activity. In intermediate lobe AChE activity high in dioestrus, otherwise low	Short paper in French. No illustrations. Enzyme activity related to reproductive state

TABLE 6.12 (continued)

Reference	Area studied	Method	Result	Comment
Phillis (1965b)	Cerebellum	Optical: Gerebtzoff (1959)	AChE very strong in molecular layer, slightly less strong in granular layer. Purkinje cells and dendrites weakly stained. **Middle peduncle + ve**	Comparison with other species. Includes some biochemistry
Storm-Mathisen and Blackstad (1964)	Hippocampus	Optical: essentially Couteaux (1951)	AChE present in some layers but absent from mossy fibre endings	Paper primarily deals with rat; one guinea-pig included for comparison

PERIPHERAL NERVES ETC.

Reference	Area studied	Method	Result	Comment
Bell (1969)	Nerve terminals on uterine artery	EM: Karnovsky and Roots (1964)	AChE on surface of axons. In late pregnancy, but not in virgins, strong postsynaptic staining observed	See ch. 7 § 2.3. **See also Bell (1971);** (this includes dog, cat, rat, rabbit, **sheep pig and cow**)
Bell and McLean (1967)	Hypogastric ganglion	Optical: Karnovsky and Roots (1964)	Fluorescence studies and AChE staining on same sections. AChE activity strong only in non-fluorescent cells; weak in fluorescent cells. Some cells negative with both procedures	Results suggest the cholinergic and adrenergic fibres to vas deferens are separate (cf. Jacobowitz and Koelle 1965)
Bulmer (1965)	Ovary	Optical: Koelle (1951)	Walls of arterioles stain for AChE; stained nerve fibres very rare. BuChE in granulosa cells of some follicles and in extensive network of thecal fibres. Interstitial cells – ve	Comparison with rat, rabbit and monkey. Results for BuChE not wholly in agreement with Gerebtzoff (1959)

TABLE 6.12 (continued)

Reference	Area studied	Method	Result	Comment
Cauna and Naik (1963)	Various sensory ganglia including dorsal root ganglia	Optical: Koelle (1951)	AChE in cytoplasm of all neurones; activity varied from high to very low. BuChE also present in some neurones but activity weaker. Satellite cells and nerve fibres − ve for both enzymes	Comparison with man, cat, rat and mole
Dixon and Gosling (1971)	Innervation of ureter	Optical: Gomori (1952); Karnovsky and Roots (1964)	AChE + ve fibres in ureteric muscle coat and adjacent to basal layer of epithelium	Fluorescent fibres had a similar distribution. Comparison with rat and rabbit
Goutier-Pirotte and Gerebtzoff (1955)	Placenta	Optical: Gerebtzoff (1953)	Very strong AChE activity (virtually no BuChE) in marginal and interlobular syncitium in early pregnancy. Increase in activity and change in localization, to labyrinth, in late pregnancy. At term, enzyme attached to foetal erythrocytes	Paper in French; mainly biochemical
Jacobowitz and Koelle (1965)	Vas deferens	Optical: Koelle (1951, 1955)	Dense network of AChE-containing fibres in muscle layers with same distribution as fluorescent fibres. Lamina propria showed strong AChE activity but no fluorescence	cf. Bell and McLean (1967). Comparison with cat and rabbit. Burn and Rand hypothesis discussed in light of species differences. See also Furness and Iwayama (1972)
Navaratnam (1965)	Heart and cardiac ganglia	Optical: Lewis (1961)	AChE in cardiac ganglion cells about mid-term; many AChE-containing fibres in node etc. by birth. BuChE around ganglion cells by 46 days gestation	Developmental study. Comparison with man, rat and rabbit

TABLE 6.12 (continued)

Reference	Area studied	Method	Result	Comment
Pecot-Dechavassine (1962)	Diaphragm	Optical: Couteaux and Taxi (1952)	Both AChE and BuChE present at end-plates but activity towards AcThCh 10 × that towards BuThCh	See below for biochemical results which indicate relatively high **BuChE** activity
Perrotta and Lewis (1958)	Placenta	Optical: Gerebtzoff (1953); Lewis (1958)	AChE activity high. Localization changes in 6th week of gestation from perilobular partitions to lobular labyrinth. Reaction with BuThCh too slight for assessment	Comparison with man and rat
Robinson (1969)	Vas deferens	EM: Karnovsky (1964)	AChE on plasma membrane of some axons in muscle, also on plasma membrane of associated Schwann cells and on muscle membrane in vicinity of + ve fibres. BuChE on plasma membrane of all axons and smooth muscle cells	
Silver (1963)	End-plates of tensor fasciae latae and superior rectus muscles	Optical: Lewis (1961)	End-plates reacted for AChE and BuChE. AChE hydrolysed PrThCh in addition to AcThCh	Comparison with rabbit, goat, monkey and hen

TABLE 6.12.1

The distribution of AChE in the hypothalamus of the female guinea-pig. (From Cottle and Silver 1970b.)

Nuclei with strong cellular staining

Supraoptic ⎫
Paraventricular ⎬ intensity of reaction varies with reproductive state
Infundibular ⎭

Mamillary (including magnocellular prefascicular nucleus)

Group dorsomedial to fornix ⎫ not distinguished as discrete nuclei
Small group ventral to paraventricular nucleus ⎬ by histological stains

Other areas with stained cells
Lateral hypothalamic area
Perifornical region

Areas with stained neuropil
Suprachiasmatic nucleus
Lateral border of paraventricular nucleus
Dorsomedial nucleus
Medial mamillary nucleus

TABLE 6.13

GUINEA-PIG BIOCHEMISTRY

CNS and peripheral tissue

Reference	Area studied	Method/substrate	Result	Comment
Ambache et al. (1971)	Small intestine and colon. Auerbach's plexus separately analysed	Manometric: Warburg; ACh, BuCh, MeCh and PrCh with inhibitors	In small intestine most of AChE, and 10% BuChE, in Auerbach's plexus. In colon, plexus contained about 57% of total AChE but no BuChE. Values for colon lower than for small intestine	
Augustinsson (1948)	Blood, liver, kidney and intestine	Manometric: Warburg; ACh, MeCh, BzCh, TB etc.	Most important result is atypical nature of nonspecific enzyme in liver; this hydrolyses BzCh very much faster than ACh. In kidney activity is low but again BzCh is preferred to ACh	Various kinetic studies etc. See also Sawyer (1945) re liver enzyme
Austin and Phillis (1965)	Cerebellum	Manometric: ACh, MeCh and BuCh	Separate analysis of molecular and granular layers, white matter and peduncles. AChE in granular layer slightly greater than in molecular layer	Note: values for peduncles expressed per g wet wt, other values per mg protein. Comparison with several other species. See also Phillis (1965b)
Esila (1963)	Retina	Starch gel electrophoresis		See table 6.32.2
Goldberg and McCaman (1967)	Cerebellum	Spectrophotometric: Ellman et al. (1961); AcThCh	AChE in molecular layer > granule layer > white matter and nuclei. See table 6.33	Comparison with cat, rabbit, rat and pigeon. Conclude no correlation between AChE and ChAc activity

TABLE 6.13 (continued)

Reference	Area studied	Method/substrate	Result	Comment
Kàsa and Silver (1969)	Cerebellum	Spectrophotometric: Ellman et al. (1961); AcThCh plus ethopropazine	AChE level in archi- and paleocortex almost equal. Peduncles, white matter and nuclei also examined	ChAc also measured. Comparison with rat
Kavaler and Kimel (1952)	Motor cortex	Manometric: Warburg; ACh, MeCh and BzCh. No inhibitor	ChE detectable at 28 days gestation, sharp rise in activity at 35 days. Level at term almost equal to adult value	Authors assumed activity due to AChE. Developmental study. Increase in activity precedes differentiation of neuroblasts to neurones
McCaman et al. (1965)	Cerebral hemisphere	Spectrophotometric: Ellman et al. (1961); AcThCh, on subcellular fractions	Subcellular distribution shown in table 6.34	ChAc, 5-HTPD and MAO also studied. Comparison with rat, rabbit and pigeon
Namba (1971)	Muscle fibres and end-plates in intercostal and other muscles	Colorimetric: Hestrin (1949); ACh and BuCh on subcellular fractions	Ratio of microsomal ChE to total ChE was low compared with that in other species	Comparison with man, rat, mouse, dog and chicken. Some histochemistry
Pecot-Dechavassine (1962)	Diaphragm muscle	Manometric: Warburg; ACh and BuCh	See table 6.35. BuChE level is higher than suggested by histochemistry	Comparison with various species
Sawyer (1946)	Sciatic nerve (intact and degenerating)	Microtitrimetric: ACh, MeCh and BzCh	Concludes from degeneration experiments that 60% AChE is in axons and $\simeq 30\%$ in sheath. PseudoChE apparently confined to sheath	See chs. 4,10 for discussion

TABLE 6.14

HORSE HISTOCHEMISTRY

Various sites

Reference	Area studied	Method	Result	Comment
Esila (1963)	Retina			See table 6.32.1
Cheng (1964a)	Extraocular and retractor bulbi muscles	Optical: Koelle (1951); Crevier and Bélanger (1955)	No distinction between AChE and BuChE. ChE at end-plates, grape-like endings and musculotendinous junctions of extraocular muscles. No grape-like endings in retractor bulbi muscles	In Japanese but with English summary and figure legends. Comparison with several other species
Hebb and Hill (1955b)	Pancreas	Optical: Koelle (1950)	AChE in nerve fibres and some un-identified cells. BuChE absent	Comparison with several other species

TABLE 6.15

HORSE BIOCHEMISTRY

CNS and blood

Reference	Area studied	Method/substrate	Result	Comment
Augustinsson (1948)	Whole blood, serum and erythrocytes	Manometric: Warburg; ACh, MeCh, BzCh, TB etc.	Relatively low AChE in erythrocytes; high level of pseudoChE in serum	Horse serum used very frequently in kinetic studies. See ch. 10 for additional references
Esila (1963)	Retina	Starch gel electrophoresis		See table 6.32.2
Ruckebusch and Ruckebusch (1959)	Whole blood, serum and erythrocytes	Potentiometric: Michel (1949) with ACh	Total ChE in serum very high	Paper in French. Comparison with dog, cow, sheep, goat, pig, rabbit and fowl
Sperti et al. (1960a)	Cerebellum	Manometric: Warburg; ACh, MeCh and BuCh	Activity in cortex due almost solely to AChE, in nuclei to BuChE and in peduncles both enzymes equally active	17 cortical areas analysed separately (see table 6.15.1a). For nuclei and peduncles see table 6.15.1b

TABLE 6.15.1
Cholinesterase activity in the cerebellum of horse. (From Sperti et al. 1960a.)

a) Cortex

Area		Q ACh	Q MeCh	Q BuCh	$\dfrac{\text{Q MeCh}}{\text{Q BuCh}}$	*ChEII%
Culmen vermis		3465	1510	451		
Culmen vermis	(23)	100.0%	100.0%	100.0%	3.35	6.0
Lingula	(4)	96.5	92.6	101.3	3.06	6.5
Centralis vermis	(5)	100.0	98.0	104.5	3.14	6.3
Centralis lateralis	(4)	106.5	114.2	103.6	3.69	5.4
Culmen lateralis	(5)	101.5	99.6	112.2	2.97	6.7
Simplex vermis	(7)	100.1	102.1	101.9	3.35	6.0
Simplex lateralis	(4)	116.9	112.4	110.5	3.41	5.8
Folium vermis	(6)	97.5	100.7	102.9	3.28	6.1
Tuber vermis	(5)	102.4	99.3	101.8	3.26	6.1
Crus I	(7)	105.1	104.5	103.3	3.38	5.9
Crus II	(5)	98.0	99.6	81.9	4.07	5.0
Paramedianus	(5)	67.3	70.9	88.5	2.68	7.3
Pyramis	(5)	76.9	75.1	95.1	2.64	7.4
Uvula	(6)	60.5	52.8	103.3	1.71	11.0
Paraflocculus	(5)	69.1	76.3	115.1	2.22	8.7
Nodulus	(8)	91.2	92.8	104.3	2.98	6.6
Flocculus	(4)	99.7	99.1	110.6	3.00	6.6

Activity expressed as a % of the absolute value for culmen vermis. Q ACh, Q MeCh, Q BuCh =
$\mu l\ CO_2$ evolved/g wet wt of tissue/10 min with respectively ACh, MeCh and BuCh as substrate.
*% of the total hydrolysis of ACh that is due to ChEII (BuChE) is derived thus:
(46 Q BuCh)/(2.17 Q MeCh + 0.46 Q BuCh)
Number of animals used given in brackets

b) Nuclei and peduncles

Area		Q ACh ± S.D.	Q MeCh ± S.D.	Q BuCh ± S.D.	ChEII%	Q ACh%	QBuCh%
Nucleus							
Dentate	(5)	557 ± 101	29 ± 40	1057 ± 193	88.5	100.0	100.0
Interpositus	(5)	311 ± 87	30 ± 5	617 ± 196	81.3	55.9	58.4
Fastigial	(5)	357 ± 42	47 ± 13	698 ± 81	75.9	64.1	66.1
Peduncle							
Superior	(4)	230 ± 77	42 ± 19	319 ± 71	61.7	100.0	100.0
Middle	(4)	146 ± 21	42 ± 4	157 ± 25	44.2	63.4	49.2
Inferior	(4)	165 ± 69	33 ± 5	245 ± 69	61.2	71.8	76.8

Q ACh, Q MeCh, Q BuCh, ChEII are as defined above.
Q ACh% and Q BuCh% give the activity as a % of the absolute value for either dentate nucleus
or superior peduncle

TABLE 6.16

MAN HISTOCHEMISTRY

CNS

Reference	Area studied	Method	Result	Comment
Brightman and Albers (1959)	Brain and spinal cord	Optical: Koelle (1951)	AChE in neurones (no details). Weak BuChE activity in white matter, too diffuse to be clearly identifiable with neuroglia	Comparison with several species
Duckett and Pearse (1967)	Basal ganglia, thalamus etc.	Optical: Gerebtzoff (1959)	AChE appears in basal ganglia and thalamus in 6th month of gestation. BuChE present only in subthalamic nucleus and near globus pallidus	Developmental study. Other enzymes examined. Oxidative activity appears long before cholinesterase activity. See Duckett and Pearse (1968) re AChE in Cajal-Retzius cells of developing cerebral cortex
Duckett and Pearse (1969)	Spinal cord	Optical: Gerebtzoff (1959)	AChE appears in neuronal cytoplasm of anterior horn cells of lumbar cord at 8–10 weeks gestation. Caudal segments show this activity before rostral ones. No such gradient in posterior horns	Developmental study. Some discussion of time of appearance of AChE in relation to muscle movement. Some other enzymes examined
Esila (1963)	Retina			See table 6.32 and 6.32.1 for Esila's own results and other references
Foldes et al. (1962)	Cortex, thalamus globus pallidus, caudate nucleus, cerebellum etc.	Optical: Koelle (1951)	AChE highest in caudate and globus pallidus. In cerebellum, molecular layer stained more strongly than granular layer. BuChE in white matter, also in cells of inferior olive	Paper primarily biochemical; see table 6.17.1

TABLE 6.16 (continued)

Reference	Area studied	Method	Result	Comment
Friede (1967)	Brain and spinal cord	Optical: modified **Gerebtzoff** (1959)	Emphasis on BuChE. Present in blood vessels and in oligodendroglia and axons in white **matter**. Dorsal vagal nucleus, subependymal tissue in region of nucleus of solitary tract and tissue round spinal canal also + ve	Comparison with rat, cat and monkey. Identification of stained elements is in some respects unorthodox (see ch. 10)
Friede and Fleming (1964)	Cerebellum	Optical: **Koelle** (1954); **Gerebtzoff** (1959)	AChE present in molecular layer. Little or no reaction in granular layer. Purkinje cell **membrane** + ve. BuChE activity very strong in deeper part of molecular layer. Granular layer virtually − ve	Comparison with other species
Ishii and Friede (1967)	Brain and spinal cord	Optical: modified **Gerebtzoff** (1959)	Comprehensive data with illustrations of most major structures except cerebellum. BuChE not studied	Compares distribution of AChE and NAD-diaphorase
Okinaka et al. (1961)	Cerebral cortex and basal ganglia	Optical: Koelle (1951)	AcThCh and BuThCh used without inhibitors. Variable activity towards AcThCh present in cells in all cortical layers but particularly in Betz cells. Strong activity in basal ganglia (including glia). Hydrolysis of BuThCh by glia in cortex, by some neurones, blood vessel walls and white matter	Paper probably more valuable for biochemical results (see table 6.17.1) as photomicrographs poorly reproduced

TABLE 6.16 (continued)

Reference	Area studied	Method	Result	Comment
Papp (1968)	Reticular formation, lower brain stem	Optical: Gerebtzoff (1959)	Cells showed wide range of AChE activity. Those in dorsal vagal nucleus all very strongly stained. Activity in different nuclei and fibres listed. BuChE not studied	
Robinson (1966)	Brain and spinal cord	Optical: Snell (1959)	In frontal cortex some AChE +ve pyramidal cells; weaker reaction in neuropil and small cells. Oligodendrocytes +ve in subcortical white matter and in cord. Cerebellar molecular layer +ve. Small number of active nuclei in medulla. Weak BuChE activity in some cortical and medullary cells. White matter +ve. Cerebellar molecular layer more +ve than granular	Normal material described and compared with that from patients with Friedreich's ataxia
			PERIPHERAL NERVES ETC.	
Cauna (1960)	Cutaneous receptors	Optical: Snell (1959)	Reaction with AcThCh and BuThCh on membranes of laminar cells and in fibres associated with hairs. Meissner's and Merkel's corpuscles and free nerve endings – ve. Some myelinated fibres and some blood vessels +ve with BuThCh	See also Cauna (1961); this includes other species
Cauna and Naik (1963)	Various sensory ganglia including dorsal root ganglia	Optical: Koelle (1951)	Variable AChE activity in neuronal cytoplasm and fibres. No BuChE in adult but in foetus is present, with AChE, in nerve cells and fibres	Comparison with cat, guinea-pig, rat and mole. See also Cauna et al. (1961)

TABLE 6.16 (continued)

Reference	Area studied	Method	Result	Comment
Cheng (1963)	Extraocular muscles	Optical: Koelle (1951)	No distinction made between AChE and BuChE. Cholinesterase present in en plaque and en grappe endings and at musculotendinous junctions and on intrafusal fibres	Care! 'A-ChE' and 'B-ChE' refer to morphologically different endings and not to AChE and BuChE
Chokroverty et al. (1971)	Striated muscle	Optical: Karnovsky (1964). Also naphthylacetate for non-specific esterases	In addition to AChE and BuChE, non-specific esterases were present in end-plates. BuChE staining in nerve fibres	
D'Agostini and Rossatti (1961)	Lymphoid tissue (tonsils)	Optical: Gerebtzoff (1959)	AChE in blood vessel walls. Weaker activity in lymphoid cells and histiocytes of germinal centre	Some comparison with cat, rabbit and guinea-pig
Ellison (1971)	Nerves of umbilical cord	Optical: Coupland and Holmes (1957)	AChE +ve fibres present at term in foetal, but not maternal, part of cord	Comparison with rat
El-Rakhawy and Bourne (1961)	Tongue	Optical: Gomori (1952)	AChE in network of sensory fibres associated with taste buds. Parasympathetic ganglia, fibres to blood vessels, and end-plates also + ve. Some nerve endings, but not fibres, stain for BuChE. Blood vessel walls, ganglia, end-plates and connective tissue show some activity	
James and Spence (1966)	Sinus and auriculo-ventricular node and ventricles of heart	Optical: Carbonell (1956); no inhibitor	ChE in conducting tissue, right atrial myocardium, and cells (? pacemaker cells) in sinus node. + ve endings in sinus node exceed those in AV node. Sinus node artery, and ventricles − ve	See also Taylor and Anderson (1973)

TABLE 6.16 (continued)

Reference	Area studied	Method	Result	Comment
Kanagasuntheram et al. (1969)	Innervation of nasopharynx	Optical: unspecified thiocholine method	AChE + ve neurones appear in ganglia of nasopharynx at early embryonic stage. Very few unstained ganglion cells. In addition to fibres from + ve cells, AChE-containing fibres reach nasopharynx from IXth nerve	Developmental study but includes adults too. See also Cauna et al. (1972) re innervation of nasal glands
Naik and Cauna (1971)	Sympathetic ganglia and vagal ganglia from stomach and gut	Optical: modified Koelle	Cervical, thoracic and lumbar *sympathetic* ganglia all similar: neurones showed a range of AChE activity in satellite cells; neurones and nerve fibres – ve. Weak BuCh activity but none – ve. In *parasympathetic* ganglia neuronal reaction for AChE stronger and satellite cells + ve. No BuChE in neurones; satellite cells strongly stained	Superior cervical ganglion in foetus (125 mmC-R) had similar pattern
Namba and Grob (1970)	End-plates in intercostal muscle	Optical: modified Koelle	AChE activity at end-plates was not very strong and there was appreciable staining of areas outside the end-plate region	Paper deals mainly with the biochemistry. Some comparison with rat
Navaratnam (1965)	Heart and cardiac ganglia	Optical: Lewis (1961)	AChE first observed in cardiac ganglion cells at 250 mm stage. Ganglia and conducting tissue – ve for BuChE at all stages examined (up to 2 days after birth)	Developmental study. Comparison with rat, rabbit and guinea-pig. Taylor and Smith (1971) found somewhat different results with very fresh foetuses and a high pH (6.0)

TABLE 6.16 (continued)

Reference	Area studied	Method	Result	Comment
Nomura and Schuknecht (1965)	Cochlea	Optical: Gomori (1952)	Cholinesterase staining (AcThCh without inhibitors) used to trace efferent fibres	Cat also examined
Perrotta and Lewis (1958)	Placenta	Optical: Gerebtzoff (1953); Lewis (1958)	Virtually no reaction for AChE. Some BuChE (probably) around foetal blood vessels	Comparison with guinea-pig and rat. The biochemically demonstrable AChE (Ord and Thompson 1950) is apparently mainly due to blood together with tissue enzyme, the activity of which is too low for histochemical detection
Taylor and Anderson (1973)	Heart: atrioventricular node and bundle (of foetuses)	Optical: Gomori (1952)	No enzyme in upper part of node, but AChE in nerve fibres. Bundle and lower part of node contain pseudoChE	Authors discuss possibility that the type of ChE may change during development. Note: identification of enzyme based on AcThCh plus inhibitors. BuThCh not used
Taylor and Smith (1971)	Heart and cardiac ganglia (of foetuses)	Optical: Gomori (1952)	AChE and pseudoChE present in cardiac muscle, ganglia and nerves in foetus of 35 mm C-R. AChE predominating in nervous tissue	cf. Navaratnam (1965). Identification of enzymes based on AcThCh plus inhibitors. BuThCh not used. For results of electrophoresis see table 6.17

TABLE 6.17

MAN BIOCHEMISTRY

CNS and peripheral tissue

Reference	Area studied	Method/substrate	Result	Comment
Augustinsson (1948)	Blood: serum and erythrocytes	Manometric: Warburg; ACh, MeCh, BzCh, TB etc.	Erythrocytes rich in AChE, serum contains pseudoChE	Kinetics studied. N.B. human serum has been used in many subsequent studies of this type
Esila (1963)	Retina	Starch gel electrophoresis		See table 6.32.2
Foldes et al. (1962)	Large number of different areas of brain	Manometric: Warburg; ACh, MeCh and BuCh. Various inhibitors	AChE in caudate nucleus > globus pallidus > thalamus > cerebellar cortex > cerebral cortex. BuChE greatest in white matter. See table 6.17.1 for selected values	Histochemical results paralleled biochemical findings; cf. Ord and Thompson (1952) re relative activity of thalamus and cerebellum
Himwich et al. (1955)	Brain; various areas; also cervical cord	Titrimetric: Aprison et al. (1954); ACh	Results include observation that in globus pallidus, putamen and, most particularly, the caudate, the ChE activity in right and left sides is different. See table 6.17.1 for selected values	Includes some foetal brains. On dry, but not wet, wt basis levels at 26–40 weeks gestation for areas other than caudate exceed adult values
Namba (1971)	Muscle fibres and end-plates in intercostal and other muscles	Colorimetric: Hestrin (1949); ACh and BuCh, subcellular fractions	Total ChE/mg muscle, and also ratio of activity in microsomal fraction and homogenate, was higher than in other species	Comparison with dog, mouse, guinea-pig, rat and chicken. See also Namba and Grob (1970)
Okinaka et al. (1961)	Different areas of cortex, basal ganglia etc.	Manometric: Warburg; ACh	See table 6.17.1 for selected values. N.B. no distinction between AChE and BuChE	For histochemistry see table 6.16

TABLE 6.17 (continued)

Reference	Area studied	Method/substrate	Result	Comment
Ord and Thompson (1950)	Placenta	Manometric: Warburg; ACh, MeCh and BzCh. DFP as inhibitor	Showed that the ChE activity in placental tissue free of blood was entirely due to AChE	
Ord and Thompson (1952)	Brain: various areas	Manometric: Warburg; ACh, MeCh, BzCh, BuCh, PrCh, TB. **DFP**, eserine and Nu 1250 as inhibitors	See table 6.17.1 for selected values	See also Cavanagh et al. (1954) which deals mainly with pseudoChE. Some values for rabbit, rat, **guinea-pig, cat and dog**
Pope et al. (1952)	Frontal cortex: sections of the different layers	Ultramicrotitrimetric, Glick (1938); ACh	ChE activity varied within and between layers. It was highest in layer I, low in layer VI and minimal in the white matter	See also Pope (1967) who studied other enzymes as well as ChE
Taylor and Smith (1971)	Foetal heart and cardiac ganglia, also foetal and adult erythrocytes and plasma	Electrophoresis: method described in paper; AcThCh and BuThCh	Confirmed histochemical findings that both AChE and BuChE present from 35 mm (C-R) onwards. **Plasma** ChE gave one band, and erythrocytes two bands, of AChE in both foetuses and adults	For histochemistry see table 6.16
Youngstrom (1941)	Foetal brain: various parts; spinal cord, liver and skeletal muscle	Bioassay of residual substrate, ACh (Bernheim and Bernheim 1936)	Total ChE determined. Rate of acquisition of enzyme activity varies in different areas of CNS. Greatest rate shown by spinal cord. Liver activity reaches peak early in foetal life and then declines	

The biology of cholinesterases

TABLE 6.17.1
ChE in various parts of human brain

Authors	Ord and Thompson (1952)	Okinaka et al. (1961)	Foldes et al. (1962)	Himwich et al. (1955)
	Sex and age not specified	Sex and age not specified	Males (33–54 yr)	Both sexes (61–87 yr)
Caudate nucleus	*100.0	[†]100.0	*100.0	[†]100.0
Cerebellum	28.9	–	–	25.3
Globus pallidus	–	32.5	87.8	47.0
Thalamus	9.6	–	44.0	17.7
Putamen	–	104.5	–	74.7

All values expressed as a percentage of the value for caudate nucleus.
* AChE activity.
† Total ChE activity.

TABLE 6.18

MOUSE HISTOCHEMISTRY

CNS

Reference	Area studied	Method	Result	Comment
Capurro et al. (1960)	Cerebellum	**Optical: thiocholine,** no details	No AChE activity. PseudoChE in granule and Purkinje cell layers and in white matter	Paper in French. Comparison with various species. Failure to find AChE could be due to the particular folium (-ia) examined (see Friede and Fleming 1964) but authors also report absence in guinea-pig (cf. various + ve results in table 6.12) which casts doubt on method
Friede and Fleming (1964)	Cerebellum	Optical: Koelle (1954); Gerebtzoff (1959)	AChE activity slight or absent in molecular layer of all folia and in granular layer of most folia but in some, activity strong. Purkinje cell membrane + ve. BuCh in granular layer and in glia of subcortical white matter	Comparison with various species
Ishii (1957a)	Brain and spinal cord	**Optical: Koelle** (1951)	AChE + ve and – ve structures listed very fully. Although BuThCh mentioned in method, BuChE not given in results	Paper in Japanese but comprehensive table of results, figure legends and summary in English. Comparison with rat, rabbit and guinea-pig

PERIPHERAL NERVES

Novikoff et al. (1966)	Dorsal root ganglia and sciatic nerve	EM and optical: Karnovsky (1964); Karnovsky and Roots (1964)	Reaction for 'AChE'* at axolemma of myelinated fibres much stronger than in rat	Paper mainly on rat. Passing reference to rabbit, cat, frog and toad. *Inhibitors not listed

TABLE 6.19

MOUSE BIOCHEMISTRY

CNS and peripheral nerves etc.

Reference	Area studied	Method/substrate	Result	Comment
Austin and Phillis (1965)	Cerebellum	Manometric: Warburg; ACh, MeCh and BuCh. No inhibitors	Total ChE activity low compared with that in most species examined except rat but proportion due to BuCh relatively high	Comparison with cat, duck, guinea-pig, sheep, monkey, rabbit, dog and rat
Ebel et al. (1973)	Brains from two strains of mice showing different performance in avoidance learning tests	Radiochemical: McCaman et al. (1968); 1-[^{14}C]-ACh with Iso-OMPA	Values given for AChE and ChAc in 5 cortical areas, olfactory bulb and cerebellum. Significant strain differences in the levels of both enzymes in temporal (but not other) cortical tissue	ChAc also measured. Authors discuss possibility that difference in behaviour-patterns of the two strains is a reflection of differences in cholinergic system in temporal lobe
Namba (1971)	Muscle fibres and end-plates in intercostal and other muscles	Colorimetric: Hestrin (1949); ACh and BuCh on subcellular fractions	ChE activity in fibres high compared with that in other species examined (except man) but relatively low at end-plates	Comparison with man, rat, guinea-pig, dog and chicken. Some histochemistry
Pryor et al. (1966)	Whole brain	Spectrophotometric: Ellman et al. (1961)	**AChE and BuChE levels listed for 5 different strains of mice. In strains with heavier brain BuChE greater, but AChE lower, than that in the other strains**	MAO, aromatic L-amino acid decarboxylase (AAD) and L-glutamic acid decarboxylase also studied. See also Pryor (1968) for a similar study on two strains during development

TABLE 6.20

PIG HISTOCHEMISTRY

CNS

Reference	Area studied	Method	Result	Comment
Arvy (1961)	Pineal gland	Optical: method not specified	No AChE activity. BuChE very feeble	Paper in French. Comparison with cow and sheep. See also Arvy (1965)
Esila (1963)	Retina			See table 6.32 and 6.32.1 for Esila's own results and other references

PERIPHERAL NERVES ETC.

Cheng (1964a)	Extraocular muscles	Optical: Koelle (1951); Crevier and Bélanger (1955)	ChE at motor end-plates, grape-like endings and musculotendinous junctions	In Japanese but with English summary and figure legends. Comparison with several other species
Mann (1971)	Bronchial innervation	Optical: Lewis (1961; see Hebb and Silver 1970)	AChE +ve fibres in smooth muscle, mucosa and blood vessels; ganglia also +ve. BuChE activity in smooth muscles and ganglia	Comparison with rabbit, sheep and calves. Catecholamines also studied
Yamauchi and Lever (1971)	Superior cervical ganglia	Optical: Lewis and Shute (1966) following fluorescence technique	High AChE activity in cells showing minimal fluorescence; moderate activity in moderately fluorescent cells and minimal activity in intensely fluorescing cells	Comparison with rat and sheep

TABLE 6.21
PIG BIOCHEMISTRY

Reference	Samples analysed	Method/substrate	Result	Comment
Augustinsson & Olsson (1959a, b)	Milk and colostrum	Manometric: Warburg; ACh, PrCh, BuCh and various aliphatic and aromatic esters. Electrophoresis on cellulose	BuChE present in milk. Concentration 25 times that in plasma. Colostrum contains the same enzyme, plus an electrophoretically slower type.	Specificity pattern atypical (see Augustinsson 1959a, c). Level of activity in milk changes during lactation
Esila (1963)	Retina	Starch gel electrophoresis		See table 6.32.2
Miller et al. (1969b)	CNS	Electrotitrimetric: modified Jensen-Holm et al. (1959); ACh, no inhibitors	ChE activity listed for many different areas of brain, also for spinal cord.	The animals used in this work were miniature pigs
Ruckebusch & Ruckebusch (1959)	Whole blood, erythrocytes and serum	Potentiometric: Michel (1949); ACh	Total ChE very high in erythrocytes but low in serum	Paper in French. Comparison with horse, cow, sheep, goat, dog, rabbit and fowl
Sekine (1951)	Spermatozoa and seminal fluid	Manometric: Warburg; ACh, BuCh, BzCh, TB	AChE activity in ejaculated and epididymal spermatozoa comparable to that in mouse brain. Much less activity in seminal fluid	Author found ACh increased, and eserine decreased, the motility of the spermatozoa (see ch. 8 §4)

TABLE 6.22
PRIMATES (excluding man) HISTOCHEMISTRY
CNS

	Reference	Area studied	Method	Result	Comment
Rhesus monkey (*Macaca mulatta*)	Brightman & Albers (1959)	Brain and spinal cord	Optical: Koelle (1951)	AChE in neurones. Weak BuChE activity in white matter. Too diffuse to be clearly identifiable with neuroglia	Comparison with other species
	Friede (1967)	Brain and spinal cord	Optical: modification of Gerebtzoff (1959)	Paper primarily concerned with BuChE which is present in blood vessels and in oligodendroglia and axons in white matter. Strong staining also of a number of cell groups	Comparison with man, cat and rat. Identification of stained elements is in some respects at variance with that in other studies. See also Shute & Lewis (1969), re BuChE in cingulate cortex
	Holmes (1961a)	Hypothalamo-hypophysial system	Optical: Coupland & Holmes (1957)	Hydrolysis of AcThCh confined to paraventricular and supraoptic nuclei and to general neuropil. Reaction with BuThCh weaker in nuclei and very slight in neuropil. In addition, some diffuse reaction in pituitary stalk and round blood vessels of neurohypophysis	Alkaline phosphatase examined too. See also Holmes (1961b) for comparative data from other animals

TABLE 6.22 (continued)

Reference	Area studied	Method	Result	Comment
Rhesus monkey (continued)				
Krnjević & Silver (1965)	Cerebral cortex	Optical: Lewis (1961)	AChE present in sub-cortical fibres, mainly round sulci. No data on BuChE	Paper deals mainly with cat; section on monkey very brief
Krnjević & Silver (unpublished, see Silver 1967)	Cerebellum	Optical: Lewis (1961)	AChE activity ex-tremely variable from folium to folium. In some, both the mole-cular and the granular layer were +ve; in others the deep, but not superficial, part of the granular layer stained, and some folia were unstained. No ac-tivity towards BuThCh	The strong activity in some folia contrasts with reports of weak ac-tivity in the Squirrel monkey (see Shantha et al. 1967)
Manocha & Shantha (1970)	Brain & spinal cord, eye including extra- and intraocular muscles, dorsal root ganglia	Optical: Coupland & Holmes (1957); Karnovsky & Roots (1964)	Results too numerous to summarize but one finding of note is ap-parent lack of AChE activity in fasciculus retroflexus of Meynert (habenulo-interpedun-cular tract) which stains so strongly in some other species (e.g. cat, Krnjević & Silver 1965)	The whole book is de-voted to comparative histochemistry of en-zymes of monkey nervous system

TABLE 6.22 (continued)

	Reference	Area studied	Method	Result	Comment
Rhesus monkey (continued)	Olivier et al. (1970a)	Thalamic nuclei	Optical: Gomori (1952)	Wide variation from nuclei to nuclei in reaction with AcThCh. Negligible reaction with BuThCh	Paper mainly concerned with anatomical delineation of nuclei rather than with cholinesterase per se
	Shute & Lewis (1969)	Brain	Optical: no details given	The ascending dorsal and ventral tegmental pathways contain AChE, as in the rat, but BuChE is also present. Distribution of ChE in thalamus and cingulate cortex differs in some respects from that in rat, particularly with regard to BuChE	Brief communication
Crab-eating monkey (*Macaca cynomolgus* = *Macaca irus*)	Phillis (1965b)	Cerebellum	Optical: Gerebtzoff (1959)	AChE present in molecular layer but activity stronger in granular layer	No details given, various species compared
Squirrel monkey (*Saimiri sciureus*)	Iijima et al. (1967)	Hypothalamus	Optical: Coupland & Holmes (1957)	AChE in neurones, neuropil and blood vessels of paraventricular nucleus. Weaker activity with similar distribution in supraoptic nucleus. In other nuclei, neurones and neuropil 'very mildly	Various other enzymes studied

TABLE 6.22 (continued)

Reference	Area studied	Method	Result	Comment
Squirrel monkey (continued)			positive' for BuChE, as were glia and ependymal cells. Some blood vessels in magnocellular nuclei stained	
Iijima & Bourne (1968)	Area postrema	Optical: Coupland & Holmes (1957)	Moderate or strong AChE activity in neurones, *glialoid cells, neuropil, and blood sinusoids. BuChE is distributed similarly but activity in sinusoids is weaker. It is present, in addition, in the ependymal cells and the glia are strongly positive	Reproduction of photomicrographs is poor but they suggest diffusion may have occurred. Various other enzymes studied. *For definition of glialoid cells see Morest (1960)
Shanthaveerappa & Bourne (1965)	Olfactory bulb	Optical: Coupland & Holmes (1957); Gerebtzoff (1959)	AChE activity moderate in glomeruli and nerve fibre layer. In external plexiform layer processes of mitral and tufted cells weakly stained. Moderate activity in inner plexiform layer, granule cells negative. BuChE activity strong in periglomerular blood vessels, weak in nerve fibre layer and doubtful in glomeruli	Various enzymes compared

TABLE 6.22 (continued)

Reference	Area studied	Method	Result	Comment
Squirrel monkey (continued)				
Shantha et al. (1967)	Cerebellum	Optical: Coupland & Holmes (1957); Gerebtzoff (1959)	Weak AChE in neuropil of molecular layer but all cell bodies negative. Purkinje cells also – ve. In granular layer the cells were – ve but there was diffuse stain in glomeruli. Fibres in white matter and also cells of deep nuclei moderately stained, as were occasional blood vessels. BuChE only in blood vessels	Various other enzymes studied
Capuchin monkey (*Cebus cebus**)	Friede & Fleming (1964) * see comment	Cerebellum	Optical: Gerebtzoff (1959)	AChE activity very weak in molecular and granular layer. **Purkinje cell** membrane stained. BuChE present in Bergmann glia of molecular layer but little or no reaction in granular layer	Comparison with various animals. *Nomenclature of monkeys is controversial. Authors describe *Cebus cebus* as a Squirrel monkey but it is generally listed as a Capuchin
Green monkey (*Cercopithecus aethiops*)	Girgis (1968)	Basal rhinencephalon	Optical: Lewis (1961)	Olfactory tubercle, caudate and putamen stain heavily but the septum and hippocampus react only weakly and most of the amygdala, and also the pyriform cortex do not stain	Author compares results with his findings in bush baby (see below) and coypu (table 6.31) Phylogenetic implications discussed

TABLE 6.22 (continued)

	Reference	Area studied	Method	Result	Comment
Green monkey (as above) Patas monkey (*Erythrocebus patos*) Anubis baboon (*Papio anubis*)	Odutola (1972)	Spinal cord	Optical: Karnovsky & Roots (1964) as modified by El-Badawi & Schenk (1967b)	AChE activity strong in many areas of grey matter including part of dorsal horn. Stain not prominent in white matter except after lesions. BuChE activity very weak in all parts	Distribution very similar in many respects to that found in cat (Silver & Wolstencroft 1971)
Monkey (*species not given*)	Olivier et al. (1970b)	Brain, with particular reference to the corpus striatum and substantia nigra	Optical: Gomori (1952)	AChE-containing fibres from corpus striatum were traced to globus pallidus and substantia nigra. Most cells of globus pallidus were unstained except in the basal part where activity was high	Monkey and cat compared (see table 6.4). Telencephalic and diencephalic lesions were made to trace the course of AChE-containing striatal efferent fibres
Bush baby (*Galago senegalensis*)	Girgis (1969)	Basal rhinencephalon	Optical: Lewis (1961)	Olfactory tubercle and septum stain strongly and parts of hippocampus are also well stained. Olfactory bulbs and lateral amygdala generally lack activity	Comments on similarities and differences compared with other mammals

TABLE 6.22 (continued)

	Reference	Area studied	Method	Result	Comment
		PERIPHERAL NERVES ETC.			•
Crab-eating monkey (*Macaca cynomolgus* = *Macaca irus*)	Hess (1962)	Motor end-plats in extraocular and buccinator muscles	Optical: modification of Koelle's method, no details	En grappe and en plaque endings contained both AChE and BuChE	Results conflict with those of Häggqvist (1960) Some muscles from rat and cat also studied
* See comment	Häggqvist (1960)	Motor end-plates in extraocular and other muscles	Optical: Holmstedt (1957b)	In general, AChE confined to en plaque endings and BuChE to en grappe endings	Separate occurrence of AChE and BuChE has been questioned (see Hess 1962; Silver 1963). * Species studied is given as gynomolgus but this is assumed to be a misprint for cynomolgus
Rhesus monkey (*Macaca mulatta*)	Silver (1963)	Motor end-plates in superior rectus, gastrocnemius and soleus muscles	Optical: Lewis (1961)	AChE present at endings in all muscles. Reaction for both enzymes seemed equally strong	Comparison with goat, guinea-pig, rabbit and hen. cf. Häggqvist
Java monkey (*Macaca fascicularis*)	Leela et al. (1971)	Nerves to nasopharynx	Optical: Koelle (1951)	Some fibres reacted only for AChE, others for AChE and BuChE. Both enzymes present in some ganglion cells. BuChE strong at base of taste buds	

TABLE 6.22 (continued)

	Reference	Area studied	Method	Result	Comment
Yellow baboon (*Papio cyno-cephalus*)	Pecot-Dechavassine (1962)	Motor end-plates in extraocular and other muscles	Optical: Couteaux & Taxi (1952)	AChE & BuChE present at endings in all muscles but reaction for BuChE was slower than that for AChE	French thesis. Comparison with other species
Monkey (species not given)	Bulmer (1965)	Ovary	Optical: Koelle, as in Pearse (1960)	Some AChE-containing fibres present (more than in other species studied)	Pattern in monkey rather different from that in rat, rabbit and guinea-pig

TABLE 6.23

PRIMATES (excluding man) BIOCHEMISTRY

	Reference	Area studied	Method/substrate	Result	Comment
Rhesus monkey (*Macaca mulatta*)	Pokrovskii & Ponomareva (1961)	Brain: 52 areas separately analysed	Colorimetric method, bromothymol blue as indicator. Details in paper. ACh & BuCh as substrates	Results obtained with ACh & BuCh given for each area	
	Robins & Smith (1953)	Cerebellum	Microdissection used. Conditions of analysis given but method (for total ChE) not specified	ChE values for granular and molecular layer and for white matter expressed separately. No significant difference in values for hemisphere and vermis. Value for molecular layer only 20% of granular	Other enzymes studied
Monkey (species not given)	Austin & Phillis (1965)	Cerebellum including deep nuclei and penduncles	Manometric: Warburg with ACh, MeCh & BuCh	AChE predominated. BuChE relatively lower than in some other species. Of the 3 peduncles the middle one had the highest activity. In this, monkey differs from other species tested	Biochemical values compared with histochemical staining. Various species compared

TABLE 6.24
RABBIT HISTOCHEMISTRY
CNS

Reference	Area studied	Method	Result	Comment
Arvy (1962)	Hypothalamus and neuro-hypophysis	Optical: Gerebtzoff (1959)	Strong AChE activity. BuChE present. No details of localization	Comparison with ox, sheep, pig, fowl and pigeon
Drukker & Schadé (1965b)	Cerebral cortex	Not specified	At birth AChE confined to neuropil	Developmental study of several enzymes but paper very brief and undetailed
Duffy et al. (1967)	Hypothalamus	EM: modified Koelle (see Brzin et al. 1966)	AChE present in nuclear envelope by 11th day of gestation. Adult picture with AChE in ER, by 17th day. In adults, AChE in nuclear envelope, ER, synapses, on axon surface and within axons of magnocellular nuclei	Adult and developing animals
Esila (1963)	Retina			See tables 6.32 & 6.32.1 for Esila's own results and other references
Friede & Fleming (1964)	Cerebellum	Optical: Koelle (1954); Gerebtzoff (1959)	In molecular layer weak AChE activity; in granular layer activity weak in most folia but strong in some, Purkinje cell membrane +ve. BuChE in granular layer in all folia	Comparison with large number of other species

TABLE 6.24 (continued)

Reference	Area studied	Method	Result	Comment
Hanson (1966)	Retinal cultures	Optical: Holmstedt (1957b), slightly modified for tissue culture	AChE present, after some days of culture, in multipolar and small nerve cells. No activity in visual cells nor neurologia	Culture from 1–2 day-old animals. Other hydrolytic enzymes studied. Rat also examined
Ishii (1957a)	Brain and anterior horn cells	Optical: Koelle (1951)	AChE +ve and −ve structures listed very fully. Although BuThCh mentioned in method, BuChE not given in results	Text in Japanese but summary, figure legends and table in English. Comparison with mouse, rat, guinea-pig. Rabbit least active species
Nichols & Koelle (1968)	Retina	Optical: Koelle (1950, 1955) or Koelle & Gromadski (1966)	Diffuse AChE activity in inner plexiform layer. Ganglion cells stained. No BuChE activity	Stained amacrine cells were not identifiable in normal material but were revealed during recovery from inhibition by in vivo DFP. Comparison with cat, rat, ground squirrel and pigeon
Papp & Bozsik (1966)	Reticular formation, caudal brain stem	Optical: Gerebtzoff (1959)	See table 6.4.4 for AChE +ve and −ve sites. BuChE too weak for evaluation (cf. cat)	Cat and rabbit compared
Phillis (1965b)	Cerebellum	Optical: Gerebtzoff (1959)	AChE moderately strong in molecular and granular layers. Purkinje cells weakly +ve. Middle peduncle +ve	cf. Friede & Fleming (1964). Various species compared. Includes some biochemistry
Reale et al. (1971)	Retina and optic nerve	EM: Karnovsky & Roots (1964)	AChE in ganglion cells, amacrine cells and some horizontal cells. Fibres (? efferent) stain in internal plexiform layer. Very few +ve fibres in optic nerve	

TABLE 6.24 (continued)

Reference	Area studied	Method	Result	Comment
Wender & Kozik (1970)	Hippocampus	Optical: Gerebtzoff (1953)	AChE + ve neurocytes present in 25-day foetuses. Neuropil activity develops later. BuChE very weak in postnatal animals; reaches adult levels at 48 days. Occurs in neurocytes and neuropil	Developmental study from 25 days gestation. Other enzymes also examined
PERIPHERAL NERVES ETC.				
Arvy (1960)	Ovary	Optical: Gerebtzoff (1959?) pH 6.8	AChE in granulosa cells, follicular fluid and corpora lutea. BuChE in interstitial cells, particularly strong in pregnancy	Paper in French. See also Bulmer (1965)
Ballantyne (1966a, b)	Liver	Optical: Coupland & Holmes (1957)	AChE in cells lining sinusoids. BuChE in parenchymal cells and nerves	See also Ballantyne (1967)
Bulmer (1965)	Ovary	Optical: Koelle (1951)	AChE in very rare fibres; strong activity in walls of arterioles in stroma. High level of BuChE in interstitial cells and in granulosa of some medium sized follicles. Large follicles —ve	Comparison with rat, guinea-pig and monkey
Cheng (1964a)	Extraocular and some skeletal muscles	Optical: Koelle (1951); Crevier & Bélanger (1955)	End-plates stained for ChE (no inhibitors). Musculo-tendinous junctions + ve. No en grappe endings in skeletal muscle	In Japanese but with English summary and figure legends. Comparison with other species

TABLE 6.24 (continued)

Reference	Area studied	Method	Result	Comment
Cheng (1964b)	Extraocular muscles	Optical: Koelle (1951)	With BuThCh, activity at en plaque end-plates weaker than with AcThCh but en grappe endings and musculotendinous junctions showed reverse	In Japanese, but English summary and figure legends
Coupland (1958)	Pancreas	Optical: Coupland & Holmes (1957), full details in paper	AChE and BuChE in nerves and Islet cells. Acinar cells − ve. BuChE also in smooth muscle cells of ducts and vessels	cf. Hebb & Hill (1955b) who did not observe any glandular activity. Cat and rat also studied
D'Agostini & Rossatti (1961)	Lymphoid tissue	Optical: Gerebtzoff (1953)	AChE mainly in margins of follicles but, in germinal centres, some cells and capillaries + ve	See also D'Agostini & Rossatti (1959). Comparison with man, cat and guinea-pig
Dixon & Gosling (1971)	Innervation of ureter	Optical: Gomori (1952); Karnovsky & Roots (1964)	AChE + ve fibres in ureteric muscle coat and adjacent to basal layer of epithelium	Fluorescent fibres similarly distributed. Comparison with rat and guinea-pig
El-Badawi & Schenk (1966)	Innervation of urinary bladder	Optical: modified Karnovsky & Roots (1964)	AChE + ve fibres and ganglion cells described. Less numerous than in cat in some areas	Comparison with other species. Noradrenergic fibres also studied. See comment on table 6.4 (cat)
El-Badawi & Schenk (1967a)	Innervation of epididymis	Optical: modified Karnovsky & Roots (1964)	AChE + ve fibres distributed only to blood vessels. Richness of innervation varies in different areas. No + ve ganglion cells	Comparison with other species. Noradrenergic fibres also studied. See comment on table 6.4 (cat)

TABLE 6.24 (continued)

Reference	Area studied	Method	Result	Comment
Grant et al. (1967)	Innervation of ear perichondrium	Optical: Karnovsky & Roots (1964); Gomori (1952) (whole mounts and isolated fibres examined)	Unmyelinated fibres contain BuChE. Myelinated fibres contain both AChE and BuChE	Rat cremaster muscle also studied. Some discussion of technique, particularly the problem of penetration of substrate etc
Hagopian et al. (1970)	Embryonic cardiac muscle	EM: See Brzin et al. (1966)	Abundant AChE in RER of myoblast of 9 day embryos; occasional reaction in nuclear envelope and Golgi complex	Discuss significance of enzymes in uninnervated muscle cells
Hebb & Hill (1955b)	Pancreas	Optical: Koelle (1950)	AChE in nerve fibres. No BuChE activity	Comparison with sheep, cat, horse and goat. cf. Coupland (1958)
Hebb & Linzell (1970)	Innervation of mammary glands	Optical: Koelle as adapted by Hebb et al. (1966)	No AChE +ve fibres to gland tissue but some +ve fibres associated with skin and muscle. BuChE +ve fibres supply smooth muscle of teat and form network round blood vessels	Study also involved fluorescence techniques and denervation experiments
Jacobowitz & Koelle (1965)	Nictitating membrane	Optical: Koelle (1955)	AChE in fibres innervating muscle. Removal of SCG reduced but did not abolish activity	Fluorescent fibres similar in distribution and number to AChE-containing fibres (cf. cat, table 6.4). Fluorescence lost after removal of SCG
Koelle (1955)	Autonomic and sensory ganglia. Anterior horn cells, Auerbach's plexus	Optical: method in paper	Variable AChE in cells of all ganglia. In general, activity higher than in cat	Comparison with cat and Rhesus monkey

TABLE 6.24 (continued)

Reference	Area studied	Method	Result	Comment
Mann (1971)	Innervation of bronchial muscle	Optical: Lewis (1961); see Hebb & Silver (1970)	AChE +ve fibres supply smooth muscle and bronchial blood vessels but not the mucosa. Ganglion cells +ve in extrachondrial plexus. BuChE +ve fibres and ganglia sparse	Fluorescence studies included. Comparison with several other species
Navaratnam (1965)	Heart and cardiac ganglia	Optical: Lewis (1961)	AChE appears in cardiac nerve cells between 24th and 27th day of gestation; thereafter, present in sinus node and atrioventricular bundle. BuChE present extracellularly, in cardiac ganglia by 10th day after birth	Developmental study. Comparison with man, rat and guinea-pig
Navaratnam & Palkama (1966)	Arteries, including coronary arteries	Optical: thiocholine, no details	AChE present in nerve fibres in wall of coronary arteries. Diffuse BuChE activity in tunica media of aorta and pulmonary arteries	Rat also studied
Novikoff et al. (1966)	Dorsal root ganglia and sciatic nerve	EM & optical: Karnovsky (1964); Karnovsky & Roots (1964)	Reaction for *'AChE' at axolemma of myelinated fibres much stronger than in rat	Paper mainly concerned with rat. Passing reference to mouse, cat, frog and toad. * Inhibitors not tested
Silver (1963)	End-plates in gastrocnemius and superior rectus muscles	Optical: Lewis (1961)	End-plates reacted for AChE and BuChE. AChE hydrolysed PrThCh in addition to AcThCh	Comparison with guinea-pig, monkey, goat and hen

TABLE 6.24 (continued)

Reference	Area studied	Method	Result	Comment
Tennyson & Brzin (1970)	Dorsal root neuroblasts	EM: method in paper; microgasometric analysis also used	In 9-day embryo AChE in reticulum of some neural crest cells. Axolemmal AChE appears in fibres in posterior fasciculus before it appears in dorsal root	Paper gives references to other related papers by Tennyson and co-workers (see ch. 4 re distal-proximal gradient in fibres)
Waterson et al. (1970)	Innervation of ear artery	Optical: Karnovsky & Roots (1964)	Considerable BuChE activity apparently associated with noradrenergic fibres in artery. Much weaker AChE activity in same region	Staining abolished by sympathetic denervation but reserpine had no effect

TABLE 6.25
RABBIT BIOCHEMISTRY
CNS and peripheral tissue

Reference	Area studied	Method/substrate	Result	Comment
Aprison & Himwich (1954)	Frontal cortex, superior colliculus, medulla and caudate nucleus	Titrimetric: see Aprison et al. (1954). ACh with no inhibitors	Time-course of acquisition of ChE activity described pre- and postnatally. Maximum levels reached at different times. Caudate and frontal cortex matured the slowest	See also Himwich & Aprison (1955)
Austin & Phillis (1965)	Cerebellum	Manometric: ACh, MeCh, BuCh	Moderate AChE activity in cortex but level in peduncles highest of any species examined. Low BuChE in cortex	Comparison with several species. See also Phillis (1965b)
Esila (1963)	Retina			See table 6.32.1
Goldberg & McCaman (1967)	Cerebellum	Spectrophotometric: Ellman et al. (1961) with AcThCh	See table 6.33 for levels of AChE in different layers	Comparison with cat, guinea-pig, rat and pigeon
Lederis & Livingston (1969)	Anterior hypothalamus and posterior pituitary	Spectrophotometric: Ellman et al. (1961) AcThCh & BuThCh with BW284C51	AChE activity of hypothalamus twice that of posterior pituitary. In both regions BuChE activity was approximately 10% of AChE	ChAc & ACh also measured
Lüdtke & Ohnesorge (1966)	Brain, muscle and gut	Manometric: Warburg; ACh, PrCh, BzCh and BuCh with neo- and physostigmine	Examined specificity, K_M and temperature optimum	Paper in German. Comparison with tench (table 6.11)

TABLE 6.25 (continued)

Reference	Area studied	Method/substrate	Result	Comment
McCaman et al. (1965)	Cerebral hemisphere	Spectrophotometric: Ellman et al. (1961); AcThCh	Subcellular distribution. See table 6.34	Comparison with guinea pig, pigeon and rat
Pecot-Dechavassine (1962)	Gemelli & diaphragm muscle	Manometric: Warburg; ACh and BuCh	See table 6.35	Comparison with several other species
Ruckebusch & Ruckebusch (1959)	Whole blood, serum and erythrocytes	Potentiometric: Michel (1949) with ACh	Total cholinesterase in plasma about equal to that in hen; exceeded that in ruminants and pig. Erythrocyte level relatively low	Paper in French. Comparison with horse, cow, sheep, goat, dog, pig and fowl

TABLE 6.26
RAT HISTOCHEMISTRY
CNS

Reference	Area studied	Method	Result	Comment
Altman & Das (1970)	Cerebellum	Optical: Gomori (1952)	No distinction made between AChE and BuChE. ChE appear-ed in granular layer and white matter at different rates in different folia. Molecular layer virtually – ve. Tran-sient staining of Purkinje cells. Cells of cerebellar nuclei + ve	Developmental study, rats 0–30 days. Good pictures, time-course of changes tabulated
Brightman & Albers (1959)	Brain and spinal cord	Optical: Koelle (1951)	AChE in neurones (not speci-fied). BuChE activity very strong in blood vessels. No staining of glia but elements (? Schwann cells) stain in dorsal roots	Comparison with several species
Brown & Howlett (1968)	Hind brain with par-ticular reference to superior salivatory nucleus	Optical: Lewis (1961); Karnovsky & Roots (1964)	Reaction to AcThCh intense in motor facial, motor tri-geminal and supragenual nu-clei and in medial nucleus of trapezoid body and pathways identified as the superior salivatory outflow and oli-vocerebellar tract. Other structures giving moderate or weak reaction listed	Results comparable to those of Shute & Lewis (1960) but anatomical interpretation different. See also Brown & Howlett (1970) for some EM studies

TABLE 6.26 (continued)

Reference	Area studied	Method	Result	Comment
Brown & Howlett (1971)	Hind brain with particular reference to inferior salivatory nucleus	Optical: Lewis (1961); Karnovsky & Roots (1964)	Reaction to AcThCh demonstrated in various nuclei including nucleus ambiguus and facial nucleus. ChE +ve area ventral to Deiters nucleus identified as inferior salivatory nucleus	Paper mainly concerned with anatomical identification of inferior salivatory nucleus
Csillik et al. (1964)	Cerebellum	Optical: Gerebtzoff (1953)	No AChE in cortex at birth. Cells which appear to be Purkinje cells stain in increasing numbers from 3rd day but later become inactive. Mossy fibres in uvula and nodule react from 7th day	Developmental study. The transient acquisition and loss of AChE by Purkinje cells has subsequently been observed in other species
Danilova (1971)	Hypothalamus and pituitary	Optical: Karnovsky & Roots (1964)	Foetal and adult tissue examined. Time-course of changes in AChE in supra-optic and paraventricular nuclei followed. Median eminence – ve. BuChE in blood vessels. AChE and BuChE +ve structures in posterior pituitary	Developmental study; cf. Hyyppä (1969) re median eminence. See also Robinson (1972)
Eränkö et al. (1967a)	Ventral horn cells	EM: modified Karnovsky & Roots (1964)	AChE in RER, nuclear envelope, some Golgi vesicles and in synapses. Synaptic vesicles – ve	Paper primarily concerned with establishing conditions for EM studies

TABLE 6.26 (continued)

Reference	Area studied	Method	Result	Comment
Eränkö et al. (1970)	Pineal gland	Optical: Gomori (1952). EM: modified Karnovsky & Roots (1964)	AChE +ve fibres and fluorescent fibres identically distributed. EM showed AChE end-product on sympathetic fibres containing dense core vesicles. BuChE in nerve trunks outside gland but after sympathectomy fine fibres within gland become +ve for BuChE and −ve for AChE	Some biochemical results mentioned. See also Eränkö & Eränkö (1971) re loss of * AChE stain after chemical sympathectomy. Results discussed in terms of Burn & Rand hypothesis. * cf. Rodrigues de Lores Arnaiz & Pellegrino de Iraldi (1972) re retention of 50% AChE & 100% ChAc in decentralized pineal
Esila (1963)	Retina			See table 6.32 & 6.32.1 for Esila's own results and for further references
Friede (1967)	Brain and spinal cord	Optical: modified Gerebtzoff (1959)	Deals mainly with BuChE. BuChE +ve glia identified as oligodendroglia. In some tracts glia − ve and axons + ve. Strong neuronal stain in some nuclei but, in general, rat had fewer + ve groups than cat	Comparison with cat, man and monkey. Details of distribution of BuChE at cellular level are in some respects unorthodox (see ch. 10)
Friede & Fleming (1964)	Cerebellum	Optical: Koelle (1954); Gerebtzoff (1959)	Little or no reaction for AChE in most folia but in some, the granular layer reacted strongly. Purkinje cell membrane +ve. Diffuse reaction for BuChE in granular layer of all folia and blood vessels stained strongly	Comparison with several species. See also Csillik et al. (1963) re limitation of staining to archicerebellum

TABLE 6.26 (continued)

Reference	Area studied	Method	Result	Comment
Gwyn (1971b)	Red nucleus	Optical: Lewis (1961)	Some cells in magnocellular portion stain heavily or moderately for AChE. Lighter stain in some cells elsewhere. Rostrally, neuropil stains strongly	Includes effects of axotomy which causes loss of stain from the most active cells. See also Gwyn (1971a); Waldron & Gwyn (1969)
Gwyn & Waldron (1968)	Spinal cord (lateral cervical nucleus)	Optical: * Lewis (1961)	Study designed to identify lateral cervical nucleus. Column of AChE + ve cells extends whole length of cord in dorsolateral funiculus	* Method designated 'Koelle & Friedenwald' but was Lewis's modification (personal communication)
Hajós et al. (1970)	Brain: caudate, amygdala, supraoptic nucleus and cerebellar glomeruli	EM: Karnovsky & Roots (1964)	End-product at 1) axon terminals and on dendrites in caudate and amygdala 2) mossy fibre-granule cell synapses 3) axodendritic and, to some extent, axosomatic synapses in supraoptic nucleus	Results discussed in terms of Koelle's amplifier role for ACh. Reaction in supraoptic nucleus and cerebellum slower than in caudate and amygdala
Hösli & Hösli (1970, 1971)	Cultures: cerebellum and brain stem (1970); spinal cord (1971)	Optical: modified Karnovsky & Roots (1964)	AChE in soma and processes of many neurones. BuChE in glia	
Hyyppä (1969)	Hypothalamus	Optical: Gomori (1952)	In adult, supraoptic and paraventricular nuclei AChE + ve; some activity in other nuclei and median eminence. Suprachiasmatic nucleus − ve.	Developmental studies included some foetuses. Nonspecific esterases also examined. cf. Danilova (1971) above, re median eminence

TABLE 6.26 (continued)

Reference	Area studied	Method	Result	Comment
Hyyppä (1969) (continued)			BuChE in capillaries, glia and ependymal cells. During development AChE activity in arcuate and ventromedial nucleus exceeds that in adult	
Ishii (1957a)	Brain and anterior horn cells	Optical: Koelle (1951)	AChE + ve and − ve structures listed very fully. BuThCh mentioned in method but BuChE activity not listed in results	Text in Japanese but summary, figure legends and table which lists distribution in rat and other rodents are in English
Ishii (1957b)	Brain and anterior horn cells	Optical: Koelle (1951)	Describes time-course of changes in AChE activity in brain structures from 15 days gestation to adulthood	Text in Japanese but English summary etc. as above. Photomicrographs of foetal material are of good quality
Kása et al. (1966)	Cerebellum	Optical: Gerebtzoff (1959)	In the intact nodule AChE present in white matter and glomeruli of granular layer. Molecular layer − ve. Capillaries in both layers contain BuChE. Purkinje cells stain transiently after lesions which isolate the archicerebellum	Effects of lesions indicate glomerular AChE is located in mossy fibre endings and Golgi cells. For some EM histochemical studies on cerebellum see Kása (1968a, 1969); Kása & Csillik (1966). For biochemical studies on isolated folia see Kása & Silver (1969) table 7.5.
Koelle (1963)	Brain and spinal cord	Optical: method given in paper	See table 6.26.1	This article covers results published in many previous papers including Koelle (1954)

TABLE 6.26 (continued)

Lewis, Shute & co-workers: For convenience, a number of papers by this group are considered together instead of being listed in strict alphabetical or chronological order.

Reference	Area studied	Method	Result	Comment
a) Shute & Lewis (1967b)	Brain: projections from brain stem to neocortex, olfactory cortex and sub-cortical areas of mid- and fore brain	Optical: Lewis (1961)	AChE + ve cells and tracts listed (table 7.1) BuChE as well as AChE present in dorsal (but not ventral tegmental pathway) See fig. 7.1	Key paper summarizing authors' previous studies and extending idea that ascending AChE-containing fibres are the anatomical basis of the ascending reticular system to forebrain (Shute & Lewis 1963) and to cerebellum and cochlear nuclei (Shute & Lewis 1965)
b) Lewis & Shute (1967)	Brain: limbic system (hippocampus, medial cortex, subfornical organ and supraoptic crest)	Optical: Lewis (1961)	AChE-containing cells and their projections are indicated in fig. 7.2 and summarized in table 7.3	Summarizes previous studies (see also Lewis et al. 1964, 1967 in which hippocampal ChAc and AChE were studied in parallel following lesions in the fimbria)
c) Shute & Lewis (1966a)	Hippocampus	EM: thiocholine, method given in paper	AChE varies in different layers. In some Golgi cells, RER and nuclear envelope + ve. Some axonal and synaptic membranes + ve; + ve synapses mainly axodendritic	The strongly stained layers seen with optical microscopy represent dense neuropil and not cells. See also Shute & Lewis (1967a) for ultrastructural changes following lesion
d) Shute & Lewis (1966b)	Hypothalamus and subthalamus	Optical: thiocholine, method not specified	Distribution of AChE broadly similar to that in cat (see table 6.4.4) but activity weaker	Paper contains discussion of monoaminergic pathways too. Brief reference to EM studies

TABLE 6.26 (continued)

Reference	Area studied	Method	Result	Comment
e) Shute & Lewis (1960)	Medulla: salivatory nuclei	Optical: thiocholine method in paper	AChE activity used to trace salivatory pathway. Other AChE +ve structures noted in mid- and hindbrain. BuChE +ve cells in part of sensory nucleus of trigeminal nerve	cf. Brown & Howlett (1968) above re difference in interpretation
f) Lewis et al. (1970)	Medulla: dorsal motor vagal nucleus and nucleus ambiguus	Optical: thiocholine, see Navaratnam et al. (1968)	Vagal nucleus contains AChE and BuChE in cells; AChE in neuropil. Vagotomy causes loss of activity from majority of cells. Nucleus ambiguus contains AChE in cells and neuropil	Results of abdominal and cervical vagotomy indicate that cells projecting to a specific area are not discretely localized in the nucleus. See also Navaratnam et al. (1968)
g) Lewis et al. (1971)	Medulla: hypoglossal nucleus	Optical: thiocholine, see Navaratnam et al. (1968); EM: thiocholine, see Lewis & Shute (1969)	Observation of the neurones in which AChE activity decreased following selective lesions of hypoglossal nerve indicated that part of the nucleus different branches arise	References given to other studies on effects of axotomy on AChE activity. See also Lewis & Flumerfelt (1970) re BuChE in some neurones
h) Lewis & Shute (1966)	Ventral horn cells, hypoglossal nucleus, dorsal motor vagal nucleus and several areas of brain with 'cholinergic' afferents e.g. caudate nucleus, hippocampus	EM: thiocholine, method in paper	In neurones reaction for AChE strong within sheets of RER and occasionally round nuclear envelope. SER, mitochondria, lysosomes & most of cell membrane − ve	Useful description of ultrastructural localization of endproduct in known 'cholinergic' sites. See also Lewis & Shute (1965b) for EM study on retina and optic nerve

TABLE 6.26 (continued)

Reference	Area studied	Method	Result	Comment
Machado & Lemos (1971)	Pineal gland	Optical: Karnovsky & Roots (1964)	AChE mainly in fibres associated with blood vessels but some fibres to parenchymal cells +ve. Stain abolished by removal of SCG. No BuChE in adult. Staining of * unusual specificity present in parenchymal cells for 2 weeks after birth	Foetal, postnatal and adult material used. * Authors attribute this activity to BuChE but it is not inhibited by 10^{-5} M DFP
McGeer et al. (1971b)	Caudate nucleus and putamen	Optical: Karnovsky & Roots (1964)	Massive lesions of thalamus cortex, ventral tegmentum and globus pallidus had no significant effect on AChE in caudate-putamen	Biochemical measurements of AChE confirm histochemical findings. Authors suggest AChE-containing elements arise within caudate-putamen (cf. Shute & Lewis 1963, 1967b; see ch. 7. § 3.2)
Navaratnam & Lewis (1970)	Spinal cord	Optical: Lewis (1961). EM: Lewis & Shute (1966)	AChE in motoneurones, intermedio-medial & -lateral columns & in lateral funiculus near substantia gelatinosa. Some activity in neuropil of dorsal horn. BuChE activity strong in blood vessels; neurones in sacral intermedio-lateral column & in medial part of ventral horn +ve. EM shows that the ER in BuChE +ve cells is unevenly distributed in the cytoplasm	Staining pattern very different from that in cat (see Silver & Wolstencroft 1971, table 6.4.5) Sciatic nerve section reduces AChE in circumscribed group of motoneurones

TABLE 6.20 (continued)

Reference	Area studied	Method	Result	Comment
Nichols & Koelle (1968)	Retina	Optical: Koelle (1955); Koelle & Gromadski (1966)	Diffuse stain for AChE in inner plexiform layer attributed to amacrine cells. BuChE in capillaries only	Amacrine cells show up clearly only during resynthesis of enzyme following inhibition by DFP. Comparison with pigeon, ground squirrel, rabbit and cat
Odutola (1970)	Cerebellum	Optical: Karnovsky & Roots (1964)	Results confirm previous reports re strong AChE activity in granular layer but definite activity also reported in * molecular layer. Certain folia, notably those of declivus, react more strongly than previously described	cf. Csillik et al. (1963); Friede & Fleming (1964); Shute & Lewis (1965). * See also Phillis (1965b)
Palkama & Sipponen (1968)	Optic nerve	Optical & EM: Palkama (1967)	After nerve section AChE accumulates in some centrifugal fibres visible at optical level. EM shows enzyme in axonal sheath. BuChE in walls of blood vessels running in septa of normal and sectioned nerves	The accumulated AChE was visible only in sections incubated at high pH (7.0–7.5). See also Lewis & Shute (1965)
Pepler & Pearse (1957)	Brain, with special emphasis on hypothalamus	Optical: Gomori (1952)	AChE activity in magnocellular hypothalamic nuclei was increased in lactation and by administration of saline. Activity in supraoptic nuclei slightly increased by castration. BuChE activity strong in habenular neurones. Blood vessels + ve, neuropil slightly stained. Astrocytes – ve	Nonspecific esterases also studied. For other experimentally-induced changes in hypothalamic AChE see Kivalo et al. (1958); Okamoto et al. (1966)

TABLE 6.26 (continued)

Reference	Area studied	Method	Result	Comment
Sharma (1968)	Olfactory bulb	Optical: Gomori (1952); Gerebtzoff (1959)	Strong AChE activity in glomeruli and moderate to strong stain round mitral cells. Elsewhere weak or negligible activity. BuChE activity moderate in olfactory fibres and glomeruli, strong in blood vessels	Nonspecific esterases also studied
Shute – see under Lewis, Shute & co-workers				
Storm-Mathisen & Blackstad (1964)	Hippocampus	Optical: modified Couteaux (1951)	AChE concentrated in some layers. Basket cell axons and areas containing fibres from cingulum particularly +ve; reaction weak in areas receiving commissural fibres. Reaction with BuThCh heavy in blood vessels but nervous tissue – ve except for very faint reaction in areas rich in AChE	Thorough, well-illustrated study (see also Storm-Mathisen 1970 for quantitative histochemical treatment)
Torack & Barrnett (1962)	Brain stem	EM: thiolacetic acid, method in paper	AChE (?) in synaptic vesicles* and some axodendritic synapses and in RER. BuChE (?) mainly in glia, particularly fibrous astrocytes, and in blood vessels	*This is one of the early EM studies. The method is not highly specific but use of inhibitors increased reliability of results. *Staining of synaptic vesicles is considered to be an artifact. Mori et al. (1964) used the same method (striatum, hypothalamus etc.) but without inhibitor

TABLE 6.26 (continued)

Reference	Area studied	Method	Result	Comment
Vanha-Perttula (1966)	Anterior pituitary gland	Optical: Gomori (1952).	No glandular cells stained for AChE. In fixed but not fresh tissue thyrotrophic cells reacted weakly for BuChE; epithelium of cleft + ve for BuChE in fresh and fixed tissue. Capillary endothelium hydrolysed AcThCh & BuThCh but activity not prevented by eserine or selective inhibitors	Other esterases studied, with a variety of inhibitors, histochemically and biochemically
Waldron (1967)	Spinal cord	Optical: thiocholine, see note under Gwyn & Waldron (1968)	Descending AChE-containing fibres present in lateral and dorsal funiculi at all levels and in ventral funiculi at some levels. Ascending AChE + ve fibres in dorsal, ventral and lateral funiculi below T_8 and in lateral only above T_8	Lesion technique used to determine polarity of fibres
		PERIPHERAL NERVES ETC.		
Adams et al. (1967)	Sciatic and genitofemoral nerve	Optical (including polarized light): Karnovsky & Roots (1964)	In genitofemoral nerve (motor) axons contained only AChE. In the sciatic nerve (mixed) the unmyelinated fibres contained BuChE (see 6.1.2 re specificity) in addition to AChE	Inhibitors used in incubation but not preincubation medium. Great auricular nerve (sensory) of rabbit used for comparison (AChE in 5% of myelinated fibres, BuChE confined to unmyelinated fibres). Useful references to other studies. See also Grant et al. (1967); Schlaepfer & Torack (1966)

TABLE 6.26 (continued)

Reference	Area studied	Method	Result	Comment
Bogart (1970, 1971)	Submandibular salivary gland (1970); parotid and sublingual salivary glands (1971)	EM: Karnovsky (1964)	AChE in axolemma of fibres associated with ducts and blood vessels but other fibres – ve. In myoepithelial cells (most prominent in sublingual) end-product in RER, nuclear envelope, surface vesicles and pits. Innervation sparse in sublingual gland compared with others	
Bulmer (1965)	Ovary	Optical: Gomori (1952)	With AcThCh, a few fibres stain but activity abolished by BuChE inhibitors. Marked BuChE activity in granulosa cells and theca externa	Comparison with guinea-pig, rabbit and monkey. Author questions whether apparent lack of a cholinergic innervation is a true result. Jordan (1971) confirmed lack of fibre staining
Cauna et al. (1961)	Superior cervical, stellate and terminal visceral ganglia	Optical: Koelle (1955); Snell (1958a)	In *sympathetic* ganglia AChE levels varied in different neurones; satellite cells – ve. BuChE in neurones also variable. In *parasympathetic* ganglia in gut, neurones and satellite cells contained AChE. Most neurones – ve for BuChE; satellite cells weakly + ve	Several species compared

TABLE 6.26 (continued)

Reference	Area studied	Method	Result	Comment
Cauna & Naik (1963)	Sensory ganglia	Optical: Naik (1963)	AChE in variable amounts in neuronal cytoplasm. Nerve fibres moderately or weakly + ve. Satellite cells − ve. No BuChE activity	Comparison with guinea-pig, man, cat and mole
Coupland (1958)	Innervation of the pancreas	Optical: Coupland & Holmes (1957), full details in paper	AChE & BuChE in nerve fibres & associated neurones. BuChE also in muscle of ducts and large arteries. Acinar and Islet cells − ve	Comparison with cat and rabbit
Csillik et al. (1968)	Iris	EM & Optical: thiocholine methods given in paper	Dense network of AChE + ve fibres throughout iris. EM shows AChE + ve and − ve fibres in close association	Authors say + ve and − ve axons cannot be distinguished by optical microscopy (cf. Csillik & Koelle 1965; Eränkö & Räisänen 1965)
Dixon & Gosling (1971)	Innervation of ureter	Optical: Gomori (1952); Karnovsky & Roots (1964)	AChE in large and small fibres in ureteric muscle and in fine fibres adjacent to basal part of epithelium	Comparison with rabbit and guinea-pig. Fluorescent fibres had a similar distribution and EM (non-histochemical) confirms that the two types run close together
El-Badawi & Schenk (1966)	Innervation of bladder	Optical: modified Karnovsky & Roots (1964)	AChE + ve nerves in sub-epithelial layer and in muscles	Fluorescence studies too. Comparison with cat, dog and rabbit

TABLE 6.26 (continued)

Reference	Area studied	Method	Result	Comment
El-Badawi & Schenk (1970)	Innervation of epididymis	Optical: modified Karnovsky & Roots (1964)	AChE + ve fibres are richest in caudal part. They form perivascular plexuses	Fluorescence studies too. Comparison with cat, dog and rabbit. No photomicrographs of AChE for rat tissue
Ellison (1971)	Innervation of umbilical cord	Optical: Coupland & Holmes (1957)	AChE + ve fibres in vitelline but not allantoic, part of cord	Comparison with man
Eränkö et al. (1959)	Adrenal gland	Optical: Koelle (1951)	AChE + ve structures include nerve trunks, fibres in cortex, fine myelinated fibres round NA-containing cells, ganglion cell bodies and processes, and satellite cells. BuChE in capsule, glomerulosa fibres and in all structures listed above except the fine fibres	Splanchnic nerve section abolished AChE activity in all structures except ganglion cells and some cortical fibres. BuChE unchanged except for slight decrease in satellite cells and some fibres
Eränkö & Härkönen (1964)	Superior cervical ganglia	Optical: Gomori (1952)	AChE activity varied from cell to cell. Both AChE staining and NA fluorescence could occur in same cell	For results of pre- and post-ganglionic nerve section see Eränkö & Härkönen (1965)
Eränkö & Teräväinen (1967)	End-plate in diaphragm, gastrocnemius and anterior tibial muscle	Optical: Gomori (1952)	AChE in synaptic folds only. BuChE in synaptic folds and in teloglia	Eserine-resistant nonspecific esterases also present in folds and teloglia
Gwyn & Flumerfelt (1971)	Dorsal root ganglion	Optical: Lewis (1961). EM: Lewis & Shute (1966)	Variable AChE activity in RER and nuclear envelope. AChE in axolemma of some myelinated fibres. Unmyelinated fibres − ve	Dorsal root section caused accumulation of AChE in fibres on cell-body side of lesion

TABLE 6.26 (continued)

Reference	Area studied	Method	Result	Comment
Huikuri (1966)	Ciliary ganglion	Optical: Gomori (1952)	AChE variable in neurones but in general, activity only moderate. Preganglionic fibres mainly − ve but terminals + ve. Postganglionic denervation abolished stain in cells *and* preganglionic structures. BuChE in fibres, satellite cells, capillaries and some neurones. Activity reduced by pre- and postganglionic nerve section	Long paper. Deals with a number of enzymes and effects of pre- and postganglionic nerve section. Author suggests that rather moderate, graded AChE activity in ganglion cells is a special feature of rat
Joó et al. (1971)	Superior cervical ganglia	EM: Lewis & Shute (1966)	AChE + ve fibres made both axosomatic and axodendritic synapses on ganglion cells	Effects of acute and chronic preganglionic denervation examined
Karnovsky (1964)	Cardiac muscle	EM: thiocholine-ferricyanide method described in paper	EM indicates that diffuse stain seen at optical level may represent nonspecific ChE activity in sarcoplasmic reticulum and in A bands. Latter activity may be due to a myosincholinesterase	
Lewis & Shute (1969)	Adrenal gland	EM: thiocholine method described in paper	AChE in RER and isolated areas of nuclear envelope of ganglion cells. Axolemma of most myelinated and un-myelinated fibres + ve as were presynaptic endings on NA cells. Schwann cells − ve. BuChE in nuclear envelope	cf. Palkama (1967) who observed some BuChE activity in axons

TABLE 6.26 (continued)

Reference	Area studied	Method	Result	Comment
Lewis & Shute (1969) (continued)			and RER of Schwann cells but rarely in axolemma of fibres. Variable activity at nerve endings. Some very faint staining of parenchymal cells	
Mazza et al. (1973)	Trigeminal ganglion	EM: Lewis & Shute (1966)	Reaction product for AChE in neuronal nuclear envelope, RER, Golgi apparatus, mitochondria and between neurones and satellite cells. BuChE in mitochondria of nerve cells and in mitochondria, ER and nuclear envelope of satellite cells. Also between neurones and satellite cells	See ch. 4 §2.3 re activity in mitochondria
Navaratnam (1965)	Heart & cardiac ganglia	Optical: Lewis (1961)	AChE appears in ganglia 4 days after birth and +ve fibres appear in sinus node and AV node and bundle by 15th day. Extracellularly located BuChE appears in ganglia between 20th and 30th day	Developmental study. Comparison with man, rabbit and guinea-pig
Navaratnam et al. (1968)	Innervation of heart	Optical: Lewis (1961)	Dorsal motor vagal nucleus contains AChE & BuChE. Vagotomy causes loss of stain in cells of nucleus and *also* loss of BuChE but not AChE from cardiac ganglia	Effects of vagotomy suggest that BuChE in the cardiac ganglia is associated with preganglionic terminals

TABLE 6.26 (continued)

Reference	Area studied	Method	Result	Comment
Navaratnam & Palkama (1966)	Aorta, pulmonary trunk & coronary arteries	Optical: thiocholine method not specified	BuChE diffusely distributed in tunica media of aorta and pulmonary trunk. In coronary artery AChE present in nerve fibres and their cells. No BuChE	Same results in rabbit. See also El-Bermani et al. (1970) re paucity of AChE-containing fibres in pulmonary vasculature
Novikoff et al. (1966)	Dorsal root ganglia & sciatic nerve	EM: Karnovsky (1964); Karnovsky & Roots (1964)	AChE in ER and in axolemma of myelinated fibres. BuChE at the Schwann cell-axon interface of unmyelinated fibres and interfaces of satellite cells	Mouse, rabbit, cat, frog and toad also studied but paper mainly on rat. Nucleotide phosphatase and BuChE similarly distributed. See comment § 6.1.2
Pecot-Dechavassine (1962)	End-plates in diaphragm & gemelli muscle	Optical: Couteaux & Taxi (1952)	AChE & BuChE present but BuChE staining is weaker than expected on basis of biochemical analyses	Comparison with other species. See table 6.35 for biochemistry
Perrotta & Lewis (1958)·	Placenta	Optical: Gerebtzoff (1953); Lewis (1958)	Syncytiotrophoblast virtually – ve for AChE & BuChE. Some yolk-sac components + ve for AChE & – ve for BuChE	Comparison with man and guinea-pig
Schlaepfer (1968)	Anterior and posterior spinal roots and ganglia	EM & Optical: Karnovsky & Roots (1964)	In anterior root, AChE associated with extra- and intrafusal fibres and (?) preganglionic sympathetic	See also Schlaepfer & Torack (1966) for EM of sciatic nerve

TABLE 6.26 (continued)

Reference	Area studied	Method	Result	Comment
Schlaepfer (1968) (continued)			fibres. In posterior root, myelinated fibres − ve; some reaction in small unmyelinated fibres. In ganglia, neurones showed range of activity in ER. BuChE in ER of satellite cells only	
Snell (1958b)	Submandibular and sublingual salivary glands	Optical: Snell (1958a)	AChE in nerve fibres to glands. Preganglionic parasympathectomy does not affect staining. Weak activity in secretory cells and in duct cytoplasm	See also Snell & Garrett (1957, 1958)
Snell (1959)	Parotid salivary gland	Optical: method in paper	Glands stain more strongly than submandibular and sublingual. Considerable reaction product in secretory cell cytoplasm with AcThCh & BuThCh. Nerves on ducts and blood vessels also + ve	Author discounts possibility of artifacts and discusses significance of secretory cell stain. Postganglionic sympathectomy had no effect on cell or nerve stain
Tewari & Bourne (1963)	Trigeminal ganglion	Optical: Gomori (1952); Gerebtzoff (1959)	AChE activity varied from cell to cell, some well stained, some − ve. BuChE confined to cell periphery	pH not specified but in previous paper on cerebellum (Tewari & Bourne 1962) the pH was 6.7 and gave misleading results
Yamauchi & Lever (1971)	Superior cervical ganglion	Optical: Lewis & Shute (1966) following fluorescence technique	High AChE in cells with minimal fluorescence, moderate activity in moderately fluorescent cells and minimal activity in intensely fluorescent ones	Comparison with sheep and pig

TABLE 6.26.1
AChE staining in the CNS of rat. (From Koelle 1963.)

Region	Stained structures
Areas of intense staining	
Reticular formation (medulla)	Scattered neurones and fibres throughout
Nuc. of lateral funiculus (medulla)	Larger neurones most heavily stained
Nuc. of Roller	Scattered neurones and fibres
Ventral nuc. of reticular formation	Neurones and fibres show fairly consistent heavy staining
Nuc. of trapezoid body	Numerous scattered neurones and fibres
Dorsal and median nuc. of the raphe (midbrain)	Small clusters of heavily stained neurones and fibres
Dorsal, caudal ventral and rostral ventral nuc. of lateral lemnicus (midbrain)	Small, scattered, heavily stained neurones and fibres
Cerebellar cortex	Scattered cells in granular layer, mostly of large type; few stellate cells in molecular layer; numerous fibres of medulla
Pontine nuc.	Numerous scattered neurones and fibres very heavily stained
Nuc. interpeduncularis	Thickly clustered very heavily stained neurones and fibres of pars lateralis; occasional neurones of pars medialis
Superior colliculus	Densely packed, moderately heavily stained neurones of stratum griseum; scattered fibres of non-optic layer of stratum opticum
Fasciculus retroflexus of Meynert	Most fibres heavily stained in cross section
Zona incerta	Few neurones and scattered fibres
Lateral habenular nuc.	Few small neurones and fibres; majority moderately stained
Anterior nuc. of thalamus	Few scattered neurones and numerous fibres of ventro-lateral portion
Lateral nuc. of thalamus	Heavily stained neurones and fibres interspersed with unstained tracts
Caudate nuc.	Very heavily stained, closely-packed neurones and fibres; interspersed bundles of internal capsule largely unstained
Putamen	Densely packed, heavily stained neurones and fibres
Central nuc. of amygdala	Scattered, very heavily stained fusiform neurones and fibres
Lateral nuc. of amygdala	Scattered, very heavily stained smaller globular neurones and fibres
Nuc. of lateral olfactory tract	Occasional neurones and fibres (majority unstained)
Tuberculum olfactorium	Densely packed small neurones and fibres
Diagonal band of Broca	Scattered neurones and fibres, many unstained
Nuc. accumbens septi	Numerous neurones and fibres

TABLE 6.26.1 (continued)

Region	Stained structures
Areas of moderate staining	
Dorsal gray column of cord	Occasional neurones and numerous fibres of dorsal columns and substantia gelatinosa
Nuc. gracilis and cuneatus	Most neurones and fibres, including internal arcuate fibres
Nuc. of lateral funiculus (medulla)	Numerous neurones and fibres
* Olivary nuc.	Great variation in intensity; few neurones and fibres moderately stained, majority lightly stained or unstained
Nuc. of fasciculus solitarius	Great variation; some neurones and fibres moderately stained, remainder lightly stained or unstained
Nuc. of Deiters	Numerous large neurones and fibres
Chief vestibular nuc.	Diffuse staining of most neurones and fibres; few unstained
Nuc. emboliformis	Occasional neurones; majority lightly stained or unstained
Dentate nuc.	Diffuse staining of most neurones and fibres; few unstained
Ventral cochlear nuc.	Larger neurones moderately stained
Dorsal and laterodorsal tegmental nuc.	Numerous small, globular neurones stained with varying intensity
Locus coeruleus	Small cluster of small, moderately stained neurones
Pontine nuc.	Occasional neurones, numerous fibres
Inferior colliculus	Few neurones and fibres in nuc. of inf. col.; occasional fibres elsewhere
Posterior nuc. of thalamus	Few small, scattered neurones; majority of fibres
Mammillo-thalamic tract	Most fibres in cross section
Medial lemnicus	Most fibres in cross section
Medial geniculate body	Few scattered small neurones and fibres
Substantia nigra	Variable intensity of staining; few neurones and fibres moderate, remainder light
Lateral geniculate body	Few neurones of dorsal, more of ventral nucleus; scattered fibres
Lateral habenular nuc.	Majority of neurones, scattered fibres
Commissure of Meynert	Few fibre bundles
Medial habenular nuc.	Few neurones and fibres; remainder light
Supraoptic nuc.	Globular neurones and fibres show uniform moderate staining; highly vascular
Anterior nuc. of thalamus	Numerous neurones and fibres of dorso-medial portion
Medial nuc. of thalamus	Occasional neurones and fibres
Nuc. reticularis of thalamus	Occasional small neurones, numerous scattered fibres
Medial forebrain bundle	Numerous fibres in cross section of lateral division
Lateral olfactory tract	Most fibres in cross section
Fimbria	Occasional neurones and fibres
Dorsal hippocampal commissure	Occasional neurones and fibres
Neocortex	Rare neurones, majority unstained; scattered fibres

TABLE 6.26.1 (continued)

Region	Stained structures
Areas of light staining	
Dorsal root ganglia of cord	Neurones faintly stained, fibres practically unstained
Dorsal gray column of cord and medulla	Majority of neurones and fibres
* Olivary nuc.	Majority of neurones and fibres
Nuc. of fasciculus solitarius	Great variation; most neurones lightly stained or unstained
Chief sensory nuc. of C.N. V	Most neurones and fibres; few more heavily stained neurones near surface
Cerebellar cortex	Scattered fibres in molecular layer
Dorsal longitudinal bundle of Schütz	Scattered small neurones and fibres
Mesencephalic nuc. of C.N. V	Large, globular, very lightly stained neurones
Red nuc.	Large neurones of magnocellular division
Substantia nigra	Majority of neurones and fibres
Posterior commissure (midbrain)	Most fibres lightly stained in cross section
Globus pallidus	Scattered neurones and fibres
Nuc. reuniens	Few lightly stained neurones
Internal capsule	Most fibres lightly stained or unstained, few moderately or heavily
Areas of slight or negligible staining	
Cervical cord	Columns of Goll and Burdach, pyramidal tract, ventral funiculus (most fibres), lateral funiculus, tract of Lissauer
Root of C.N. V	Cross section of fibres
Cerebellar cortex	Purkinje cells, numerous small granular cells
Reticular formation of medulla	Numerous small neurones and fibres
* Olivary nuc.	Numerous small neurones and fibres
Fasciculus solitarius	Numerous neurones and fibres
Inferior colliculus	Majority of fibres
Posterior nuc. of thalamus	Majority of neurones
Diencephalon	Neurones and fibres of intermediate mass and tuber cinereum
Optic chiasma	Majority of fibres (few lightly stained)
Corpus callosum	Majority of fibres
Lateral olfactory tract	Majority of fibres
Anterior commissure	Most fibres of anterior limb
Medial forebrain bundle	Most fibres of lateral division
Neocortex	Majority of neurones and fibres; few lightly or moderately stained

* No indication of whether superior or inferior.

TABLE 6.27
RAT BIOCHEMISTRY
CNS and peripheral tissue

Reference	Area studied	Method/substrate	Result	Comment
Austin and Phillis (1965)	Cerebellum	Manometric: ACh, MeCh and BuCh	AChE activity in cortex lowest of any species examined but values for peduncles among the highest	Comparison with several other species
Bennett et al. (1966)	Brain: 15 separate areas	Titrimetric: Rosenzweig et al. (1958); ACh	AChE activity per unit weight decreased in rats between 100 and 150 days of age. Comprehensive tables of values for different areas	Comparison of 6 strains of rat (see also Bennett et al. 1958b)
Elkes and Todrick (1955)	Brain: various areas; also spinal cord	Manometric: Warburg; MeCh and BzCh. BW62C47 as AChE inhibitor	Tabulates both AChE and pseudoChE in the various parts of the brain at ages between 8 and 77 days. Rate of change in activity of enzymes also studied in whole brain and spinal cord	Developmental study (see also Bayliss and Todrick 1953b)
Esila (1963)	Retina	Starch gel electrophoresis; AcThCh and BuThCh etc. plus inhibitors		See table 6.32.2
Giacobini and Holmstedt (1958)	Spinal cord	Electrometric: Tammelin (1953). Cartesian diver: Giacobini (1956b); AcThCh. Mipafox as inhibitor	Measurements of AChE in single anterior horn cells suggest * 2 populations, one with high and one with low activity	Comparison with histochemical findings. * Later work (Brzin et al. 1966) suggests this may reflect differences in substrate penetration rather than real differences in activity
Goldberg and McCaman (1967)	Cerebellum and other brain areas	Spectrophotometric: Ellman et al. (1961); AcThCh	For analysis of different layers of the cerebellum see table 6.33; for other tissues see table 6.27.1	Cerebellar results compared with those from cat, guinea-pig, rabbit and pigeon. ChAc also studied
Hobbiger and Lancaster (1971)	Slices of temporoparietal region and whole brain homogenates	Spectrophotometric: Ellman et al. (1961) with AcThCh. Titrimetric: Jensen-Holm et al. (1959); MeCh, BuCh plus iso-OMPA	Comparison of values for homogenates and slices show that diffusion of substrate into slices is limited	Authors conclude that AChE activity of slices does not indicate the extracellular AChE. See also Lancaster (1971)

TABLE 6.27 (continued)

Reference	Area studied	Method/substrate	Result	Comment
Kása and Silver (1969)	Cerebellum	Spectrophotometric: Ellman et al. (1961); AcThCh plus ethopropazine as inhibitor	Results discussed in detail in ch. 7 (tables 7.4, 5)	ChAc measured and effect of lesions on activity of both enzymes examined. Comparison with guinea-pig (see table 7.4)
Klingman et al. (1968)	Sympathetic ganglia; also brain and heart	Spectrophotometric: Ellman et al. (1961); AcThCh and BuThCh; BW284C51 and ethopropazine as inhibitors	See table 6.27.2	Some cat tissues included
McCaman et al. (1965)	Cerebral hemispheres	Spectrophotometric: Ellman et al. (1961); AcThCh on subcellular fractions	Subcellular distribution shown in table 6.34	ChAc, 5-HTPD and MAO also studied. Comparison with guinea-pig, rabbit and pigeon
McGeer et al. (1971a)	Caudate nucleus	Spectrophotometric; see McGeer et al. (1969); ACh	AChE at 4 days of age is 70% of the adult value which is reached at 15–16 days	Developmental study, 4 days postnatal to adulthood. Tyrosine hydroxylase, glutamic dehydrogenase and ChAc also studied
McLennan (1954)	Superior cervical ganglion	Manometric: Warburg; ACh	Axotomy reduced ChE activity by 50%	Normal and axotomized ganglia. ChAc levels unaffected by axotomy
Metzler and Humm (1951)	Whole brain	Microtitrimetric: Glick (1938); ACh	ChE activity reaches peak in 4th postnatal week, then declines. Adult level only slightly more than half peak value	Developmental study; 14-day foetus to adults. Some data from chick and salamander
Namba (1971)	Muscle fibres and endplates in intercostal and other muscles	Colorimetric: Hestrin (1949) ACh and BuCh on subcellular fractions	ChE in membrane fraction exceeded that in microsomal fraction. Activity per end-plate was higher than that in other species	Comparison with guinea-pig, man, mouse, dog and chicken. EM showed end-product in transverse tubules and sarcoplasmic reticulum

TABLE 6.27 (continued)

Reference	Area studied	Method/substrate	Result	Comment
Pavlin (1966)	Individual neurones, glial clumps and brain homogenates	Magnetic diver: Brzin et al. (1964); ACh, MeCh, BuCh and PrCh, BW284C51 and iso-OMPA used in some experiments	ACh hydrolysed to greater or lesser degree by nearly all reticular neurones. Low activity towards BuCh also shown by most cells but 50% inactive towards MeCh. Glia and neurones also active towards PrCh	Note, particularly, presence of BuChE in most neurones. Includes some values for man. See also Pavlin (1965)
Pecot-Dechavassine (1962)	Diaphragm and gemelli muscle	Manometric: Warburg; ACh and BuCh	See table 6.35	Comparison with several species
Piras et al. (1970)	Brain homogenates and subcellular fractions	Colorimetric: modified Hestrin (1949); ACh and BW284C51	Specific activity of AChE increased with age in fraction containing synaptic debris and nerve endings but fell in mitochondrial fraction	Postnatal developmental study. Other enzymes examined
Sinha and Rose (1972)	Cerebral cortex: neurones and neuropil separated by density gradient	Spectrophotometric: Ellman et al. (1961); AcThCh and BuThCh. Inhibitors used but not specified	Main finding is that AChE and BuChE are present in both types of tissue fraction so cannot be used as differential markers	MAO also measured
Sperti and Sperti (1959a, b)	Cerebellum	Manometric: Warburg; MeCh and BuCh	Values given. Lesions reduce AChE but not BuChE	Object of experiment is to see effect of peduncular section or of splitting the cerebellum or of cerebral cortical lesions on enzymes
Vlk and Tuček (1962a)	Heart	Colorimetric: Hestrin (1949); ACh, MeCh and BzCh	ChE activity varies in different parts, being lowest in left ventricle	Data also given for dog, cat and rabbit. Only in rat is there any parallelism between value for AChE and ACh

TABLE 6.27.1

Comparison of ChAc and AChE activity in certain areas of rat brain. (From Goldberg and McCaman 1967.)

Tissue	* Choline acetylase	‡ Acetylcholinesterase
Mamillary body	52.5 ± 4.93 (5)	3498 ± 638 (4)
Caudate nucleus (Head)	22.0 ± 3.24 (7)	3866 ± 479 (5)
Dorsal Grey composed of Gracilis, Cuneatus and Spinal 5th Nuclei	12.7 ± 0.21 (5)	1760 ± 75 (3)
Olfactory bulb (Undissected)	11.7 ± 1.20 (8)	1693 ± 316 (4)
Thalamic nuclei (Caudal portion)	10.6 ± 0.98 (10)	2010 ± 258 (4)
Hippocampus	9.88 ± 1.15 (5)	2692 ± 303 (4)
Thalamic nuclei (Rostral portion)	4.60 ± 0.89 (9)	1037 ± 144 (4)

* Expressed as μmoles of acetylcholine synthesised/g dry wt/hr. ± Standard error (number of sections).

‡ Expressed as μmoles of acetylthiocholine hydrolysed/g dry wt/hr. ± Standard error (number of sections).

The biology of cholinesterases

TABLE 6.27.2
Cholinesterase activity in [†]rat sympathetic ganglia. (From Klingman et al. 1968.)

Ganglion	Wt. ganglion mg	Cholinesterases			Number of ganglia
		Total	Acetyl-	Pseudo-	
Right superior cervical	1.69 ± 0.19	28.5 ± 2.6	16.4 ± 1.7	9.7 ± 1.0	8
Left superior cervical	1.55 ± 0.15	34.3 ± 5.0	20.2 ± 3.6	12.7 ± 1.4	9
Paired superior cervical, right	1.58 ± 0.15	27.7 ± 2.0	16.1 ± 1.3	9.6 ± 0.6	11
Paired superior cervical, left	1.44 ± 0.11	34.6 ± 4.8	20.4 ± 3.2	11.4 ± 1.1	11
Cervical trunk	1.52 ± 0.14	11.0 ± 2.1	7.5 ± 1.6	3.0 ± 0.4	7
Right stellate	1.99 ± 0.23	36.7 ± 5.5	23.1 ± 4.5	12.6 ± 1.4	8
Left stellate	2.06 ± 0.21	35.7 ± 2.3	19.9 ± 1.7	12.6 ± 1.1	10
Upper thoracic chain *	1.99 ± 0.22	19.7 ± 2.2	11.1 ± 1.2	7.3 ± 0.8	7
Left cardiac (abdominal)	1.65 ± 0.29	17.3 ± 1.8	10.1 ± 0.9	5.3 ± 1.0	6
Coeliac	1.93 ± 0.44	23.5 ± 0.7	13.9 ± 1.1	9.1 ± 1.2	8
Superior mesenteric	2.26 ± 0.14	32.5 ± 5.1	18.1 ± 2.2	12.1 ± 2.9	7
Cat, superior cervical	13.8 ± 0.6	27.6 ± 1.1	17.4 ± 0.6	8.6 ± 0.8	6
Cat, cervical trunk	4.4 ± 0.7	8.7 ± 1.7	6.6 ± 1.4	1.2 ± 0.0	3

* Three ganglia from the right and left thoracic chain caudally to the stellate ganglion were pooled; the thoracic sympathetic trunk was included.

Activity expressed as μmoles substrate hydrolysed/min/g ganglion.
$\pm$ Standard error
[†] Some cat tissues included.

TABLE 6.28
REPLILES

Histochemistry and biochemistry: CNS and peripheral tissue

Species	Reference	Area studied	Method	Result	Comment
LIZARDS					
Green lizard (*Lacerta viridis*)	Capurro et al. (1959b)	Retina	Histochemical		See table 6.32
Sand lizard (*Lacerta agilis*)					
Gecko (*Hemidactylus bowringi*)	Liu and Maneely (1968)	Motor end-plates in embryonic and regenerating tail	Histochemical: modification of Couteaux (1951)	ChE-containing end-plates appear first at end of myotubules near intersegmental zone but later occupy middle of neurotubules. In early stages of development and regeneration intersegmental connective tissue also reacts	
SNAKES					
Grass snake (*Natrix natrix*)	Hebb (1957b)	Retina	Histochemical: Koelle (1950)	AChE present. No details given	Comparison with other species
Horned viper (*Vipera aspis*)	Bastide et al. (1967)	Brain, kidney etc.	Biochemical: modification of Hestrin (1949)	High level of pseudoChE in brain and kidney	Other enzymes studied
Desert Cobra (*Naje haje*)	Mohamed et al. (1969)	Venom	Biochemical: modification of Hestrin (1949)	Venoms contained cholinesterase and proteinase	Venom from a number of other snakes had no ChE activity, neither did that of scorpion, *Buthus quinquestriatus*
Egyptian Cobra (*Walterinnesia aegyptia*)					

TABLE 6.28 (continued)

Species	Reference	Area studied	Method	Result	Comment
TURTLES					
Common Greek or European turtle (*Testudo graeca*)	Augustinsson (1959b, c)	Plasma	Biochemical: Warburg with various substrates and inhibitors	Plasma contains an eserine-sensitive esterase which hydrolyses choline esters less rapidly than non-choline esters	Augustinsson suggests that phylogenetically this enzyme is intermediate between the cholinesterases and ali-esterases (see ch. 10 § 3.1)
Box turtle (*Terrapene carolina*)	Brightman and Albers (1959)	Brain and spinal cord	Histochemistry: Koelle (1951)	AChE activity strong in ventral horn cells and substantia gelatinosa. Neuroglia unstained. No BuChE activity	Comparison with several other species
Cumberland turtle (*Pseudemys scripta troosti*) European turtle (*?Testudo graeca*)	Capurro et al. (1959b)	Retina	Histochemical		See table 6.32

288

TABLE 6.29

SHEEP HISTOCHEMISTRY

CNS

Reference	Area studied	Method	Result	Comment
Arvy (1961)	Pineal gland	Optical: method not specified	AChE – ve. BuChE very feeble, no details of localization	Paper in French. Comparison with pig and cow. See also Arvy (1965)
Arvy (1962)	Hypothalamoneuro-hypophysial system	Optical: Gerebtzoff (1959)	Strong AChE activity and some BuChE present. No details of localization	3 foetuses (youngest 112 days gestation) gave results as for adult. Comparison with several species. Other enzymes studied
Brightman and Albers (1959)	Brain and spinal cord	Optical: Koelle (1951)	Weak BuChE activity, too diffuse to be definitely assigned to neuroglia. Activity towards AcThCh not tested in sheep	Various species compared
Esila (1963)	Retina			See table 6.32 and 6.32.1 for Esila's own results and further references
Palmer and Ellerker (1961)	Brain stem	Optical: modified Koelle	AChE and BuChE + ve structures listed. BuChE present in a number of neuronal groups as well as in neuro-pil. Vessels and glia of area postrema showed AChE reaction	Includes an appendix (Hardwick and Palmer 1961) on the effect of formaldehyde fixation
Phillis (1965b)	Cerebellum	Optical: Gerebtzoff (1959)	Some AChE in molecular layer. Much stronger activity in granular layer. No data for BuChE	Various species compared
Silver (1967, 1969, 1971)	Cortex, cerebellum and midbrain	Optical: Lewis (1961)	AChE demonstrated in developing cells including Purkinje cells. Some loss of activity during maturation	Papers mainly concerned with foetal tissue

TABLE 6.29 (continued)

PERIPHERAL NERVES ETC.

Reference	Area studied	Method	Result	Comment
Hebb and Hill (1955b)	Pancreas and pancreatic juice	Optical: Koelle (1950). For juice the histochemical medium was used and production of a precipitate noted. Also ACh hydrolysis (bioassay)	AChE confined to nerves and some unidentified cells. No BuChE activity. Juice unreactive	Comparison with several species
Mann (1971)	Innervation of bronchial muscle	Optical: Lewis (1961; see Hebb and Silver 1970)	AChE + ve fibres supplied smooth muscle and bronchial glands and blood vessels. Ganglia in extrachondrial plexus + ve. BuChE present in smooth muscle and ganglia	Fluorescence studies included. Comparison with cow, rabbit, piglet and goat
Yamauchi and Lever (1971)	Superior cervical ganglion	Optical: Lewis and Shute (1966) following fluorescence technique	AChE activity high in cells showing minimal fluorescence; moderate activity in moderately fluorescent cells and minimal activity in intensely fluorescing cells	Comparison with pig and rat

TABLE 6.30

SHEEP BIOCHEMISTRY

Reference	Area studied	Method/substrate	Result	Comment
Austin and Phillis (1965)	Cerebellum	Manometric: ACh, MeCh, BuCh	All 3 substrates hydrolysed by cortex. Values also given for nuclei and peduncles. In sheep, all 3 peduncles give fairly * similar result	Comparison with several other species. In some of these, the different peduncles show * different activity
Davies et al. (1953)	Kidney and blood and other tissues (not specified)	Specificity towards ACh, MeCh, BzCh, BuCh and PrCh examined. Method not given	Results indicate variability of specificity patterns in ruminant tissue compared with rat	Use of BuCh as well as BzCh proved existence of pseudoChE in ruminant tissue (see ch. 10 § 3.1 for earlier refs suggesting its absence)
Nachmansohn (1940)	Brain and spinal cord	Manometric: Warburg; ACh	Total ChE activity (fresh wt) changes at different rate in different parts of CNS. Enzyme present in muscle at nerve endings before formation of end-plates	Foetuses of 60–138 days used
Ruckebusch and Ruckebusch (1959)	Whole blood, serum and erythrocytes	Potentiometric: Michel (1949) with ACh	Total ChE in serum higher than in other ruminants but lower than non-ruminants. Erythrocyte value exceeded only by pig	Paper in French. Comparison with horse, cow, rabbit, goat, dog, pig and fowl

TABLE 6.31

MISCELLANEOUS SPECIES: HISTOCHEMISTRY and BIOCHEMISTRY
CNS and peripheral tissue

	Reference	Area studied	Method	Result	Comment
BEAR	Augustinsson (1948)	Cortex and caudate nucleus	Biochemical: Manometric; Warburg with ACh, MeCh and BzCh	Activity towards ACh high in caudate and low in cortex. No detectable hydrolysis of BzCh in cortex	Comparison with other species
COYPU (*Myocastor coypus*)	Girgis (1967)	Basal rhinencephalon	Histochemical: Lewis (1961)	AcThCh without inhibitors showed many + ve structures	Authors comment that AChE-containing structures are well developed in phylogenetically old areas
DEER (Red) (*Cervus elaphus*)	Jenkinson and Maloiy (1969)	Skin	Histochemical: Naik (1963)	AChE and BuChE present in same nerve fibres. BuChE widely distributed in other structures including sebaceous and sweat glands. Slight reaction for AChE in myoepithelium of latter	Monoamineoxidase in nerves devoid of cholinesterases
ELEPHANT	Augustinsson (1948)	Whole brain	Biochemical: Manometric; Warburg with ACh, MeCh, BzCh	Activity towards ACh and MeCh but very little towards BzCh	Comparison with other species
FERRET	Holmes (1961b)	Hypothalamus and pituitary gland	Histochemical: Coupland and Holmes (1957)	AChE but no BuChE in neurones of magnocellular nuclei. Neuropil + ve for AChE. Slightly + ve for BuChE. Variable reaction towards AcThCh and BuThCh in pituitary stalk and infundibular process	Comparison with other species. Non-specific esterases in glial cells
	Trueman and Herbert (1970)	Pineal gland and habenula	Histochemical: Koelle and Friedenwald (1949)	AChE in fibres and neurones of ganglion in posterior pineal; also in fibres connecting posterior and anterior part. Neurones and fibres of medial habenula strongly + ve for AChE, staining in lateral habenula mainly con-	Paper also deals, in some detail, with monoamines

TABLE 6.31 (continued)

	Reference	Area studied	Method	Result	Comment
GOAT	Hebb and Hill (1955b)	Pancreas	Histochemical: Koelle (1950)	AChE in nerves and some un-identified cells. No BuChE detected	Comparison with other species
	Mann (1971)	Innervation of bronchial muscle	Histochemical: Lewis (1961, see Hebb and Silver 1970)	AChE + ve nerve fibres in smooth muscle and bron-chial glands. Some fibres to blood vessels + ve. Ganglia in extrachondrial plexus + ve. BuChE in smooth muscle and ganglia	Fluorescence studies included. Comparison with several other species
	Ruckebusch and Ruckebusch (1959)	Whole blood, serum and erythrocytes	Biochemical: Potentio-metric; Michel (1949) with ACh	Activity in erythrocytes lower than that in pig and sheep, slightly greater than in cow. Serum level very low	Comparison also with dog, horse, rabbit and hen. N.B. AChE and pseudoChE not dis-tinguished
	Silver (1963)	End-plates in tensor fasciae latae and superior rectus muscles	Histochemical: Lewis (1961)	End-plates reacted for AChE and BuChE. AChE hydrolys-ed PrThCh in addition to AcThCh	Comparison with rabbit, guinea-pig, monkey and hen
HAMSTER (Mesocricetus auratus)	Friede and Fleming (1964)	Cerebellum	Histochemical: Koelle (1954); Gerebtzoff (1959)	Reaction for AChE only very weak; granular layer slight-ly more + ve than white matter. Diffuse BuChE ac-tivity in granular layer	Comparison with several other species
HEDGEHOG (Erinaceus europaeus)	Cauna (1961)	Skin of nose, pads and nipple	Histochemical: Koelle (1955); Snell (1959)	Nerves associated with sense organs contained AChE. Myelinated stem-fibres as well as nerve endings + ve in hedgehog but − ve in most other species	Comparison with man, cat, rat and guinea-pig

TABLE 6.31 (continued)

	Reference	Area studied	Method	Result	Comment
HEDGEHOG (continued)	Holmes (1961b)	Hypothalamus and pituitary gland	Histochemical: Coupland and Holmes (1957)	AChE and BuChE in magnocellular nuclei. These cells also contained non-specific esterase. In the infundibular process reaction towards AcThCh intense; less reaction with BuThCh. In some animals eserine-resistant activity found	Comparison with other species
	Waldron (1969)	Spinal cord (lateral cervical nucleus)	Histochemical: Lewis (1961)	AChE activity variable in different cells of nucleus which extends whole length of cord	
KANGAROO	Winkelman (1966)	Skin of nose, mouth etc.	Histochemical: Gomori (1952)	AChE-containing fibres sparse, mainly sympathetic to blood vessels and glands. Sensory endings contain BuChE, particularly those on hair follicles. Some sebaceous glands BuChE + ve	Alkaline phosphatase also studied
MOLE (Talpa europaea)	Cauna and Alberti (1961)	Skin of nose	Histochemical: Koelle (1951, 1955); Snell (1959)	AChE in various nerve trunks and free endings. Plexuses round hairs + ve; reaction very strong in fibres and cells at base of Eimer's sense organs. Sebaceous glands − ve. No reaction for BuChE	Silver staining as well. Role of various sense organs and free nerve endings discussed
	Cauna and Naik (1963)	Sensory ganglia (semilunar, nodose, and dorsal root)	Histochemical: Naik (1963)	AChE in neuronal cytoplasm and fibres. Satellite cells − ve. No BuChE activity	Comparison with man, cat, guinea-pig and rat

TABLE 6.31 (continued)

	Reference	Area studied	Method	Result	Comment
POSSUM (*Trichosurus vulpecula*)	Harrex (1971)	Carotid body	Histochemical (EM?): Karnovsky and Roots (1964)	AChE round blood vessels and in axons. Weak activity in cytoplasm of type I cells. BuChE activity more pronounced than AChE, mainly in cytoplasm of type II cells	
SQUIRRELS a) Grey *(Neosciurus carolinensis)*	Hebb (1957b)	Retina	Histochemical: Koelle (1950)	AChE present. No details given	Pure cone retina
b) Ground *(Citellus mexicanus)*	Nichols and Koelle (1968)	Retina	Histochemical: Koelle (1951, 1955); Koelle and Gromadski (1966)	AChE activity in inner plexiform layer and cells which are probably amacrine. Other cells of ganglion cell layer – ve. BuChE in some horizontal cells	Comparison with pigeon, cat, rat and rabbit. Pattern in pigeon (cone retina) is similar
c) Red *(Sciurus ruber)*	Friede and Fleming (1964)	Cerebellum	Histochemical: Koelle (1954); Gerebtzoff (1959)	AChE activity strong in both molecular and granular layer. Purkinje cell membrane + ve. BuChE in granular layer	Comparison with several other species

TABLE 6.32

Reported histochemical distribution of cholinesterases in the retina of various species.
(Adapted from Esila 1963.)

SPECIES	SUBSTRATE	INHIBITOR	NATURE OF ENZYME	Nerve fibre layer	Ganglion cell layer	Inner plexiform layer	Amacrine cells	Inner nuclear layer	Bipolar cells	Outer plexiform layer	Outer nuclear layer	Inner segments of visual cells	Outer segments of visual cells	AUTHOR
BIRDS														
Domestic fowl	AcThCh		ChE's	+	+	+	+							Capurro et al. (1958)
Domestic fowl (chick)	AcThCh	DFP 10^{-7} M	AChE	+	+	+	+			±		±		Shen et al. (1956)
Goldfinch	AcThCh (?)		ChE's			+								
Greenfinch	AcThCh (?)		ChE's		+	+	+							Capurro et al. (1958)
Hawfinch	AcThCh (?)		ChE's		+	+	+							
Italian sparrow	AcThCh (?)		ChE's			+								
Parrot	AcThCh (?)		ChE's			+								Francis (1953)
Pigeon	AcThCh (?)		ChE's			+								
Rock dove	AcThCh (?)		ChE's			+	+							Capurro et al. (1958)
Songthrush	AcThCh (?)		ChE's		+	+	+							
CAT	AcThCh (?)		ChE's			+								Eichner (1959)
	AcThCh	DFP 10^{-7} M	AChE				+			±				Koelle et al. (1952)
	AcThCh	DFP 10^{-7} M	AChE			+								Leplat & Gerebtzoff (1956)
FISH														
Angelfish	AcThCh	DFP 10^{-7} M	AChE		+	+	+			+				Capurro et al. (1959a)
	BuThCh		PseudoChE		+	+	+							
Gambusi	AcThCh	DFP 10^{-7} M	AChE		+	+	+							
	BuThCh		Pseudo ChE			+	+							
Goldbream	AcThCh	DFP 10^{-7} M	AChE		+	+	+							Capurro et al. (1959a)
	BuThCh		PseudoChE		+		+							
Goldfish	AcThCh	DFP 10^{-7} M	AChE		+	+	+			+				
	BuThCh		PseudoChE		+		+				+			
Herring	AcThCh	DFP 10^{-7} M	AChE		+	+	+							
	BuThCh		PseudoChE	+		+								
Minnow	AcThCh		ChE's			+								Francis (1953)

TABLE 6.32 (continued)

SPECIES	SUBSTRATE	INHIBITOR	NATURE OF ENZYME	Nerve fibre layer	Ganglion cell layer	Inner plexiform layer	Amacrine cells	Inner nuclear layer	Bipolar cells	Outer plexiform layer	Outer nuclear layer	Inner segments of visual cells	Outer segments of visual cells	AUTHOR
FROG	AcThCh	DFP 10^{-7} M	AChE		±	+	±			+				Boell et al. (1955)
	AcThCh	ISO-OMPA 10^{-6} M	AChE			+	±			±	±			Eränkö et al. (1961)
	BuThCh	62C47 10^{-5} M	PseudoChE			±				±				Francis (1953)
	AcThCh		ChE's											
GUINEA-PIG	AcThCh	DFP 10^{-7} M	AChE		+	+								Capurro et al. (1959c)
	AcThCh	ISO-OMPA 10^{-6} M	AChE			+	+			±				Eränkö et al. (1961)
	BuThCh	62C47 10^{-5} M	PseudoChE			±								Francis (1953)
	AcThCh		ChE's			+								
MAN	AcThCh		ChE's			+				±	±			Eichner (1956)
	AcThCh	DFP 10^{-7} M	AChE			+				+		+		Eichner (1957)
	AcThCh	DFP 10^{-7} M	AChE	±	+	+				+		+		Eichner (1958)
	AcThCh (?)		ChE's			+				+		+		Eichner (1959)
	AcThCh	DFP 10^{-7} M	AChE		+	+	±			+		±		Viale and Apponi (1961)
	BuThCh		Pseudo ChE		+	+	±					±		
MOUSE	AcThCh	ISO-OMPA 10^{-6} M	AChE			+				±				Eränkö et al. (1961)
	BuThCh	62C47 10^{-5} M	PseudoChE			±								
OX * (Cow)	AcThCh	DFP 10^{-6} or 10^{-7} M	AChE		+									Ábrahám (1960)
	BuThCh		PseudoChE		+	+								Eichner (1955a)
	AcThCh		ChE's			+		+						Eichner (1955b)
	AcThCh		ChE's			+								
PIG	AcThCh		ChE's			+								Francis (1953)
	AcThCh	DFP 10^{-8} M	AChE		+	±	+							Hebb et al. (1953)
	BuThCh		PseudoChE		+	+	+							

TABLE 6.32 (continued)

SPECIES	SUBSTRATE	INHIBITOR	NATURE OF ENZYME	Nerve fibre layer	Ganglion cell layer	Inner plexiform layer	Amacrine cells	Inner nuclear layer	Bipolar cells	Outer plexiform layer	Outer nuclear layer	Inner segments of visual cells	Outer segments of visual cells	AUTHOR
RABBIT	AcThCh	ISO-OMPA 10^{-6} M	AChE			+	±			±				Eränkö et al. (1961)
	BuThCh	62C47 10^{-5} M	PseudoChE			±	±			±		±		Francis (1953)
	AcThCh		ChE's		+	±	±							
	AcThCh	DFP 10^{-8} M	AChE		+	±	+			+				Hebb et al. (1953)
	BuThCh		PseudoChE		+	+	+		+					
	AcThCh		ChE's		+	+	+		+					Koelle & Friedenwald (1950)
	AcThCh	DFP 10^{-8} M	AChE		±	+	±							Raviola & Raviola (1961)
	AcThCh		ChE's			+								Raviola & Raviola (1962)
RAT	AcThCh	ISO-OMPA 10^{-6} M	AChE			+								Eränkö et al. (1961)
	BuThCh	62C47 10^{-5} M	PseudoChE			±				±				Francis (1953)
	AcThCh		ChE's			+								
REPTILES														
Cumberland turtle	AcThCh	DFP 10^{-7} M	AChE		+	+	±							
	BuThCh		PseudoChE											
European turtle	AcThCh	DFP 10^{-7} M	AChE		+	+	±		+	±	+			Capurro et al. (1959b)
	BuThCh		PseudoChE											
Green lizard	AcThCh	DFP 10^{-7} M	AChE		±	+								
	BuThCh		PseudoChE											
Sand lizard	AcThCh	DFP 10^{-7} M	AChE		+	+	±			±				
	BuThCh		PseudoChE											
SHEEP	AcThCh	DFP 10^{-7} M	AChE			+	±			+				Francis (1953)

* Esila lists Anfinsen (1944). This was a microchemical study and has been included in table 6.7.

298

TABLE 6.32.1

Histochemical localization of cholinesterases in the retina of different species. (Modified from Esila 1963.)

Retinal layer	SPECIES									
	Cat		Cow		Dog		Guinea-pig		Horse	
	AChE	BuChE	AChE	BuChE	AChE	BuChE	AChE	BuChE	AChE	BuChE
Müller's cells						1				
Nerve fibre layer	$\frac{1}{2}$			1		$2\frac{1}{2}$				
Ganglion cell layer	$\frac{1}{2}$		1 *	3 *	*	3	*	1 *		*
Inner plexiform layer	3		3	$2\frac{1}{2}$ *	3	3	2	$\frac{1}{2}$	$2\frac{1}{2}$ *	2 *
Amacrine cells				$\frac{1}{2}$	*	$2\frac{1}{2}$ *	*	N	N	N
Inner nuclear layer				$\leqslant\frac{1}{2}$		$\leqslant\frac{1}{2}$ *				
Outer plexiform layer		$\leqslant\frac{1}{2}$		$1\frac{1}{2}$	1 *	1		1 *		2
Outer nuclear layer				$\leqslant\frac{1}{2}$		$\frac{1}{2}$				
Inner segments of visual cells				$\leqslant\frac{1}{2}$ *		2		$\frac{1}{2}$		$\leqslant\frac{1}{2}$ *
Outer segments of visual cells										

(cont.)

TABLE 6.32.1 (continued)

Retinal layer	SPECIES									
	Man		Pig		Rabbit		Rat		Sheep	
	AChE	BuChE	AChE	BuChE	AChE	BuChE	AChE	BuChE	AChE	BuChE
Müller's cells		*								$\frac{1}{2}$ *
Nerve fibre layer		1	$\leqslant\frac{1}{2}$ *	$\leqslant\frac{1}{2}$ *		*			*	$1\frac{1}{2}$ *
Ganglion cell layer		3 *	$\leqslant\frac{1}{2}$ *	2 *	*	*	$2\frac{1}{2}$ *		*	3 *
Inner plexiform layer	$2\frac{1}{2}$ *	3 *	3 *	1	$2\frac{1}{2}$ *	*	$2\frac{1}{2}$ *		$2\frac{1}{2}$	3 *
Amacrine cells	$\leqslant\frac{1}{2}$	N	$\leqslant\frac{1}{2}$ *		N	N	N	N	$\leqslant\frac{1}{2}$	N
Inner nuclear layer		$\leqslant\frac{1}{2}$		$\leqslant\frac{1}{2}$						$\leqslant\frac{1}{2}$
Outer plexiform layer		$1\frac{1}{2}$ *	1 *	$\frac{1}{2}$ *			$\leqslant\frac{1}{2}$ *			3
Outer nuclear layer		$\frac{1}{2}$								$\frac{1}{2}$
Inner segments of visual cells		1		*						$\frac{1}{2}$ *
Outer segments of visual cells										

A blank denotes the total absence of staining
$\leqslant\frac{1}{2}$ denotes little or no staining
3 indicates maximal intensity
N, no result reported
Fixed and fresh tissues gave similar results except in some cases (*) where fresh tissue gave a value which differs from that shown for fixed material
Method: Gomori (1952).

TABLE 6.32.2

Starch gel electrophoresis of retinal cholinesterases (Esila 1963).

Species	Number of bands		Comment
	AcThCh + 10^{-6} M ISO-OMPA	BuThCh + 10^{-5} M BW284C51	
Cat	2	0	Histochemical sections also − ve for BuChE
Cow	2	0	Lack of reaction with BuThCh at variance with marked histochemical activity
Dog	3 (1 weak and immobile)	3 (1 weak and immobile)	
Guinea-pig	2	0	Histochemical sections also − ve for BuChE
Horse	2	3 (weak)	
Man	2	2	
Pig	2 (or 3)	1 (weak)	
Rabbit	2 (both very weak)	0	Histochemical sections also − ve for BuChE
Rat	2 (both weak)	0	Histochemical sections also − ve for BuChE
Sheep	3 (1 immobile)	2	

The method was that of Markert and Hunter (1959); non-specific esterases were also studied.

TABLE 6.33

Activity * of acetylcholinesterase in the cerebellum. (From Goldberg and McCaman 1967.)

	Rat	Pigeon	Guinea-pig	Rabbit	Cat
Molecular layer	1000 ± 123 (4)	4293 ± 316 (6)	3757 ± 285 (7)	2087 ± 216 (6)	2330 ± 288 (7)
Granular layer	1472 ± 89 (5)	2399 ± 190 (8)	2390 ± 131 (6)	3702 ± 643 (6)	3694 ± 282 (5)
White	1620 ± 223 (4)	775 ± 132 (7)	1192 ± 69 (7)	1270 ± 126 (5)	1193 ± 211 (6)
Nuclei	1525 ± 232 (5)	610 ± 58 (4)	1354 ± 189 (5)	−	1790 ± 276 (9)

* The activity is expressed as μmoles of acetylthiocholine hydrolysed/g dry wt/hr.
  Each value represents the mean ± standard error (number of sections).

TABLE 6.34
AChE in brain subcellular fractions of four species. (From McCaman et al. 1965.)

Fraction	Rat		Rabbit		Guinea-pig		Pigeon	
	P (%)	Enz (%)	P (%)	Enz (%)	P (%)	Enz (%)	P (%)	Enz (%)
Nuclear	13	8	16	7	15	9	11	2
Mitochondrial	42	32	38	32	39	34	37	32
Microsomal	24	51	24	49	24	48	28	60
Supernatant	21	9	22	12	22	9	24	6
% Recovered	114	107	101	99	108	104	111	104
M_1	70	73	67	60	72	69	71	70
M_2 submitochondrial	10	22	15	36	11	23	10	26
M_3	20	5	19	4	17	8	19	4
% Recovered	95	86	109	97	97	115	103	55
Absolute values								
Enzyme (Enz) μmoles/g/hr		343		623		223		1104
Protein (P) mg/g	85		116		85		75	

TABLE 6.35
Hydrolysis of acetylcholine and butyrylcholine by muscles of fishes, frog and mammals. (From Pecot-Dechavassi 1962.)

Species	Muscle	Hydrolysis of ACh μl CO_2/hr/200 mg	Hydrolysis of BuCh μl CO_2/hr/200 mg	Hydrolysis of BuC as % of hydrolysis ACh
Fish				
Catfish	Lateral superficial	1,598.4	29.9	1.8
	Lateral deep	930.6	13.2	1.4
Gnathonemus	Electric organ	5,096	115.2	2.8
Malapterurus	Electric organ	1,156.2	31.5	2.6
Frog				
Rana esculenta	Rectus abdominis	91.1	29.9	32
Mammals				
Guinea-pig	Diaphragm	116.5	106.9	91.7
Rat	Gemelli	74.5	22.7	30.4
	Diaphragm	119.2	69.9	58.6
Cat	Anterior tibial	57.6	8.5	14.7
	Soleus	71.7	25.2	35.1
	Diaphragm	115.3	66.6	58.6
Rabbit	Gemelli	36.4	17.5	48
	Diaphragm	74.3	35.3	47

CHAPTER 7

Acetylcholinesterase in the context of nervous transmission

7.1 Introduction

The last chapter indicated the widespread distribution of AChE in the nervous system of various vertebrates. The next point to consider is the functional significance of this localization. Krnjević (1969) described as a much repeated truism the statement that 'the presence of acetylcholinesterase is not by itself sufficient evidence of the existence of a cholinergic pathway'. I make no apology for reiterating these words yet again since many authors will persist in regarding 'cholinesterase-containing' as synonymous with 'cholinergic'. This uncritical equating of morphological sites of AChE activity with physiological sites of cholinergic transmission has had two effects. First, it has led to some probably erroneous conclusions (see for instance §7.3.3 re cerebellum) and second, it has caused doubts to be shed on the value of AChE histochemistry as a research tool. The data to be discussed in this chapter should show to what extent the presence or absence of AChE can be taken as an indicator — an indicator, *not* a proof — of the presence or absence of cholinergic mechanisms in a particular pathway.

The central nervous system and the peripheral nervous system, both somatic and autonomic, properly merit the same attention but this forms so large a field of study that some arbitrary limit is unavoidable. I have chosen to start with a rather general treatment of peripheral motor systems (see §7.3.7 for sensory fibres) and then to consider the CNS in more detail.

7.2 The peripheral motor system

7.2.1 Somatic fibres

Motor nerves which innervate skeletal muscle arise from neurones situated in certain of the cranial nuclei and in the ventral horns of the spinal cord. Since it can be directly demonstrated that ACh is released from their axon terminals these cells fulfil Dale's criterion (Dale 1934, 1935) for a cholinergic neurone. AChE is present not only in the cell body and axon but also at the motor end-plate, the point of contact between nerve and muscle. The function of AChE at the end-plate seems clear-cut. Most of the AChE is situated post-synaptically on the surface of the junctional folds in the sarcoplasm (for references see Guth 1968; Friedenberg and Seligman 1972). It would thus appear to be in a good position for hydrolysing ACh released by the pre-synaptic nerve endings. Brown et al. (1936) showed in cat gastrocnemius muscle that if the hydrolysis of ACh by cholinesterase was blocked with eserine the muscle no longer gave a single response to a single stimulus to the nerve but twitched repetitively. This and much subsequent work (see Hebb and Krnjević 1962; Werner and Kuperman 1963; Katz and Miledi 1973) has indicated that by terminating the action of ACh on the post-junctional membrane, cholinesterase plays a vital part in the ordered transmission of impulses at the neuromuscular junction.

Although there can be little doubt about the function of AChE at the end-plate it is still not clear whether BuChE plays any part in the reaction. It was at one time suggested (Häggqvist 1960) that endings of the en grappe type in slow muscle might contain BuChE rather than AChE but there is evidence from many muscles that where BuChE is found AChE is also present (Denz 1953; Hess 1961b; Pecot-Dechavassine 1962; Silver 1963).

The idea that AChE might be part of the acetylcholine receptor molecule was discussed in ch. 2. Studies on denervated muscle add to the reasons for rejecting this hypothesis. It has long been recognized that after a muscle is denervated, its sensitivity to ACh is no longer confined to the region of the end-plate but spreads throughout the length of the fibres (Axelsson and Thesleff 1959). Waser and Nichel (1969) reported experiments on denervated mouse diaphragm in which AChE activity was demonstrated histochemically and the newly developed ACh receptors were labelled for autoradiography with [^{14}C]-decamethonium. Their results showed that AChE activity was detectable in the end-plate region but not at the sites binding decamethonium outside the region.

7.2.2 The autonomic nervous system – ganglia

ACh is generally accepted as the transmitter at sympathetic and parasympathetic ganglia and also at terminals of parasympathetic postganglionic fibres. AChE is present at all these sites but certain experimental findings, more particularly at ganglia in the adrenergic system, have cast doubts on its function. The two most important points which suggest that the situation at the sympathetic ganglion cell may not be analogous to that at the end-plate concern first, the localization of AChE and second, the effects of anticholinesterases. Knowledge about the precise localization of AChE within a ganglion is obviously important to the understanding of its function and the question is whether it is present pre-synaptically, post-synaptically or at both these sites. In the case of parasympathetic ganglia, for example the ciliary ganglion, there is no problem: AChE seems to be associated with both the pre- and postganglionic cholinergic elements (Koelle and Koelle 1959). With sympathetic ganglia the localization has been more difficult to establish. Sawyer and Hollinshead (1945) showed in the cat superior cervical ganglion that preganglionic nerve section caused a marked reduction in biochemically detectable AChE. Their conclusion, that most of the enzyme was localized in preganglionic structures, was supported by subsequent histochemical studies which showed that the staining occurred predominantly in the neuropil, most cell bodies giving a weak or negligible reaction (see Koelle 1951; Holmstedt and Sjöqvist 1959a; Fredricsson and Sjöqvist 1963; Koelle 1963). Biochemical analyses of single cells confirmed, in the main, the impression gained from histochemistry. Using the Cartesian diver technique, Giacobini (1956b) found that apart from a few highly reactive cells which could be cholinergic, the majority of cells showed only low levels of AChE activity. In some cells no reaction was detectable but later work by Brzin et al. (1966) indicates that these negative results were probably due to non-penetration of substrate. These experiments together with further histochemical work (see Eränkö 1972) suggested that probably all the neurones, adrenergic as well as cholinergic, in the superior cervical ganglion (at least of certain species) possess some AChE. Nevertheless there seemed no reason to question the well-established belief that the major part of the ganglionic AChE is localized pre-synaptically. Now, however, evidence is available which casts serious doubt on this view. In the course of developing an electron microscopic technique for use on cat autonomic ganglia Koelle et al. (1971) came to the conclusion that the reason why methods which worked satisfactorily at the end-plate would not work on ganglia was because of the greater solubility of ganglionic AChE. To overcome this problem they modified their gold-thiolacetic acid method (ch. 3) and fixed the ganglion in 4% formaldehyde for 8 hr. Under these conditions the end-product at axodendritic synapses was deposited predominantly on the post-synaptic membranes, the pre-synaptic membranes showing much less reaction. The authors suggest two ways in which these results might be reconciled with the earlier evidence for a pre-synaptic localization. The first is that the pre-synaptic endings contain considerable amounts of AChE (as originally believed) but this is more soluble than the enzyme at post-synaptic sites. Under the conditions of the experiment it is lost while the less concentrated, but more insoluble, post-synaptic enzyme remains. The second and, according to the authors, the more likely, explanation is that the true localization of the enzyme is predominantly post-synaptic, as electron microscopy suggests, but its presence depends on the integrity of the pre-synaptic fibres. The loss of some necessary trophic influence would account for the decrease in activity following section of the preganglionic trunk. These results of Koelle et al. have two important implications, one general and one specific. First, the demonstration that the histochemical picture can be 'manipulated' by changes in conditions emphasizes the need for extreme caution in the functional interpretation of histochemical experiments. Second, if the site of deposition of end-product is taken as evidence for the site of AChE activity, then the apparent difference between the neuromuscular and the ganglionic junction disappears. This means that it would no longer be necessary to speculate on the significance of a mainly preganglionic localization of enzyme. However, until it is quite certain (cf. Koelle et al. 1974) that the new technique shows the true localization of the AChE, the original idea that it is the pre-synaptic enzyme which is functionally important cannot be wholly abandoned. It is thus still appropriate to outline the earlier theories of what that enzyme is doing there.

Because anticholinesterases do not affect ganglionic transmission as dramatically as they affect neuromuscular transmission (Emmelin and MacIntosh 1956; see also Koelle 1963; Zaimis 1963) it was suggested that in the ganglion the termination of transmitter action is not primarily dependent on the hydrolysis of ACh by AChE but depends more on the diffusion of ACh from the synapse (Ogston 1955). According to this scheme AChE might have a secondary role in the eventual hydrolysis of ACh outside the synaptic area. Another, different, idea (see Koelle 1962, 1971) is that AChE is involved in an 'amplifier' mechanism. Koelle suggested that the impulse in the pre-synaptic fibre releases only a small amount of ACh which is insufficient to excite the post-synaptic membrane but adequate to stimulate the pre-synaptic ending to release further ACh. In this scheme (see also Jacobowitz and Koelle 1965; McKinstry and Koelle 1967) the role of the AChE would be to prevent perpetual restimulation of the pre-synaptic terminals. This mechanism requires that the pre-synaptic membrane is more sensitive to ACh than is the post-synaptic membrane but work by Collier and Katz (1970) suggests the reverse is true. These authors used radiochemically labelled choline to mark the ACh of either the 'transmitter' pool or the 'surplus' pool (see ch. 4) in the perfused superior cervical ganglion of cat. They showed that while low concentrations of ACh perfused through the ganglion were capable of exciting the postganglionic cells they did not bring about a release of labelled ACh from the transmitter pool in the pre-synaptic endings. This argues against ACh having physiologically significant effects pre-synaptically. Furthermore, by distinguishing between transmitter and surplus ACh, Collier and Katz were able to show that the earlier demonstration that carbachol could release ACh (McKinstry and Koelle 1967) was probably related to ACh which had accumulated under non-physiological conditions and which did not represent the ACh involved in transmission.

Another function which has been suggested for the pre-synaptic enzyme is that it prevents ACh accumulating in the resting ganglion. Some findings which appeared to support this idea were reported by Volle and Koelle (1961). The same authors were unable to find evidence for a different proposal, that hydrolysis of ACh by AChE was essential to supply choline for the synthesis of more ACh.

As mentioned already, the doubts about the role of AChE at the ganglion stemmed not only from its apparently inappropriate localization but also from the equivocal effects of anticholinesterases. Even if the question of localization proves to be resolved, the problem of anticholinesterases remains. Emmelin and MacIntosh

(1956) perfused the cat superior cervical ganglion with Locke's solution and monitored ganglionic function by recording the response of the nictitating membrane to stimulation of the preganglionic fibres. They found that the speed of relaxation of the membrane at the end of stimulation was the same in the presence or absence of eserine. These results were interpreted in favour of the idea that AChE acted only secondarily, after ACh had diffused away from the synapse. In other words inhibition of AChE does not prevent the disappearance of ACh from the synaptic region. A different explanation, but one which does not entirely exclude the idea just discussed, was outlined by Zaimis (1963). She suggested that inhibition of AChE in a ganglion causes ACh to accumulate, just as at the neuromuscular junction, but the *effect* of the accumulation may be masked by simultaneous non-specific actions of the drugs (see ch. 11, also Karczmar 1967) which may be relatively more important at the ganglion.

In the course of early work with anticholinesterases Koelle and his co-workers (see McIsaac and Koelle 1959) established that ganglionic enzyme was not all equally accessible to inhibitors. Using lipid-soluble and lipid-insoluble inhibitors alone or in combination they found that 'external' AChE, the enzyme accessible to lipid-insoluble inhibitors was functional whereas the 'internal' enzyme, i.e. that inhibited by lipid-soluble compounds seemed to be in reserve. The two components were taken to represent, respectively, enzyme on the surface of the terminal and enzyme inside the preganglionic fibres. In the light of the more recent idea that the enzyme is largely post-synaptic this interpretation would need revision.

7.2.3 The autonomic system – peripheral fibres and endings

Autonomic fibres in smooth muscle do not form discrete end-plates of the type found in voluntary muscle but instead form what are called 'en passage' synapses. At these synapses the varicosities of the axon are closely applied to the membrane of the muscle cell and are not separated from it, as elsewhere, by Schwann cells or other elements. The ultrastructural localization of AChE in the intramuscular portions of autonomic fibres has been examined by various authors. Studies include those of Imagawa (1969) on cat ciliary muscle and of Matsuda (1970) and Ivens et al. (1973) on the iris dilator muscle of rabbit and rat respectively. In most cases the end-product was predominantly localized in the space between the axon and Schwann cell; adrenergic as well as cholinergic fibres gave this reaction. Robinson and Bell (1967) examined the toad bladder and noted staining of the muscle membrane but only after rather prolonged incubation. Since the authors found evidence that the AChE was liable to diffuse under certain experimental conditions it is difficult to assess the true functional significance of the reported localization. It could be argued that the staining of the muscle which occurs after the longer periods of incubation is purely a diffusion artifact and the genuine localization is pre-synaptic. On the other hand, the localization at endings in somatic muscle is predominantly post-synaptic and an analogous situation might be expected in smooth muscle. A possible explanation for the paucity of post-synaptic end-product, after incubation periods which are adequate to demonstrate the axonal reaction, may be that the enzyme in the muscle is more soluble than that in the axon and most has thus diffused away. This idea gains some support from the rather different picture in mammalian tissue. Robinson (1969) examined the ultra-structural localization of AChE at nerve junctions in guinea-pig vas deferens. The innervation is predominantly noradrenergic but some fibres are very rich in AChE and may well be cholinergic. Where such fibres pass close to smooth muscle cells the muscle membrane was consistently stained.

The localization of AChE has also been studied at synapses between cholinergic vasodilator fibres and smooth muscle cells in the uterine artery of the guinea-pig (Bell 1969). These vessels have both a noradrenergic and a cholinergic innervation and this was apparently reflected in the staining patterns of the axons. In one type the axonal membrane was heavily stained for AChE whereas in the other type, which are presumed to be noradrenergic, the axonal reaction was much weaker. The membrane of the Schwann cells was stained lightly regardless of the type of axon or axons with which the cell was associated, and the fibroblasts stained similarly. Observations made on the AChE activity in the membrane of the muscle cell are particularly interesting. In vessels examined in late pregnancy areas of the muscle membrane which were closely apposed to strongly stained axons showed a marked reaction. With virgin animals strong post-synaptic activity was seen at only 20% of the junctions, most others were weakly stained but some did not stain at all. Since the degree of axonal staining is similar in pregnant and virgin guinea-pigs the post-synaptic activity seen in the former is unlikely to be a diffusion artifact. Previous work (Bell 1968) indicated that during late pregnancy the uterine artery shows a striking increase in sensitivity to the periarterial injection of ACh. Bell (1968, 1969; see also 1972) has suggested that this change in sensitivity is involved in the maintenance of uterine hyperaemia and it seems probable that the increase in AChE activity postjunctionally is part of the same process. No other system in the body is subjected to the same massive physiological adjustments as the female reproductive organs and it remains to be established whether the fluctuation of post-synaptic AChE is confined to arteries in this system or whether a degree of lability is a characteristic of other peripheral autonomic synapses.

7.2.4 Controversial theories

Brief reference must be made to two controversial ideas which have attracted a number of adherents but which are not widely accepted. The first is Nachmansohn's belief that acetylcholine has a role not only at the synapse but also in the conduction of the impulse along the axon. The second is Burn and Rand's hypothesis that acetylcholine is involved in the release of transmitter from adrenergic terminals. The principles embodied in these ideas are applicable to the CNS as well as to the peripheral nervous system but experiments to establish or refute them have been done almost entirely on peripheral nerves.

The idea that ACh plays a part in the propagation of axonal action potentials was implicit in Calabro's (1933) discussion of 'cardiac neurohumour' but it was due to the trenchant advocacy of Nachmansohn and his co-workers (Fulton and Nachmansohn 1943; Nachmansohn 1946; see also Nachmansohn 1959, 1963, 1970) that it gained such prominence and evoked correspondingly trenchant rebuttals (Dale 1948; Feldberg 1945, 1957; Koelle 1963). Nachmansohn's contention has been that the release of ACh together with its rapid removal by AChE is essential to axonal conduction. Moreover, this mechanism is held to occur not only in cholinergic nerves but in non-cholinergic ones as well and, according to Nachmansohn, ACh, ChAc and AChE are invariably present in all types of conducting tissue regardless of the nature of their terminal transmitter. Although it is true – and undoubtedly puzzling – that reputedly non-cholinergic fibres may contain appreciable AChE activity (see Burgen and Chipman 1951; Gruber and Zenker 1973) the claim that ChAc and ACh are universally present in functionally significant amounts is not supported by experimental data. Some tracts, in particular those of the optic system and also the dorsal roots, contain scarcely any ACh (MacIntosh 1941) or ChAc (Hebb and Silver 1956) and it may well be that the little that is present is concentrated in a few intermingled cholinergic fibres (Hebb 1957a). Apart from these 'biochemical' objections to the theory there are physiological ones. Thus Lorente de Nó (1944) showed that when ACh was applied together with eserine to the sciatic nerve of bullfrog, the fibres were not depolarized nor was conduction blocked. This criticism Nachmansohn rejected on the grounds of non-penetration of ACh. Similar claims and counter claims have been made with regard to the effects of anticholinesterases. If AChE activity is essential to conduction then anticholinesterases would be expected to alter the action potential. Bullock et al. (1946a, b) found that eserine and DFP were both capable of abolishing the action potential in squid axons and assumed this to be the direct result of the inhibition of AChE. This view was immediately disputed (Crescitelli et al. 1946; Toman et al. 1947) on the grounds that the conduction block could be a non-specific effect unrelated to any anticholinesterase action. One of the strongest pieces of evidence that the block is not caused by an uncontrolled accumulation of ACh is that the membrane is not necessarily depolarized under conditions in which block occurs (Toman et al. 1947; see also Ritchie 1967). One of the possible ways in which anticholinesterases could block conduction other than by inhibiting cholinesterase is by causing pH changes (Hoskin 1971).

Despite its vigorous defence by Nachmansohn's group the theory cannot be reconciled with the experimental findings and, as discussed in ch. 4, the general concensus is that most of the ChAc and AChE in the axons of cholinergic nerves is in transit to its site of action at the terminals. The ACh present could be the product of accidental synthesis. The possibility that it is functionally important, for instance in permeability changes associated with metabolism or transport, seems remote in view of its virtual absence from certain types of fibre. Rejection of Nachmansohn's hypothesis does leave us without an explanation for the presence of AChE in non-cholinergic nerves but no case can be made for retaining it on these grounds since AChE is also present in many non-conducting tissues (ch. 8).

Burn and Rand put forward their hypothesis in 1959 (see also Burn and Rand 1965; Burn 1971) and like that of Nachmansohn it has continued to attract both support and criticism (see Ferry 1966; Jaju 1969; Campbell 1970, also various entries in ch. 6). In brief the theory is this: the nerve impulse arriving at the 'adrenergic' terminal releases ACh which alters membrane permeability in such a way that calcium enters the terminal and causes release of noradrenaline from the storage granules. The hypothesis was advanced to account for various pharmacological phenomena that could be demonstrated under certain conditions. For example, if a cat is treated with reserpine to deplete its stores of noradrenaline, stimulation of sympathetic fibres causes effects normally considered to be due to ACh.

The so-called cholinergic link hypothesis implies that all sympathetic fibres, postganglionic as well as preganglionic should be classed as cholinergic. Here again is a theory which might explain the unexpected presence of cholinesterase in adrenergic neurones but accumulating evidence suggests the idea is not tenable. If the release of noradrenaline involves a cholinergic process then adrenergic fibres should contain AChE, ChAc and ACh. As discussed earlier, most if not all adrenergic neurones do have some AChE both in their perikarya and axons but there is as yet no clear biochemical evidence to show that any contain significant amounts of either ChAc or ACh (Nordenfelt 1965; Ehinger et al. 1966, 1970; Buckley et al. 1967). Unequivocal proof that a material is absent is always hard to provide since it can be argued that methods of detection are inadequate. In this instance, however, the negative analytical results are supported by certain observations made with the electron microscope. For instance it has been shown in cat iris (Tranzer and Thoenen 1967) and cat spleen (Fillenz 1970) that the adrenergic terminals are devoid of the cholinergic type of synaptic vesicle. The clear vesicles sometimes seen in adrenergic terminals and originally thought to contain ACh have been identified as empty adrenergic vesicles (see Tranzer et al. 1969). The absence of detectable levels of ChAc and ACh and the lack of synaptic vesicles in terminals render unlikely the possibility that impulses in adrenergic axons release ACh as a preliminary to the release of noradrenaline. A proposed modification to the cholinergic link hypothesis has been that although the adrenergic fibres do not themselves release ACh they are influenced by ACh released from contiguous cholinergic fibres (see Fray and Leaders 1967; Ehinger et al. 1970). Again there is considerable evidence against this suggestion. Perhaps the most telling argument is that in a number of tissues, for example the longitudinal muscle of the guinea-pig vas deferens (Furness and Iwayama 1972), the renal arteries of the rat (Lorez et al. 1973) and the rat mesenteric blood vessels (Finch et al. 1973; Furness 1973) the adrenergic axons are not accompanied by cholinergic fibres. Another piece of evidence comes from the iris muscle of cat which receives adrenergic fibres from the superior cervical ganglion and all or nearly all its cholinergic fibres from the ciliary ganglion (Ehinger 1967). This means that it is possible to stimulate the fibres separately. Langley and Anderson (1892) found that stimulation of the superior cervical nerve (which would be expected to activate only the adrenergic axons to the iris) caused contraction of the dilator muscle. Furthermore, Anderson (1905) showed that this effect could still be evoked after extirpation of the ciliary ganglion.

7.3 The central nervous system

7.3.1 Is ACh a central nervous transmitter?

The question we have to answer is whether the AChE which is present in the CNS is fulfilling a role similar to that at the motor end-plate. In other words, is it there to hydrolyse ACh released at cholinergic synapses? AChE is generally found both pre- and post-synaptically (ch. 4) hence the localization does not pose particular difficulties. The real problem concerns ACh. Is it or is it not acting as a transmitter in the CNS? The major bar to answering this question is that the identification of transmitters at central synapses is less straightforward than it is at peripheral junctions. Werman (1966) has proposed eight criteria which should be met in the identification of a substance as a natural central nervous transmitter. Among these is the collection of the putative transmitter from the stimulated junction. With peripheral structures such as muscle or ganglia, ACh can be collected relatively easily following nervous stimulation and there is little doubt that it has come from the nerve terminals. Direct confirmation of this sort cannot be obtained so readily for reputed cholinergic synapses in the CNS and proof that a pathway is or is not cholinergic must rest on the accumulation of indirect evidence obtained in a number of ways. Extrapolation from the situation at the periphery suggests that no junction is likely to be cholinergic unless it contains AChE but it is unsafe to carry this analogy further and to assume that because an area contains AChE, transmission there is necessarily cholinergic.

In 1937 Dale (see Dale 1938) speculated whether the concept of cholinergic transmission, then so recently demonstrated at the periphery, could be extended to some central synapses. He made the plea that any such extension should be made 'only in the light of direct and critically scrutinized evidence'. Dale's exhortation to caution has in general been heeded and even now, 40 years later, most physiologists are hesitant about designating a pathway cholinergic. A particular difficulty in accepting ACh as a central transmitter is that when it is applied by iontophoresis to neurones in the brain (cf. §7.3.6 re Renshaw cells in the spinal cord) the effect is slow in onset and may outlast the application (Curtis and Koizumi 1961; Krnjević and Phillis 1963a, b). Such a time-course is very different from the rapid action to be expected of a transmitter in a fast conducting pathway and it has raised doubts about the role of ACh at central synapses. The ability of certain cells to respond to ACh could be an incidental and functionally unimportant property of the membrane. In their now classical paper Shute and Lewis (1963) concluded that the ACh, ChAc and AChE present in the CNS were likely to be of some functional significance and Krnjević (1969) reviewing the subject of cholinergic transmission reiterated this opinion. He suggested that although ACh apparently had no part as a 'detonator' transmitter in the major pathways it might be released from subsidiary pathways and function in a facilitatory way (see also Krnjević et al. 1971) governing the general level of CNS activity. Krnjević pointed out that the pattern of AChE-containing fibres in the brain supported the view that the cholinergic system was likely to be composed of fine, diffusely distributed fibres which were distinct from those of the major efferent, afferent and commissural pathways (plate 7.1). Analysis of the physiological significance of these AChE-containing systems is most conveniently done under 6 general headings.

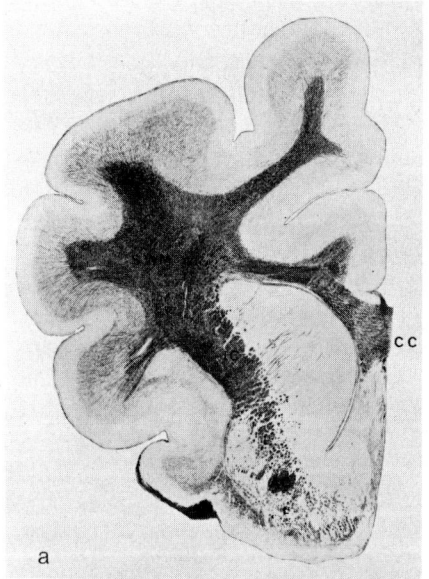 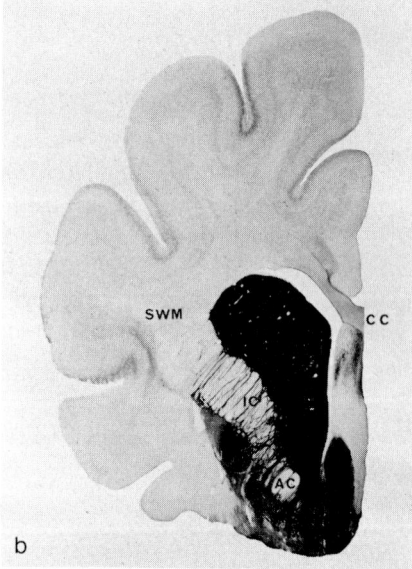

Plate 7.1. To illustrate the absence of AChE activity from the major pathways of the cat forebrain. a) Section stained to show white matter. (From Vogt and Vogt 1902.) b) Section stained for AChE. AC, anterior commissure; CC, corpus callosum; IC, internal capsule; SWM, subcortical white matter. (After plate shown by Krnjević 1969.)

7.3.2. *The cholinesterase-containing ascending reticular system*

Shute and Lewis (1963, 1965, 1967b, see also ch. 6) examined the distribution of AChE in the rat brain and by using the lesion technique (ch. 3) established the polarity of various enzyme-containing fibre systems. The results of this work, which are shown in table 7.1 and fig. 7.1, may be summarized as follows. The fibres which stain for AChE seem to come from two different areas of the brain stem. The dorsal tegmental pathway starts in the midbrain reticular formation and projects to the tectum, metathalamus (this includes the medial and lateral geniculate bodies) and the thalamus. A point to note here is that Shute and Lewis designate the midbrain

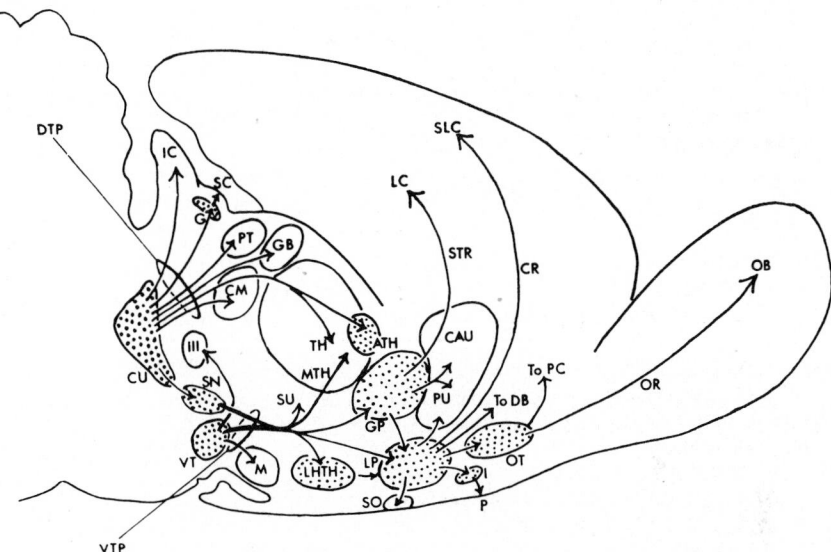

Fig. 7.1. Diagram of Shute and Lewis's 'ascending cholinergic reticular system'. Stippling shows constituent AChE-containing nuclei of the system in the midbrain and forebrain, with projections to the cerebellum, tectum, thalamus, hypothalamus, striatum, lateral cortex and olfactory bulb. Abbreviations: ATH, anteroventral and anterodorsal thalamic nuclei; CAU, caudate; CM, centromedian (parafascicular) nucleus; CR, cingulate radiation; CU, nucleus cuneiformis; DB, diagonal band; DTP, dorsal tegmental pathway; G, stratum griseum intermediale of superior colliculus; GB, medial and lateral geniculate bodies; GP, globus pallidus and entopeduncular nucleus; I, islets of Calleja; IC, inferior colliculus; III, oculomotor nucleus; LC, lateral cortex; LHTH, lateral hypothalamic area; LP, lateral preoptic area; M, mamillary body; MTH, mamillo-thalamic tract; OB, olfactory bulb; OR, olfactory radiation; OT, olfactory tubercle; P, plexi-form layer of olfactory tubercle; PC, precallosal cells; PT, pretectal nuclei; PU, putamen; SC, superior colliculus; SLC, supero-lateral cortex; SN, substantia nigra pars compacta; SO, supra-optic nucleus; STR, striatal radiation; SU, subthalamus; TH, thalamus; TP, nucleus reticularis tegmenti pontis (of Bechterew); VT, ventral tegmental area and nucleus of basal optic root; VTP, ventral tegmental pathway. (From Shute and Lewis 1967b.)

TABLE 7.1

Summary of Shute and Lewis's 'ascending cholinergic reticular system' (cf. fig. 7.1). Sites of cholinesterase-containing cells in capitals, sites of dense cholinergic neuropil in italics. (From Shute and Lewis 1967b.)

Dorsal tegmental pathway
  from NUCLEUS CUNEIFORMIS to tectum, metathalamus, thalamus

---

1) direct fibres, supplying:
    posterior colliculus
    anterior colliculus via STRATUM GRISEUM INTERMEDIALE
    dorsal and *deep pretectal nuclei*
    medial and lateral geniculate bodies, especially *ventral nucleus of lateral geniculate body*
    centromedian and *intralaminar thalamic nuclei*
    specific thalamic nuclei
    anterior thalamic nuclei, especially *antero-ventral thalamic nucleus*

2) indirect fibres, via ventral supra-optic decussation, supplying:
    anterior colliculus
    pretectal nuclei

3) indirect fibres, via medial strial bundle, supplying:
    lateral geniculate body

Ventral tegmental pathway
  from VENTRAL TEGMENTAL AREA and SUBSTANTIA NIGRA to subthalamus, hypothalamus and basal fore-brain areas

---

1) direct fibres, supplying:
    oculomotor nucleus via NUCLEUS OF BASAL OPTIC ROOT
mamillary bodies via SUPRAMAMILLARY NUCLEI
subthalamic nucleus
      ENTOPEDUNCULAR NUCLEUS and GLOBUS PALLIDUS
      POSTERIOR and LATERAL HYPOTHALAMIC AREAS
      LATERAL PREOPTIC AREA
      paraventricular and supra-optic hypothalamic nuclei
      OLFACTORY TUBERCLE (polymorph layer)
      olfactory tubercle (plexiform layer) via ISLETS OF CALLEJA

2) indirect fibres, via lateral strial bundle, supplying:
    amygdaloid nuclei

3) corticopetal radiations, from ENTOPEDUNCULAR NUCLEUS, GLOBUS PALLIDUS, LATERAL PREOPTIC AREA, OLFACTORY TUBERCLE, supplying:
    caudate-putamen
    amygdaloid nuclei, especially *pars ventralis of lateral amygdaloid nucleus*
    nucleus of lateral olfactory tract
    supero-lateral, lateral and ventral (including olfactory) cortex
    olfactory bulb
    nucleus accumbens

4) connexions, via LATERAL PREOPTIC AREA, with DIAGONAL BAND and SEPTUM
  (origin of cholinergic hippocampal afferents)

reticular formation 'nucleus cuneiformis' which is in accordance with Valverde's (1962) terminology for the rat. In cat, however, nucleus cuneiformis denotes a much more restricted area, and I found (unpublished) that its cells do not stain for AChE. The second pathway recognized by Shute and Lewis, the ventral tegmental pathway, arises from neurones in the ventral tegmental area and the substantia nigra and sends fibres to the subthalamus, hypothalamus and basal forebrain areas. Fibres to the neocortex and striatum project from reticular cells in various nuclei, for example the globus pallidus and entopeduncular nucleus, situated at the rostral end of the ventral tegmental pathway. The cell groups projecting to the cortex are not only in continuity with each other and with the neurones of the ventral tegmental pathway but are composed of large neurones with long dendrites. All of these features suggest that the cells may be activated from many different sources. This is in keeping with the view that the cholinergic system is likely to be involved in diffuse rather than specific activation and Shute and Lewis proposed that their 'ascending reticular system' is the morphological basis of the 'ascending reticular activating system' recognized physiologically (Starzl et al. 1951). That cholinergic mechanisms are involved in this latter system seems probable from the demonstration in the cat that the desynchronization of the EEG which results from high-frequency stimulation of the medial reticular formation is accompanied by increases in the output of ACh from the cortex (Kanai and Szerb 1965; see also Beani et al. 1968). Although Szerb (1967) obtained some evidence that the pathway responsible for arousal and that responsible for ACh release may not be one and the same throughout their length, Shute and Lewis's proposal remains a valuable working hypothesis. The system Shute and Lewis described was recognized on the basis of AChE content; we now have to see what other evidence there is to suggest the presence of cholinergic synapses in the different areas included in table 7.1.

7.3.2.1 Geniculate bodies and other thalamic nuclei   Shute and Lewis showed that in the rat cholinesterase-containing fibres projected from the midbrain reticular formation to the medial and lateral geniculate bodies and to various other thalamic nuclei. AChE activity was particularly marked in the neuropil of the ventral nucleus of the lateral geniculate body. A strong histochemical reaction for AChE has also been reported in the neuropil of the lateral geniculate of cat (Phillis et al. 1967). Earlier, Feldberg and Vogt (1948) and also Hebb and Silver (1956) had demonstrated choline acetylase activity in the geniculate bodies of several species including man and dog but rat and cat tissues were not analysed. Iontophoretic studies on the cat bear out the possibility that cholinoceptive neurones are present in the lateral (Phillis et al. 1967) and medial (Tebēcis 1970a,b, 1972) geniculate bodies. The results suggest, however, that the visual and auditory fibres synapsing respectively in the lateral and medial geniculate bodies do not release ACh and that any cholinergic fibres which may innervate the receptive cells are likely to have come from the mesencephalic region. Deffenu et al. (1967) who measured the levels of ACh in the lateral geniculate bodies of cats following various lesions came to the same conclusion. The controversial question (see Hebb 1970; Pepeu 1972) of whether there is a specific cholinergic link between the lateral geniculate body and the visual cortex will be taken up later (§ § 7.3.2.5, 7.3.7).

Evidence suggests that cholinergic synapses are present in other parts of the thalamus. ACh release has been detected by the Gaddum (1961) 'push-pull cannula' technique in both the ventrobasal complex and dorsal thalamus of cat (Phillis et al. 1968). Iontophoretic observations on the ventrobasal complex (Andersen and Curtis 1964; McCance et al. 1968a) showed the thalamocortical neurones to be particularly sensitive to ACh but it seems clear that the main ascending sensory fibres do not release ACh. McCance et al. (1968a, b) concluded that the cholinergic input to the cholinoceptive cells comes from the midbrain reticular formation and also from the cerebellum via the superior peduncle. For further references see Phillis (1971).

7.3.2.2 Substantia nigra   Two features serve to distinguish the substantia nigra from most other parts of the brain. First, a large proportion of the cells are dopaminergic and second, in primates and some subprimates including cat and dog (Brown 1943) the cells contain the pigment melanin. It is this pigment which gives the structure its slightly dark appearance and hence its name. Where it occurs, this natural coloration in no way masks the histochemical reaction for AChE and the substantia nigra stains strongly in various species including monkey, cat and rat. The nucleus sends fibres to, and receives fibres from, a number of different areas. Afferent pathways include the striato- and corticonigral tracts while the major efferent path is the nigrostriatal or comb bundle. As yet the connections are not entirely certain and species differences are likely. Some of the findings of Shute and Lewis (see 1967b) for rat differ from those of Olivier et al. (1970b) for cat and monkey.

On the basis of the histochemically demonstrable pattern of AChE activity and from the effects of lesions on staining, Olivier et al. (1970b) showed that the AChE is not exclusively attributable to striatonigral fibres since considerable staining persisted in the substantia nigra of cat after the tract had been cut. In addition to any activity which may be associated with afferent fibres from other sources, some of the nigral cells would seem to contain enzyme since staining was lost (in the monkey) when cells degenerated as a result of a lesion in the ventromedial part of the tegmentum. In the rat, Shute and Lewis (see 1967b) found marked AChE activity in the cells of the region of the substantia nigra termed the pars compacta. They suggested these cells contribute to the AChE-containing fibres in the ventral tegmental pathway projecting to the subthalamus, hypothalamus and basal forebrain areas. The histochemical picture in different species thus indicates that both cholinergic synapses and cholinergic cell bodies may be present in the substantia nigra and some results from physiological experiments support this view (Smelik

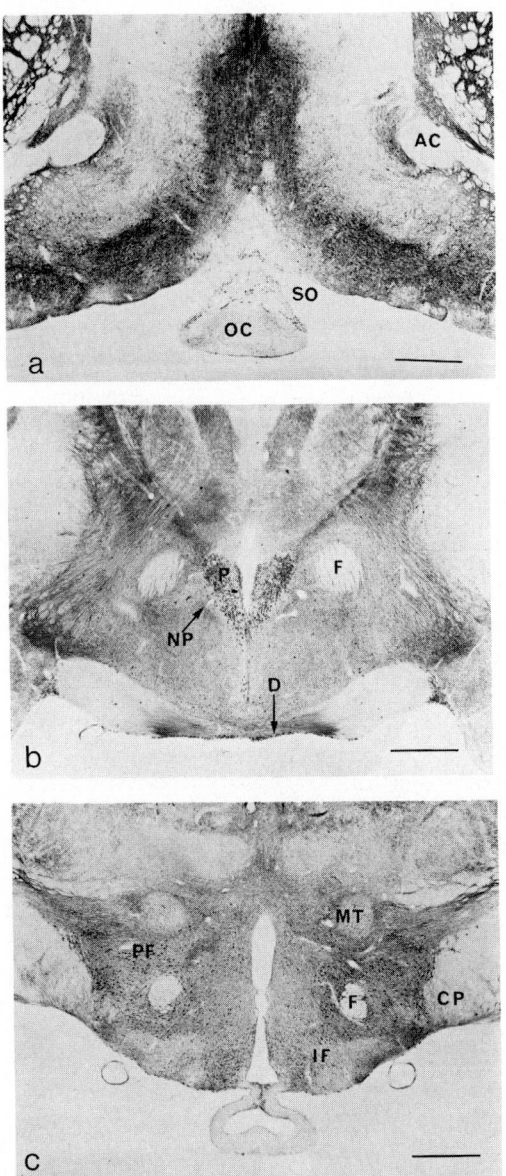

Plate 7.2. To show AChE activity in the guinea-pig hypothalamus sectioned at different levels. AC, anterior commissure; CP, cerebral peduncle; D, supraoptic nucleus (diffusus); F, columns of fornix; IF, infundibular nucleus; MT, mamillothalamic tract; NP, neuropil ventral to para-ventricular nucleus; OC, optic chiasma; P, paraventricular nucleus; PF, perifornical scatter; SO, supraoptic nucleus. Scale bar: 1 mm. (From Cottle and Silver 1970b.)

and Ernst 1966; Costall and Olley 1971; Costall et al. 1972). On the other hand, biochemical analyses of choline acetyltransferase raise certain difficulties. Both in cat and man the level of ChAc is disproportionately low in comparison to the level of AChE. This discrepancy raises the possibility that the esterase may be doing something more than hydrolysing ACh released at synapses. This will be discussed again in ch. 8 §3.1.2.

7.3.2.3 Hypothalamus   The hypothalamus is responsible for a variety of physiolo-gical functions and exerts its effects both directly through nervous pathways and indirectly via the pituitary hormones. Many of the actions of the hypothalamus involve catecholaminergic links but various pieces of evidence show that ACh also has a place in some of these activities. This evidence includes the presence of ChAc (Hebb and Silver 1956; Lederis and Livingston 1969; Bull et al. 1970) and ACh (MacIntosh 1941; Lederis and Livingston 1969) as well as AChE (see Shute and Lewis 1966b; Cottle and Silver 1970a, b). As yet the exact location and importance of cholinergic synapses is poorly documented hence the problem is best tackled from a functional rather than an anatomical angle. It will become clear from what follows that the hypothalamic pathways may differ pharmacologically from species to species. For this reason it is unsafe to assume that data obtained from one species are necessarily relevant to another. This point is well illustrated by the distribution of AChE. As shown in table 6.12.1, Cottle and Silver (1970b) demon-

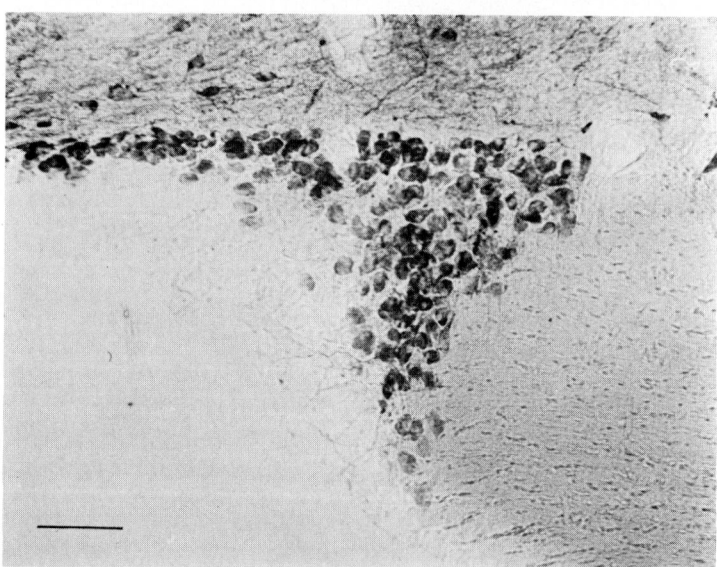

Plate 7.3. AChE activity in the cells of the supraoptic nucleus of a female guinea-pig. Note absence of stained neuropil. Scale bar: 100 μ.

strated AChE activity in cells or neuropil, or in both cells and neuropil, in many areas of the guinea-pig hypothalamus (plate 7.2). On the other hand, Abrahams et al. (1957) working on dog found AChE in only 3 nuclei and in each it was confined to the cells. One of the nuclei with positive cells was the suprachiasmatic. In the guinea-pig, it is only the neuropil of this nucleus which stains.

One of the best known of the hypothalamic functions is the production of the hormones vasopressin (also called antidiuretic hormone, ADH) and oxytocin. These are elaborated in the neurones of the supraoptic and paraventricular nuclei and are released from the endings of these cells in the pars nervosa of the pituitary gland. In 1939 Pickford showed that water diuresis in dogs could be temporarily inhibited by the intravenous injection of ACh and she concluded that ACh was involved in the release of ADH. Subsequent work has included the recording of effects of intra-carotid injections of ACh on the unit activity of neurones in the supraoptic and paraventricular nuclei (Koizumi et al. 1964; Brooks et al. 1966; Dyball and Koizumi 1969). The results of these and other experiments suggest that ACh stimu-lates the release of oxytocin and of ADH by acting on the cell bodies in the two nuclei. Despite the convincing experimental demonstration that ACh influences the release of these hormones, there is one difficulty about accepting this as evidence that ACh is the major transmitter at synapses on the supraoptic and paraventricular neurones. The problem is that the AChE seems to be in the wrong place. Strong staining has been observed in the cells of both nuclei in dog (Abrahams et al. 1957), cat (Shute and Lewis 1966b) and guinea-pig (Cottle and Silver 1970b), but staining of the neuropil was found to be weak or negligible. This distribution (plate 7.3) is not typical of an area of cholinoceptive cells: what is generally found in such an area (ch. 6) is that the cholinergic neuropil is well stained but the cell bodies, though stained to some extent are not intensely reactive. Koelle (1961) has suggest-ed that the supraoptic and paraventricular neurones are cholinergic rather than cholinoceptive and that ACh is involved in the release of transmitter from the terminals and not in the synaptic excitation of the cells themselves. This idea has not been widely accepted although it has recently been revived, very tentatively, by Gosbee and Lederis (1972). Among the reasons for its rejection (see also Heller and Ginsburg 1966) is that in the regions where the axons terminate, i.e. in the median eminence and in the pars nervosa of the pituitary gland, AChE activity is generally low. In summary, the dilemma in explaining the distribution of AChE is this. The lack of a cholinesterase-containing neuropil is difficult to reconcile with the un-doubtedly strong evidence that the cells are cholinoceptive; on the other hand the paucity of AChE in the cell terminals (despite the strong staining of the cell bodies) is difficult to reconcile with the view that they are cholinergic. This problem merits further investigation. In particular it is important to establish that the lack of neuropil staining in the magnocellular nuclei is not a technical artifact due, perhaps, to non-penetration of substrate or to atypical solubility of the enzyme.

Another important function of the hypothalamus is the control of body temper-

ature. The pathways responsible for this regulation have not been individually
identified but experimental evidence implicates structures in the preoptic, anterior
and posterior hypothalamic areas. These structures apparently lie close to the ventral part of the IIIrd ventricle and the intraventricular injection of appropriate drugs
can induce changes in body temperature. Cholinomimetics, with and without anticholinesterases, have been injected systemically, intraventricularly or directly into
the hypothalamus, or have been applied to hypothalamic cells iontophoretically.
The results (see Bligh 1973) indicate that cholinergic synapses may exist in pathways between temperature sensors and temperature regulatory effectors. A particularly interesting feature of these experiments is that different species give different
results. In rat and mouse the administration of ACh or cholinomimetics causes
body temperature to fall (Lomax and Jenden 1966; Lomax et al. 1969; Kirkpatrick
and Lomax 1970). This suggests that ACh could be a transmitter somewhere on the
pathway between the warm sensors and the heat loss effectors. In contrast to this
situation in rat and mouse, where ACh lowers body temperature, in sheep, goat and
rabbit it produces a rise. Bligh et al. (1971) took this rise as evidence that in these
latter species ACh acts not on the warm sensor-to-heat loss effector pathway but on
that between the cold sensors and heat production effectors. In the model which
they have developed on the basis of their findings, the transmitter proposed for the
warm sensor-to-heat loss effector path is 5-HT. Myers and Yaksh (1969) have
suggested that in the Rhesus monkey ACh has a dual role. They postulated that it
could be involved at synapses between cold sensors and heat production effectors in
both the anterior and posterior hypothalamus but that in the latter it might also be
acting on the heat loss pathway (see also Hall and Myers 1972). So far the distribution of AChE, 5-HT and NA has not been systematically mapped and compared in
the various species which seem to use the same putative transmitters in different
pathways. A study of this type might give more information about the anatomical
organization of the structures responsible for temperature control but it is likely to
prove difficult to decide which system subserves which particular hypothalamic
function. As plate 7.2 shows, AChE-containing fibres are diffusely distributed in
the hypothalamus and do not belong to any obvious tracts. Unit activity studies on
thermosensitive neurones (see Bligh 1973; Knox et al. 1973) have indicated that the
'hypothalamic temperature regulator' is likely to be composed of scattered cells
rather than discrete nuclei. That being so, it is perhaps to be expected that the
fibres associated with these cells would have a similarly irregular distribution. However, the point has been made repeatedly that the diffuse distribution of AChE-
containing fibres in many parts of the CNS is in keeping with the idea that ACh is
acting not as a transmitter of a specific signal but in a facilitatory way. The possibility that this could also be true in the hypothalamus cannot be eliminated. Indeed,
the idea that certain putative hypothalamic transmitters should be regarded as
'modulators' (Myers 1970) has already been mooted in connection with 5-HT and
NA (see Bligh 1973). At present the type of response of hypothalamic neurones to

ACh is not well documented (see Jell 1973). Beckman and Eisenman (1970) have reported that the iontophoretic application of ACh to a warm-sensitive cell in the rat hypothalamus produced a 'sharp rise' in firing rate but so far as can be judged from their published records, the latency of the response is in the order of seconds rather than milliseconds. Rather more detailed data are necessary before it will be possible to decide whether ACh should be regarded as a signal transmitter, a modulator, both, or neither.

Yet another important role of the hypothalamus is the regulation of the reproductive cycle. This aspect of hypothalamic activity has formed a major field of research but details of transmission processes in the areas concerned with reproduction are still largely unknown. Sawyer et al. (1949) found that when rabbits were injected intravenously with atropine (30 mg/kg) within 10–40 sec of copulation post-coital ovulation was suppressed in 69% of the animals. If the injection was withheld until 5 min after mating, ovulation was inhibited in only one of 11 animals. The authors concluded that there was a cholinergic link somewhere on the pathway responsible for triggering the release of luteinizing hormone. Other evidence that cholinergic mechanisms are involved in reproductive events includes reports that in the rat hypothalamus the levels of an acetylcholine-like substance (Kuwashima 1957) and of ChAc (Kato and Minaguchi 1964; Kobayashi et al. 1966) fluctuate during the oestrous cycle. Cottle and Silver (1970b) made a histochemical study of AChE activity in hypothalami taken from guinea-pigs in various reproductive states including pregnancy and lactation. Since histochemical material cannot be reliably analysed in quantitative terms we assessed the number of stained elements as well as the depth of staining in various nuclei and were able to discern certain trends in the different reproductive conditions. Fluctuations were most marked in the infundibular nucleus, a structure which has been implicated in the control of pituitary gonadotrophin activity (see Szentagothai et al. 1968). In lactating animals (10 days postpartum) almost all the neurones were very strongly stained but, in contrast, virtually no staining was found in the infundibular nuclei of females examined immediately after parturition. Infundibular staining of ovariectomized and hysterectomized guinea-pigs was in general weaker than that in intact animals. Taken together the separate pieces of evidence support the concept that AChE is involved in the hypothalamic regulation of reproductive processes but, again, we are in the dark about the details of this involvement. As with temperature control and with the activation of the supraoptic and paraventricular neurones it is uncertain whether ACh is responsible for transmitting the signal or is modulating the response to the signal. A further question is prompted by the particularly marked fluctuations in AChE activity. Can changes of this order be attributed solely to variable levels of nervous activity or do they indicate that AChE has another role in addition to hydrolysing ACh at synapses? This point will be taken up again in the next chapter.

The division of the hypothalamus into its component nuclei has been based on

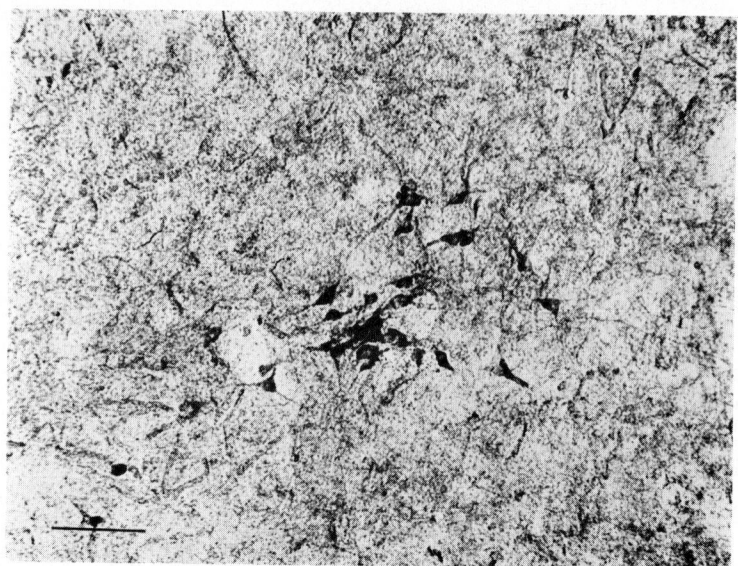

Plate 7.4. Strong AChE activity in a nest of cells lateral to the IIIrd ventricle in the guinea-pig hypothalamus. Scale bar: 100 μ. (From Cottle and Silver 1970b.)

differences in cell types revealed by conventional histological stains (see for example Aus der Mühlen 1966). One of the merits of the histochemical technique for AChE is that it may show up hitherto unrecognized subgroups. An example of this comes from some work on the guinea-pig hypothalamus (Cottle and Silver 1970b). We found that in the ventral part, just lateral to the IIIrd ventricle, a small clump of cells (plate 7.4) stood out from the rest by virtue of their intense reaction for AChE. The type of staining suggests that the cells could be cholinergic but the system to which they belong, and their function, is as yet unknown.

7.3.2.4 The basal ganglia   The term 'basal ganglia' embraces the subcortical nuclei of the telencephalon and usually includes the caudate nucleus, putamen, globus pallidus, the substantia innominata, the amygdala and the claustrum. Other terminology used to denote these structures is more complicated and may be inconsistent from author to author. The term corpus striatum can include the caudate, putamen, globus pallidus, claustrum and amygdala but sometimes it refers to the first 3 structures only. Neostriatum covers the caudate and putamen in those species such as rat where the two cannot be separately distinguished. With this terminology, the globus pallidus is designated the paleostriatum and the amygdala is known as the archistriatum. The name lenticular or lentiform nucleus is given, particularly in man, to the putamen plus globus pallidus. The imprecise use in the

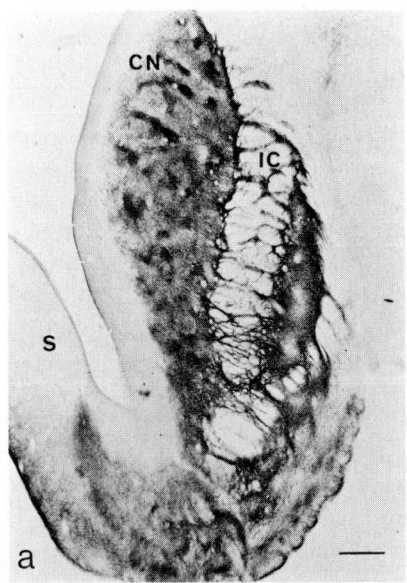

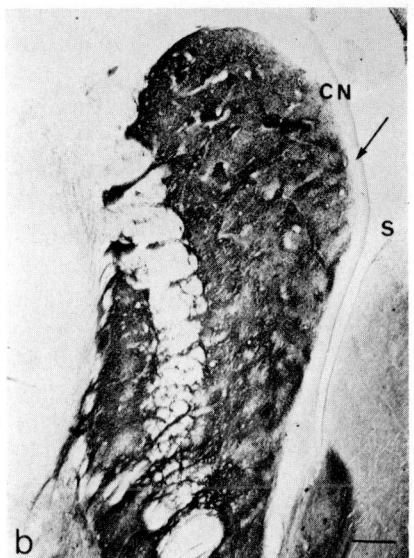

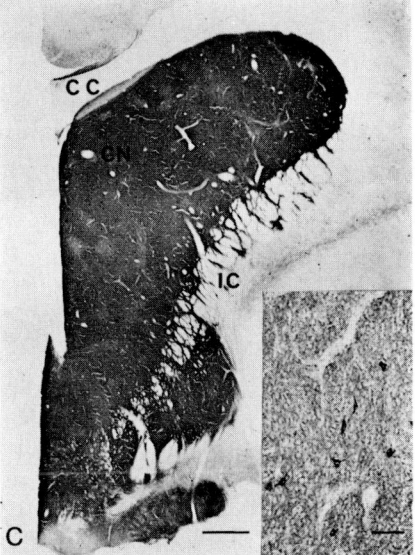

Plate 7.5. To show the development of AChE activity in the cat caudate nucleus. a) Foetus, 53 days gestation. b) Newborn kitten. Note that staining is still absent from the medial edge (arrowed) of the caudate. c) 40-day old kitten. The entire caudate nucleus now shows activity in the neuropil. A photographic filter was used to show up the scattered neurones, seen at a higher magnification in the inset. CC, corpus callosum; CN, caudate nucleus; IC, internal capsule; S, septum. Scale bars: a), b) 500 μ; c) 1 mm (inset 100 μ).

literature of these various designations can cause confusion and makes difficult the interpretation of different experiments.

The basal ganglia are particularly rich in a number of substances thought to be involved in nervous transmission. These include ACh, ChAc and AChE as well as dopamine, DOPA decarboxylase, 5-HT and monoamine oxidase. The AChE activity in the caudate-putamen is characteristically intense (plate 7.1) the level being among the highest of any brain area. With optical microscopy Shute and Lewis (1963) detected very few AChE-containing cells in the rat 'caudate' nucleus, most of the activity being associated with neuropil. This observation was later confirmed at the electron microscopic level (Lewis and Shute 1964). Using the lesion technique to determine the source of AChE-containing fibres Shute and Lewis (see 1967b) concluded that these came mainly from the ventral tegmental pathway·and globus pallidus. The reduction of staining resulting from lesions in the ventral tegmental pathway rendered more visible a few scattered AChE-containing cells, the processes of which also contributed to the reaction in the neuropil. In the adult cat, as in the rat, the caudate and putamen react so strongly for AChE that it is difficult to identify the elements with which the enzyme is associated. The situation in the developing foetus is, however, somewhat clearer. Krnjević and I (1966) found that the putamen and globus pallidus of foetal cats gave a fairly intense reaction for AChE from an early stage but the reaction in the neuropil of the caudate nucleus remained weak (plate 7.5a, b, c) for a considerable part of prenatal and early postnatal development and thus did not mask any AChE-containing cells that were present. As plate 7.5c shows, such cells were not numerous and were scattered. In the claustrum and in the region of the globus pallidus, stained cells were more frequent. If, as these results suggest (but see findings of McGeer et al. 1971b, 1973 below), the enzyme is predominantly associated with the neuropil, then the majority of cells in the caudate are more likely to be cholinoceptive than cholinergic. Iontophoretic studies have indicated that in the cat a proportion of cells respond to ACh, some neurones being excited and others depressed (Bloom et al. 1965; McLennan and York 1966). Similar results have been found in rabbit (Herz and Zieglgänsberger 1968). The pattern of the response to ACh is the same as that already described for other structures: the effect appears after an appreciable latency and persists after the application of ACh ends. McLennan and York (1967) found that dopamine depressed ACh-sensitive neurones in the cat caudate nucleus. A rigorous survey was not made to establish whether all the cholinoceptive cells detected were sensitive to dopamine but it was established that dopamine had a depressant action both on cells which were excited by ACh and on those which were inhibited.

Mitchell and Szerb (1962) demonstrated the spontaneous release of ACh from the caudate nucleus of the anaesthetized cat and showed that the release increased when part of the anterior sigmoid gyrus of the cortex was stimulated. Using a similar technique to collect effluent from the caudate, McLennan (1964) was able to show that stimulation of the anteroventral nucleus of the thalamus increased the output of ACh but stimulation of the centromedial nucleus caused a release of dopamine. The observation that the anteroventral nucleus influences the release of ACh in the caudate gained added significance from the subsequent work of McLennan and York (1966). They found that the cells of the caudate nucleus which are either excited or depressed by iontophoretically applied ACh are similarly excited or depressed by stimulation of the anteroventral thalamic nucleus. Taken all together, these various pieces of evidence support the idea that in the cat the caudate nucleus receives an appreciable input of cholinergic fibres.

Shute and Lewis's results, referred to earlier, suggest that in the rat too, much of the AChE in the neuropil of the neostriatum is associated with extrinsic fibres. Somewhat conflicting findings have, however, been reported by McGeer et al. (1971b; see also McGeer et al. 1973). These workers sectioned the main afferent pathways from the cortex, thalamus and brain stem to the striatum but could not detect any significant change in the striatal level of either AChE or ChAc. Their conclusion was that the majority of cholinergic elements in the rat caudate-putamen are intrinsic rather than extrinsic. A similar view has been taken by Lynch et al. (1972) who injected DFP into the rat neostriatum and followed the reappearance of histochemical staining. This technique showed up AChE-containing cells dispersed throughout the caudate. The authors describe these cells as being 'numerous' which contrasts with Shute and Lewis's (1963, 1967b) report of 'occasional' AChE-containing cells. Lynch et al. concluded that the cells were cholinergic inter-neurones and that the profuse branching of their axons as well as their dendrites accounted for the particularly marked AChE activity in the caudate-putamen. Obviously, more work is necessary to establish the relative contributions of extrinsic and intrinsic elements to cholinergic systems in this structure and in any interpretation it is particularly important to integrate biochemical and physiological data. As McLennan (1970) points out, the very high levels of AChE, ChAc and ACh in the caudate nucleus might lead one to expect plenty of physiological evidence of cholinergic synapses. In fact, McLennan and York (1966) found only 30% of the cells were cholinoceptive in the caudate of the cat decerebrated by midcollicular section. Bloom et al. (1965) give a figure of 54% for the anaesthetized cat and 83% for the decerebrate animal (midcollicular section). McLennan and York (1966) suggest that the difference between their results and those of Bloom et al. may be an indication that when the cortex overlying the caudate is removed (as in their own experiments) neurones in the caudate are depressed. With many factors of this type to contend with it will not be easy to map the cholinergic mechanisms operating in the caudate nucleus but until this is done the full explanation of the exceptionally high content of AChE and ChAc must wait.

The amygdaloid nucleus is rich in ChAc although the levels are not as high as in the caudate nucleus (Hebb and Silver 1956); it is also well endowed with AChE. Shute and Lewis (1967b) suggest that in the rat some AChE-containing fibres terminate on cells in the nucleus while others pass through to the entorhinal cortex. In the cat, iontophoretic studies revealed cholinoceptive cells scattered throughout the amygdaloid complex (Straughan and Legge 1965). Since they were not concentrated in any one area it is probable (see Silver 1967) that the particularly marked histochemical reaction for AChE in certain parts such as the central and basal nuclei (Krnjević and Silver 1965) is attributable to fibres travelling on elsewhere. Hall and Geneser-Jensen (1971) have pointed out that the central nucleus in guinea-pig, in contrast to that in cat, is almost devoid of histochemically detectable AChE.

7.3.2.5 Cerebral and hippocampal cortex ACh is generally thought to have some part in cortical function. Biochemical and physiological data supporting this view have been accumulating over the years but as with so many other regions, details are still controversial. In 1941 MacIntosh extracted ACh from the cerebral cortex of dog and the presence of ChAc was later demonstrated by Feldberg and Vogt (1948). Using a more sensitive technique Hebb and I measured the ChAc activity in a number of species including man and found the levels to be inversely related to the degree of cortical development of the different species (Hebb and Silver 1956). In most species examined there was also considerable variation between the activity in different cortical areas although this was not true of the rabbit or guinea-pig

TABLE 7.2

Choline acetylase activity in the cerebral cortex of mammals. Values expressed as μg ACh hydrolysed/g dry wt/hr. Figures in brackets are results obtained from tissues extracted 24—50 hr after death. (From Hebb and Silver 1956.)

	Cortical areas				
Species	17	4	28	51	Ammon's horn
Man	100—140* (100)	(290)	(580)	–	(675—1050)
Pig	520—950	720—1125 (990)	1800	2450—2600	–
Cat	800—950	1350	2100	4500	3400
Sheep	1000—1350	1500—1700	1600	5200	2800—2900
Dog	1250—1400 (1300)	3000	3800	3750	2600—4200
Rabbit	4000—5200*	3750—4800*	3500	4500	4125
Guinea-pig	4000*	4000*	3750[†]		

* Includes some cortical tissue from adjacent areas.
[†] Areas 28 and 51 pooled.

which showed similarly high levels in all the areas we examined (table 7.2). Quantitative values for AChE in the cortex have been reported for various species (see ch. 6 for references) but it is the histochemical results which are probably most useful when it comes to discussing functional implications and these will be considered more fully below.

Other techniques used to investigate the existence of cholinergic synapses in the cortex have included the collection of ACh from the surface of the cortex either at rest or during stimulation (MacIntosh and Oborin 1953; Mitchell 1963; Szerb 1964, 1967; Phillis and Chong 1965; Collier and Mitchell 1967; Bèani et al. 1968; Phillis 1968a; Hemsworth and Mitchell 1969; Szerb et al. 1970; Domino and Bartolini 1972). It is generally accepted that ACh collected in such experiments could have been released as a result of activity in the diffuse ascending reticulocortical pathway responsible for cortical arousal. What is controversial (see Hebb 1970; Spehlmann 1971; Spehlmann et al. 1971) is whether there are, in addition, more restricted cholinergic systems serving specific sensory areas. Results obtained in the rabbit suggest that cholinergic fibres may run from the lateral geniculate body to the visual cortex (Collier and Mitchell 1966, 1967) and from the medial geniculate body to the auditory cortex (Hemsworth and Mitchell 1969). Mitchell and his co-workers, while rejecting the possibility that these fibres were primary thalamocortical afferents (cf. §7.3.7), proposed that they might belong to the system originating in the thalamus which is responsible for augmenting and repetitive responses in the cortex. These conclusions have been questioned by Phillis (1968a) who was unable to demonstrate that ACh release in cats could be selectively increased in limited cortical areas by the stimulation of specific thalamic structures. He took the view that the diffuse reticular activating system was the only ascending cholinergic system to the cortex but that some ACh could stem from intracortical cholinergic cells (see below) activated by thalamocortical fibres. Hemsworth and Mitchell (1969) considered that latter possibility in the rabbit but came to the conclusion that these cells were unlikely to contribute significantly to the ACh collected.

In 1963 Krnjević and Phillis used the iontophoretic technique to look for cholinoceptive cells in the cerebral cortex of cat, rabbit and monkey. Cells that responded to ACh were in general found in the motor cortex at a depth of between 0.8 and 1.3 mm from the surface. A number of characteristics suggested that these responsive cells were deep pyramidal (or Betz) cells of cortical layer V. Further confirmation for this view came from histochemical studies. Krnjević and I (Krnjević and Silver 1963) found that in the cat cortex the Betz cells and associated neuropil stained for AChE in a way which suggested that these cells could be cholinoceptive (plate 7.6). To determine the source of the AChE-containing fibres innervating these cells we made an extensive histochemical study of the distribution of AChE in the forebrain of adult, young, neonatal and foetal cats (Krnjević and Silver 1965, 1966). We used the lesion technique (ch. 3) to determine the polarity of the fibres, and the ChAc levels in partially isolated cortical slabs were also examined (Hebb et al. 1963). The loss of fibre staining and the marked reduction of ChAc activity in isolated slabs suggested that the majority of 'cholinergic' elements were subcortical in origin. The histochemical picture indicated that the fibres innervating the medial part of the cortex came mainly from the septum and the, innervating the lateral regions from the basal ganglia. Although many of the AChE-containing fibres seemed to originate outside the cortex the intracortical system of U- or arcuate fibres which travelled round the base of the sulci to connect adjacent gyri was also strongly stained. Observations on developing cats (Krnjević and Silver 1966) provided evidence that the cells giving rise to these fibres may be the spindle or polymorph cells of layer VI (Meynert 1867). These cells which migrate into the cortex from some undetermined site relatively late in foetal development stain very strongly for AChE (plate 7.7). The staining reaction is not lost when the cortex is undercut; on the contrary, it may be intensified (see plate 7.9). These features indicate that the cells could be cholinergic and this raises the question of what neurones are innervated by their processes, some of which may be represented by the stained horizontal fibres that we found (Krnjević and Silver 1965) in layer I (plate 7.8). In view of the apparent lack of cholinoceptive cells elsewhere than in layer V, we suggested that the stained fibres present in superficial layers could terminate on apical dendrites belonging to cholinoceptive cells situated more deeply.

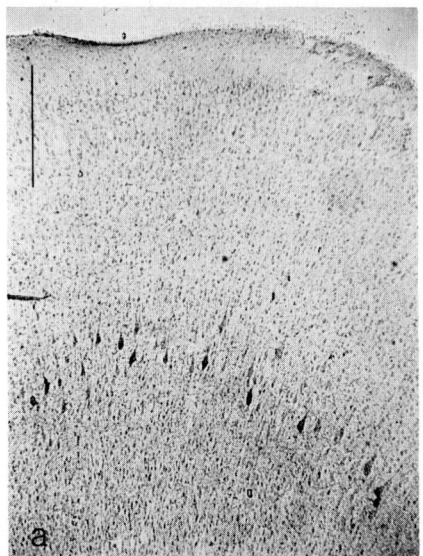 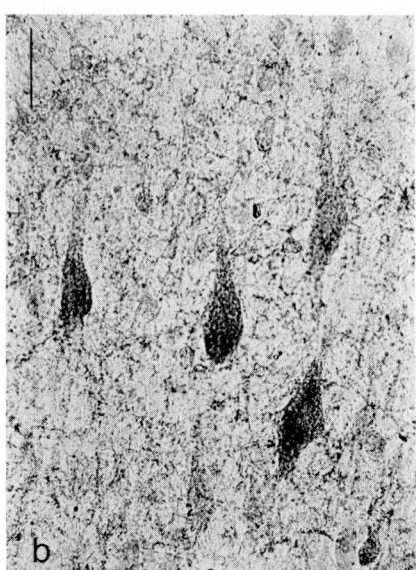

Plate 7.6. AChE activity in Betz cells in the anterior sigmoid gyrus of cat cerebral cortex. Scale bars: a) 500 μ; b) 50 μ. (From Krnjević and Phillis 1963a.)

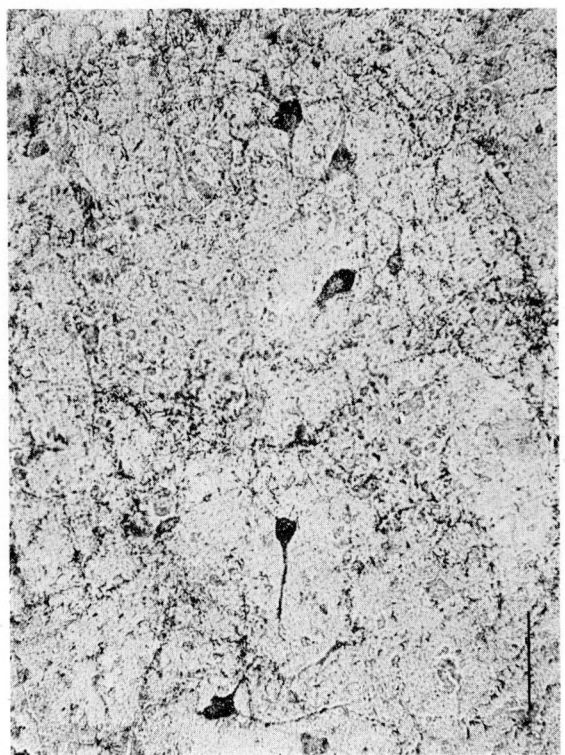

Plate 7.7. Strong AChE activity in small polymorph cells of layer VI in cat cerebral cortex. Scale bar: 50 μ. (From Krnjević and Silver 1965.)

Subsequently, however, Phillis and York (1967, 1968a, b) have detected cholino-ceptive cells in all except the most superficial layers, more especially in II, III and IV. In contrast to the Betz cells which are excited by ACh, these cells are inhibited and hence escape detection if their activity is already depressed by deep anaesthesia. Phillis and York suggest that these cells are innervated by cholinergic interneurones lying wholly within the cortex and it has been proposed (Hebb 1970, see also Jordan and Phillis 1972) that these interneurones could indeed be the AChE-containing polymorph cells of layer VI. This view is supported by the finding that 'cholinergic' inhibition can be induced by surface stimulation of isolated cortical slabs (Phillis and York 1968a) from which most of the AChE activity has been lost except for that associated with the polymorph cells (see plate 7.9).

An area of the forebrain which deserves special mention is the hippocampus. This structure has been the subject of numerous investigations including iontopho-retic studies (Biscoe and Straughan 1966) and histochemical staining (Storm-

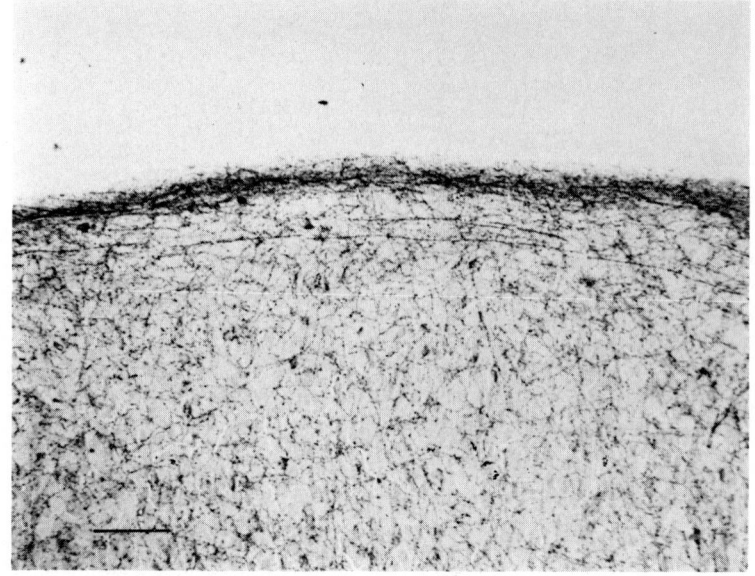

Plate 7.8. Horizontally-orientated AChE-containing fibres in the plexiform layer of the cat cerebral cortex. Scale bar: 100 μ.

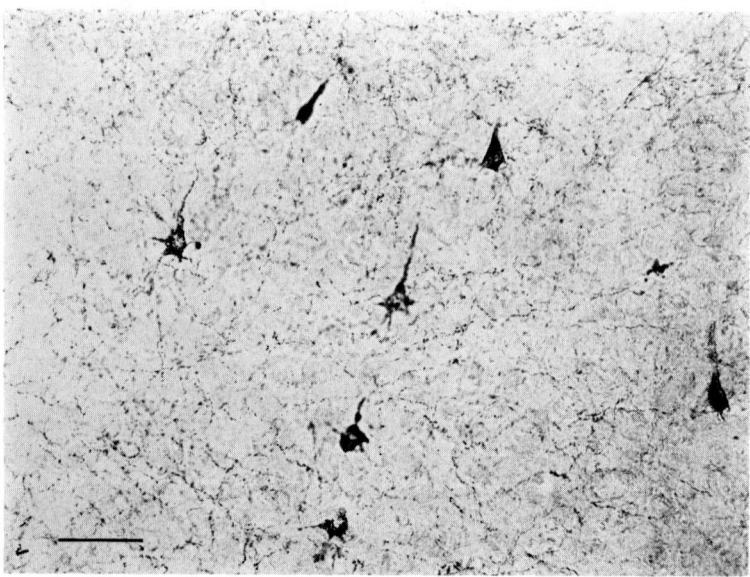

Plate 7.9. Retention of AChE activity in polymorph cells of suprasylvian gyrus of cat cerebral cortex which had been partly undercut 31 days previously. Scale bar: 50 μ. (From Krnjević and Silver 1965.)

Mathisen and Blackstad 1964; Lewis and Shute 1967; Lewis et al. 1967; Storm-Mathisen 1970; Geneser-Jensen and Blackstad 1971; Geneser-Jensen 1972a, b; Mellgren 1973; Mellgren and Srebro 1973). Fonnum (1970) showed that the correspondence between the level of AChE and ChAc in different laminae of the rat hippocampus was good, and Biscoe and Straughan (1966) found a reasonable fit between the laminated distribution of AChE and of cholinoceptive cells in that of the cat. These various results support the concept that the hippocampus receives a considerable cholinergic input and Lewis et al. (1967) showed that in the rat this probably comes from the medial septum and the nucleus of the diagonal band. Whether the same is true for other species is not established, and it is worth noting

TABLE 7.3

Summary of Lewis and Shute's 'cholinergic limbic system' (cf. fig. 7.2). (From Lewis and Shute 1967.)

Group 1 neurones: afferent to the hippocampal formation

nucleus of diagonal band ⎤ → ⎧ dentate gyrus
medial septal nucleus ⎦ ⎩ dorsal and ventral hippocampus

Group 2a neurones: afferent to medial cortex

anterodorsal thalamic nucleus → retrosplenial area
anteroventral thalamic nucleus → cingular area
interstitial nucleus of ventral ⎤
    hippocampal commissure ⎥ → ⎧ anterior limbic area and area infralimbica
precallosal cells ⎦ ⎩ olfactory bulb
nucleus accumbens → olfactory tubercle

Group 2b neurones: afferent to the cerebellum

                          ⎧ laterodorsal tegmental nucleus
                          ⎪ (l.d.t.n.) → brachium conjunctivum
dorsal tegmental nucleus (d.t.n.) → ⎨
                          ⎪ nucleus reticularis tegmenti pontis
                          ⎩ (n.r.t.p.) → brachium pontis

Group 2c neurones: afferent to the ascending cholinergic reticular system

1) to the dorsal tegmental pathway:
        d.t.n. → l.d.t.n. → nucleus cuneiformis
2) to the ventral tegmental pathway:
        d.t.n. and deep tegmental nucleus ⎫
        dorsal and median raphe nuclei   ⎬ → interpeduncular nucleus
        habenular nuclei   ⎭
        d.t.n. → l.d.t.n. → nucleus cuneiformis → substantia nigra
        d.t.n. → n.r.t.p. → ventral tegmental area

Group 2d neurones: afferent to non-neural structures

cells in dorsal fornix ⎤ → ⎧ subfornical organ
cells in septal raphe ⎦ ⎩ supraoptic crest

---

that the pattern of AChE activity in the dentate region of guinea-pig differs considerably from that in rat (see Geneser-Jensen 1972b).

Lewis and Shute (1967) examined the hippocampal efferent system as well as the afferent fibres. They have shown that although probably non-cholinergic themselves, the hippocampal efferent fibres relay to various nuclei which in turn send AChE-containing fibres to the cingulate cortex, the olfactory bulb, the cerebellum and structures associated directly or indirectly with the ascending reticular system (table 7.3). As fig. 7.2 shows, the latter projects on to the nucleus of the diagonal band and the septum via the lateral preoptic area, hence the hippocampus is linked to this activating system by both afferent and efferent paths. On the basis of these observations, Lewis and Shute have proposed that the hippocampus forms part of a 'cholinergic limbic system' which could be involved not only in arousal and attention but possibly in memory and learning as well. Clearly this hypothesis needs rigorous physiological and pharmacological investigation. Experiments designed to test the validity of the idea as a whole may also throw light on certain features of the cholinesterase-containing system which at present are a mystery. Thus in the rat many of the neurones in the nuclei on to which the hippocampal efferent fibres project contain varying amounts of pseudocholinesterase in addition to AChE (Lewis and Shute 1967). Since pseudocholinesterase acts best at high concentrations of substrate it has been mooted (see ch. 10) that it can take over the hydrolysis of ACh if AChE suffers substrate inhibition. Although theoretically reasonable this mechanism has not been unequivocally demonstrated (ch. 2).

7.3.2.6 Some other structures of the AChE-containing ascending and limbic systems   Of the various other AChE-containing areas which Shute and Lewis identified in rat brain, three systems deserve some mention. The first is the interpeduncular nucleus which according to Lewis and Shute's (1967) scheme is a link in the chain (hippocampus − habenular nucleus − interpeduncular nucleus) that connects the hippocampus with the ascending reticular system (see fig. 7.2).

Plate 7,10a,d show the marked AChE activity in the interpeduncular nucleus and habenulo-interpeduncular tract (fasciculus retroflexus of Meynert) of adult and foetal cats respectively. In agreement with this indication that the system might be cholinergic Lewis et al. (1967) found the ChAc level in the rat interpeduncular nucleus to be extremely high. They obtained a value of 55,000 μg ACh/g wet wt/hr which compared with values of 13,000 for the hypoglossal and facial nuclei and 6200 for the caudate nucleus. The question is why it should be so exceptionally rich in these enzymes. According to Lewis and Shute (1967) the AChE activity in the rat interpeduncular nucleus is associated with both nerve terminals and cell bodies. It was shown by Cajal (1911) that the terminal parts of the fibres in the habenulo-interpeduncular tract cross and recross the interpeduncular nucleus and it is possible that it is this anatomical peculiarity rather than any special functional requirement which accounts for this plethora of AChE and ChAc.

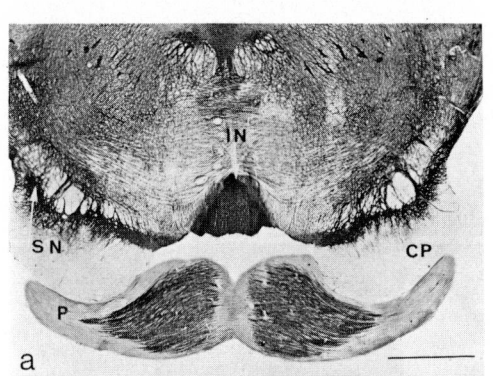
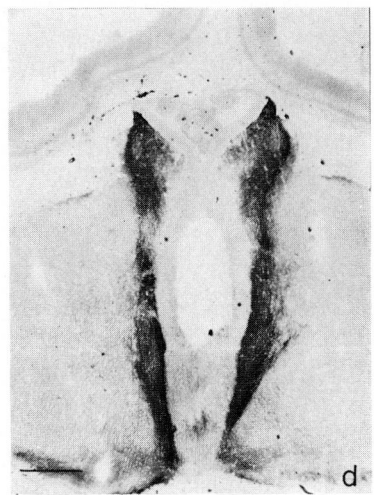

Plate 7.10. a) Intense AChE activity in the interpeduncular nucleus (IN) of adult cat. CP, cerebral peduncles; P, pons; SN, substantia nigra. Scale bar: 2 mm. d) Intense AChE activity in the habenulo-interpeduncular tract of foetal cat, 36 days gestation. Frontal section. Scale bar: 500 µ. (From Krnjević and Silver 1966.)

Two other structures with high levels of AChE were described by Lewis and Shute (1967) as 'problematical'. These are the subfornical organ and the supraoptic crest. The subfornical organ is a vascular structure found beneath the ventral hippocampal commissure. The enzyme is associated with incoming fibres from the dorsal fornix but since its function is unknown the reason for this massive innervation is not understood. The same is true of the supraoptic crest which is histologically similar and lies just rostral to the optic chiasma. Its very dense reaction for AChE (see Shute and Lewis 1966b fig. A) is due to fibres reaching it from the septal raphe.

7.3.3 The red nucleus and superior colliculus

The red nucleus is anatomically closely related to the deep tegmental grey but since Shute and Lewis (1967b) did not include it in their investigation of the ascending system it is considered separately. Although the red nucleus has been the subject of several studies its pharmacology is still uncertain. McLennan (1969) found that in the cat some of the rubral neurones, both large and small, showed a fairly marked histochemical reaction for AChE but the neuropil was much less reactive. This picture which argues against a cholinergic innervation of the nucleus is in keeping with results of iontophoretic experiments (Davis and Vaughan 1969). It was suggested by McLennan and by Davis and Vaughan that some of the rubral neurones could be cholinergic and this idea is supported by the presence of a small amount of ChAc in samples of red nucleus from both cat (A. Silver and J.H. Wolstencroft unpublished) and man (Bull et al. 1970). Waldron and Gwyn (1969) found that in rat (see Waldron 1970 re cat and guinea-pig) hemisection of the cervical spinal cord failed to cause an accumulation of enzyme proximal to a lesion in the rubrospinal tract but it did produce a loss of AChE in the neurones of the contralateral red nucleus (see also Gwyn 1971b). While the loss of staining suggests that the enzyme-containing cells are those which project to the cord, the absence of any accumulation of AChE is at variance with the idea that they are cholinergic (see Lewis et al. 1967). A possible explanation for these somewhat contradictory findings is that the ChAc detected in samples of red nucleus is not attributable to its neurónes but represents contamination from adjacent structures. The red nucleus lies close to the IIIrd nerve and the habenulo-interpeduncular tract both of which contain ChAc.

Another midbrain structure not yet discussed is the superior colliculus. Burgen and Chipman (1951) found relatively high AChE activity in the dog, and Ramon-Moliner (1972) showed that in the cat colliculus histochemical staining for AChE occurred in layers. The strongest activity was in the most superficial part, i.e. the zonal layer and superficial grey. The underlying optic layer was very pale as was the deep grey and deep white but the intermediate layer showed patches of strong activity interspersed with areas which stained only weakly. That the AChE in the superior colliculus is associated with cholinergic mechanisms would seem likely since the area contains appreciable ChAc (Hebb and Silver 1956). On the other hand, Straschill and Perwein (1971) found that only 7% of the collicular cells tested in the cat were cholinoceptive. Until experiments are done to see whether or not there is any correlation between the distribution of cholinoceptive cells and ChAc in particular layers it is unwise to speculate on the significance of the superficial AChE.

7.3.4 Cerebellum

In many species the cerebellum is one of the anomalous areas of the brain in which the AChE level is high but the levels of ChAc and ACh are disproportionately low (for review see Silver 1967). The reason for this disparity is not understood (ch. 8 §3.1.1) but the cerebellum exemplifies the risk of accepting the presence of AChE as firm evidence for the occurrence of cholinergic transmission. The problem of deciding whether cholinergic synapses do exist in the cerebellum is complicated by the marked differences in the distribution of AChE (chs. 6 and 8) not only between species and between different areas of the cerebellum within a species but even between the deep and superficial parts of the same cell layer within a single folium. The structure of the cerebellum is shown diagrammatically in fig. 7.3.

The difficulties involved in trying to make sense of cerebellar pharmacology are

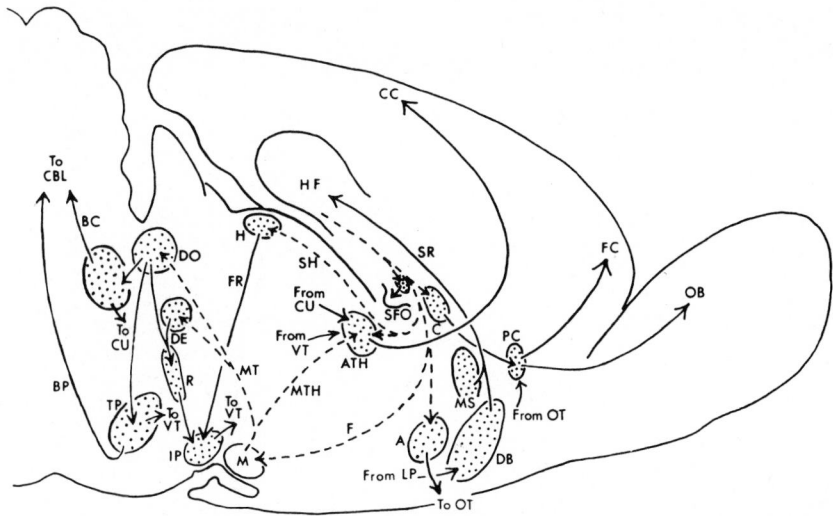

Fig. 7.2. Diagram of Lewis and Shute's 'cholinergic limbic system'. Stippling shows AChE-containing nuclei of the midbrain and forebrain connected with the hippocampus, their projections to the medial cortex, and their connexions with the ascending cholinergic reticular system. Abbreviations: A, nucleus accumbens; ATH, anteroventral and anterodorsal thalamic nuclei; BC, brachium conjunctivum; BP, brachium pontis; C, interstitial nucleus of the ventral hippocampal commissure; CBL, cerebellum; CC, cingulate cortex (cingular and retrosplenial areas); CU, nucleus cuneiformis; DB, diagonal band; DE, deep tegmental nucleus (ventral tegmental nucleus of Gudden); DO, dorsal tegmental nucleus; F, fornix; FC, frontal cortex (area infralimbica and anterior limbic area); FR, fasciculus retroflexus (habenulo-interpeduncular tract); H, habenular nuclei; HF, hippocampal formation; IP, interpeduncular nucleus; LD, laterodorsal tegmental nucleus; LP, lateral preoptic area; M, mamillary body; MS, medial septal nucleus; MT, mamillo-tegmental tract; MTH, mamillothalamic tract; OB, olfactory bulb; OT, olfactory tubercle; PC, precallosal cells; R, dorsal and median nuclei of raphe (nucleus centralis superior); SFO, subfornical organ; SH, stria habenularis; SR, septal radiation; TP, nucleus reticularis tegmenti pontis (of Bechterew); VT, ventral tegmental area. (From Lewis and Shute 1967.)

well illustrated by a comparison of rat and guinea-pig. In the rat, histochemically demonstrable AChE activity (plate 7.11) is more pronounced in the archicerebellar part of the vermal cortex — this is the part which receives its input from the vestibular system — than in the paleocerebellar folia (Csillik et al. 1963; Shute and Lewis 1965; Kása and Silver 1969). The reaction is largely confined to the granule cell layer. In contrast, in the guinea-pig the entire cortex reacts to much the same extent (plate 7.12) and furthermore, marked staining occurs in the molecular as well as the granular layer. Kása and I (1969) analysed samples of rat and guinea-pig archi- and paleocerebellum for AChE and ChAc; the results are shown in table 7.4. The first point to notice is that the quantitative data for AChE confirm the impres-

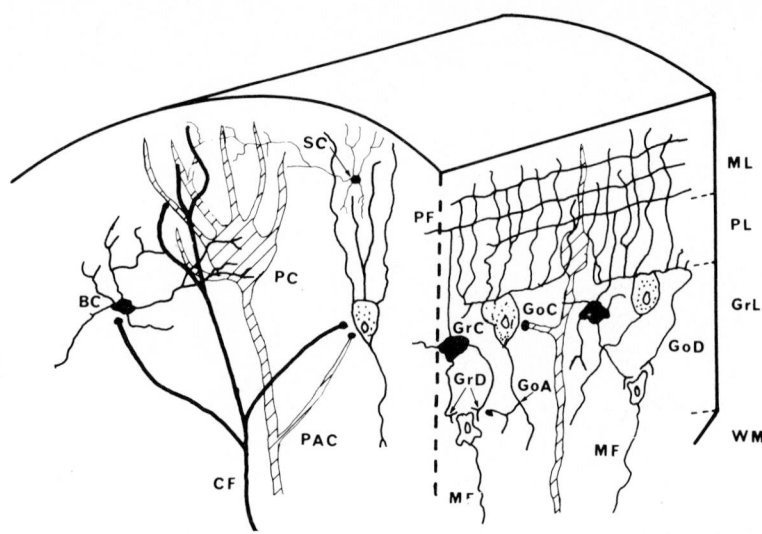

Fig. 7.3. Diagram of neuronal interrelations in the cerebellum. BC, basket cell; CF, climbing fibre; GoA, Golgi cell axon; GoC, Golgi cell; GoD, Golgi cell dendrite; GrC, granule cell; GrD, granule cell dendrite; GrL, granular layer; MF, mossy fibre; ML, molecular layer; PAC, Purkinje cell axon collateral; PC, Purkinje cell; PF, parallel fibre; PL, Purkinje cell layer; SC, stellate cell; WM, white matter. (Redrawn from Silver 1967.)

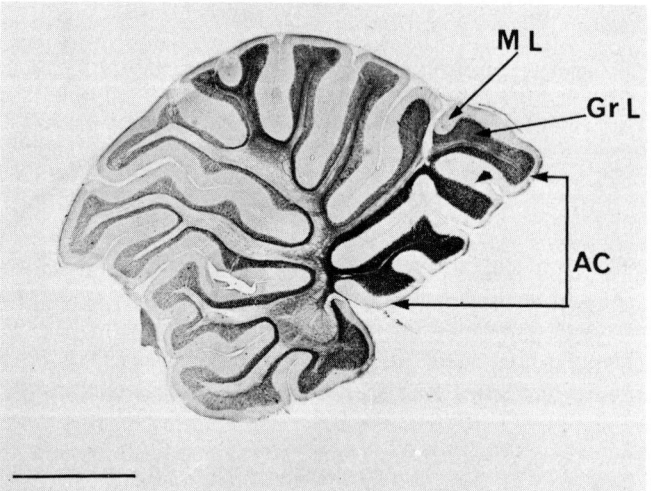

Plate 7.11. Rat cerebellum stained for AChE. Note that reaction is strongest in the archi-cerebellar cortex (AC). GrL, granule cell layer; ML, molecular layer. Scale bar: 2 mm.

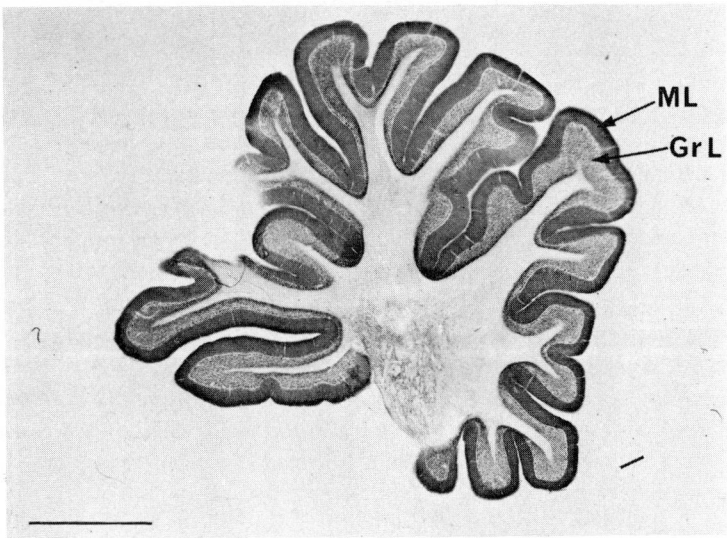

Plate 7.12. Guinea-pig cerebellum stained for AChE. Note the reaction is uniform in different folia and occurs in both the granular (GrL) and molecular layer (ML). Scale bar: 2 mm.

sion given by the histochemical pictures: in rat the activity in the archicerebellum is 3.6 times that in the paleocerebellum while in the guinea-pig the values for the two areas are almost equal and exceed the maximum value for the rat. The ratio of the AChE levels in the rat archi- and paleocerebellum is very closely paralleled by the ratio of the ChAc levels, this figure being 3.7:1. When AChE and ChAc are directly compared the ratio of activities for archicerebellum is 166:1 and for paleocerebellum 170:1. In the guinea-pig the relatively higher values for AChE in both the archi- and paleocerebellum are not matched by a correspondingly higher figure for ChAc. On the contrary, as table 7.4 shows, the choline acetylase activities are far lower

TABLE 7.4

Comparison of AChE and ChAc activity in the archi- and paleocerebellar cortex of rat and guinea-pig. (Data from Kása and Silver 1969, and unpublished.)

	AChE μmole/g wet wt/hr		ChAc μmole/g wet wt/hr		AChE : ChAc	
	Rat	Guinea-pig	Rat	Guinea-pig	Rat	Guinea-pig
Archicerbellum (Ac)	762	1110	4.6	0.48	166 : 1	2312 : 1
Paleocerebellum (Pc)	213	1140	1.25	0.45	170 : 1	2533 : 1
Ratio Ac : Pc	3.6 : 1	0.97 : 1	3.7 : 1	0.97 : 1		

than those in rat and this is reflected in AChE:ChAc ratios of approximately 2300:1 and 2500:1 for the archi- and paleocerebellum respectively. In addition to analysing intact tissue Kása and I measured enzyme activity in folia of rat archi- or paleocerebellum which had been undercut in such a way that the blood supply was left intact but the area was neurally isolated. The rats were killed between 3 and 30 days after operation but since the results were similar regardless of survival time the figures shown in table 7.5 represent average values. The lesion caused a pronounced fall in both ChAc and AChE in the archicerebellum but in the paleocerebellum the decrease was less dramatic. Conclusions can be drawn both from the loss of enzyme activity and from the fact that this loss was never total. The decrease in ChAc and AChE would suggest that afferent fibres make an appreciable contribution to the enzyme levels in intact tissue and that this contribution — and hence the consequence of its loss — is greater in the archicerebellum than in the paleocerebellum. Since the lack of an appreciable histochemical reaction for AChE in the molecular layer indicates that the majority of climbing fibres which synapse with Purkinje cell dendrites are probably non-cholinergic (but see further discussion below), it seems most likely that the enzyme-rich afferents are mossy fibres. These fibres, which terminate in the glomeruli of the granule cell layer, arise from various parts of the brain stem and cord (Jansen and Brodal 1954; Shute and Lewis 1965; Ha and Liu 1966; Gwyn et al. 1972). Although the fibres reaching different areas of the cerebellum are morphologically similar it looks as if chemically they may be less homogeneous. Kása and I suggested that the difference in ChAc activity in rat archi- and paleocerebellar cortex could mean that the latter received fewer of the cholinergic type of fibre. If ChAc and AChE are associated only with incoming fibres then deafferentation could be expected to cause an almost total loss of enzyme but, as already mentioned, this was not so (see table 7.5). Choline acetylase levels remained at about 16% of normal in the archicerebellum and 45% in the paleocerebellum; the corresponding figures for AChE were 26 and 65%. We proposed that the elements most likely to be responsible for the persisting ChAc activity were the Golgi cells which are known to retain histochemically detectable AChE following deafferenta-

TABLE 7.5

Comparison of AChE and ChAc activity in intact and isolated folia of rat archi- and paleo-cerebellar cortex. (Data from Kása and Silver 1969, and unpublished.)

	AChE			ChAc		
	μmoles/g wet wt/hr		% retained in	μmoles/g wet wt/hr		% retained in
	Intact	Isolated	isolated areas	Intact	Isolated	isolated areas
Archicerebellum	762	198	26	4.6	0.72	16
Paleocerebellum	213	138	65	1.25	0.56	45

tion (Shute and Lewis 1965; Kása et al. 1966). Although Golgi cells are more numerous in the archicerebellum the total activity retained was much the same as that in the paleocerebellum. We took this to mean that only a proportion of the Golgi cells are cholinergic and that the total number of this type is similar in both parts. It may be wrong to assume that the Golgi cells are alone responsible for the residual activity especially in view of evidence that supportive elements associated with cholinergic fibres may show enzyme activity when the axons degenerate (Birks et al. 1960; Bevan et al. 1973, see ch. 9). A significant contribution from supportive elements does not, however, seem likely in our experiments since the archicerebellum which appears to contain a greater number of cholinergic fibres is scarcely more active, when deafferentated, than the paleocerebellum. In summary, this series of biochemical experiments suggests that in rat a proportion of mossy fibres, more particularly some of those innervating the archicerebellum, are cholinergic. In addition to these afferent fibres some intrinsic cerebellar elements, possibly certain Golgi cells, may also be cholinergic. The situation in the guinea-pig seems quite different. Despite a high level of AChE activity, the ChAc level is very low suggesting that cholinergic synapses may be sparse in this species.

In the cat, the species most often used for neurophysiological experiments, the biochemical evidence for the existence of cholinergic synapses includes the presence in the cerebellar cortex and peduncles of AChE together with small but significant amounts of ChAc (Hebb and Silver 1956; Goldberg and McCaman 1967) and of ACh (MacIntosh 1941; Crossland and Merrick 1954; Israël and Whittaker 1965). It has also been found that ACh can be collected from the cerebellar surface (Phillis and Chong 1965). Unfortunately the iontophoretic studies aimed at elucidating cerebellar pharmacology have given conflicting results. Initially Crawford et al. (1963) found evidence indicating that Purkinje cells were cholinoceptive and they suggested that the cholinergic input was provided by the parallel fibres from granule cells. After further investigation (Curtis and Crawford 1965), this idea was abandoned and it was concluded that granule cells were unlikely to be either cholinergic or cholinoceptive and that the apparent excitatory effect of ACh on Purkinje cells could be due to an action on non-synaptic areas of the cell membrane. This conclusion was strengthened by the finding (Crawford et al. 1966) that intravenous injections of atropine and dihydro-β-erythroidine had no effect on the field potential that was generated when a number of Purkinje cells responded to electrical stimulation of various pathways. Against this evidence which suggests that ACh plays little part in cerebellar transmission must be set the observations of McCance and Phillis (1964a, b, 1968). They found that granule cells, more particularly those in the deeper parts of the folia, could be stimulated by iontophoretically applied ACh and that the mossy fibres innervating areas with responsive cells were especially rich in AChE. Although some of the Purkinje cells responded to ACh, the excitation did not apparently involve the dendritic receptors and McCance and Phillis (1968) suggested it might result indirectly from stimulation of granule cells or directly from an action on receptors on the Purkinje cell soma. The task of reconciling the evidence against the existence of cholinergic synapses with the evidence for their existence is possibly less difficult than at first appearance. Although it is always unwise to extrapolate from one species to another, and particularly so when dealing with the cerebellum, the biochemical results for rat showed very clearly that morphologically homogeneous structures might be chemically heterogenous. The same probably holds true for the cat and this may well explain why McCance and Phillis (1968) found cholinoceptive granule cells while Crawford et al. (1966) did not. The latter group examined cells only in the superficial parts of the folia where, according to McCance and Phillis, cholinoceptive cells are relatively rare. As more becomes known about the source, and thus the possible nature, of the mossy fibres which project to specific areas of the cerebellum, cholinergic synapses can be sought on a reasoned basis rather than by trial and error. McCance and Phillis (1968) made this sort of approach by taking account of the difference in AChE activity in the cerebellar peduncles. They argued that the pontine relay nuclei (Jansen and Brodal 1954) which project to the cerebellum via the AChE-rich middle peduncle were a more likely source of cholinergic afferents than were the medullary nuclei which project through the inferior peduncle, a tract relatively low in enzyme. This idea gains support from their findings that stimulation of the pontine nuclei, but not the medullary nuclei, resulted in activation of ACh-sensitive granule cells.

So far little attention has been given to the climbing fibres. These are generally thought to be non-cholinergic (see Crawford et al. 1966) but, again, it may be a mistake to assume that all the fibres are pharmacologically identical. Climbing fibres arise from the inferior olive and although in the cat the great majority of its cells do not stain for AChE, reactive cells can occasionally be distinguished in the positive neuropil (plate 7.13). Furthermore, there is biochemical evidence of some cholinergic elements in the molecular layer where the fibres end, both AChE (Austin and Phillis 1965; Goldberg and McCaman 1967; Phillis 1968b) and ChAc (Goldberg and McCaman 1967) being present. Israël and Whittaker (1965) found that the ACh-containing synaptosomes isolated from the cat cerebellum were apparently derived from the molecular layer and not, as in rat, from the mossy fibre endings. This result is in keeping with the suggestion that some cholinergic fibres do reach the molecular layer but the lack of ACh-containing synaptosomes of mossy fibre origin is hard to reconcile with the evidence that a proportion of these fibres appears cholinergic. Climbing fibres are not, of course, the only possible source of the ChAc found in the molecular layer, some intrinsic elements may contribute. Phillis (1968b) pointed out that in the cat, in contrast to the rat (Kása et al. 1965), the AChE in the granular layer is not confined to the mossy fibres but is present in some granule cell bodies. This type of cell, Phillis suggests, could be sending cholinergic parallel fibres to the molecular layer. Other elements which might be responsible for the ChAc content of the molecular layer are certain Golgi cell dendrites. As already discussed, some of the Golgi cells in the rat may be cholinergic and this

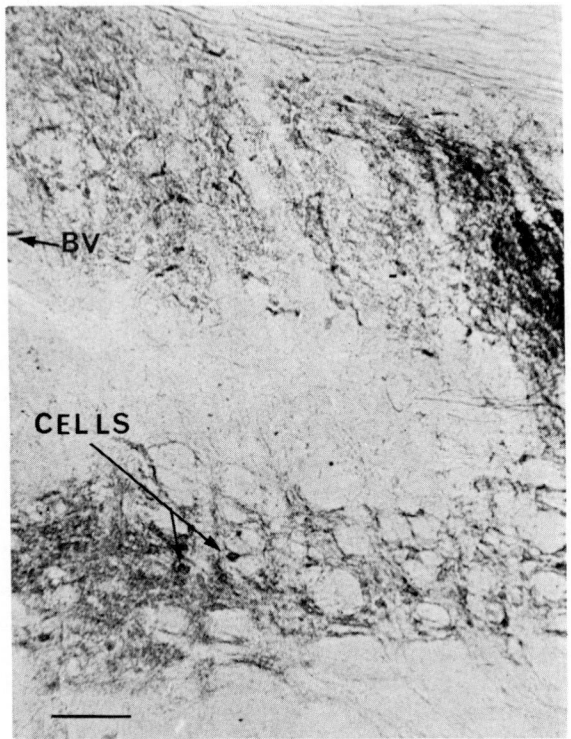

Plate 7.13. To show occasional AChE-containing cells (arrowed) in the inferior olive of the cat. Note staining of blood vessels (BV). Scale bar: 100 μ.

·could perhaps be true of the cat too. A finding which is in keeping with the existence of intrinsic cholinergic structures, whether in the granular or molecular layer, is that about 70% of the biochemically detectable AChE activity persists in the cat cerebellar cortex following deafferentation (Austin and Phillis 1965). Until ChAc is studied in similar experiments the true significance of this observation remains in doubt.

In considering the cerebellum, some attention must be given to the nuclei which are situated within it. Iontophoretic studies on cat (Chapman and McCance 1967) have shown that cells in the cerebellar nuclei can be excited with ACh. In all three nuclei the cells showed a similar range of reaction, from inexcitable to highly excitable. The presence of AChE in the neuropil indicates that some cells may receive a cholinergic input either from interneurones within the nuclei or from afferent fibres. A cholinergic input from the Purkinje cells of the cerebellar cortex seems unlikely both because of the presumed non-cholinergic nature of these cells and because their action on the nuclei is inhibitory rather than excitatory (Ito et al. 1964). The extrinsic afferents to the nuclei come from a number of sources (see Dow 1936; Jansen and Brodal 1954) which include the inferior olive, the vestibular nuclei and the reticular formation. AChE-containing cells in these regions could be a source of a cholinergic input.

Evidence suggests that some cells of the cerebellar nuclei in the cat may be cholinergic. Phillis (1965a, 1968b) found that when the superior cerebellar peduncle (also termed the brachium conjunctivum) was cut AChE accumulated predominantly on the side of the lesion nearest to the cerebellum thus showing up the presence of AChE-containing efferent fibres. In contrast, the accumulation caused by section of the middle peduncle was confined to the side away from the cerebellum. This bears out Flood and Jansen's (1966) view that the middle peduncle in the cat is a purely afferent pathway. If the accumulation of AChE on the cerebellar side of a lesion in the superior peduncle (and the similar but less marked accumulation seen in the inferior peduncle) can be taken as an indication of a cholinergic projection from the cerebellum where does it go? McCance et al. (1968a, b) have evidence that a proportion of the fibres travelling in the superior peduncle make cholinergic synapses in the thalamus (see also Phillis 1971) but McLennan (1970) has questioned this view. It must be emphasized that Phillis's evidence for an AChE-containing efferent projection applies to cats and contrasts with Shute and Lewis's (1965) finding that in rat the accumulation of AChE following peduncular lesions is confined to afferent fibres. Another indication that the cerebellar nuclei differ from species to species is that the enzyme present in the nuclei in the horse is mainly pseudocholinesterase (Sperti et al. 1960a).

These data, so often conflicting, demand a summary. The difficulty is that when for brevity and clarity the many 'ifs' and 'buts' are omitted statements appear misleadingly dogmatic. To counter this I would re-emphasize two points. First, that data from one species are often inapplicable to another and second, that within a species morphologically similar structures may be pharmacologically quite different. Evidence for the existence of some cholinergic synapses in the cerebellum includes the presence of ACh and ChAc as well as AChE and also the demonstration that cells are cholinoceptive. The relatively low level of ACh would suggest, however, that it is not a major transmitter. The case for cholinergic mossy fibres ending in the granule cell layer seems stronger than that for cholinergic endings in the molecular layer but the presence of ChAc in this layer cannot be ignored. It may mean that a proportion of one, or some, of the various elements are cholinergic. Climbing fibres, Golgi cell dendrites and parallel fibres have all been proffered as candidates. The realization that similar cells may be pharmacologically different goes some way to explain the conflicting results reported in the literature. What remains a mystery is the cause of the disparity between the high values for AChE and the low values for ChAc, found in so many species. This problem will be taken up again in the next chapter.

7.3.5 Medulla and pons

The medulla and pons contain the nuclei of the majority of cholinergic cranial nerves but the distribution of AChE suggests that other cholinergic structures may be present as well (Holmes and Wolstencroft 1964; Papp and Bozsik 1966; Ramon-Moliner 1972). This possibility is supported by various iontophoretic studies which have demonstrated the existence of cholinoceptive cells in cat (Salmoiraghi and Steiner 1963; Bradley et al. 1966) and in rat (Couch 1970; Bradley and Dray 1972). Salmoiraghi and Steiner had evidence that most of the responsive cells which they detected belonged to the medullary reticular system. Although cholinoceptive cells have not been systematically mapped in the lower brain stem other work has endorsed this view. In general the cells are scattered through the reticular formation, dispersed among non-cholinoceptive cells, but Avanzino et al. (1966) obtained a response from practically all of the cells they examined in the paramedian reticular nuclei of cat. Similarly Couch (1970) found 81% of cells tested in the pontine raphe nuclei of rat to be cholinoceptive. Effects have also been obtained from the midline pontine area of cats as the result of intracerebral injection of cholinergic drugs (Kostowski 1971). Contrary to earlier ideas (see Metz 1958) that ACh is a transmitter in the respiratory centre, Salmoiraghi and Steiner (1963) showed that in the case of neurones with respiratory characteristics, all except two of those tested were unresponsive to iontophoretically applied ACh. Both the vestibular (Steiner and Weber 1965) and the cochlear (Whitfield and Comis 1968) nucleus have been shown to contain cholinoceptive neurones. The possibility that here, and in other systems, cholinergic mechanisms could be involved in regulating the passage of sensory information will be considered in §7.3.7. It is, however, relevant to mention that Iwata et al. (1971) found that systemic injections of physostigmine depressed monosynaptic activation of trigeminal neurones and enhanced the inhibitory effect, on these neurones, of cutaneous nerve stimulation. They tentatively suggested that physostigmine could be causing stimulation of cholinoceptive inhibitory interneurones.

As yet little is known of the source of fibres which might be innervating the cholinoceptive cells but ascending and descending fibres may be involved, as well as fibres originating within the medulla and pons. According to Klemm (1972) the medullary reticular formation projects both to the cord and to more rostral levels of the brain. Certain evidence suggests that this system, like the midbrain reticular formation may in part comprise cholinergic cell bodies. Plate 7.14 shows that in cat a number of cells stain for AChE in a manner which suggests that they are cholinergic rather than cholinoceptive and, as will be discussed in §7.3.6 the grey matter of the cord is well stained in many of the areas where medullary axons terminate. In some unpublished experiments, J.H. Wolstencroft and I found that ChAc was present in the medullary reticular formation of the cat. The levels in the various areas examined were generally only 5 or 10% of those in known cholinergic nuclei.

Such relatively low values need not argue against the presence of cholinergic neurones since the histochemical picture suggests that the samples analysed are likely
to include a considerable amount of non-cholinergic tissue as well.

Other structures of interest in the medulla and pons include the cuneate and
gracile nuclei and the inferior and superior olives. In cat, both the cuneate and the
gracile nucleus show a strong reaction for AChE in the neuropil (Silver and Wolstencroft 1971; Ramon-Moliner 1972) but this does not fit with the results of iontophoretic studies which have revealed a marked lack of cholinoceptive cells (Steiner
and Meyer 1966; Galindo et al. 1967). The enzyme in these nuclei may be involved
in some non-synaptic role of the type to be discussed in the next chapter.

The neuropil of the inferior olive gives, in general, a striking reaction for AChE
but the intensity varies somewhat in its different parts. Ramon-Moliner (1972)
found that in cat the staining was strongest in the dorsal accessory olive. As mentioned earlier, stained cells are rare and this is one of the reasons for suggesting that
the climbing fibres, which arise in the inferior olive, are unlikely to be cholinergic.

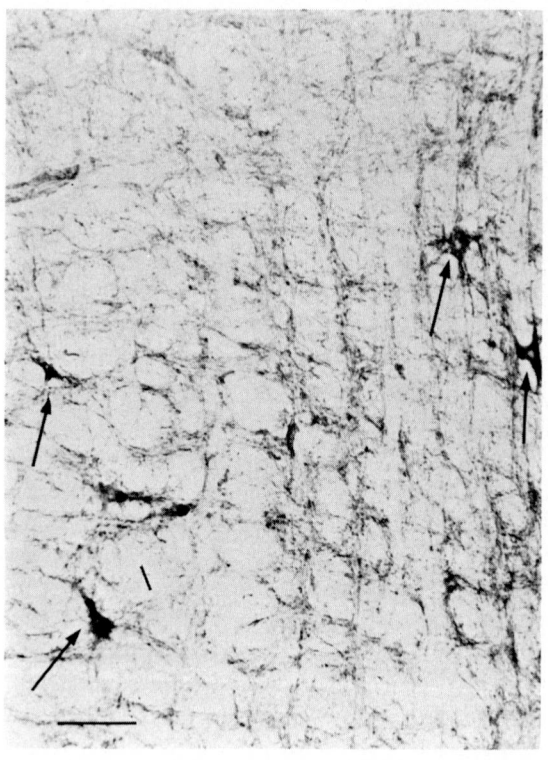

Plate 7.14. Cells (arrowed) in the medullary reticular formation of cat, showing strong AChE
activity. Scale bar: 100 μ.

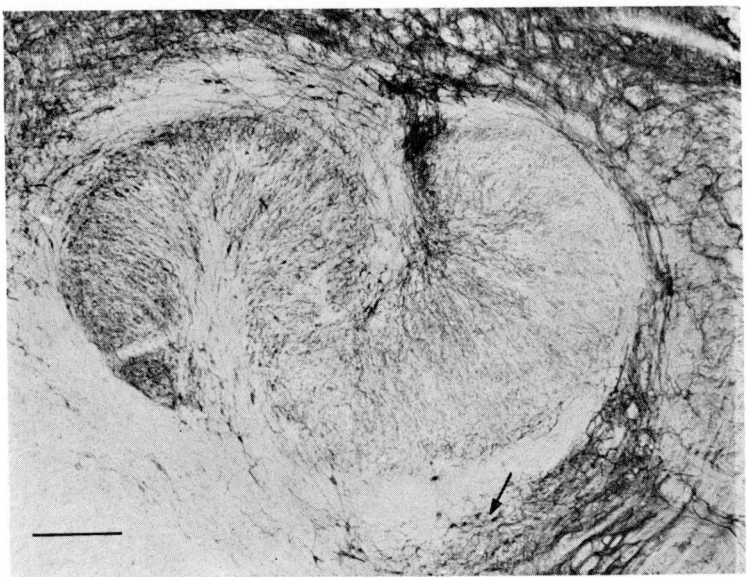

Plate 7.15. Superior olive of cat stained for AChE. Note scarcity of reactive cells in the nucleus itself. Some strongly positive reticular cells are present in the preolivary nucleus (arrowed). Scale bar: 250 μ.

A certain confusion is apparent in the literature over the superior olivary complex. The whole structure is often referred to loosely as the superior olive and is described as being the source of the olivocochlear bundle (OCB). This implication, that the olive is a single entity, is misleading. In most mammals the complex consists of at least 5 different cell groups and strictly speaking only the prominent S-shaped lateral group should be called the superior olive (see Rasmussen 1946). Plate 7.15 shows that few of the neurones in the olivary complex react for AChE, a result which would seem to conflict (see Ramon-Moliner 1972) with the steadily growing evidence (§7.3.7) that the olivocochlear bundle is cholinergic. The explanation is that the cells of origin of the OCB are not situated in the superior olive itself but are confined to the small retro-olivary and pre-olivary nuclei (Rasmussen 1946; Shute and Lewis 1965; Kingsley and Barnes 1973). In keeping with the proposed cholinergic nature of the bundle, the reticular cells of these groups do contain AChE.

7.3.6 Spinal cord

Motor neurones in the ventral horns of the spinal cord release ACh from their peripheral endings on muscle and are thus by definition, cholinergic. The same is

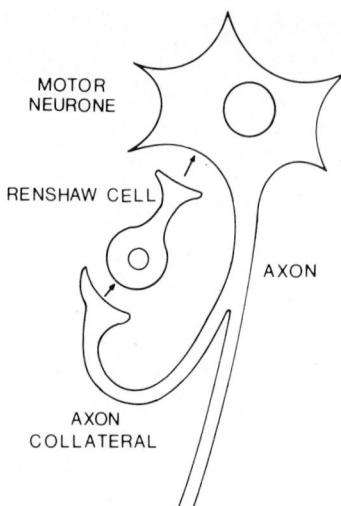

MOTOR
NEURONE

RENSHAW CELL

AXON

AXON
COLLATERAL

Fig. 7.4. Diagram showing proposed relation of Renshaw cell to motor neurone.

true of the cells in the intermediolateral nucleus which give rise to preganglionic autonomic fibres. The pharmacology of elements which do not project to the periphery cannot be examined so easily, hence the extent of cholinergic transmission within the cord itself is still uncertain.

The best authenticated cholinergic synapse in the cord is that between motor axon collaterals (fig. 7.4) and Renshaw cells. These are inhibitory cells (see below) which project on to the motor neurones (Eccles et al. 1954, 1956; Curtis et al. 1961). Since Dale's (1935) principle states that all the terminals of a particular cell employ the same transmitter, it is to be expected that the collaterals, like the parent motor fibres, will release ACh. In keeping with this theoretical concept, numerous experiments in cats have established that the Renshaw cells are cholinoceptive. It has also been shown that antidromic stimulation of peripheral nerves results in a marked increase in the amount of ACh present in the eserinized effluent from the cat's perfused lumbar cord (Kuno and Rudomin 1966; see also Edery and Levinger 1971). Although some of the receptors on the Renshaw cells appear to be muscarinic or 'intermediate' in type the majority are nicotinic (Curtis and Ryall 1964, 1966a, b) and the response to ACh closely resembles that at the motor end-plate. It is rapid in onset, it is blocked by dihydro-β-erythroidine but not atropine and is prolonged by anticholinesterases (Eccles et al. 1956). To some extent it was the failure to identify cerebral neurones with similar characteristics which cast doubt on the existence of cholinergic transmission in the brain (see Curtis 1963).

An interesting point about the study of Renshaw cells is that for a long time these were identified purely by their physiological effects and not as morphological entities. Renshaw first demonstrated that central inhibition could be produced by antidromic impulses in motor fibres in 1941 but it was not until 1954 that Eccles et al. established that the action was mediated via interneurones. They designated these interneurones 'Renshaw' cells although at this stage they had not been seen. Morphological reality was given to the cells in 1958 when Szentagothai using the 'isolated horn preparation' of dog cord found that the axon collaterals of motor neurones ended on small cells situated in seemingly appropriate positions in the ventral horn. In subsequent work by Erulkar et al. (1968) on cats, cells which had been identified as Renshaw cells on the basis of their response to ventral root stimulation were marked with dye injected via the recording electrode. The cord was then stained for AChE and it was found that the 'body' of the marked structures gave a slight to moderate reaction but their surfaces stained more strongly. AChE activity on the surfaces of structures thought to be dendritic bulbs of Renshaw cells has also been described in cat and rat by Csillik and Tóth (1972). This pattern of staining is in keeping with the pharmacological evidence that Renshaw cells are cholinoceptive but not cholinergic. A tentative identification of Renshaw cells in material stained for AChE has also been made in monkey (Odutola 1972). Navaratnam and Lewis (1970) working on rat found some pseudocholinesterase-containing cells which from size and position might be Renshaw cells but the authors could see no connection with motor axons.

Although physiological, pharmacological and histochemical findings appear to support the idea that Renshaw cells are probably cholinoceptive interneurones which release an inhibitory transmitter such as glycine (Curtis et al. 1971) their existence has been doubted by some authors (Weight 1968; Scheibel and Scheibel 1969). Weight has suggested that the so-called cell is nothing more than the terminal bouton of the axon collateral. This does not seem likely. Unless ACh itself is the inhibitory transmitter — and there is no evidence that it is — such a scheme would involve the release of ACh from the axon going to the muscle and of some different transmitter from the branch to the motor neurone. This is counter to Dale's law. A scheme whereby ACh was involved in the release of inhibitory transmitter from pre-synaptic terminals was proposed by McKinstry and Koelle (1967) but has been discounted (Curtis et al. 1971).

Duggan and Curtis (1972) take the view that the evidence for ACh as the transmitter from motor axon collaterals to the Renshaw cell is almost complete but this cannot be said of any other synapse within the cord. Although Curtis et al. (1966) showed iontophoretically that some interneurones apart from Renshaw cells respond to ACh, they were careful to point out that these responses were not necessarily related to synaptic effects. In attempts to get more information about the possibility of other cholinergic systems in the cord a number of histochemical studies have been made. The distribution of AChE in the rat (Navaratnam and Lewis 1970), cat (Silver and Wolstencroft 1971) and monkey (Odutola 1972) has

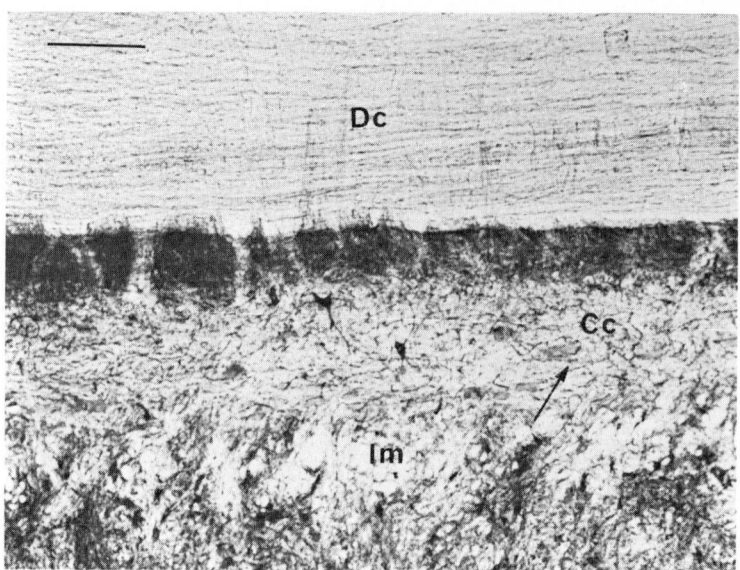

Plate 7.16. Sagittal section of cat spinal cord in plane of Clarke's columns (Cc). Note strong AChE activity in small cells in contrast to the weak reaction in the larger neurones (arrowed). Dc, dorsal columns; Im, intermediomedial nucleus. Scale bar: 250 μ. (From Silver and Wolstencroft 1971.)

been examined in considerable detail. The two latter species show a very similar staining pattern. Wolstencroft and I were hesitant about interpreting our data in terms of function but we noted a number of cells which reacted as if they might be cholinergic. In addition to motor horn and autonomic preganglionic cells other strongly stained cells were found in the commissural, spinal reticular and lateral cervical nuclei (see also Gwyn and Waldron 1968 re rat). Some large stained cells which were scattered in the dorsal part of the ventral horn were tentatively identified as the cells of origin of the ventral spinocerebellar tract (VSCT; see Ha and Liu 1968). Supporting evidence that at least some of the fibres in the cat VSCT are cholinergic was obtained by Gwyn et al. (1972). The lesion technique (ch. 3) showed that caudal to a hemisection both ChAc and AChE accumulated in the lateral part of the white matter in the region occupied by the tract. The dorsal spinocerebellar tract (DSCT), on the other hand, seems unlikely to be cholinergic since its cells of origin, the large neurones in Clarke's columns, react only slightly for AChE (Silver and Wolstencroft 1971). The weak reaction in these cells was in marked contrast to the strong staining in some small cells which are also present in the column (plate 7.16). Odutola (1972) concluded from his experiments on monkeys with cord section that in this species the fibres of the DSCT contain considerable amounts of AChE and he suggested that they may contribute to the AChE-positive mossy fibre endings in the cerebellar glomeruli. In view of the controversy associated with transmission in the cerebellum this should be investigated further.

In discussing the question of whether Renshaw cells were the only type of spinal interneurone to be cholinoceptive, Curtis et al. (1966) pointed out that since the pharmacology of the descending spinal pathways was unknown it would be unwise to make such an assumption. Subsequently Gwyn et al. (1972) showed that some descending fibres may be cholinergic. When a chronic hemisection is made in the cat spinal cord the accumulation of ChAc is not confined to the caudal segment discussed earlier but also occurs rostral to the lesion in the medial part of the white matter. As judged from their position, the fibres which show this reaction could belong to a number of descending tracts which include the raphe-spinal and lateral reticulospinal pathways. Gwyn et al. also found that the level of choline acetylase in the dorsal horns was unexpectedly high. In experiments in which the mean value for the whole ventral horn was 6640 μg ACh/g/hr the corresponding value for the dorsal horn was 5860. This adds weight to the idea that the dorsal horns contain cholinergic endings or cells or both. No attempt was made to measure the ChAc in the separate laminae (Rexed 1954) of the dorsal horn but such experiments might be informative since the AChE activity varies from lamina to lamina (Silver and Wolstencroft 1971). One of the laminae which reacts particularly strongly is lamina III. This possesses very few AChE-containing cells and we have suggested that the staining could be associated either with descending fibres or with the processes of the strongly stained cells in lamina IV or V. Another AChE-rich pathway which merits special attention is Lissauer's tract. This is composed of the ascending and descending branches of some of the fine fibres in the dorsal root (see Szentagothai 1964) and it is surprising that this tract stains so heavily (plate 7.17) when the dorsal root itself does not. One possibility is that the fibres which are concentrated in this tract arise from those neurones in the dorsal root ganglion which stain strongly for AChE.

Neuropharmacological experiments made some time ago indicated that there might be cholinergic synapses in the spinal cord of amphibia (Kiraly and Phillis 1961; Mitchell and Phillis 1962). The latter authors found that in the frog, antidromic stimulation of ventral (but not dorsal) roots resulted in an increase in the amount of ACh that collected in the fluid bathing the cord. Eserine potentiated and dihydro-β-erythroidine depressed the response that could be recorded in the adjacent dorsal roots. More recently Koketsu et al. (1969) working with toad and bullfrog examined the mechanism by which ACh reduced the dorsal root potential produced by stimulation of dorsal or ventral roots. They proposed that ACh acted either indirectly on a cholinoceptive interneurone, which in turn depolarized the dorsal root terminals, or directly on cholinergic receptors on the terminals themselves. Both ideas can be only tentative. So far no one has identified a Renshaw-type of cholinoceptive interneurone in the amphibian cord nor is there morphological evidence for a cholinergic synapse on the dorsal root terminals.

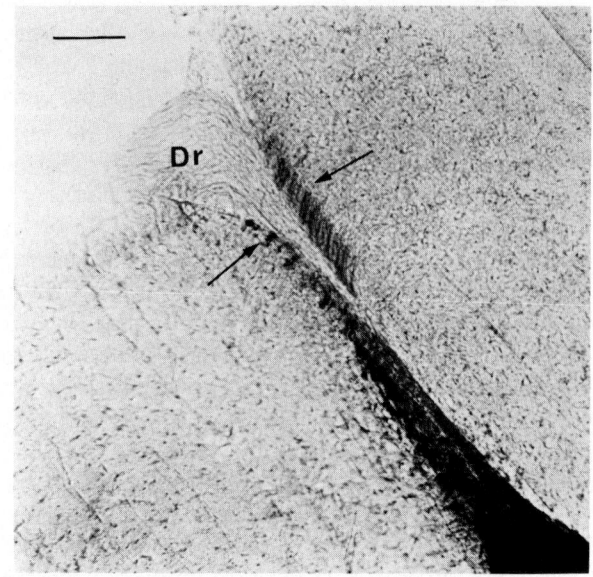

Plate 7.17. AChE in fibres in Lissauer's tract (arrowed). Dr, dorsal root. Scale bar: 100 μ. (From Silver and Wolstencroft 1971.)

7.3.7 The role of AChE in sensory pathways

The afferent pathways which link peripheral receptors to the CNS contain extremely low levels of ChAc and ACh and relatively little AChE. For these reasons sensory fibres, for instance those in the dorsal roots, and the optic and auditory nerves are regarded as non-cholinergic (Feldberg and Vogt 1948; Hebb 1955, 1957a; Hebb and Silver 1956). This does not mean, however, that the whole of the sensory system is operated by something other than ACh. On the contrary, there are indications that ACh and hence AChE, could be involved in one of two ways in the transmission of information from the periphery to the brain. Evidence suggests it may act as the transmitter at some synapses interposed on the direct route from the receptor to the cortex or it may be released from neurones which impinge on, and so modify activity in, these pathways. The original proposal that the sensory pathways might comprise cholinergic as well as non-cholinergic neurones, possibly arranged alternately, was made by Feldberg and Vogt (1948) on the basis of the variation in ChAc levels in different parts of the CNS. They noted that in the optic system of the dog the retina and lateral geniculate body are rich in ChAc while the optic nerve and tract which joins them is not. Using a more sensitive assay for ChAc, Hebb and I later obtained qualitatively similar results for a number of species

(Hebb and Silver 1956). Feldberg and Vogt postulated that cholinergic elements in the retina might include the bipolar cells. These are interposed between the receptors (i.e. the rods and cones) and the ganglion cells which give rise to the fibres of the apparently non-cholinergic optic nerve. The optic nerve, which on entering the brain stem is designated the optic tract, ends in the lateral geniculate body and in the superior colliculus. Since both these structures contain considerable amounts of ChAc and AChE, Feldberg and Vogt suggested that this might mean that the cells on which the tract ended were cholinergic. They made the reservation, however, that the ChAc and AChE could be associated with systems other than the optic system. As discussed in § § 7.3.2.1 and 7.3.2.5, Mitchell and his colleagues (Collier and Mitchell 1966, 1967; Hemsworth and Mitchell 1969) take the view that in the rabbit there is a cholinergic path between the lateral geniculate body and the visual cortex and between the medial geniculate body and the auditory cortex but that these pathways are not part of the 'primary' afferent chain. In other words these authors do not subscribe to the idea that the final neurones in either the optic or the auditory pathway are cholinergic. Hebb (1970), on the other hand, considers that this possibility has not been excluded. To support her view she cites results obtained by Miller et al. (1969a) on the rabbit. These authors found that removal of one eye caused a reduction of 23% in the concentration and of 27% in the content of ACh in the contralateral lateral geniculate body. Since the optic tract is non-cholinergic the fall in ACh cannot be attributed to disappearance of terminals and a possible explanation is that it represents the loss, due to transynaptic degeneration, of cholinergic cells. In addition to this demonstration that the fibres of the optic tract may end on cholinergic cells, Miller et al. also showed that removal of the visual cortex produced retrograde degeneration of the lateral geniculate body accompanied by a reduction in ACh content of 28%. Together, these findings could be interpreted in favour of the argument that the optic tract fibres end on cholinergic cells which in turn project to the cortex. It remains to be seen if this view can be reconciled with that of Mitchell's group.

Another sensory pathway which may be cholinergic is the ventral spinocerebellar tract. Evidence for this has been mentioned earlier and includes the finding that in the cat both ChAc and AChE accumulate on the caudal side of a lesion which transects this tract in the spinal cord. The tract carries to the cerebellum tactile and proprioceptive impulses which it receives from the periphery via the non-cholinergic dorsal root fibres. What is interesting is that the dorsal spinal cerebellar tract which apparently performs much the same function appears to be non-cholinergic, at least in the cat. Its cells of origin, the large neurones in Clarke's columns possess very little histochemically detectable AChE activity (Silver and Wolstencroft 1971; see also plate 7.16).

The next point to consider is the extent to which cholinergic neurones that are not part of a direct chain between sensor and effector may modify events occurring in such pathways. The probable participation of cholinergic mechanisms in a general ascending activating system which may alter the level of CNS activity (see

Krnjević 1969) has already been discussed in this chapter. What we come to now is the evidence that cholinergic neurones may also have a specific role in controlling the level of activity in the retina and in the auditory system.

As mentioned earlier, the retina appears to possess cholinergic synapses since it is particularly rich in ChAc. Hebb (1955, 1957b) found that the ChAc activity in the retina of birds was higher than any then reported for other vertebrate tissues. Furthermore, because ChAc levels in the optic nerve are extremely low in almost all species examined, it seems likely that the retinal enzyme is associated with intrinsic elements and not with the limited number of efferent fibres which travel in the nerve. The pigeon would seem to be atypical since its optic nerve contains an appreciable number of efferent fibres (Cowan et al. 1961; Cowan and Powell 1962) and the ChAc activity is considerably greater than that in other species (Hebb 1955, 1957b). Histochemical studies of AChE and BuChE have been made on various species but, as table 6.32 shows, results of different authors are not always in agreement. There are two reasons for this. First, the retina is particularly prone to diffusion artifacts and second, the very close packing of cells and processes makes accurate identification of stained elements difficult. At one stage for instance, there was doubt about the presence of AChE in ganglion cells. Hebb et al. (1953) reported staining in the ganglion cells in unfixed sections of rabbit retina but this localization was not seen by others (Eränkö et al. 1961). Subsequent work with the electron microscope has, however, shown AChE in some of the ganglion cells of rat (Lewis and Shute 1965b) and rabbit (Reale et al. 1971). There is almost unanimous agreement that the amacrine cells in the inner nuclear layer react for AChE (see Nichols and Koelle 1968; Nichols et al. 1972). These cells, which are believed to be inhibitory, send their processes into the inner plexiform, molecular or synaptic layer where they make contact with the dendritic processes of the ganglion cells and also with the bipolar cells. Koelle (1969) has suggested that ACh might evoke the release of the inhibitory transmitter or that it could itself be this transmitter. In view of the doubts about ACh acting in an 'amplifying' capacity (see § 7.2.2), and taking account of the exceptionally high level of ChAc in the retina, the second alternative is the more attractive. One other point should be made about the amacrine cells. Nichols and Koelle (1968) found that in a pure cone retina (e.g. in pigeon) both the soma and the processes of the amacrine cells stained for AChE. In pure rod retinae (e.g. of rabbit, rat and cat) the cell processes were stained but the cell bodies did not react. Only during the resynthesis of enzyme, following its inhibition by DFP, was it possible to detect AChE in the soma, and to trace its passage into the processes in the inner plexiform layer. The functional implications of this difference between rod and cone retinae remain to be clarified.

We come next to the auditory system. Electrophysiological experiments (see Guth and Bobbin 1971 for references) have shown that the potentials which can be recorded from the inner ear in response to a sound are inhibited by stimulation of the olivocochlear bundle (OCB). As mentioned earlier (§7.3.5) this tract is composed of fibres which arise from cells in the retro-olivary nucleus. It terminates in the cochlea on the hair cells of the organ of Corti and on other nerve endings (Smith and Sjöstrand 1961). Since many of the fibres are rich in AChE (Churchill et al. 1956; Shute and Lewis 1965; see also comprehensive review by Iurato et al. 1971), and since hair cells respond to iontophoretic application of ACh, it was suggested that ACh might be the inhibitory transmitter. When this idea was first investigated (see Tanaka and Katsuki 1966; Guth and Amaro 1969) it appeared unlikely that ACh could be the final mediator of the inhibitory effect since this was not blocked by intra-arterial injections of atropine or dihydro-β-erythroidine but was blocked by strychnine. Subsequent work has shown, however, that when injected directly into the cochlea, cholinergic antagonists including atropine and d-tubocurarine (Konishi 1972) and gallamine (Galley et al. 1971) do block the inhibitory action of the OCB. This evidence which strongly suggests that ACh is, indeed, the transmitter is strengthened by three other findings. First, ACh is released into cochlear fluid when the OCB is stimulated (Norris and Guth 1974) second, choline acetyltransferase is present in the OCB of guinea-pig (Jasser and Guth 1973) and third, the effect of stimulating the OCB can be mimicked by perfusing the scala tympani with ACh plus eserine (Bobbin and Konishi 1971).

When auditory impulses are set up in the cochlea they are relayed to the cochlear nucleus in the brain stem. Certain evidence suggests that here they come under the influence of a controlling mechanism operated by ACh. The nucleus contains AChE (see for instance work on cat by Osen and Roth 1969) and according to McDonald and Rasmussen (1971), nerve endings reacting for AChE seem to be of three types as judged from their ultrastructural characteristics. Some of the fibres innervating the ventral cochlear nucleus are collaterals from the olivocochlear bundle and Comis and Guth (1974) have demonstrated a release of ACh from the nucleus following electrical stimulation of the OCB. Whitfield (1968) and Whitfield and Comis (1968) have detected cholinoceptive cells in the cochlear nucleus and, on the basis of neuropharmacological experiments (see also Comis and Whitfield 1968; Comis and Davies 1969) together with the histochemical demonstration of AChE, these authors have suggested that ACh operates a so-called gating mechanism whereby impulses set up in the ipsilateral superior olive reduce the sound threshold in the ventral cochlear nucleus. If this is so, then the cholinergic system is in this case acting in a facilitatory manner rather than in the inhibitory way postulated for the retina and the cochlea itself. It may be added here that Bloom et al. (1964) proposed that ACh might be the transmitter in a pathway responsible for inhibiting the mitral cells in the olfactory bulb of rabbit.

7.4 Concluding remarks

Certain anomalous areas such as the cerebellum (see ch. 8) still justify the maxim that 'AChE-containing' is not synonymous with 'cholinergic'. On the other hand, more and more areas tentatively designated cholinergic, purely on the basis of their reaction for AChE have been found to fulfil some of the more rigorous criteria needed for this classification. Among questions which remain to be resolved is the source of the AChE activity in the corpus striatum. Is it associated mainly with extrinsic elements or with intrinsic structures? Another question concerns the hypothalamus. If the magnocellular nuclei are cholinoceptive, why do the fibres innervating the cells give such a poor reaction for AChE?

Where evidence of cholinergic transmission is sound, we no longer need to ask whether AChE is present to hydrolyse ACh or whether it performs some other task. Instead we can consider whether the enzyme has any part in determining the different effects of ACh at different sites. ACh seems to act as a 'detonator' transmitter on Renshaw cells and peripheral nicotinic junctions but elsewhere it appears to function more as a modulator. Presumably the difference is a reflection of differences in the type of ACh receptor (Dale 1938; see Krnjević and Phillis 1963b). Since there is evidence (Chothia 1970; Smythies 1974; see ch. 2) that at nicotinic sites ACh is bound upside down relative to its binding at muscarinic sites, it will be interesting to discover whether or not the molecular arrangement of the associated AChE is identical in both types of post-synaptic membrane. At present virtually all the work on the behaviour of membrane-bound AChE has been done on diaphragm or electroplax which represent nicotinic sites only.

CHAPTER 8

Do cholinesterases have a function other than in transmission?

8.1 Introduction

The presence of acetylcholinesterase at cholinergic junctions, both in the CNS and in the periphery, is one of the corner-stones on which our understanding of cholinergic transmission is based (ch. 7). There is generally little difficulty about assigning a function to the enzyme present at a particular site in nervous tissue: except in certain cases — and some of these will be discussed later in the chapter — the AChE seems to be where it is in order to hydrolyse ACh released as the result of traffic in cholinergic fibres. Difficulties do arise, however, when it comes to explaining the presence of cholinesterases at sites where transmitter mechanisms are not operating. What, for instance, is the role of AChE in the erythrocyte? The main purpose of this chapter is to consider some of these 'non-transmission' functions which can be grouped under four headings. The first section tackles the question of whether cholinesterases play any part in the mechanisms controlling permeability and transport processes of non-neuronal membranes. The question will be considered in general terms and also in the particular context of the erythrocyte and the blood-brain barrier. The next part deals with the apparent anomalies in the level of cholinesterase activity in certain areas of the nervous system — areas such as the cerebellum which may be rich in AChE but in which the number of cholinergic synapses as judged from other criteria seems to be low. The third section covers the cholinesterases in developing animals and considers the appearance and disappearance of enzyme from certain regions of the nervous system at different stages. The final part accommodates miscellaneous structures not considered elsewhere.

8.2 Cholinesterase in permeable membranes

8.2.1 General considerations

Cholinesterases, both AChE and pseudoChE, are associated with many membranes across which transport of water or ions is taking place (see chs. 6 and 10). Because this association occurs so frequently and in so many different tissues and animals the idea has grown up that cholinesterases must have some involvement in the process of permeability control and of transport – in particular, the transport of sodium – across membranes. Proposals of this sort have been made in the case of the erythrocyte and the blood-brain barrier (see § § 8.2.2, 3); frog skin (Kirschner 1953; Koblick 1959; Koblick et al. 1962); frog sartorius muscle (Van der Kloot 1956); the gills of the crab, *Eriocheir sinensis* (Koch 1954); the anal papillae of the larva of *Chironomus plumosus*, an insect of the order Diptera (Koch 1954); the hydatid cyst of *Echinococcus* (Schwabe 1959; Schwabe et al. 1961); the cuticle of the cestode worms, *Hymenolepis* and *Hydatigera* (Lee et al. 1963); the kidney of the crayfish (Kamemoto 1961) and of the rat (Fourman 1966a), and the salt-gland of the duck (see Fourman 1969). Faced with the fact that cholinesterases are present in so many places where materials are crossing membranes, it is difficult to believe that the enzymes are of no functional significance at these sites. Attempts to establish a link between cholinesterase activity and membrane transport have, however, met with little success. Obviously the presence of cholinesterase on permeable membranes is itself suggestive of a role but supporting evidence is sparse. Most claims that cholinesterase is actively involved in events at membranes are based on the observation that the normal permeability characteristics are affected by drugs which are best known for their ability to inhibit cholinesterases. When evidence of this sort is analysed it is clear that the doses of drug used are often far in excess of the dose necessary to inhibit cholinesterases. In other words, the drugs could well be producing their effect by something other than the mere inhibition of cholinesterase. This point is well illustrated in Koch's work (Koch 1954) on the crab gill. To inhibit sodium absorption it was necessary to use the anticholinesterase DFP (see ch. 11) in a concentration of 1.9×10^{-3} M. A concentration of 10^{-3} M was ineffective although this is probably 100–1000 times greater than that needed to knock out the cholinesterases. Duncan (1967) advanced the theory mentioned in ch. 5 that some sort of complex of ATPase and pseudocholinesterase could play an important part in controlling cation permeability. Again, the effects of anticholinesterases are cited as providing some support for this idea but Duncan makes the very important point that ATPase, being a serine-containing enzyme, is itself susceptible to phosphorylation by anticholinesterases of the organophosphorus type (see Hokin and Yoda 1964). A non-specific action of this sort on various membrane systems could well be responsible for some of the permeability changes produced by anticholinesterases. Other factors, however, might also be involved. Membrane permeability is influenced by pH and it is possible that, in some instances, the pH of the system is altered when anticholinesterases are administered. Hoskin (1971) found that squid axon contains an enzyme capable of hydrolysing DFP and he has suggested that the fall in pH which results from this reaction could be sufficient to account for the block of nervous conduction caused by injection of DFP. An induced fall of pH of this type will occur only if enzymes which can hydrolyse the anticholinesterase are present but there is also the possibility that the pH of the compound administered may, itself, produce an effect in certain experiments on small animals. In Kamemoto's (1961) studies on the kidneys of the crayfish, *Orconectes virilis*, the final concentration of eserine in the blood was 2×10^{-4} M. Kamemoto did not specify which eserine salt he injected but since eserine itself is only slightly soluble it is probable that the sulphate or salicylate was used. At 2×10^{-4} M the pH of the sulphate is approximately 5.0. The amount of eserine injected into a 30-g animal was only 20 μg but since the buffering activity of the blood of invertebrates is often less efficient than that of vertebrates (see Prosser 1950), it is possible that the injection would lower the pH sufficiently to affect sodium uptake by the kidneys. Another point to consider is that the administration of a high dose of inhibitor involves the injection not only of the anticholinesterase moiety but also, in certain cases, of a radicle which could have some effect of its own. Salicylate is commonly employed as the salt in preparations of eserine, and it has been established that salicylates do affect membranes. Barker and Levitan (1971) found that when molluscan ganglion cells were perfused with media containing sodium salicylate (from 1 to 30 mM) the permeability to potassium increased and that to chloride decreased. Salicylates produce a qualitatively similar change, an increase in cation permeability and a decrease in anion permeability, in the erythrocyte membrane (Wieth 1970a, b). In experiments with anticholinesterases on whole animals it is difficult to be sure that an observed effect is related to a specific action on the particular system being studied and is not an indication of more generalized poisoning. Roşca and Dordea (1971) showed that anticholinesterases disturb the osmotic balance in the leech, *Hirudo medicinalis*. At first sight this could be taken as yet another example of a system in which cholinesterase activity is apparently concerned in permeability processes. Earlier work (Roşca et al. 1958) had shown, however, that the osmotic balance is similarly disturbed when the leech is anaesthetized with ether or chloral hydrate and also when the ganglionic chain is removed surgically. These experiments indicate that the nervous system is important in the maintenance of osmotic equilibrium and the possibility remains that the effects of anticholinesterases are mediated indirectly via enzyme associated with nervous tissue rather than directly via enzyme associated with the permeable membranes themselves.

Amniotic and allantoic membranes deserve special mention. Cuthbert (1963) found that the chick amnion contained a substance indistinguishable from ACh and that ChE was histochemically demonstrable in muscle cells. The reaction was particularly marked in the 'cross' figures and Cuthbert suggested these might represent ACh-sensitive pace-maker cells which initiate the amniotic contractions. Burt et al. (1970) have shown that the rabbit allantoic membrane contains both AChE and ChAc and that the electrical potential across the membrane is sensitive to ACh. Injection of ACh into the allantoic sac caused a fall in potential whereas application to the exocoelomic surface caused a rise. The action of ACh was blocked by the anticholinesterase reagent, phospholine iodide (Echothiophate) but this drug had no effect on the normal resting potential. This latter finding suggests that AChE is not an essential part of the mechanism which generates the potential.

Numerous references are given in chs. 6 and 10 to the presence of cholinesterases in secretory glands. The type of enzyme varies with the species as does the exact location, which may also be influenced by the state of glandular activity. Secretory tissues which in some species contain cholinesterases include salivary glands (where enzyme may be localized in alveolar cells, ducts or both), sweat glands and the glands of the alimentary tract, prostate and pancreas. Welsch and Pearse (1969) showed the rat has BuChE and the rabbit AChE, in the C cells of the thyroid and parathyroid glands. Cholinesterases are found in some secretions including the colostrum and milk of dog and pig (ch. 10 §7.5), dog sweat (Aoki 1966) and pancreatic secretion (McCance et al. 1951; Hebb and Hill 1955b) and rat tears (Bingham and Buckley 1973). Large amounts of AChE are present in secretions produced by the parasitic stages of some nematode worms (Sanderson and Ogilvie 1971; see ch. 5 §4.1). Pseudocholinesterase generally predominates in glands but in some species, notably rabbit and cat, the main enzyme present is AChE. In addition to its occurrence in secretory cells of glandular tissue, cholinesterase also occurs in the specialized secreting cells — the neurosecretory cells — of nervous tissue. Thus AChE is found in the paraventricular and supraoptic nuclei of the hypothalamus where the hormones oxytocin and vasopressin are synthesised (see ch. 7, and Cottle and Silver 1970b). It is also present in cells of a similar type in the caudal neurosecretory system of the spinal cord of carp (Luppa et al. 1968). As with all the examples discussed so far, the association of cholinesterase with glandular and neurosecretory systems has yet to be satisfactorily explained. The problem of the role of cholinesterases in non-neuronal sites has probably been most rigorously pursued in the case of the blood cells and vascular tissue and these two aspects warrant separate consideration. The placenta, which presents special problems, is also dealt with separately.

8.2.2 Acetylcholinesterase in erythrocytes and platelets

There has been no lack of ideas about possible roles for AChE in the erythrocyte. On the contrary, these have been so numerous that AChE seems as much associated with red herrings as with red blood cells. It is best to deal first with facts and to reserve examination of theory until later.

Bovine red cells are well known as the starting material for commercial preparations of AChE and as a result there is a tendency to assume that the possession of AChE on the red cell stroma is a characteristic common to all species. This is not so. Table 8.1 shows the strikingly variable levels in different mammalian species reported by Zajicek (1957). Man is at one end of the range, cat is at the other and the difference between them is almost 100-fold.

Much the same sort of spectrum has been found by other workers although absolute values have been expressed in a variety of ways. Not only do the different methods of measurement yield different units but these units may be related to a

TABLE 8.1

AChE activity of erythrocyte and platelet stroma from different animals. (From Zajicek 1957.)

| Species | AChE activity (μl CO_2/30 min/mg N) | | | |
	Erythrocyte stroma	Average	Platelet stroma	Average
Man ♂	2400		0	
♂	2050		0	
♀	2100	2200	0	0
Cow	1880		60	
	1900		70	
	1960	1930	30	50
Guinea-pig ♂	730		630	
♂	630		430	
♀	780	710	420	490
Horse ♀	820		1240	
♀	640		1740	
♂	420	660	1540	1500
Rabbit ♂	530		3700	
♀	550		4100	
♂	340	470	4000	3930
Rat ♂	290		3700	
♂	305		5050	
♂	320		3000	
♂	180	260	5200	4240
Cat ♂	36		5200	
♀	26	30	5700	5450

given number of erythrocytes (say 10^8 or 10^9), or to a volume of suspended cells or to mg haemoglobin or protein nitrogen. Table 8.2 shows results obtained by Callahan and Kruckenberg (1967) expressed as a percentage of their value for man.

Neither birds (Mendel et al. 1943; Augustinsson 1948) nor most fish (Augustinsson 1948; Clos et al. 1957, but see Dupé and Bockelée-Morvan 1968 re lungfish) possess AChE in their erythrocytes and Augustinsson (1950) suggested there could be some sort of reciprocity between AChE levels in erythrocytes and in plasma. For example, the cow with one of the highest levels of AChE in its erythrocytes has little AChE in its plasma while the domestic hen has a relatively high level in plasma and none in the erythrocytes (table 8.3). Augustinsson's idea needs some qualification. In birds, much of the activity shown by the plasma towards MeCh is not due to AChE but to pseudocholinesterase of an atypical specificity (Earl and Thompson 1952a; Myers 1953) while in those fish (teleosts and elasmobranchs) which lack erythrocytic AChE (see ch. 6, table 6.11) the plasma level is low and of the same order as that found in the cow with AChE-rich erythrocytes. In addition to species

TABLE 8.2

Erythrocyte cholinesterase activities for nine species of animals expressed as a % of the normal value for man. (From Callahan and Kruckenberg 1967.)

Species	No. of animals	Percentage of value for man*
Monkeys	80	64.0
Chimpanzees	10	70.7
Horses	9	11.7
Goats	33	12.4
Pigs	135	16.4
Guinea-pigs	45	13.7
Dogs	115	7.7
Rabbits	204	5.2
Cats	15	2.0

* Based on a value for man of 0.75 delta pH units/hr (Michel 1949.)

differences, variations with age have been found in some cases. The AChE level in the human red cell is lower in foetuses and newborn babies than in adults (Kaplan et al. 1964) but in sheep no such difference is seen (Herz et al. 1967). Erythrocytes of man and sheep do, however, show a similar pattern of AChE activity in relation to the age of individual cells. When red cells are separated by centrifugation into older and younger populations the level of AChE is higher in the latter fraction which is rich in reticulocytes. Augustinsson (1955a) found no difference in the AChE activity of erythrocytes from men and women nor was there any significant fluctuation in the level of enzyme activity in one male subject studied weekly for 2¼ years. In contrast, pseudocholinesterase levels of serum were higher in males than females and weekly variations of ± 6% were found in the long-term study.

White cells do not apparently possess AChE (see Zajicek 1957) but platelets, on

TABLE 8.3

Hydrolysis of ACh and MeCh by blood from different animals. (From Augustinsson 1948.)

Species	Erythrocytes		Plasma	
	*ACh	[†]MeCh	*ACh	[†]MeCh
Cow	163	70	5	3
Hen	0	1	28	11
Dogfish (*Scyllium*)	0	0	9	5
Ballan wrasse (*Labrus berggylta*)	0	0	5	2

* 1.10×10^{-2}M
[†] 1.02×10^{-2}M

Results expressed in μl CO_2 evolved/30 min, calculated for 100 μl of serum or suspended erythrocytes.

the other hand, exhibit high levels of activity in certain species. As table 8.1 shows, Zajicek found an inverse relation between the AChE content of erythrocytes and platelets; decreasing activity of AChE in erythrocytes was accompanied by increasing activity in platelets. Zajicek's negative finding for human platelets was confirmed by Aster and Enright (1969) but it has been questioned by Saba and Mason (1970) who obtained a value of 7.71×10^{-4} μmole ACh/min/100 μg membrane protein. This is of the same order of activity as that reported by Zajicek for the cow. His value of 50 μl CO_2/30 min/mg N, when converted to the same units (assuming 22.4 μl $CO_2 \equiv 1$ μmole ACh and 1 mg platelet stroma N $\equiv$ 6.25 mg plasma membrane protein), gives a figure of 11×10^{-4} μmole ACh/min/100 μg protein.

The comparative studies cited above show that AChE is not an indispensable constituent of all erythrocytes and platelets and this is one of the factors that must be considered when it comes to assigning a role for the enzyme in these cells. The first question to ask is whether AChE is related to an ACh-mediated mechanism of the type present at neuronal synapses. The characteristic action of ACh is to alter permeability of membranes. Since the proper functioning of the erythrocyte depends on the ordered exchange of materials across its surrounding membrane some involvement of ACh would not seem unreasonable. The major snag about such an idea is that although the membrane possesses the hydrolytic enzyme, AChE, it does not possess the enzyme for ACh synthesis, choline acetyltransferase. Early evidence that low ChAc activity could be detected in human red cells (Korey cited by Nachmansohn and Wilson 1951; Holland and Greig 1952) was later discounted (see Hebb 1957a) and it is generally held that erythrocytes lack both ChAc and ACh. Before this had been established, Greig and her co-workers published a series of papers describing experiments on human and dog erythrocytes designed to demonstrate the importance of AChE in the control of red cell permeability (see Greig and Holland 1949a; Lindvig et al. 1951; Greig et al. 1953). Among other results reported by the group was the finding (Greig et al. 1953) that if cells which had lost part of their potassium were incubated with ACh, potassium would be taken into the cell against a concentration gradient. Since this process was blocked by physostigmine it was concluded that AChE was involved. Almost the same uptake and a similar blocking effect by eserine were still seen, however, when ACh was replaced with glucose. The authors' explanation for this latter result was based on the now untenable theory that glucose was needed for the synthesis, inside the erythrocyte, of ACh which then provided the substrate for AChE as before.

In contrast to the results of Greig's group a number of other workers have failed to find evidence that AChE plays any significant part in the normal processes controlling ionic permeability. Thompson and Whittaker (1952) showed that inhibition of AChE did not interfere with active sodium transport in human red cells. A similar conclusion was reached by Martin (1970) who found that choline transport, like sodium transport, was unaltered when AChE was removed from red cells by papain. In the case of potassium, Christensen and Riggs (1951) reported that although treatment with eserine led to some loss of potassium from human erythrocytes the concentration necessary to give this result was at least ten times greater than that which inhibited cholinesterases. Subsequently Giberman et al. (1973) showed in monkey erythrocytes that while eserine (again in high concentration) had some inhibitory action on potassium influx, DFP and neostigmine did not. These findings suggest that non-specific actions could be responsible for some of the changes which were detected by Greig's group in the presence of ACh or absence of AChE. The influence of pH on membrane properties has been mentioned already (§8.2.1) and it is possible that pH changes in the vicinity of the erythrocyte membrane might explain certain of the actions caused by the addition of ACh. Parpart and Hoffman (1952) found that the changes in potassium permeability produced by ACh could be mimicked by the addition of acetic acid (10^{-2} to 10^{-4} M) in place of ACh. When AChE was inhibited with eserine the effect of ACh was lost but the effect of acid remained. More recently Gilboa-Garber and Mizrahi (1972) have provided evidence that ACh-induced permeability changes, as indicated by haemolysis, can also be mimicked by acid. Under the conditions of their experiments, butyric acid, produced in the suspension by the reaction of plasma cholinesterase with BuCh, or the addition of ACh or acetic acid all caused haemolysis of human cells. Like Parpart and Hoffman (1952) they found that inactivation of AChE prevented the effect of ACh. It must be emphasized that the doses of ACh used were high (up to 1.4×10^{-1} M) but some increase in osmotic fragility was apparent with doses of only 10^{-3} M. It seems paradoxical that the cell is better protected against the haemolytic effects of ACh when the hydrolytic enzyme is inactivated than when it is functional. But is there really a paradox? The phenomenon demonstrable under experimental conditions may have little bearing on what is happening in circulating blood. Is it likely, for instance, that a red cell would ever encounter ACh in millimolar concentrations? So far as is known, sites at which ACh is released are well supplied with AChE. Even a transient local spill-over of appreciable amounts of ACh into the blood does not seem very probable. In 1933 Dudley, working in England, attempted to detect ACh in ox blood saturated with oxalic acid but he was unable to find more than 0.1 mg/l of a material with ACh-like properties. Because this finding was contrary to results obtained in Germany by Bischoff et al. (1931), Dudley went to Freiburg to repeat his experiments. There he found values of up to 40 mg/l which were in agreement with the results already published by the German workers. This discrepancy was never satisfactorily resolved but it is generally held that vertebrate blood contains little or no ACh. Douglas and Paton (1951) showed that even when massive doses (10 mg/kg) of the anticholinesterase TEPP (ch. 11) were given intravenously to cats, the ACh level in the blood did not exceed 0.04 mg/l. If the concentration of ACh in blood remains so low (about 2×10^{-7} M) under conditions in which there is likely to be a drastic reduction in enzyme activity, at least at peripheral sites (TEPP does not readily enter the CNS), it is difficult to believe that a red cell would ever be exposed to a damaging concentration under physiological conditions.

When the correct type of experiment has been devised to reveal the role of erythrocyte and platelet AChE it may well prove to be unconnected with permeability control. Nevertheless, the long-standing suspicion that there is a connection has led to various attempts to implicate some facet of AChE activity (or inactivity) in several blood diseases characterized by haemolysis. Among these is paroxysmal nocturnal haemaglobinuria (PNH). Erythrocytes from patients with this illness can be separated into two types on the basis of their susceptibility to lysis by complement (for a definition of complement see Müller-Eberhard 1972). The complement-sensitive cells are totally devoid of AChE while the activity of the so-called 'complement-insensitive' cells may be slightly below normal. Even the complement-insensitive cells may be more susceptible to lysis than are cells from normal subjects and the degree of increased sensitivity parallels the decrease of AChE in the membrane. If the AChE of normal cells is inhibited, susceptibility to lysis by complement is not, however, increased (Kunstling and Rosse 1969). Similarly, in a family whose erythrocytes contained only 20—40% of the normal amount of AChE there was no evidence of spontaneous haemolysis or of abnormal sensitivity to complement (Johns 1962). The lack of association between AChE levels and complement-sensitivity in these last two examples suggests that the absence of AChE from some red cells of patients with PNH is a symptom rather than a cause. As Kunstling and Rosse put it 'the membrane defect which increases the susceptibility of PNH red cells to lysis by c′ [complement] also makes them deficient in AChE activity......'. Results of investigations of platelets and granulocytes from PNH patients support this view. Aster and Enright (1969) found such platelets to be up to 250 times more sensitive to antibody than were those from normal subjects or subjects with other blood disorders. Like Zajicek (1957), these authors were unable to detect any AChE in either normal or abnormal human platelets. Since mixed white cells were also unusually sensitive to antibody Aster and Enright concluded that PNH is a manifestation of a membrane defect common to erythrocytes, granulocytes and platelets. The basic defect probably results from a somatic mutation in a primitive cell capable of differentiating into an erythroblast, myeloblast or megakaryoblast.

Decreased levels of erythrocyte AChE are found in certain haemolytic anaemias (see Sirchia et al. 1970). Although the reduction is not of the same magnitude in all types, Palek et al. (1969) regard the decrease as a single syndrome and attribute the variability in absolute levels to the variability in the age of the cell population in the different diseases. Younger erythrocytes carry more AChE than do older ones hence in patients with hyperplastic anaemia, a disease characterized by many young cells, the enzyme deficit per cell may be masked in measurements made on whole blood. In newborn babies with haemolytic disease due to ABO blood group incompatibility the AChE levels are reduced (Gerlini et al. 1968). In contrast, babies with haemolysis due to Rh factor incompatibility have normal levels (Kaplan et al. 1964). This indication that the decreased AChE activity in the former disease is not a direct effect of circulating antibody is strengthened by the failure of Kaplan et al. to induce any change in the AChE activity of normal erythrocytes treated either in vitro or in vivo with autoantibody.

All the available evidence suggests that in anaemias, as in PNH, the decrease in erythrocyte AChE is simply a manifestation of some basic lesion of the cell membrane. Studies of these conditions are therefore unlikely to shed any light on why the erythrocytes of certain species are supplied with the enzyme. Zajicek (1956, 1957) believed that neither the erythrocyte nor the platelet needed AChE for its own integrity but were merely 'carriers of this enzyme, the presence of which in the circulation is doubtless of importance'. Is the word 'doubtless' still justifiable? In the time that has passed since that was written little has come to light to indicate that there are any mechanisms in the body which depend for their correct function on circulating AChE. One system which may possibly be affected by blood-borne AChE is the erythropoietic system. Davis (1960) found that subcutaneous injections of bovine AChE (200 Ammon units) in rats resulted in an increase in reticulocytosis with a rise in circulating erythrocytes. Since the haematocrit did not change, the new cells were apparently smaller than usual. Rather more complicated experiments were done on mice by Samuels et al. (1968). Some of the mice were made polycytaemic by the injection of blood. When these animals were given 1 unit of sheep erythropoietin, the incorporation of ^{59}Fe into red cells was increased but daily subcutaneous injections of AChE (25–200 units) had no such effect. If, however, AChE was given together with erythropoietin, the effect of the latter was reduced. In another series of mice which were only mildly polycytaemic, injection of 200 units of AChE did result in an increased incorporation of ^{59}Fe into the red blood cells but again the effect of AChE plus erythropoietin was less than the effect of erythropoietin alone. Samuels et al. suggested that AChE, like testosterone, might stimulate the production of endogenous erythropoietin; they also postulated that the decreased effect of sheep erythropoietin when combined with AChE was due to some 'interfering effect' of the latter. Once again we have to ask whether this experimentally induced phenomenon is significant physiologically. Several steps are involved in the production of new blood cells. First a reaction occurs between a renal factor (REF) and a plasma-borne globulin produced in the liver. This results in the formation in the blood of erythropoietin which in turn stimulates erythropoiesis in bone marrow. Whether AChE, either in situ on the circulating erythrocytes or during the breakdown of cells in the reticuloendothelial system, can take part in or influence any stage of these reactions must remain conjectural.

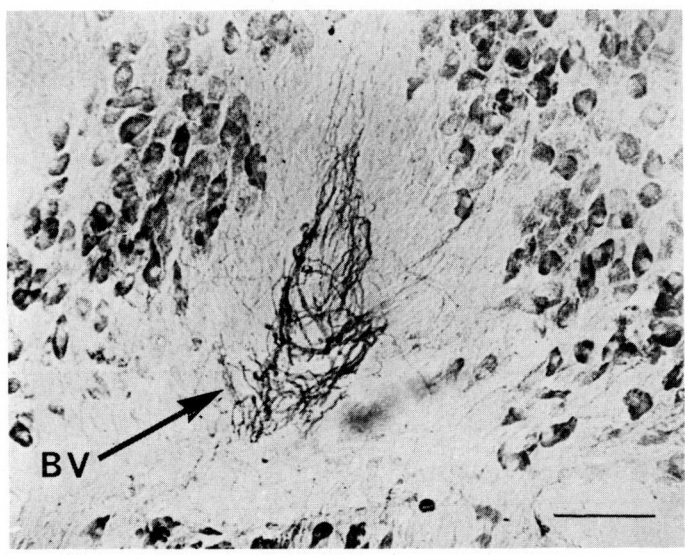

Plate 8.1. Strong AChE activity in a tuft of blood vessels (BV) in the guinea-pig hypothalamus at the level of the optic chiasma. Scale bar: 100 μ. (From Cottle and Silver 1970a.)

8.2.3 Cholinesterases in vascular tissue

Cholinesterases are present in the walls of some blood vessels in the nervous system and other tissues. Not all the reports of histochemical reactions for cholinesterases in blood vessels make a clear distinction between the staining attributable to enzyme in nerve fibres innervating the vessels and that due to enzyme associated with the vascular tissue itself. The present discussion is concerned with the latter case and entries in table 8.4 are restricted to reports of unequivocal vascular staining. One of the peculiar features of this localization is the difference in the occurrence of stained vessels, not only in different species but within different areas of the same species. Furthermore, as table 8.4 shows, the type of enzyme is variable: in some vessels it is AChE, in some BuChE, while yet other vessels contain both together. Species differences in vascular staining are well illustrated by results from the spinal cord. In the rat cord treated for BuChE the blood vessels stain strikingly (Navaratnam and Lewis 1970). In the cat, on the other hand, vessels in the cord react neither for AChE nor BuChE (Silver and Wolstencroft 1971). The cat also provides a good example of regional differences. In contrast to the absence of enzymes from vessels in the spinal cord, both AChE and BuChE are present in the capillaries of the brain stem structure which Gwyn and Wolstencroft (1968) have designated the area subpostrema. As its name suggests this is situated immediately ventral to the area postrema. In the area postrema itself, a few capillaries stain

TABLE 8.4
Cholinesterase activity in blood vessels of the CNS.

Species	Area	Enzyme	Reference
Rat	Brain and cord in general	BuChE	Brightman and Albers (1959)
	Brain	BuChE and AChE	Flumerfelt et al. (1973)
	Brain	BuChE and AChE	Joó (1969); Joó and Várkonyi (1969)
	Cerebellum	BuChE	Friede and Fleming (1964); Kása et al. (1965); Kása and Csillik (1966)
	Hippocampus	BuChE	Storm-Mathisen and Blackstad (1964)
	Hypothalamus	BuChE	Hyyppä (1969); Danilova (1971)
	Olfactory bulb	BuChE	Sharma (1968)
	Optic nerve	BuChE	Palkama and Sipponen (1968
	Retina	BuChE	Nichols and Koelle (1968)
	Spinal cord	BuChE	Navaratnam and Lewis (1970
Monkey (Squirrel monkey)	Area postrema	AChE and weak BuChE activity in lining of sinusoids	Iijima and Bourne (1968)
(Rhesus monkey)	Brain stem	BuChE	Friede (1967)
(Squirrel monkey)	Cerebellum	BuChE (a few vessels react for AChE)	Shanthaveerappa et al. (1967
(Squirrel monkey)	Hypothalamus	AChE (some vessels react for BuChE)	Iijima et al. (1967)
	Olfactory bulb	BuChE	Shanthaveerappa and Bourne (1965)
Cat	Area subpostrema	AChE and BuChE	Gwyn and Wolstencroft (196
	Cerebellum	? Cholinesterase of atypical specificity. No activity towards BuThCh. Activity with AcThCh inhibited by both Mipafox and BW248C51 (ch. 11)	Kása et al. (1965) see also Ká and Csillik (1966) re electro microscopy
	Inferior olive	AChE only	Gwyn and Wolstencroft (196
Man	Brain in general	BuChE	Friede (1967)
	Cerebral cortex	Hydrolysis of BuThCh	Okinaka et al. (1961)
Rabbit	Cerebellum	AChE	Crook (1963)
Guinea-pig	Hypothalamus	AChE	Cottle and Silver (1970a,b)
Sheep	Area postrema	AChE	Palmer and Ellerker (1961)
Frog	Brain especially optic lobes	ChE (type not specified)	Shen et al. (1955) see also Boell et al. (1955)

TABLE 8.4 (continued)

Species	Area	Enzyme	Reference
Toad	Spinal cord	AChE	Brightman and Albers (1959); Phillis and Tebēcis (1967)
Canary, pigeon and parakeet	Cerebellum	BuChE and some AChE	Friede and Fleming (1964)
Chicken	Brain stem and spinal cord	AChE; slight reaction for BuChE	Cavanagh and Holland (1961b)
Sparrow (white crowned)	Hypothalamus	BuChE	Kobayashi and Farner (1964)
Goldfish	Cerebrum, cerebellum and cord	AChE and BuChE	Brightman and Albers (1959)

weakly for AChE; none react for BuChE. Vessels in the inferior olive possess strong AChE activity (see plate 7.13) but like those in the area postrema, they lack BuChE.

The often very striking appearance of the stained vessels (see plate 8.1) prompts the question of function. It was suggested many years ago that cholinesterases had a role in the maintenance of the blood—brain barrier (Greig and Holland 1949b; Greig and Mayberry 1951). This hypothesis was based mainly on the observation that administration of anticholinesterase agents such as eserine increased the penetration into the nervous system of acid fuchsin (Greig and Holland, working with frog cord) and of barbital and chloralose (Greig and Mayberry, working with mouse brain). The crucial point, whether or not the effect of eserine was attributable solely to its anticholinesterase action was not definitely established in these experiments. Subsequent work by Strickland and Thompson (1954, 1955) has shown that eserine, DFP and the neostigmine analogue NU 1250 all increase the leakage of potassium from slices of chicken brain but the doses needed for this action are much bigger than the doses which inhibit cholinesterases. Although the postulated role of cholinesterases in maintaining the blood—brain barrier has been generally ignored or rejected (see for example Rosić and Milošević 1967), the theory is periodically revived. Joó and Csillik (1966) noted that in the rat the capillaries in brain areas which lie outside the blood—brain barrier (Wislocki and Leduc 1952) did not contain BuChE. In the case of capillaries from areas within the barrier, ultrastructural examination of the histochemical end-product in sections stained for BuChE suggested the enzyme was associated with cytoplasmic particles. The authors thought these particles could be pinocytotic vesicles and on this basis proposed that 'the barrier function is correlated to the hydrolysis of some biologically active choline ester participating in the pinocytotic process'. Joó and Várkonyi (1969; see also Joó 1969) extended the electron microscope studies in rat and concluded that capillaries in tissue within the blood—brain barrier contained

TABLE 8.5
Cholinesterase activity in blood vessels in peripheral tissue.

Species	Area	Enzyme	Reference
Rat	Aorta and pulmonary arteries (N.B. coronary arteries were not reactive)	BuChE (diffuse in tunica media)	Navaratnam and Palkama (1965)
	Bronchus	ChE (in big veins of smooth muscle)	Gerebtzoff (1959)
	Ciliary ganglion	BuChE	Huikuri (1966)
	Kidney	AcThCh and BuThCh hydrolysed (specific inhibitors not used)	Marx and Carter (1963)
	Mesentery	AChE	J.B. Furness (personal communication) cf. Gerebtzoff (1959) re -ve veins
	Pancreas	BuChE (in arteries)	Coupland (1958)
Monkey (no species given)	Ovary	BuChE	Bulmer (1965)
Cat	Autonomic ganglia	AChE and BuChE	Koelle (1950)
	Bronchus	ChE (in big veins of smooth muscle)	Gerebtzoff (1959)
	Carotid body	AChE and BuChE	Biscoe and Silver (1966)
	Kidney	AcThCh and BuThCh hydrolysed (specific inhibitors not used)	Marx and Carter (1963)
	Lung	AChE and BuChE	Koelle (1950)
	Lymphoid tissue	AChE	D'Agostini and Rossatti (1961)
	Pancreas	BuChE (arteries)	Coupland (1958)
Man	Kidney	AcThCh and BuThCh hydrolysed (specific inhibitors not used)	Marx and Carter (1963)
	Skin	BuChE in adventitia of large vessels. Capillaries -ve	Cauna (1960)
	Tongue	BuChE	El-Rakhawy and Bourne (1961)
	Tonsils	AChE (in extrafollicular vessels)	D'Agostini and Rossatti (1961)
Rabbit	Aorta and pulmonary arteries	BuChE (diffuse in tunica media)	Navaratnam and Palkama (1965)
	Bronchus	AChE (in smooth muscle of big veins)	Gerebtzoff (1959)
	Kidney	AcThCh and BuThCh hydrolysed (specific inhibitors not used)	Marx and Carter (1963)
	Lymphoid tissue	AChE	D'Agostini and Rossatti (1961)
	Pancreas	BuChE	Coupland (1958)

TABLE 8.5 (continued)

Species	Area	Enzyme	Reference
Guinea-pig	Bronchus	ChE (in smooth muscle of big veins)	Gerebtzoff (1959)
	Kidney	AcThCh and BuThCh hydrolysed (specific inhibitors not used)	Marx and Carter (1963)
	Ovary	AChE (in arterioles)	Bulmer (1965)
	Placenta	AChE (in capillaries)	Gerebtzoff (1959)
	Tongue	AChE (in small arteries and arterioles)	Gerebtzoff (1959)
Mouse	Bronchus	ChE (in smooth muscle of big veins)	Gerebtzoff (1959)
	Kidney	BuChE (in arteries of hilum)	Gerebtzoff (1959)
	Kidney	AcThCh and BuThCh hydrolysed (specific inhibitors not used)	Marx and Carter (1963)
Dog	Kidney	as above	Marx and Carter (1963)
Possum	Carotid body	AChE	Harrex (1971)
Duck	Salt-gland	AChE (uncertain whether activity is in, or around, capillary endothelial cells)	Fourman (1969)
Teleost fish (*Uranoscopus scaber*)	Liver	AChE and BuChE (veins and capillaries)	Gerebtzoff (1959)

AChE as well as BuChE. The reaction for AChE was confined to the basement membranes. Since the intraperitoneal injection of eserine and of 'specific' (see ch. 11) anticholinesterases — either Mipafox, an inhibitor of BuChE, or BW 284C51, an inhibitor of AChE — increased the penetration of dye from the capillaries into the brain tissue, Joó and Várkonyi concluded that BuChE and AChE both had some part in the maintenance of the barrier. The possibility that the increased capillary permeability was a non-specific effect resulting from the grossly non-physiological state of an animal poisoned with anticholinesterases was not eliminated. Flumerfelt et al. (1973) also subscribed to the view that the presence of cholinesterases in brain capillaries is suggestive of a role in the maintenance of the blood—brain barrier but their report of the ultrastructural localization of BuChE and AChE in rat capillaries differed in some respects from that of Joó and Várkonyi (1969). In contrast to the results of the latter authors, Flumerfelt et al. found the main sites of BuChE activity to be the basement membrane and nuclear envelope.

The pinocytotic vesicles were unstained but, occasionally, isolated profiles of rough endoplasmic reticulum were positive. A reaction for AChE occurred in a small proportion of capillaries, the distribution of end-product being similar to that for BuChE. The difference in the results from the two studies is presumably due to differences in methods: Flumerfelt et al. were using copper as the capture agent while Joó and Várkonyi used copper plus lead (Kása and Csillik 1966). Because the conditions of a histochemical experiment are of necessity non-physiological it is not easy to decide whether a particular distribution of end-product mirrors the true localization of the enzyme or is artifactual. On the other hand, the fact that different methods do give different results may mean that an enzyme is liable to changes in solubility, sometimes behaving as a cytoplasmic enzyme and sometimes as a membrane-bound one. Whether this is the case for capillary BuChE and whether any such lability could be related to function merits further study.

Another problem in understanding the significance of capillary cholinesterases is that of the regional and species differences mentioned earlier. In the rat, the majority of capillaries in the brain react for cholinesterases but stain is lacking from vessels in the area postrema, choroid plexus, pituitary gland and stalk, pineal body, subfornical organ and supraoptic crest. As has been pointed out (see Joó and Csillik 1966; Joó and Várkonyi 1969; Flumerfelt et al. 1973) the feature common to all these areas is that they lack a blood—brain barrier. In species in which capillary staining is much more limited, no such obvious character distinguishes areas with stained vessels from areas with unstained ones. In the cat, stained capillaries are the exception rather than the rule and while cholinesterase is absent from almost all vessels in areas outside the blood—brain barrier it is also absent from the majority of vessels in areas within it (see Gwyn and Wolstencroft 1968). Crook (1963) found in the rabbit that although the capillaries of the cerebellum are rich in AChE the vessels in the cerebral cortex are unstained. In guinea-pig hypothalamus the reaction for AChE is particularly strong in a tuft of vessels (plate 8.1) in the preoptic area and in vessels in the ventral part of the median eminence (Cottle and Silver 1970a; see plate 10.5b). Both areas are probably concerned in some way with hypothalamic releasing factors and while it is tempting to relate these two aspects there is as yet no experimental support for speculation along these lines. In our experiments (Cottle and Silver 1970a) the blood vessels were unstained in sections treated for BuChE but reaction product was found in some globules near the ventral surface of the median eminence (plate 10.5a). As far as we could judge the globules seemed to lie between the vessels which stained so strongly for AChE. They were not, however, found among the stained vessels in the preoptic area, and, again, their significance is uncertain.

Cholinesterase-containing capillaries are not confined to the central nervous system. Reports of activity in vessels in peripheral tissue are indicated in table 8.5. One of the striking features of the table is how the type of enzyme may vary in different vessels in the same animal. Assuming that the enzymes are present to hydrolyse some substrate, is this substrate the same in vessels with AChE and vessels with BuChE? At present there are no experimental data to help answer that question. Two findings, however, do indicate that at least in some vessels the hydrolytic activity could differ from the typical 'cholinergic' reaction between AChE and ACh. Thus Kása et al. (1965) showed that the enzyme in the cerebellar capillaries of cat hydrolysed AcThCh but not BuThCh. This suggests it is AChE yet it was inhibited by Mipafox which is specific for BuChE, as well as by BW284C51, an inhibitor of AChE. In contrast, Vanha-Perttula (1966) found that in rat anterior pituitary gland the enzyme present in the sinusoids hydrolysed both AcThCh and BuThCh but was resistant to eserine as well as to specific inhibitors of AChE and BuChE. This indication that the enzyme responsible for the histochemical reaction in these particular vessels is not a cholinesterase at all must raise doubts about any assumption that cholinesterases in vessels are necessarily involved in the hydrolysis of choline esters.

8.2.4 The placenta

The placenta is not innervated yet it contains AChE and, in the primates but not other species (see Hebb and Ratković 1962), ChAc and ACh as well. It is assumed that in primates ChAc and AChE are present respectively to synthesise and hydro-lyse acetylcholine which is playing some part in the exchange of material between mother and foetus. Why this apparent ACh-mediated step should be confined to primates is uncertain. One possibility is that the primate placenta is called upon to perform some function which is not necessary to other species. Hebb and Ratković suggested that this could, perhaps, be concerned with the passage of certain anti-bodies from mother to foetus since primates and non-primates show differences in this respect (Bangham 1960). Despite the lack of ChAc and ACh, non-primate placentae possess AChE. Gerebtzoff (1959) has described the localization of the enzyme in guinea-pig placenta in some detail. By the 16th day of gestation the capillaries contain AChE and the syncitium acquires activity by the 19th day. Gerebtzoff suggested that the initial syncitial activity represents deposits from maternal blood but that later the syncitium itself starts to synthesise AChE. About 48 hours before delivery AChE appears in the placental labyrinth surround-ing the lacunae of foetal blood. Although the syncitial activity decreases at this time the total activity in the placenta increases but falls again at delivery. Gerebt-zoff's interpretation of these observations is that AChE migrates from the syn-citium to the labyrinth and then immediately before delivery it passes onto the foetal erythrocytes in the lacunae. In Gerebtzoff's view this placental contribution at the time of delivery provides the newborn animal with erythrocytic AChE until effective synthesis of AChE begins in the haemopoietic tissue about 48 hours later. Once the significance of AChE in the erythocyte is established it may be easier to decide whether the placenta is synthesising AChE solely for export to the erythro-cytes or whether AChE has some function within the placenta itself.

8.2.5 Comments

The safest conclusion to be drawn from these investigations is that we do not know what cholinesterases are doing at the sites discussed. Furthermore, care must be taken to separate possibility from probability in any speculation. The AChE at synapses seems to be involved in terminating the ACh-induced permeability changes in post-synaptic membranes and because of the frequent occurrence of AChE or BuChE at other sites where permeability changes are important, anyone trying to explain the presence of some strangely located enzyme may be tempted to write that 'it possibly has a role in permeability control'. Such opinions, originally only tentative, are quoted by others with less reservation and the theory gains further currency. What is needed is experimental evidence that will give the hypothesis the solid foundation it requires or will dispose of it once and for all, clearing the way for other proposals. So long as the idea remains in its present form – a long-established legend with some possible truth – it will continue to be invoked, often uncritically and unsupported, as a convenient explanation for an inconvenient finding. Unfortunately, and perhaps significantly, interest in the possible involvement of cholinesterases in permeability mechanisms is more often shown by those seeking a role for cholinesterase than by those seeking to understand the mechanisms themselves. In the 860-page-long supplement to the Journal of General Physiology 51 (1968) which covers a symposium on cell membrane biophysics, only one paper (Kernan's) reports any experiments involving anticholinesterases. That the biophysicist gives cholinesterases scant consideration when it comes to defining mechanisms involved in movements across membranes is equally clear from Keynes's (1969) comprehensive review of work on salt and water transport across a wide range of multicellular structures. Apparently biophysicists can produce adequate explanations of membrane phenomena without invoking a cholinesterase-dependent step. As a result, they ignore its presence. This leaves the physiologist, pharmacologist and histochemist in a dilemma. To fall back on the idea that the enzyme is functionless is unsatisfactory but to continue in the unproven belief that it is 'likely to be involved in permeability control' is equally so. As some anonymous aphorist wrote over 300 years ago, 'Truth has no greater enemy than verisimilitude and likelihood' (see Shorter Oxford English dictionary, 3rd edition).

8.3 Acetylcholinesterase in nervous tissue: a metabolic role?

8.3.1 Mature systems

The previous sections have dealt with structures in which AChE is involved in some process which has nothing to do with the transmission of nervous impulses either from neurone to neurone or from neurone to effector. Although the role of the enzyme remains obscure the problem itself is clearly recognizable. When we turn to the nervous system it is much more difficult to decide where AChE is fulfilling its accepted function of hydrolysing the ACh released as a neurotransmitter and where it may be acting in quite a different capacity. The problem, or perhaps the recognition of the problem, is made more complex because of the increasing application of the term 'cholinergic' not only to neurones which release ACh but also to any AChE-containing system, regardless of whether or not there is evidence that ACh is the transmitter. When the proper distinction is made, it becomes clear that high levels of AChE may be present in certain regions where the major transmitter seems to be something other than ACh. Two such regions are the cerebellum and the substantia nigra.

8.3.1.1 The cerebellum   The controversy surrounding the nature of the neurotransmitter in the cerebellum was discussed in the last chapter. As indicated there, one of the reasons for rejecting ACh as a major transmitter is that although the AChE activity in most species is very high the values for choline acetyltransferase and ACh may be among the lowest found in the brain. Another peculiarity about the AChE of the cerebellum is the extraordinarily wide variation in distribution from species to species (see Silver 1967). Friede and Fleming (1964) recognized five main histochemically demonstrable patterns of localization. Their results, which are given in more detail in the appropriate tables in ch. 6, may be summarized as follows:

1) Molecular layer strongly stained, granular layer and white matter have little activity. Example: birds.

2) Granular layer strongly stained, molecular layer and white matter have little activity. Example: cat.

3) Neither the molecular nor granular layer react significantly. Example: monkey (*Cebus cebus*).

4) Both the molecular and granular layer stain strongly. Examples: guinea-pig, squirrel.

5) Pattern of stain varies from folium to folium. In some folia the granular layer stains strongly (as in Type 2) but in others the reaction is weak (as in Type 3). Examples: rat, mouse, rabbit, cow (also Rhesus monkey, see Silver 1967).

Our experiments (Kása and Silver 1969) on rat and guinea-pig cerebella were mentioned earlier (ch. 7). These showed that in the rat those folia which stained strongly for AChE (i.e. folia of the archicerebellum, Shute and Lewis 1963; Friede and Fleming 1964) contained significant amounts of ChAc associated mainly with mossy-fibre afferents. Folia of the paleocerebellum, which stain only weakly, contained less than one-third of the activity of the archicerebellum. These findings indicate that a limited amount of cholinergic transmission may occur in the rat cerebellum and that its existence is better revealed by analysis of discrete areas than by analysis of the cerebellum as a whole. In the guinea-pig the situation was quite different. While the AChE activity was 40% higher than in the archicerebellum of rat, the ChAc level was only about 10% of the value for rat archicerebellum. If we accept this latter figure as evidence that there are far fewer cholinergic synapses in guinea-pig than in rat it further emphasizes the need to assign a role to the particularly large amount of AChE in the guinea-pig cerebellum. Any proposals for function in the cerebellum of different species have to take account of the variable patterns of activity summarized above. If the enzyme is needed for some 'metabolic function' (I use the term for want of a better) in certain cells in one species, what takes its place in those species in which the corresponding cells lack AChE activity? Kása and I showed by means of lesions that in the rat archicerebellum much of the AChE was associated with incoming fibres which were apparently cholinergic in the true physiological sense. It seems likely that in some other species a proportion of the histochemical staining may be similarly attributable to afferent fibres, and differences in the proportion of AChE-containing fibres in different areas could account, as in rat, for a variation from folium to folium. But merely assigning the enzyme activity to extrinsic elements — and this would have to be established for each species — does not alter the basic fact that in many species there is still a discrepancy between AChE and the other components of a 'cholinergic system'. Kása and Csillik (1965a) suggested that in the guinea-pig all the climbing fibres and mossy fibres (as well as the majority of intrinsic cells) contain AChE. This would account for the strong histochemical reaction and the high quantitative values for AChE but we are left with the same problem, now shifted one structure downstream, of why cells which are not apparently releasing ACh should have axons which are rich in AChE.

The cerebellar Purkinje cells pose a special problem in the context of the 'metabolic' function of AChE. Kása and his co-workers (Kása and Csillik 1964; Kása et al. 1966) made the interesting observation in the adult rat that when lesions are placed in the cerebellar white matter, transient AChE activity is histochemically demonstrable in the normally non-reactive Purkinje cells. As a result of the lesion, Purkinje cells would be both deafferentated and axotomized but the authors believed it to be the axotomy that triggered the response (however, see ch. 9 §5.1.1 re Kreutzberg's work on the morphological consequence of axotomy). Kása et al. suggested that the enzyme could be concerned in protein synthesis initiated in the cell in an attempt to repair the damage. Phillis (1968b) observed a somewhat similar effect in the cat but in this case Purkinje cell staining could be induced by lesions restricted to the cerebellar peduncles. Since the Purkinje cell axons terminate mainly in the cerebellar nuclei and would not be damaged by peduncular section, he concluded that deafferentation rather than axotomy was responsible for the effect and suggested that the disturbance in the normal pattern of activation by afferent impulses brought about 'metabolic changes'. At this stage it is not possible to define any such change in biochemical terms but one finding, which belongs more properly to §8.3.2 below, may have some bearing on the question. Histochemical studies on a number of species have shown that although the Purkinje cells in the adult may not stain for AChE, activity is demonstrable in the cells at some stage of development (Joó et al. 1963; Csillik et al. 1964; Sakharova 1966; Silver 1967, 1971). Does the reappearance of AChE following lesions represent a reversion to the immature state? In view of Phillis's (1968b) findings in the cat that deafferentation causes the return of AChE activity, it is tempting to suggest a parallel between a mature cell which has been deprived of its afferent input and an immature cell which has yet to receive it. If AChE has a metabolic role in the cell while it is developing, but not when it is mature, some mechanism must switch off enzyme synthesis at a certain stage of maturation. Perhaps the arrival of an adequate afferent innervation informs the cell that the particular process in which AChE is involved is no longer required. Should the mature cell become deafferentated, the restraint on the process would be removed and it might restart. The possible nature of this metabolic activity will be taken up again in the context of development (§8.3.2) and some ideas about its control are discussed in ch. 9 §6.

8.3.1.2 The substantia nigra   Bull et al. (1970) measured the ChAc level in a number of areas of human brain and found that in the substantia nigra, just as in the cerebellum and also the globus pallidus (see later), there was a striking disparity between the activities of ChAc and AChE. They took 100 as an arbitrary value for the activity of the two enzymes in the caudate nucleus and, on a proportional basis, the ChAc values for the cerebellum and the substantia nigra were, respectively, 2.2 and 3.3 while the corresponding figures for AChE (taken from Ord and Thompson 1952) were 26.7 and 38.2. This means that in these structures the percentage level of AChE is about 10 times higher than that of ChAc. In most of the other structures analysed the equivalent figure was not more than 2.5 while in the thalamus the percentages were equal. In some unpublished experiments on cat brain J.H. Wolstencroft and I found a qualitatively similar result. In the substantia nigra the histochemical reaction for AChE was strong (see plate 7.10) yet the figure for ChAc activity was among the lowest obtained from brain stem structures. In experiments that gave values of 47–50 μmoles ACh/g tissue/hr for cholinergic nuclei such as the hypoglossal, the value for the substantia nigra was only about 0.8 μmoles, much the same as that for the pyramids. Since the ChAc activity in other tissues was not reduced by the addition of samples of substantia nigra it seems unlikely that the low level is due to any inhibitory action on the enzyme by the pigment or some other component. In view of the physiological evidence for cholinergic mechanisms in the substantia nigra (ch. 7 §3.2.2) it may be that the specific activity of ChAc in discrete areas is considerably higher than that indicated by relatively gross sampling. While this could go some way to explaining why the figures so far available for ChAc activity are unexpectedly low, it still would not explain why the AChE levels should be disproportionately high. Olivier et al. (1970b) postulated that the 'cholinergic pathway represented by the striatonigral system may regulate the rate of striatal dopamine formation by influencing the enzymatic mechanism involved in the synthesis of dopamine'. If this is, indeed, what happens then AChE would be needed to hydrolyse synaptically-released ACh. Alternatively — or in addition — AChE may have a more direct 'metabolic' role akin to that mooted but not defined for the Purkinje cell.

8.3.1.3 The globus pallidus   The globus pallidus provides another example of a site in which there seems to be a disparity between acetylcholinesterase and choline acetylase activity. In their analysis of human brain Bull et al. (1970) found a figure of 12.4 when the choline acetylase activity in the globus pallidus was expressed in relation to an arbitrary value of 100 for the caudate nucleus. On the other hand the figure for AChE, taken from Ord and Thompson (1952) but expressed in the same way, was 100.3. The percentage level of AChE is thus almost 10 times greater than that of ChAc. As mentioned above, the cerebellar cortex and substantia nigra also give ratios of approximately 10 but whereas in these structures the high figure reflects particularly low levels of ChAc, in the globus pallidus the ChAc level is appreciable and the high ratio results from the especially pronounced AChE activity. This suggests that AChE is not limited to the presumably fairly extensive population of cholinergic neurones but is present in non-cholinergic ones as well.

8.3.1.4 The hypothalamo-neurohypophysial system   Lederis and Livingston (1969) have shown in rabbit that the values for acetylcholine content and ChAc activity in the hypothalamus are about the same as those in the neural lobe of the pituitary gland. In contrast, the AChE activity of the hypothalamus is about twice that of the neural lobe. As Cottle and I have pointed out (Cottle and Silver 1970b), this could mean that in the hypothalamus AChE is not merely hydrolysing transmitter but is doing something else as well. Additional support for this possibility comes from a number of reports (for references see Cottle and Silver 1970b, also ch. 9) that AChE levels in certain hypothalamic nuclei are prone to fluctuate. Fluctuations occur not only when different conditions are imposed experimentally (e.g. salt loading) but also in response to naturally occurring changes associated with the oestrous cycle, pregnancy and lactation (ch. 7 § 3.2.3). The proposition that these quite considerable variations in enzyme levels reflect changes in nervous activity seems dubious. To postulate that they reflect changes in metabolism may seem more reasonable but experimental proof is still lacking. Furthermore, as will be discussed in the next chapter, variations in AChE activity do not necessarily mean that the enzyme is directly involved in producing metabolic changes: it may fluctuate as a consequence of them.

8.3.1.5 Glia, Schwann cells and sense organs   The presence of AChE together with ChAc and ACh throughout the length of a cholinergic fibre has been cited by Nachmansohn as evidence for his theory that ACh has a role in axonal conduction.

Rejection of Nachmansohn's idea for reasons discussed in the previous chapter means that we still have to consider what AChE is doing in the fibres. The most widely accepted view (ch. 4) is that axonal enzyme is probably doing nothing and is merely in transit from its site of production in the soma to its site of action at the terminal. It is less easy to account for the AChE that is found in some of the supportive elements in neuronal structures. The tables in ch. 6 show that where a cholinesterase is present in glial and satellite cells of vertebrates this is generally BuChE but, in a few places, AChE is found. For instance, Abrahams and Edery (1964) demonstrated a strong reaction for AChE in astroglia in the cat hypothalamus and hippocampus, Hebb et al. (1953) reported AChE activity in glia of pigeon optic nerve and Palmer and Ellerker (1961) found AChE-containing glia in the area postrema of sheep. Similarly, Cauna et al. (1961) found AChE in the satellite cells of parasympathetic ganglia in the gut of rat. The role of this AChE like that of BuChE (see ch. 10) is at present unknown. While AChE-containing glia are not plentiful in vertebrates they are common in invertebrates. For example, Hess (1972) found them in the cockroach and I found them in the leech (Silver 1972, see ch. 5). Cholinergic mechanisms are not fully documented in these species and it is possible that the enzyme in the glia could be involved in the hydrolysis of ACh released at the synapses which, in both animals, are present in the neuropil rather than on the cell bodies. On the other hand, the function of the glial enzyme may be unconnected with transmission. The idea of a non-transmitter role for AChE in the invertebrates gains some support from observations on the mollusc *Aplysia* (see ch. 5). Only 4 cells in the abdominal and pleural ganglia contain significant amounts of ChAc yet all the neurones, or possibly their associated glia, contain AChE (McCaman and Dewhurst 1971). Furthermore, although the AChE activity varies from cell to cell, there is no consistent difference between the levels in cholinergic and non-cholinergic cells.

The relation of AChE to Schwann cells is controversial (see ch. 4). Early evidence that Schwann cells contained AChE came from Sawyer's (1946) demonstration that when the sciatic nerve was cut in a guinea-pig, AChE activity persisted in the peripheral stump after the axons had degenerated. It was shown subsequently by Cavanagh et al. (1954) that in the chicken, AChE activity was not retained in the degenerating stump and they concluded that Schwann cells were devoid of enzyme. Lubińska et al. (1963) cited results which indicate that the response to nerve section in the dog is qualitatively the same as that in the guinea-pig and they suggested that the ability to synthesise AChE might be acquired by the Schwann cell· only as a result of axonal disruption. This idea, which will be considered again (ch. 9 §6), is to some extent supported by electron microscopical findings on normal nerve (chs. 4 and 6). In such studies a reaction for AChE in the Schwann cell cytoplasm is seen only rarely (e.g. Brzin et al. 1966) and although reaction product is frequently observed between the axolemma and the Schwann cell membrane, staining of the membrane itself is not an invariable finding (see Reale et al.

1971). Whether the relative paucity of staining reflects the true situation, or is due to technical inadequacies, is not clear. It may be significant that in two cases in which staining of the Schwann cell membranes was a consistent finding, the Karnovsky and Roots technique, rather than the Koelle method, was used (see Bell 1969; Robinson 1969). On the other hand, both studies were made on autonomic nerves of the guinea-pig, hence the positive result could be characteristic of the tissue rather than the method.

The subject of cholinergic transmission in the vertebrate sensory system was touched on in the previous chapter. It was pointed out that, although the transmitter released by primary afferent fibres terminating in the CNS is not ACh, it is very probable that cholinergic cells may form a part of the neuronal chain linking the peripheral receptors with the higher centres (see Feldberg and Vogt 1948; Hebb 1955, 1970 and other references in ch. 7). This means that AChE found at some levels of the sensory pathway is probably doing no more than hydrolysing ACh released at cholinergic synapses. In a few places, however, the role of AChE is less obvious and this is particularly true of the enzyme associated with peripheral sensory receptors such as those in skin (for example, see Cauna 1960, 1961; El-Rakhawy and Bourne 1961, and other references in ch. 6). At one stage it was thought that ACh might take part in initiating impulses in sensory fibres but the theory has been discounted for most systems (see J.A.B. Gray 1959 for a general treatment). As is all too often the case in this chapter it is again not possible to make any reasoned suggestion for the function of AChE at such sites. Among points to be kept in mind when seeking an answer is, first, that BuChE is also found at sensory endings, either alone (e.g. in the Pacinian corpuscle of cat pancreas, Hebb and Hill 1955a) or together with AChE (e.g. in Meissner's corpuscles in human skin, Cauna 1960); and second, that in some receptors (e.g. Merkel's discs in human skin, Cauna 1960) neither enzyme is present.

Long after ACh had been rejected as the impulse initiator in most sensory systems it remained and, in the view of some workers still remains, a possible candidate in the carotid body. This structure contains considerable amounts of cholinesterase, with BuChE predominating, but it seems to be associated with blood vessel walls, nerve plexuses and ganglion cells and not with chemoreceptors (Biscoe and Silver 1966). While the bulk of evidence argues against ACh being involved in the production of afferent impulses (see Biscoe 1971) its exact role and that of the cholinesterases is still in dispute. The sinus nerve contains some efferent fibres (De Castro 1926) and there is evidence that these can influence chemoreceptor function (Neil and O'Regan 1969). Since the sinus nerve contains ChAc (Hebb 1968) it is possible that the efferent fibres are cholinergic. Biscoe and I did not, however, detect any change in histochemically demonstrable cholinesterases following section of the sinus nerve.

The idea of ACh as a sensory trigger has recently been revived in connection with the cornea. The cornea contains what is possibly the largest amount of ACh of any mammalian tissue (see Williams and Cooper 1965). Since denervation of the rabbit cornea reduced the level by 87–100% Fitzgerald and Cooper (1971) have suggested that it is associated with sensory nerve fibres. The point of special relevance to this chapter is that the high level of ChAc activity and ACh content is not matched by a correspondingly high value for AChE. J.R. Cooper (personal communication) found in cow cornea that the rate of synthesis by ChAc was 18.38 μg ACh/mg protein/hr and the rate of hydrolysis by AChE was 40.4 μg ACh/mg protein/hr. The ratio of AChE to ChAc activity is thus only 2.2 whereas in rat brain the figure found in this laboratory is more than 50. The cornea presents a problem which is the reverse of that in most examples in this chapter. Till now the difficulty has been to explain why AChE activity should be high in areas where ChAc activity is low. Here, the question to ask is why the level of AChE is not higher.

Another example of an as yet unexplained occurrence of AChE in sensory tissue is provided by the dorsal root ganglia. Giacobini (1956a) showed histochemically in frog, rat and cat that 10–15% of the dorsal root ganglion cells possessed a high level of AChE activity. He later confirmed this finding in experiments with the Cartesian diver (Giacobini 1959). Hebb (1957a) pointed out that the overall level of ChAc in the dorsal root ganglion and fibres is so low that the enzyme could not be present in all cells but might be confined to a few special cholinergic, sensory cells, possibly those which reacted strongly for AChE. Subsequently Brzin et al. (1966), working on frog, showed that the AChE activity is much more uniform than was originally thought, the apparent variability from cell to cell reflecting differences in substrate penetration rather than in enzyme content. These studies would suggest that AChE is probably present in most ganglion cells although ChAc may be absent from the majority. The situation thus seems very similar to that in *Aplysia*.

8.3.2 The developing nervous system

8.3.2.1 Introduction   The developing nervous system provides more examples of sites where there may be considerable AChE activity but little else to suggest cholinergic function. In 1925 Needham wrote of the egg that it 'has been largely used as a source of information about other things much more than as a source of information about itself'. To some extent the same is true of the embryo and results obtained from developing organisms can be useful in offering some pointers to the possible role of AChE in 'non-transmission' functions in mature tissues. The time-course of the appearance of cholinesterase activity has been followed in numerous studies on a variety of species. For references to biochemical and histochemical work see Karczmar (1963); Silver (1967, 1971); Burt (1968); Filogamo and Marchisio (1971) and various tables in ch. 6. Results from many of these investigations indicate that significant levels of AChE may be present in cells well before synaptic transmission is likely to be occurring. In addition, some cells which contain AChE during development may later lose it as they mature. These findings have led to the idea that AChE may be involved in metabolic processes associated with maturation.

Although the embryonic nervous system is morphologically simpler than that of the adult it still presents pitfalls in the expression and subsequent interpretation of data. During development, tissue weight and protein content are likely to be changing all the time but the rate of change may not be uniform in the different parts. As a result, any decrease in the specific activity of the enzyme (activity/mg protein) at a particular stage of development could mean one of two things: either that cells which were previously rich in enzyme are losing it or that the absolute activity in each enzyme-containing cell is unaltered but these cells are being diluted out by a disproportionate increase in the number of enzyme-free cells. A further difficulty is introduced if some cells are losing activity at the same time as others are acquiring it. It is in recognizing such simultaneous changes that histochemistry is valuable but, as with the biochemical studies, results must be interpreted with caution. In our work on the forebrain of the developing cat, Krnjević and I (1966) found that the histochemical reaction for AChE in fibres was more intense in foetuses of 50 days gestation than in newborn and older kittens (see fig. 9.1). This could mean that enzyme activity in the fibres decreases with maturation but alternatively it may mean that the method is less suitable for demonstrating AChE once myelination has begun. As mentioned in ch. 2 Krnjević and I made another observation which underlines the complications that may be involved in work on developing tissues. We found that the cholinesterase activity initially present in the forebrain of foetal cats was of an atypical specificity as judged by its reaction to different substrate and inhibitor combinations. Other examples of changes in substrate specificity and isozyme patterns during development were also given in ch. 2 (§3.2).

8.3.2.2 Specific areas   As a result of work done mainly on chick, Filogamo and Marchisio (1971) suggested that 'practically all neurons are born with an ACh system marked on their birth certificate'. This may prove to be too sweeping a generalization for species other than chick; nevertheless, as mentioned above, there are examples from various animals of cells which react for AChE during development yet are non-reactive later on. The case of developing Purkinje cells has been touched on already (§8.3.1.2). Post-natal acquisition of enzyme with subsequent loss during the next 3 weeks has been reported in rat (Joó et al. 1963; Csillik et al. 1964) and cat (Sakharova 1966). In the sheep a similar phenomenon occurs, but in utero (Silver 1967). Gestation lasts approximately 147 days and cells become reactive (plate 8.2) about half-way through this period. Activity is maximal around 90 days, the number of stained cells dwindling thereafter. In contrast to the situation in the rat, staining does persist in a few cells in the adult cerebellum. Other developing cells which exhibit AChE activity but later lose it include neuroblasts in cat cortex (Krnjević and Silver 1966), in sheep cortex (Silver 1967), in rat hippocampus (Mellgren 1973) and in the olfactory bulbs of rabbit, rat and guinea-pig

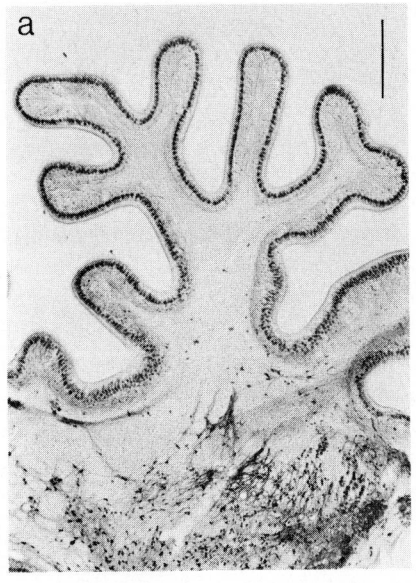

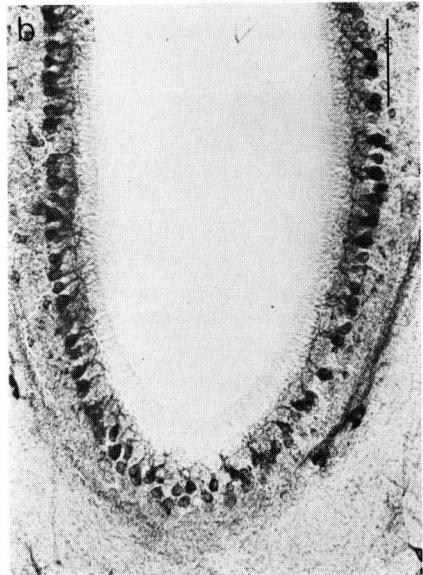
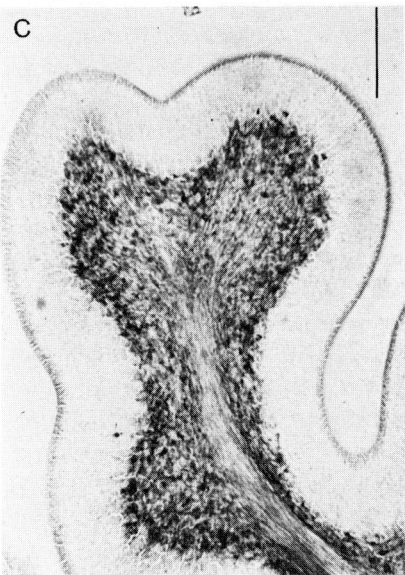

Plate 8.2. a, b) AChE activity in Purkinje cells of the cerebellum of a foetal sheep of 94 days gestation (from Silver 1967). c) Cerebellum of a foetal sheep of 112 days gestation showing that by this stage the Purkinje cells have largely lost their AChE activity. Scale bars: a) 500 μ; b) 100 μ; c) 250 μ.

(Filogamo and Robecchi 1969). Amacrine cells of the embryonic chick retina show transient AChE activity (Shen et al. 1956) and AChE is also present in the Cajal-Retzius cells of the human foetus (Duckett and Pearse 1968) which atrophy very soon after birth. Rossi (1961) found two patterns of staining in the acoustic ganglion of the guinea-pig. The cells which eventually form the cochlear ganglion exhibit only weak activity for a short time during development. In contrast, those cells which make up the vestibular ganglion stain strongly at the time when they separate from those of the cochlear ganglion. Activity decreases later but is still retained in the adult. In the spinal ganglia of chick, post-hatching levels of enzyme are very much lower than the maximum reached during development in the egg (Giacobini et al. 1970). This situation makes an interesting contrast with that in the chick sympathetic ganglion. In this case, the AChE level reaches a maximum at exactly the same stage as that in the spinal ganglion but the subsequent fall is only slight and the maximum value is regained within two days of hatching (fig. 8.1). The significance of this contrasting behaviour in the non-cholinergic spinal ganglion and the cholinergic sympathetic ganglion will be followed up a little later.

Other examples of cells in which AChE activity is higher during development than at subsequent stages can be found in the literature (see Filogamo and Marchisio 1971). Together with the cases given above these seem to provide sufficient justification for the hypothesis that the enzyme could have a function that is specifically related to maturation. That AChE may have a 'metabolic' role has been mooted already but the possible nature of this role in terms of chemical reactions has been ignored. This question can be avoided no longer and we have to decide whether there is any good experimental evidence for the assertion or whether it is mere verbiage used to veil a lack of understanding. Since AChE is a hydrolytic enzyme it is most unlikely to be catalysing any synthetic process but it could perform a catabolic role somewhere along a synthetic pathway. The most important thing to establish is whether the substrate is ACh or some other ester. As pointed out in ch. 2, AChE is able to hydrolyse a variety of compounds including non-choline esters and it may be naïve to assume that ACh is its sole substrate. Indeed Crawford et al. (1966) have even questioned whether ACh *is* the natural substrate for AChE. With the development of a histochemical method for demonstrating choline acetyltransferase it should be possible to see whether all cells possessing AChE activity during development contain ChAc as well. Biochemical analyses of the two enzymes in maturing tissue have yielded a variety of patterns. Fig. 8.2 shows the situation in the spinal and sympathetic ganglia of the chick. Since ChAc is detectable in both tissues this suggests that ACh could be the substrate for the acetylcholinesterase. In the sympathetic ganglion the intital rise in ChAc roughly parallels the rise in AChE although it reaches its maximum earlier. The second peak in the ChAc level occurs at a time when the AChE activity is falling and is probably attributable to the formation of synapses by cholinergic preganglionic fibres. The spinal ganglion lacks a cholinergic input and is, itself, formed of predominantly

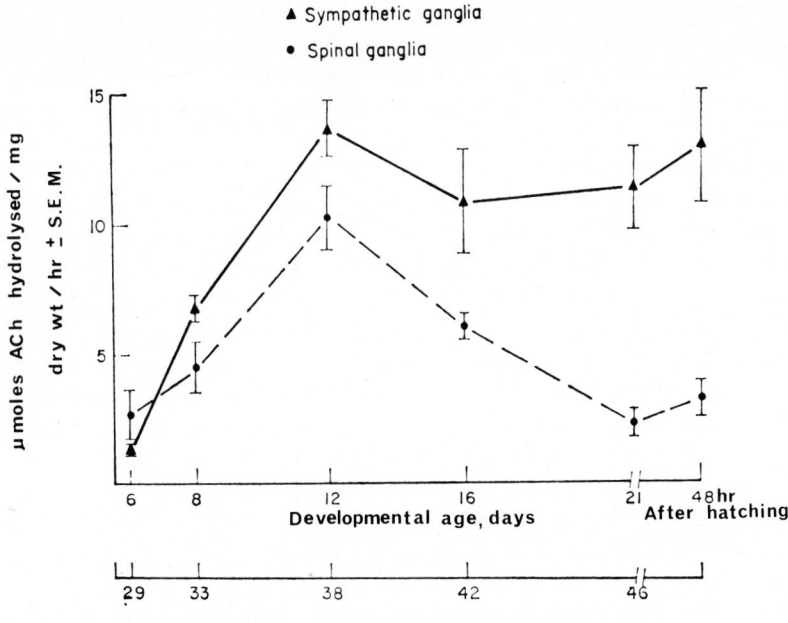

Fig. 8.1. Variations of AChE activity in chick embryo spinal and sympathetic ganglia. (From Giacobini et al. 1970.)

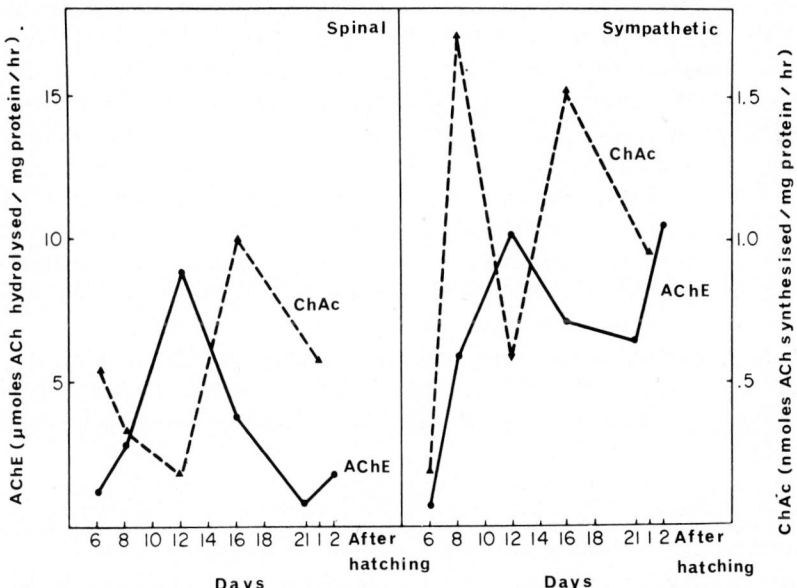

Fig. 8.2. AChE and ChAc activities in spinal and sympathetic ganglia of chick embryos. (From Filogamo and Marchisio 1971.)

non-cholinergic cells; nevertheless, it is endowed with significant ChAc activity which shows two peaks. The apparent lack of an early peak for AChE corresponding to that for ChAc may simply be due to the design of the experiment. Strumia and Baima-Bollone (1964) using optical microscopy found a strong histochemical reaction for AChE in the spinal ganglion at 4 days of incubation and an even earlier occurrence (at 3 days) was noted by Pannese et al. (1971) with the electron microscope. It is thus probable that had the quantitative analyses been done at this stage an early peak for AChE would have been found. The two periods at which ChAc activity shows a peak coincide with the two periods of maximal proliferation of neuroblasts (Levi-Montalcini and Levi 1943). This has been taken as an indication that ACh could be important in histogenesis (Marchisio and Consolo 1968). It is easier to make such an assertion in the case of a non-cholinergic ganglion than for the sympathetic ganglion in which non-cholinergic cells receive a cholinergic innervation. However, the increase in both ChAc and AChE activity well before the formation of synapses again suggests that ACh could be important in cell maturation. Burt (1968) has studied the levels of AChE and ChAc in the spinal cord of chick embryos. On the basis of his results which are indicated in fig. 8.3 he came to the conclusion that here, too, the early increase in AChE levels is related to cellular growth and differentiation while the rapid increase in ChAc activity which occurs immediately before hatching is associated with the establishment of cholinergic transmission.

8.3.2.3 Possible roles for the cholinergic triad in the developing cell   If the presence of ChAc can be taken as an indication that ACh is also present, this seems to answer the question of what AChE is hydrolysing. The basic problem, however, still remains: what is the function of this cholinergic triad, ChAc, ACh and AChE in the growing cell? There are various possibilities but each is speculative. All the known actions of ACh seem to be manifestations of its ability to change membrane permeability. The synthetic processes of the growing cell could perhaps be dependent on ACh-controlled permeability changes which ensure the efficient supply of substrate or the removal of catabolites. Evidence suggesting that ACh can influence certain types of synthesis has come from the demonstration by the Hokins and their group (Hokin and Hokin 1958 a, b, 1959; Hokin, L. 1965, 1969; also Yagihara and Hawthorne 1972) that ACh stimulates the incorporation of ^{32}P into phosphatidyl inositol and phosphatidic acid in some preparations of nervous tissue including brain slices, brain homogenates and microsomal fractions. With nerve-ending preparations, however, an inhibitory effect has been shown (Abdel-Latif 1966).

When a reaction is demonstrable in vitro the question which comes to mind is whether it also occurs in the intact system in vivo. Larrabee and his colleagues (see Larrabee 1967a) showed that in rat sympathetic ganglia, either excised or in situ with an intact blood supply, stimulation of the preganglionic trunk caused an

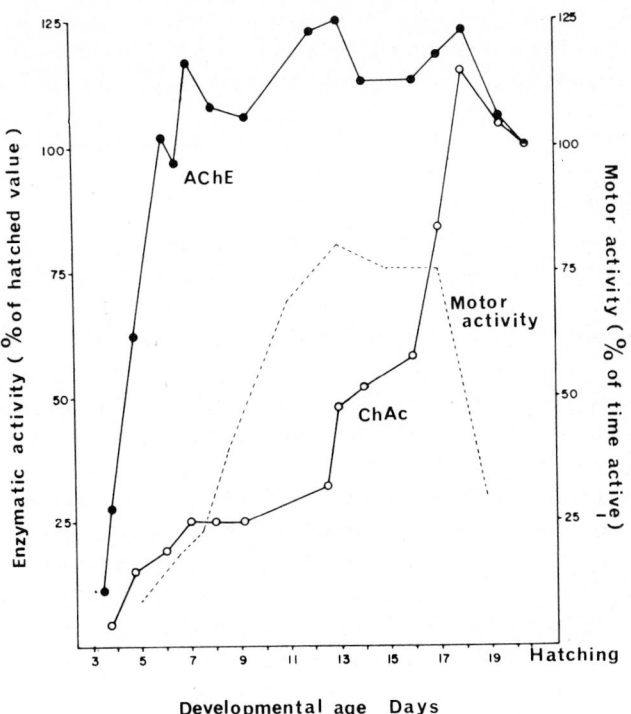

Fig. 8.3. AChE and ChAc activities in the spinal cord of the chick embryo, expressed as a percentage of the value at hatching. Spontaneous motor activity according to stages of Hamburger et al. (1965). (From Burt 1968.)

increased incorporation of ^{32}P into phosphatidyl inositol (but not into phosphatidic acid). No change in the rate of incorporation was found in the nodose ganglion nor in nerve trunks; this suggests the effect occurs post-synaptically and could be associated with the release of ACh. These experiments do provide some evidence that reactions similar to those obtained in the test-tube might be taking place in intact tissue. Larrabee (1967b) speculated that a transmitter could be responsible not only for depolarizing the post-synaptic membrane but also for triggering the mechanism which restores it. Durell et al. (1969) took a rather different view. In their model the interaction between ACh and phospholipid metabolism is merely one step in the transmitter-induced permeability changes which result in depolarization. Yet another idea (Hokin, L. 1969) is that the accelerated phospholipid turnover reflects an increase in the production of vesicle membrane which is needed to package transmitter for use at the terminals of the stimulated cell. In view of these suggestions it is interesting that other putative transmitters have subsequently been shown to have an effect on the uptake of ^{32}P. Thus Hokin, M. (1970) found that

5-HT increased the incorporation of ^{32}P into phosphatidic acid, and, to a lesser extent, into phosphatidyl inositol and diphosphoinositol in slices of guinea-pig cortex. Using noradrenaline she obtained a similar result with guinea-pig cortex (Hokin, M. 1969) and it has since been shown by Berg and Klein (1972) that noradrenaline stimulates the uptake of ^{32}P into the phosphoinositol- and phosphatidyl serine-containing fractions of post-synaptic membranes from rat pineal gland (see also Eichberg et al. 1973).

These indications that transmitters could play a part in the restoration or formation of membranes following synaptic events cannot, as yet, be accepted as evidence that ACh has a similar role in the de novo synthesis of membrane in the developing cell. Until further work is done, possibly on cell cultures, the idea that membrane synthesis in developing cells includes an ACh-dependent step can be no more than a working hypothesis. It seems reasonable to assume, however, that should this step exist, AChE might have a regulatory effect. Such a function might explain the presence of AChE both in developing cells and also in axotomized Purkinje cells attempting repair (but cf. ch. 9 §5.1.1 re Kreutzberg's work and ch. 9 §6 re ideas on the control of enzyme synthesis). Bearing in mind L. Hokin's suggestion about vesicle membrane, the AChE in the endoplasmic reticulum of non-cholinergic cholinoceptive cells might again represent part of an ACh-stimulated membrane-producing system. These speculations are of little worth without support from experimental evidence but this is difficult to provide. Of necessity, anticholinesterases (ch. 11) such as Sarin or physostigmine (see Widlund and Heilbronn 1973 re different effects) are used in most experiments to test the influence of ACh on phospholipid turnover. Any normal role of AChE is thus eliminated from the test system.

8.4 Miscellaneous structures

A wide range of structures has been included in the preceding sections but there are many other sites at which the role of cholinesterases is equally uncertain. While the AChE at the neuromuscular junction presents no problem of interpretation the significance of the enzyme at the myotendinous junction is not understood. AChE is found at the point where the fibres of the tendon are inserted into the infoldings of the muscle membrane. In this part of the muscle neither motor nor sensory endings are present (see Schwarzacher 1961; Lubińska and Zelená 1967) and as yet there is no explanation for this localization of AChE. Flood et al. (1970) have pointed out that it is unlikely to be hydrolysing ACh since the junction lacks significant ChAc activity and diffusion of ACh from the end-plates seems improbable.

Cholinesterase, predominantly AChE, has been found in the spermatozoa of a number of species including pig (Sekine 1951), ram (Mann 1954), guinea-pig

(Grieten 1956), bull (Nelson 1964; Meizel et al. 1971), dogfish (Kaswin and Serfaty 1946), trout and perch (Tibbs 1960), and the mussel *Mytilus edulis* (Applegate and Nelson 1962). In human semen the overall activity is low and most of the enzyme is in the seminal plasma rather than in the spermatozoa (Zeller and Joël 1941). The presence of cholinesterase in other motile structures, such as cilia, was mentioned in ch. 5. Whether the enzyme in spermatozoa is concerned with their movement is uncertain. Sekine (1951) reported that the motility of pig spermatozoa was increased by ACh and decreased by eserine. In contrast, Ishiwata et al. (1956, cited by Mann 1964) found that DFP had no effect on either the motility or the metabolism of spermatozoa. Furthermore, although in ram spermatozoa the enzyme is mostly associated with the tail (Legge and Mann cited by Mann 1954), in the perch it is mainly in the head.

The invertebrates provide many examples of problematical cholinesterase-positive tissue (see ch. 5). In the liver fluke, *Fasciola hepatica*, for example, AChE activity is very pronounced in the muscle of the suckers, pharynx and cirrus; the latter is an ejaculatory organ. Cholinesterase is similarly present in muscles of the nematode, *Ascaris lumbricoides*. In this worm, enzyme is found in and between the cells of the intestine but, in contrast to the situation in the rat nematode, *Nippostrongylus brasiliensis* (see Lee 1970) it is not present in the secretory cells (Lee 1962).

Cholinesterases have been examined in invertebrates during development. In developing echinoderms and ascidians AChE activity appears before neurogenesis begins and even in mature animals the enzyme seems to be associated mainly with muscle (see Karczmar 1963) although some activity is present in nerve as well (Pentreath and Cottrell 1968). A similar association of enzyme with developing musculature has been described by Rybicka (1967) in the platyhelminth, *Hymenolepis*. In insects the relation of enzyme activity to neurogenesis varies with species. According to Karczmar (1963), cholinesterase may appear before or after the beginning of neurogenesis or may be correlated with it. An example of the latter pattern is provided by *Drosophila melanogaster* (Dewhurst et al. 1970). There is some disagreement about the level of cholinesterase activity during diapause. This is a stage of insect development during which electrical activity in the nervous system disappears. Van der Kloot (1955) reported that in the silkworm, *Hyalophora cecropia,* ChE activity fell precipitously at the time when electrical activity ceased and both electrical and enzymatic activity reappeared, simultaneously, just before the end of diapause. On the other hand, Tyshtchenko and Mandelstam (1965) found only a slight reduction in histochemically demonstrable activity during diapause in the oak silk worm, *Antheraea pernyi.*

8.5 Conclusion

This chapter has failed to answer the question posed in the heading 'Do cholinesterases have a function other than in transmission?' Despite wide-ranging studies of
the problem, it is still impossible to assign a definite role to a considerable amount
of the cholinesterase present in animal tissue. This negative answer nevertheless
adds point to the warning (ch. 7) that the presence of AChE cannot be accepted as
proof of the existence of cholinergic mechanisms unless supported by other sorts of
evidence.

CHAPTER 9

The variability of cholinesterase activity

9.1 Introduction

The existence of tables giving values for cholinesterase activity in tissues from various species may lull the unwary into the belief that the level of cholinesterase in a particular organ is fixed and expressible in absolute terms. This unfortunately is an over-simplification since cholinesterase is not a 'static' constituent and its activity is affected by a variety of factors, some of which may overlap. Superimposed on the obvious variants of sex, age and strain may be others such as the season, the time of day, the nutritional and reproductive state and the health or otherwise of the animal. In addition, cholinesterase levels can be influenced by a number of experimental procedures including surgical interference, X-irradiation and training.

9.2 Natural variants

Sex differences in cholinesterase activity in serum have long been recognized. Zeller et al. (1941) measured total activity in serum from man and guinea-pig and, in both

TABLE 9.1

Acetylcholinesterase and pseudocholinesterase activities and weights of whole brain (minus cerebellum) of male and female rats. (From Woolley 1963.)

Group	Brain weights mg	Pseudocholinesterase		Acetylcholinesterase	
		Total activity*	Activity per mg fresh tissue[†]	Total activity*	Activity per mg fresh tissue[†]
Male rats					
intact	1452	13,120	9.03	216,400	149
castrated	1463	13,300	9.12	228,400	156
Female rats					
intact	1360	11,700	8.57	207,600	153
castrated	1377	11,300	8.23	218,700	159

* Moles of substrate $\times$ 10^{10} hydrolysed per min by the whole brain, minus cerebellum.
[†] Moles of substrate $\times$ 10^{10} hydrolysed per min per mg of fresh tissue.

TABLE 9.2

Acetylcholinesterase* activity in brain of male and female rats (intact and gonadectomized). (From Woolley 1963.)

Brain area	Male rats		Female rats	
	intact	castrated	intact	ovariectomized
Visual-somaesthetic cortex	62.9 ± 1.6	62.2 ± 1.5	59.4 ± 1.4	61.8 ± 1.2
Hypothalamus	119 ± 5	112 ± 3	118 ± 3	117 ± 3
Olfactory tubercles	748 ± 22	729 ± 19	697 ± 14	715 ± 18
Pyriform cortex-amygdala	144 ± 4	148 ± 5	148 ± 4	157 ± 5 (< 0.05)
Mesencephalic reticular formation	183 ± 6	179 ± 5	185 ± 6	178 ± 3
Remaining brain minus cerebellum	149 ± 3	156 ± 2	152 ± 2	158 ± 3

* Expressed as moles ACh $\times$ 10^{10} hydrolysed/min/mg fresh tissue. Values are means ± S.E. of 8–20 samples per point. Figure in parentheses is P value (t test) for significance of difference between intact males and other groups.

species, levels were higher in males than females. In rats and mice, on the other hand, serum levels of pseudocholinesterase are greater in females (Mundell 1944; Sawyer and Everett 1946; Skramstad 1956). The direction of the sex difference is not necessarily the same for all tissues in a given species, thus Woolley (1963) showed that in rat brain the value for pseudoChE in males exceeded that in females whether expressed as total activity or per unit weight of brain (table 9.1). The difference was maintained even in animals gonadectomized when 30 days old; gonadectomy was not tested at an earlier stage hence it is not clear whether the difference is directly dependent on hormones or is due to a sex-linked genetic factor. In contrast to these results for the non-specific enzyme, Woolley found no obvious sex difference in AChE activity in whole brain but AChE levels in the pyriform-amygdala region of the cortex were higher than normal in ovariectomized females (table 9.2). This suggests that in certain discrete areas AChE activity may be influenced by sex hormones. Other evidence for this view will be discussed in more detail in §9.2 which considers endocrine influences in general.

Age is another factor modifying enzyme activity and when measurements are made on developing animals the terms in which results are expressed are particularly important. As discussed in the previous chapter, enzyme activity in various parts of the immature nervous system is higher than that in the adult and a number of processes could account for this apparent reduction with maturation. First, some ChE-containing structures actually disappear after a certain stage of development. The Cajal-Retzius cells provide an example in foetal human brain (see Duckett and

Pearse 1968) and an analogous situation is encountered peripherally in rat adrenal medulla. This is richly innervated in young animals but with aging the number of fibres and terminals is reduced and there is a concomitant loss of AChE; rats 2–3½ years old show a 56% reduction in enzyme activity compared with rats of 2–3 months (Méhes and Décsi 1958). Second, individual structures within a given region may retain the same concentration of enzyme per unit area or length, but a disproportionate increase in tissue devoid of ChE will result in decreased values for activity expressed per unit weight of sample. From a functional point of view this decrease would be apparent rather than real. A different situation, which would probably be more significant in terms of function, could arise if the concentration in a specific cell were reduced. This would be the case if an increase in the dimensions were not accompanied by an appropriate increase in enzyme synthesis so that a given amount of enzyme has to be distributed over a larger area.

The danger of attempting to interpret histochemical material quantitatively has been emphasized before but it is appropriate to repeat the warning here in connection with material from young animals. As mentioned already (ch. 8 §3.2.1) Krnjević and I, working on foetal cats, observed strong staining of a degree never seen in adult cats, in certain areas of the cerebrum (see plate 9.1). The impression gained is that the enzyme activity is far greater than that in the same region of the adult brain but this may be a misinterpretation; very possibly some factor in foetal

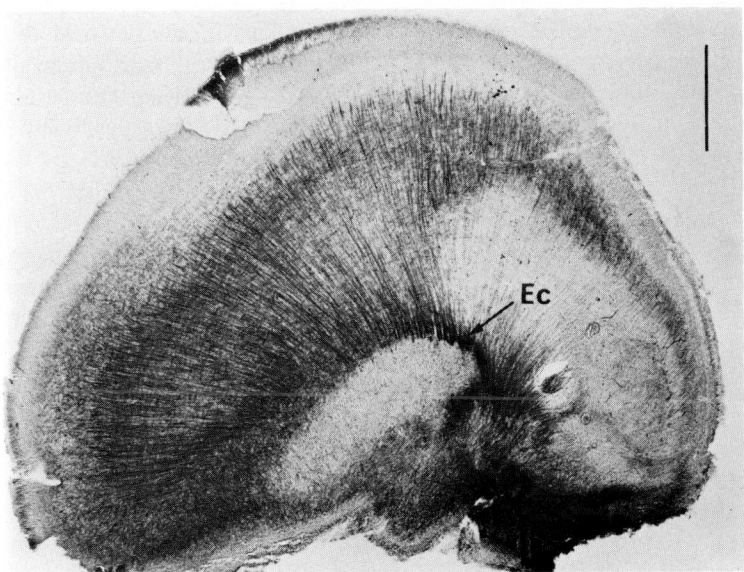

Plate 9.1. Sagittal section of cerebral hemisphere of foetal cat (50 days gestation). Note marked reaction for AChE in fibres. Ec, external capsule. Scale bar: 2 mm. (From Krnjević and Silver 1966.)

brain such as the relative lack of myelination may favour the histochemical reaction. A similar difference in histochemical results from mature and immature animals was reported by Cottle and Silver (1970b): cells in the infundibular stalk of prepubertal guinea-pigs stained for AChE much more strongly than those in the adult.

Bennett et al. (1958a) found that total cholinesterase activity in brain varied between different strains of rat. Certain observations which K. Krnjević and I made in this laboratory suggest there may be a similar strain difference in cats. Some years ago a tom-cat with 6 toes per paw was introduced into the breeding colony and in his easily recognizable polydactyl progeny the intensity of histochemical staining for AChE in the forebrain was consistently above average.

The possibility of seasonal changes in enzyme activity is often overlooked and systematic studies are rare. Nevertheless, certain evidence suggests that seasonal influences should be borne in mind in any long-term quantitative study of cholinesterases. The acetylcholine component of the ACh—AChE system is undoubtedly subject to seasonal variation and bioassayists are all too familiar with the relative insensitivity of the frog rectus abdominis muscle to ACh during the summer months. J.F. Mitchell and I (unpublished) noticed that the resting release of ACh from the isolated diaphragm of rat fell progressively between December and June; similarly, A.K. Armitage and G.H. Hall (personal communication) found a particularly low release of ACh from cat cerebral cortex during June and July. In both these cases the results were obtained incidentally in the course of other experiments and are necessarily incomplete. Experiments specifically designed to follow the variations could prove interesting. As the result of work with various pharmacological antagonists, Singh (1964) claimed that the transmitter released at vagal nerve endings on frog stomach actually changes with the seasons: depending on the time of year it may be Substance P, ACh, histamine or 5-hydroxytryptamine! This unorthodox view does make the point, albeit too forcibly, that the cholinergic system is more labile than is generally conceded.

Obviously, seasonal changes in ACh release need not necessarily be paralleled by changes in cholinesterase but direct evidence for seasonal fluctuations is slowly accumulating. Firer and Khamitov (1957) working in Russia examined total cholinesterase activity (AChE and pseudocholinesterase were not distinguished) in various organs of the frog including stomach, aorta and lung parenchyma. They found a sharp peak of activity in the spring with values starting to rise at the beginning of the second half of March and returning to 'normal' by mid-April. Values for stomach muscle, expressed as % ACh hydrolysed by a given volume of supernatant from homogenized tissue, were 3.00—6.20% in winter and 14.56% in spring. Activity in femoral muscle showed a smaller variation, rising from winter values of 3.27—4.10% to 9.37% at the end of March. It seems possible that the peak could be related to the sexual activity occurring at that time (see §9.3).

Evidence of a change in AChE (as opposed to total cholinesterases) which was apparently linked to seasonal factors comes from work by Lubińska et al. (1964)

on dogs. These authors measured the AChE which accumulated in the proximal stump of the peroneal nerve following chronic surgical section. Although the animals were kept at a reasonably constant temperature, the rate of accumulation of enzyme increased suddenly with the start of the cold weather in the notoriously severe Polish winter of 1962 and the rates returned to normal as the weather became milder. Since AChE arriving at the stump is in transit from the cell body (ch. 4), changes could be related to alterations in the rate of synthesis and/or mechanism of transport but, whatever the basis, the end result is an alteration in the concentration of enzyme peripherally.

The finding, made with J.F. Mitchell, that ACh release in the rat phrenic nerve-diaphragm preparation altered with the time of year has been mentioned above; in addition we noticed a variation with the time of day (Mitchell and Silver 1963; see also Hanin et al. 1970 re rat brain). The spontaneous release of ACh from diaphragms of rats killed between 3.00 a.m. and 5.00 a.m. was about twice that from rats killed in the early afternoon. Whether AChE in the diaphragms also showed circadian changes was not established but evidence of a circadian rhythm in cholinesterase of the nerve cord of the scorpion, *Heterometrus fulvipes*, has been reported by Venkatachari and Dass (1968). They found maximal cholinesterase levels at 4.00 p.m. with minimal levels at 4.00 a.m. and they pointed out that the electrical activity in the cord showed a similar pattern. There may, however, be no causal relation; since the scorpion is cold-blooded both parameters may be independently reflecting the effect of ambient temperature. In view of the evidence, scant though it is, that ChE levels may vary at different times of the day, comparative experiments on enzyme activity might gain in value if each were done at the same hour of the day (see Cottle and Silver 1970a).

In recent investigations of photoregulation of biological activity Bieth et al. (1970) produced a model in which the activity of AChE was influenced by the presence or absence of sunlight (or ultraviolet light). The system depended on the addition of a photochromic cholinesterase inhibitor which was reversibly converted by sunlight from the fairly strongly inhibitory *trans* form to the weaker *cis* form. Activity in preparations exposed to ultraviolet light was 30% greater than that in preparations kept in the dark. The authors suggested their system might be a model for naturally occurring photoperiodicity and cited the carotenoids as possible light-sensitive regulators of enzyme activity.

9.3 Endocrine influences

Sex differences in enzyme activity could result from sex-linked genetic factors but, as mentioned in §9.2, hormonal influences may be involved in controlling cholinesterase levels, at least in certain areas. Woolley's (1963) finding that ovariectomy increased AChE activity in the pyriform-amygdala area of rat cortex has been cited already and other evidence of an association between enzyme levels and reproductive states has come from work on the hypothalamus. Pepler and Pearse (1957) found that the depth of histochemical staining for AChE in the supraoptic and paraventricular nuclei of lactating rats was greater than in non-lactating rats. Particularly strong AChE activity during lactation has also been reported in guinea-pigs (Cottle and Silver 1970b), but in this species the increase was most noticeable in the infundibular (arcuate) nucleus. Recognizing that the quantitative evaluation of histochemical material is not reliable we were cautious in interpreting these results but we gained the impression that variations in AChE activity in the infundibular nucleus were not due to experimental artifacts or to individual differences but could be related to definite physiological states. Thus in lactating animals 10 days post-partum, the majority of neurones showed strong AChE activity which was in marked contrast to the almost complete absence of staining in infundibular cells of animals killed immediately post-partum. In prepubertal guinea-pigs activity in the infundibular nucleus was likewise low, few cells being appreciably stained. Ovariectomy and hysterectomy both *tended* to reduce staining in the infundibular nucleus, and also in the medial mamillary nucleus but results from these experiments were not entirely consistent. Mészáros et al. (1969) reported a similar histochemically demonstrable decrease in hypothalamic AChE following ovariectomy and adrenalectomy in the rat. On the other hand they saw no correlation between the degree of staining and normal reproductive function and concluded that the level of AChE in the hypothalamus reflected general cellular activity rather than specific hormonal states. According to Kobayashi et al. (1966) choline acetylase levels in rat anterior hypothalamus change during the oestrous cycle, maximum values being reached during dioestrus; in contrast, changes in ChAc activity produced by ovariectomy were confined to the posterior hypothalamus. It is difficult to interpret this apparent discrepancy in terms of physiological function and while parallel studies of AChE and ChAc in individual nuclei during different reproductive conditions might yield interesting data they would not necessarily elucidate the mechanism underlying these fluctuations. If cholinergic processes are involved in the elaboration, transport and secretion of hypothalamic releasing (or inhibitory) factors (see Cottle and Silver 1970a), then the rhythmic changes are probably causally related to the cyclic events of reproduction. On the other hand, the fluctuations may be a passive consequence — a mere side effect — of hormonal changes. Here too, one can envisage more than one route for these side effects: the hormones may influence enzyme synthesis directly, or, alternatively, enzyme synthesis might be affected indirectly following hormonally induced changes in some other condition such as water balance.

Although the mechanism may be in doubt, the fact that cholinesterase activity in the CNS is influenced by reproductive events seems to be established; conversely, cholinesterase in the reproductive tract can be influenced indirectly by events in the CNS. Sharma et al. (1970) followed histochemically the changes in ChE activity in various parts of the rat genital tract during the normal oestrous cycle and they also examined the tracts of rats from which the olfactory lobes had been removed. In the latter, ChE activity increased during the first 4 days after operation, then the levels fell and by 28 days after operation ChE had virtually disappeared. The authors do not mention the effects of the lesions on the oestrous cycle itself but it may well be abnormal. In the sow, for instance, excision of the olfactory lobes produces permanent anoestrus (see Clegg and Doyle 1967). Sharma et al. also studied male rats and found that olfactory lesions had no effect on the ChE in the vas deferens.

The effect of sex hormones on serum pseudocholinesterase has been the subject of a number of studies. In the rat the level in the serum is normally higher in females than in males. Everett and Sawyer (1946) found that after castration the activity fell in females and rose in the males, ultimately reaching a very similar level in the two sexes. When oestrogen was given to the castrated animals the enzyme level rose rapidly. In a subsequent study Omole (1972) confirmed that castration increased the serum ChE in the male but saw no elevation following oestrogen treatment.

Like sex hormones, thyroid hormone also exerts an influence on brain ChE. Much of the work showing this has been done on immature rats (Hamburgh and Flexner 1957; Geel and Timiras 1967) and results suggest that hypothyroidism during development reduces the level of AChE and pseudoChE in the brain. Geel and Timiras showed that enzyme activity (expressed per unit DNA) was restored to normal by 22 days of life if rats which had been thyroidectomized at birth received thyroxine from the 6th day. Hamburgh and Flexner sounded a necessary note of caution by pointing out that it was not clear whether the absence of thyroxine reduced the rate of enzyme synthesis or whether it lowered the activity of individual enzyme molecules; this question could be resolved by in vitro studies on enzyme from thyroxine-deficient animals. In some more recent work, Ling (1970) examined the influence of the thyroid gland on total cholinesterase in the brain of mature rats. When animals were given subcutaneous injections of thyroxine the ChE content of whole brain increased but analysis of discrete samples showed the rise was actually confined to the telencephalon. When hypothyroidism was induced by propylthiouracil, ChE levels fell, the diencephalon, midbrain, pons and medulla being specifically affected. On the other hand, if hypothyroidism was induced by the removal of the thyroid together with the parathyroid glands, the fall occurred in the midbrain, pons, medulla and cerebellum but not in the diencephalon. The difference in effects of these two procedures is yet another example of the variables involved in studies of cholinesterase.

Vaccarezza and Peltz (1960) showed that ACTH increased ChE levels in whole blood, serum and erythrocytes of man but this may be of more importance as a side-effect of the therapeutic use of ACTH than as a physiological factor. As might be expected, the action of ACTH is via the adrenal gland: in adrenalectomized rats, blood ChE falls and administration of ACTH has no effect but injections of cortisone can reverse the fall (see Vaccarezza and Willson 1965a). The effect of adrenalectomy on cholinesterases is not confined to the blood and the reduction in hypothalamic AChE which follows adrenalectomy in the rat (see earlier) is an example of the generalized response to the operation. It seems unlikely that the decreased level of enzyme is directly related to the fall in circulating adrenaline since Benson (1948) has shown that in vitro, adrenaline inhibits AChE and pseudoChE. Bajgar et al. (1972) did some experiments on young rats born to mothers that had been adrenalectomized before mating. At birth the AChE activity in the brain was similar to that in controls but after 16 days the levels were above normal in brain homogenates and particulate subfractions but not in the supernatant. The greatest increase was found in the microsomal fraction. The authors suggested that the change could be a result of the high levels of ACTH which would have been present in the plasma of the adrenalectomized mothers.

Firer and Khamitov (1957) showed that in frogs removal of the pancreas caused an increase, followed by a fall, in the total ChE in a number of organs (nervous tissue was not examined). However, since pancreatectomy is probably a rather drastic procedure this effect should perhaps be regarded as a 'clinical' rather than an endocrine variation.

9.4 Clinical conditions and drugs

Malnutrition and certain diseases have been shown to affect serum ChE (see Davies 1954). Sereni et al. (1966) examined the developing rat brain and found that levels of AChE, 5-hydroxytryptamine and noradrenaline (but not succinic dehydrogenase) were depressed by undernourishment, but all reached normal values quite quickly once ad lib feeding was allowed. Adlard and Dobbing (1971a) obtained similar results except that succinic dehydrogenase was also depressed in their animals. In a subsequent paper they showed that AChE levels in the brains of adult rats which had been undernourished during the suckling period were 14% *above* control values, on a weight basis (Adlard and Dobbing 1971b). Since most research workers can safely assume that their experimental animals are well-fed and healthy these factors are unlikely to be of practical importance in the majority of experiments on cholinesterase levels. When, however, an experimental regime is liable to reduce food intake any decrease in cholinesterase might be the result of malnutrition and not a direct result of the procedure under test. For instance, I found (Silver 1960) that hens poisoned with anticholinesterase organophosphorus compounds such as tri-orthocresyl phosphate (TOCP, ch. 11) and tri-paraethylphenyl phosphate, showed a marked loss of appetite and a considerable fall in weight, in some cases as great as 35%. Unless paired-feeding experiments are done to assess the effect of malnutrition per se it is unsafe to assume that any pathological changes observed in this type of experiment are directly attributable to the drugs.

Dikshit and Mahal (1937) reported a decrease in ChE activity in defibrinated whole blood from guinea-pigs injected with bubonic plague (*Bacillus pestis*) but injections of *Bacillus subtilis* appear to cause an increase in brain cholinesterase in the mouse (Reznik and Kutovoi 1970). Again though, it is not clear if these changes are a primary or secondary effect of the infection. Augustinsson (1948) cites a number of other clinical conditions which apparently affect cholinesterases. Among those which reduce activity in the serum are carcinoma (see also Vaccarezza and Willson 1965b, Vaccarezza et al. 1969), tuberculosis, anaemia and liver disease. Diseases which increase serum cholinesterase include diabetes, asthma (see also Vaccarazza and Peltz 1960) and, as already mentioned, hyperthyroidism. A reduction of erythrocyte, as well as serum ChE occurs in renal ischaemia (Vincent et al. 1970a) and enzyme activity is improved by haemodialysis. Since the liver is a source of the serum enzyme (see ch. 10) it is not surprising that serum pseudo-cholinesterase falls in liver disease (see also Davies 1954) but Sushko (1970) found that in infective hepatitis there was also a reduction in the level of AChE in blood. Blood cholinesterase may well be low in alcoholics as the result of liver damage but there is evidence that alcohol can decrease brain AChE as well. In a study on the rat Smyth and Beck (1969) showed that the initial effect of alcohol was to decrease the level of brain CoA; later ACh, ChAc and AChE levels also fell. They suggested that the reduction of AChE activity might be a manifestation of an adaptive process whereby the brain could maintain, or attempt to maintain, a workable concentration of ACh despite the fall in available CoA. The idea of a positive adaptive process of this type is interesting but as will be discussed again in §9.5, the difficulty is to distinguish between primary and secondary results and to recognize what is, and what is not, cause and effect; a direct action of alcohol on AChE synthesis cannot therefore be ruled out.

Chang et al. (1973) have reported that the chronic administration of nicotine reduces the AChE demonstrable by electron microscope histochemistry in the hypoglossal nucleus and hypothalamus of the rat but the doses needed for the effect are high (1 mg/kg, 5 times a day for 8 to 16 weeks). Vernadakis and Rutledge (1973) made the interesting observation that BuChE activity was above control values in the cortex and corpus striatum of rats which were killed one or 3 weeks after being anaesthetized with ether or pentobarbitone. Since the levels were normal in animals killed during or immediately after anaesthesia the elevation seems to be a delayed long-term effect which the authors tentatively attributed to a change in glial metabolism. Cortical AChE activity showed no increase as a result of previous anaesthesia but that in the corpus striatum rose slightly. The fact that anaesthesia can have this prolonged effect is of practical significance when BuChE is measured in tissues taken from animals prepared by previous surgery. Ideally, control tissue should be taken from the same animal but if this is not possible sham-operated ones should be used.

The serum cholinesterases of spastic, moronic and mongoloid children were measured by Sideman et al. (1965) using ACh and MeCh as substrates. In the spastic children activity towards both compounds was greater than that in the controls but in the morons and mongols levels were within normal limits. The authors tentatively suggest that the excessive motor activity of spastics might result in an 'overflow' of transmitter into the blood but whether this actually occurs and, if it does, whether it would cause an elevation of cholinesterase levels is mere speculation. Pandey et al. (1970) measured enzyme levels in the erythrocytes of schizophrenics. They found the ChE activity to be below normal but some other enzymes showed an increase.

A condition which cannot be omitted in a discussion of the clinical aspects of cholinesterase is myasthenia gravis. This disease is characterized by weakness of the striated muscles, particularly those innervated by the cranial nerves. Basically there is some failure of neuromuscular transmission and this has been attributed at different times to a deficiency of ACh or to an excess of AChE. Neither of these simple explanations seems to be correct, however, and the exact cause has yet to be pinpointed (for discussion see Grob 1963b; Nastuk and Plescia 1966; Barnes and Irvine 1973). The apparent association of the disease with tumours of the thymus, together with immunological data (see Adner et al. 1966) suggests that myasthenia gravis could be an auto-immune disease possibly resulting in an abnormality of the receptor protein (ch. 2). Anticholinesterase drugs (see ch. 11) such as neostigmine (Prostigmin), edrophonium (Tensilon) and ambenonium (Mytelase) are useful in the diagnosis and management of the disease but exactly how they work is not clear. Possibly all they do is to allow the accumulation of endogenous ACh, but a more complex action, perhaps one involving the receptor protein, cannot be ruled out.

Another condition of clinical importance concerns an atypical serum cholinesterase which is genetically determined and occurs in certain families (see ch. 10). The enzyme is unable to hydrolyse succinylcholine and while this deficiency is unimportant under physiological conditions, patients with the atypical enzyme run the risk of a fatal overdose if given succinylcholine (suxamethonium) as a relaxant during surgery (see Evans et al. 1952; Lehmann and Ryan 1956).

9.5 Experimental interference

9.5.1 Surgical lesions

9.5.1.1 Axotomy   Sawyer (1946) showed biochemically that when a cholinergic nerve (in his experiments, the sciatic nerve of a guinea-pig) is transected, AChE accumulates proximal to the cut (i.e. in the part still connected to the cell body) but in the isolated stump distal to the cut the enzyme activity falls. This phenomenon forms the basis of the useful histochemical method developed by Shute and Lewis (1961) for tracing the polarity of AChE-containing fibres within the CNS (see chs. 3 and 7). Sites of enzyme accumulation are easily recognizable and Gwyn and Wolstencroft (1966), for example, were able to distinguish ascending from descending cholinesterase-containing tracts in cat spinal cord by mapping areas of AChE pile-up above and below a hemisecting lesion. A little more caution is needed when considering the significance of a decrease in enzyme because the loss may not be entirely confined to the isolated fibres. Interference with the blood supply, for instance, may produce a loss which extends beyond the area to which the sectioned fibres project.

When a cell is axotomized AChE activity in the perikaryon is usually reduced and subsequently restored, the time-course of the changes approximating to the time-course of chromatolytic changes in Nissl substance (see Schwarzacher 1958; Chacko and Cerf 1960; Taxi 1961; Filogamo and Candiollo 1962; Lewis and Shute 1965a). Work by Kreutzberg and his co-workers suggests that in at least some species the changes in enzyme activity in axotomized neurones may not result directly from the damage to the cell itself. Electron microscopic studies (Blinzinger and Kreutzberg 1968) have shown that, in the rat, section of the facial nerve induces a very rapid proliferation of microglia in the facial nucleus. Within 5 days of the operation, most of the surface of the soma and dendrites of the axotomized cells are covered with glia. These glia invade the synaptic cleft and by displacing the terminal boutons they effectively cut off the afferent input. At the same time as these changes are occurring, the AChE activity of the axotomized cells is falling and by 10 days after operation, the histochemical reaction is no longer obtainable (Kreutzberg et al. 1974). The possibility that the enzyme activity is lost from these cells because they are, in effect, deafferentated is supported by the somewhat different findings in the guinea-pig. In this species section of the facial nerve is not followed by disruption of the synapses on the axotomized neurones and although there is some loss of AChE from the dendrites, the histochemical reaction in the soma of the facial neurones persists and may even be slightly intensified.

Reference has already been made (ch. 7 §3.3) to the interesting observation by Waldron and Gwyn (1969) on AChE activity in the red nucleus of rat (see also Gwyn 1971b). In keeping with other evidence that the red nucleus is probably non-cholinergic (McLennan 1969; A. Silver and J.H. Wolstencroft, unpublished; Waldron 1970), enzyme failed to accumulate in the rubrospinal tract after section, but what did occur was a complete loss of enzyme activity from the cells of the contralateral nucleus (the rubrospinal tract is crossed). It seems from this result that synthesis of AChE in non-cholinergic cells can be reduced by axotomy just as in cholinergic ones. Furthermore, pseudocholinesterase-containing cells such as those present in the vagal nucleus of rat undergo a similar loss of enzyme when axotomized (Navaratnam et al. 1964, 1968).

9.5.1.2 Deafferentation   The view that changes in AChE activity in axotomized neurones are caused by the loss of the afferent input is in accord with the results of experiments involving surgical deafferentation. Hess (1960) showed that when he removed an eye from newborn rabbits or mice or from foetal guinea-pigs the AChE in the contralateral superior colliculus was reduced and although the enzyme level gradually increased, it was still deficient in animals killed 7 months later. In contrast, removal of an eye from an adult mouse did not reduce cholinesterase activity in the superior colliculus. Hess suggested that neuronal synthesis of enzyme is influenced by the number of afferent fibres synapsing on the cell. Following removal of an eye, the input to collicular cells is severely reduced but afferents growing in from sources other than the retina (e.g. the cerebral cortex) apparently stimulate a slow synthesis of enzyme. In the adult the number of afferent fibres needed to maintain production seems less critical, hence the ineffectiveness of the operation.

Experiments in which animals are deprived of light bear out Hess's view that it is synaptic function rather than purely anatomical connections which are important in the maintenance of post-synaptic enzyme levels. Maletta and Timiras (1967) measured the effect of 21 days of complete light deprivation on both AChE and pseudoChE in visual and extra-visual areas of the nervous system of female rats and their male babies — the period of complete darkness for both groups began when the litters were born. Compared with that in control rats of the same age, the AChE activity in the superior colliculus and lateral geniculate body of the young animals was significantly reduced; in the adults, however, all values were within the control range. In neither adults nor young was pseudoChE affected but Maletta and Timiras (1968) subsequently found ChAc activity to be reduced in the same areas as AChE following light deprivation from birth. The role of pre-synaptic activity in the regulation of synthetic processes in post-synaptic cells is receiving increasing recognition (see Bloom et al. 1970) and AChE seems to be an obvious marker in investigations of this effect. This question will be considered again in §9.6.

In non-mammals (e.g. frog) removal of an eye can also cause a fall in enzyme in the optic tectum (Boell et al. 1955) but this may be attributable to an actual loss of cells rather than to a reduction of synthesis (Hess 1961a); the same is apparently true of the chick following section of the optic nerve (Filogamo 1960). In mammals, cellular loss does not seem to occur in the superior colliculus and such differences afford one more instance of the difficulty of interpreting the results of lesions. Another factor to be considered in this type of experiment is that there may well be regional differences. Thus although, in mammals, transneuronal degeneration does not occur in the superior colliculus after removal of an eye, it does occur to a marked extent in the lateral geniculate body (see Kupfer and Palmer 1964).

The reduced level of enzyme in the red nucleus and facial nucleus following axotomy, and in the superior colliculus after deafferentation, is in contrast to the situation in the cerebellum (ch. 8 §3.1.1) where lesions cause an increase rather than a decrease in the AChE activity of Purkinje cells (see also §9.6.1). Yet another situation, that in which deafferentation is apparently without any effect on AChE, has been described by Burt and Narayanan (1970). They made lesions in the neural tube of chick embryos (50–53 hr of incubation) such that the ventral half could develop in the absence of any connections with the dorsal half or with the dorsal root ganglia and more rostral levels. Over the whole period studied, AChE activity was equal to control values although ChAc levels were significantly reduced. The divergent results for the two enzymes may be explicable if, as discussed in the previous chapter, much of the AChE is associated with the metabolic processes of development rather than with synaptic aspects. Any fall in 'neuronal' AChE concomitant with the fall in ChAc might be small in proportion to the total level of 'metabolic' AChE and so be undetected.

9.5.1.3 Isolated cortical slabs   Krnjević and Silver (1965) used the technique of undercutting cortical areas to determine the source of AChE-containing fibres in the cat forebrain. In view of the marked decrease in histochemical reaction in such preparations, together with the evidence of a profound loss of ChAc (Hebb et al. 1963), they concluded that most of the fibres arose from subcortical cells – the arcuate fibres round the sulci being a possible exception (see ch. 7). Subsequently it has been shown (Duncan et al. 1968; Chu et al. 1971a) that the loss of AChE from the isolated area can be partly prevented if the undercut region is electically stimulated with currents below the threshold needed to produce seizure discharges from such slabs. It is difficult to interpret this effect, particularly since no histological studies were done to see if any AChE-containing fibres had sprouted into the slabs from the borders of the lesions. When intact cortex was subjected to similar stimulation the AChE activity did not change ipsilaterally but, surprisingly, AChE levels in the contralateral area seemed to be above normal. Even more surprising, in view of the loss of transcallosal fibres, this increase was also seen contralateral to a stimulated undercut area. Changes in enzyme activity are not the only result of the cortical stimulation: the same treatment prevents the development of supersensitivity in the isolated slab (Rutledge et al. 1967; Chu et al. 1971a, b). Whether or not these two findings are manifestations of a single mechanism must remain speculative. Krnjević et al. (1970) made the tentative suggestion that the relation of AChE to the ACh receptor might provide common ground but, as discussed in ch. 2, this relationship does not appear to be as close as was originally thought. That the nature of the receptor may change seems possible from the work of Chu et al. (1971b) who showed that the capacity to bind tubocurarine (see ch. 2) was increased in isolated cortical slabs. For a further discussion of the possible role of cholinergic mechanisms in the development of cortical hyperexcitability see Green et al. (1970) and Goldberg et al. (1972).

9.5.2 X-irradiation

Over the last few years the literature on the results of exposure to X-irradiation has proliferated enormously. Effects on function, on structure and on composition of tissues have all been documented and cholinesterase has not been neglected in these studies. Comprehensive data of this type are obviously important in the assessment of radiation hazards and in the understanding of defects resulting from exposure. Whether the results are of equal value in a narrower context, that of elucidating the mechanism of a specific enzyme, seems more doubtful. The important point to establish is whether AChE is a specific target or whether any changes produced by irradiation are part of a general interference with protein metabolism. Results from developing rats irradiated prenatally (Maletta et al. 1967) suggest that the AChE activity in parts of the CNS may be decreased specifically since overall protein levels were unaltered (the contribution of AChE to total protein is so small that its decrease would not influence the total figure). The phylogenetically older parts of the CNS such as the spinal cord were more radio-resistant than the newer areas such as the sensorimotor cortex (see also Maletta and Timiras 1966). The probable explanation is that areas which were developing at the time of irradiation (14 days gestation) were more vulnerable than the phylogenetically older areas which would have been relatively more mature (see Adlard and Dobbing 1972 re sparing of ChE-containing terminals in rats irradiated postnatally). This still leaves open the question of what exactly it is that is vulnerable: is the AChE-synthesising machinery depressed primarily or secondarily? Morphological changes causing a loss of synapses might result in a secondary reduction in enzyme as discussed previously (§9.5.1.2). Maletta et al. (1967) found that levels were low at 18 days gestation and at 2 and 9 days after birth but were normal by 23 days which indicates that the deficiency caused by irradiation is gradually made good. According to Krnjević and Silver (1965, 1966) most of the AChE in the sensorimotor cortex would be in afferent fibres and it seems possible that it is their rate of growth into the area which could be affected. If the effect were on synthesis in the cells of origin of the cortical fibres, one would expect to see a depression in subcortical areas too.

9.5.3 Electroshock

Rats born to mothers that had been subjected to daily electric shocks during part of their pregnancy were found to grow initially more slowly than rats born to un-shocked mothers; by puberty, however, the rates were reversed (Petropoulos et al. 1968). AChE levels in the brain of the experimental young differed from those of the controls but the degree of difference was not always statistically significant and the direction depended on the length of time over which mothers had been shocked. In offspring from rats treated from day 7 of pregnancy, AChE levels were lower than normal in cerebral cortex, cerebellum and hypothalamus. When the treatment started later in pregnancy, AChE activity was again below normal in the cerebral cortex but in the cerebellum and hypothalamus it was above normal. Once more, the significance of these findings is uncertain although the authors believe the changes to be associated with alterations in cholinergic processes. They make the interesting suggestion that the effects seen in the offspring result from shock-induced changes in maternal hormones. The points which emerge most decisively from these experiments and from those on X-rays are that experimental inter-ference affects AChE levels differently in different brain areas and the complexities of interpretation should not be underestimated.

9.5.4 Training and stress

Various workers have shown that the way in which animals are handled during the postnatal and weaning period can affect the cholinesterases in their nervous tissue. A good many of the experiments have been aimed at establishing whether 'behav-iour' depends on cholinergic mechanisms, and foremost among those with this motive is Rosenzweig's group at Berkeley, California (for references see Rosenzweig 1966; Bennett et al. 1970). Rosenzweig et al. (1962) reared post-weaning rats under conditions (termed an 'enriched environment') in which several animals were housed together and were given 'toys' and trained to run mazes. AChE and pseudo-cholinesterase were not distinguished in the brain analyses which were done about 105 days after birth and which were expressed per unit wet weight. Cortical ChE levels in maze-running rats were lower than those of untrained rats reared in isola-tion but subcortical levels were increased. In contrast, the cortical weights of ex-perimental animals rose while subcortical weights were unchanged. Similar experi-ments by Tapp and Markowitz (1963) produced results which were almost the reverse of these. In the cortex neither ChE activity nor weight was influenced by the enriched environment but a significant decrease in ChE and an increase in weight occurred in the subcortical samples. This pair of experiments is only one example of the contradictory results which confuse this whole field – there are many other instances. Often the effects produced by the experimental situation are very small and become statistically significant only if 'manipulated'. For instance, Kling et al. (1965) compared the AChE levels in various brain areas from control pre-weaning rats with those from similar rats which for 10 days had been subjected daily to a 15-min period of handling or exposure to a light flash. The light treat-ment caused a decrease in hypothalamic AChE which was significant by comparison with values in control rats though not significant by comparison with handled rats; levels in these latter rats were not themselves statistically different from control values. In the caudate nucleus, a significant difference was demonstrable only when results for light exposure and handling were compared; when the control values were used for comparison the effect of light was not significant.

In an article entitled 'Is the central cholinergic nervous system overexploited?'

Karczmar (1969) brings some timely scepticism to the question of interpretation of this sort of experiment. He points out that not only are the reported changes often very small, but also, because of concurrent changes in weight, the direction of the change can vary depending on how the enzyme activity is expressed. From a functional point of view ACh is of much greater relevance than AChE (Aprison et al. 1968; see also Aprison and Hingtgen 1970) and the small variations in cholinesterases may be giving no indication of what, if anything, is happening to the cholinergic system as a whole. More important still, cholinesterases seem able to work with a wide safety factor (see Russell 1969) and a number of experiments have shown that in the rat brain, cholinesterases can be severely depressed by DFP without producing any obvious behavioural change (see Glow et al. 1966). Clearly, more definitive evidence is needed before levels of brain AChE can be accepted as an index of behavioural or 'intellectual' activity. What the results to date do provide is a further indication of the lability of the AChE system. If an equivalent degree of flexibility were brought to the analysis of the data the picture might be clearer. The behaviourists would like to equate changes in enzyme with processes of learning but a broader view might be more revealing. An 'enriched environment' or an 'altered early experience' is not a single, well-defined modality like 'heat' or 'light'. Which of all the many factors, some possibly conflicting, that must be involved in these experiments are affecting the enzymes? Singh et al. (1967) have suggested that visual stimulation or deprivation may be the critical factor in alterations of cortical ChE. They postulate that the inevitably greater visual stimulus received by handled rats and rats in an enriched environment could account for the changes observed by previous workers. The universal validity of this theory must be doubted since their own experiments on rats showed that increased visual stimulation caused a rise in cortical ChE whereas Rosenzweig et al. (1962) observed a fall in cortical levels in rats from an enriched environment. Whether or not this particular idea is correct, it does emphasize the need to recognize that the factor under test (e.g. learning) is not necessarily the factor actually responsible for any change.

Kling et al. (1969) could find no consistent differences between AChE levels in various brain areas of kittens raised by their mothers and of kittens raised in isolation; they pointed out that this lack of effect of 'altered early experience' is at variance with the results for rat. This discrepancy may represent a genuine species difference but equally it could stem from the difference in experimental design: the factor or factors affecting ChE levels in the various experiments on rats may well be missing in these particular experimental conditions. In Rhesus monkeys trained to discriminate colours and to perform a more complex learned task, AChE activity was significantly reduced in the orbital cortex but in 18 other areas, both cortical and subcortical, levels were within the range for untrained control animals (Kling et al. 1970). Following bilateral ablation of the dorsolateral frontal cortex the level of AChE in the orbital cortex increased but the difference between trained and non-trained animals persisted. Křivánek and Burešová (1972) measured the AChE in the 'handedness' area of the cerebral cortex of rats trained to perform a task for which the same paw was used every time. They concluded that the small increase in AChE that was seen in rats which had been trained for an extra long time was a consequence of the prolonged *performance* of the task and was not associated with the formation of a long-term memory trace. Their argument was that if the change were correlated with the establishment of memory, the enzyme activity should increase gradually from the start of training and would be independent of the number of overlearning trials.

9.6 Control of enzyme synthesis

9.6.1 Introduction

The experiments described above show the sort of factors which influence cholinesterase activity but they give little indication of the mechanisms underlying the changes. To analyse what is involved when there is an 'increase' or 'decrease' in synthesis it is necessary to know something of the processes which normally control synthetic activity.

The point has been made earlier (chs. 5 and 7) that during phylogeny chemicals which were available to the primitive cell were not necessarily incorporated into the nervous system of different animals in the same way. Somehow, the vertebrates have ended up with ACh as their transmitter at the neuromuscular junction and something else, so far unidentified, as their sensory transmitter. In contrast, in some invertebrates (e.g. crab, lobster; see ch. 5) the neuromuscular transmitter is not ACh but the sensory transmitter may be. As yet no one has identified the specific factors which determined this state of affairs but presumably the differences will eventually be explained in terms of the genetic code which governs the synthetic activity of the cell. In their scheme for the genetic regulation of protein synthesis in bacteria Jacob and Monod (1961) proposed that the rate of protein synthesis was regulated by repressors. If the repressors are in the inactivated state, synthesis is induced, but if they are activated, synthesis is repressed. It follows from this that although a cell may have the genetic make-up to produce a certain protein it may not do so because of the inhibition imposed by a repressor. In some most elegant work on amphibia involving the transplantation of the nucleus of a differentiated cell into an enucleated unfertilized egg, Gurdon (1968, 1971) has shown that nuclei from differentiated cells (e.g. skin, intestine) possess all the genetic material necessary to form every other cell type present in the adult frog. Although there is no evidence that the expression of genetic potential in the cells of higher organisms is regulated by the type of repressor-activator system proposed for bacteria, it would seem that some form of suppression must be at work to prevent a differentiated cell manifesting the other types of activity of which it is genetically capable.

9.6.2 The suppression or stimulation of enzyme activity

Attempts were made in the last chapter (ch. 8 §3.1.1) to find some metabolic reason for the appearance of AChE in Purkinje cells which had been deafferentated. In view of Gurdon's demonstration that cells possess unexpressed potential it may perhaps be unnecessary to seek a function for this enzyme activity. AChE might appear in the deafferentated cell not because this receives some urgent signal to begin synthesis, but because it is no longer receiving the normal inhibitory signal which stops it producing the enzyme. Bevan et al. (1973) considered alternatives of this sort to explain their observations on the release of ACh from Schwann cells in degenerating frog nerves. Their evidence indicated that Schwann cells are able to synthesise ACh but do not do so unless the nerve is cut. They suggested that the degenerating axon may release something which activates synthesis or, alternatively, that the axon normally suppresses synthesis but when it degenerates this suppression is removed and synthesis can occur. Speculation of this type may be easy but to obtain experimental support is not. Evidence for or against such ideas may come with the development of a specific and sensitive method for the histochemical demonstration of choline acetyltransferase. Such a technique used in conjunction with methods for other enzymes should show whether ChAc has been selectively activated or whether it is but one of a number of enzymes from which the suppressor has been removed by denervation. Eränkö and Teräväinen (1967) and also Eränkö et al. (1970), have demonstrated an increase in BuChE activity of Schwann cells following nerve section. The suggestion that the same may be true of AChE has been mentioned already (ch. 8 §3.1.5).

In the case of Purkinje and Schwann cells, the nerves seem to have a restraining effect on synthesis and the removal of this restraint by denervation apparently allows an increase in enzyme activity. In certain other structures (see §9.5.1.2) denervation causes a fall in enzyme level. Although the fall may be in part attributable to a loss of enzyme-containing fibres there is also a decrease in the enzyme activity in the denervated cells themselves. In this case it appears that the nerves normally stimulate AChE synthesis. This phenomenon is perhaps best illustrated by reference to muscle. The influence of innervation on AChE activity at the end-plate has been a particular interest of Guth and his colleagues (see Guth and Zalewski 1963; Guth et al. 1964; Guth 1968, 1969a). Guth et al. (1966) showed that when a nerve is implanted into denervated rat muscle, AChE activity in the persisting sole-plates recovers faster than it does in a muscle without an implant. Furthermore, this occurs even though there is no direct connection between the new implant and the original sole-plates. Filogamo and Gabella (1966) did somewhat similar experiments on chick and guinea-pig. They denervated certain muscles and then gave DFP to inhibit the enzyme remaining in the subneural apparatus. If reinnervation occurred enzyme activity was restored but there was no restoration if reinnervation was prevented. What emerges from these studies is that the nerve is essential for stimulating and maintaining the synthesis of AChE in the end-plate. Various experiments have established that the nerve exerts its effect on the end-plate via a trophic factor (Drachman 1967; Filogamo and Gabella 1967; Lentz 1971) and despite earlier doubts (see Guth 1968) most current evidence supports the view that this factor could be ACh (Guth 1969b; Filogamo and Marchisio 1971; Giacobini 1972; Burt 1973). If this is so, then AChE synthesis could be triggered by the process of substrate induction (see §9.6.3).

According to Filogamo and Marchisio (1971) the sequence of events in developing muscle is as follows. On reaching the myoblasts the ingrowing nerve fibres release ACh and this acts as a 'primer' for the synthesis of AChE. To start with, the synthesis of AChE occurs throughout the fibre but as this matures, the overall activity decreases except at each end. As a final stage, the end-plate is established and AChE becomes concentrated in this region (Filogamo and Gabella 1967). Denervation causes a drop in AChE activity at the end-plates but the extent of the decrease and the exact time-course varies with the species (see Guth 1968). The finding that the loss is not total may reflect the fact that some release of ACh still occurs in denervated muscle (Mitchell and Silver 1963; Bevan et al. 1973). On the other hand, it may mean that synthesis of AChE, once established can continue, albeit at a reduced level whether or not ACh is present. There is some evidence that at the myotendinous junction (ch. 8 §4) maintenance of AChE is not dependent on the presence of ACh although the initial synthesis in the developmental stage may be. Despite the lack of nerves at the junction itself, AChE activity will not develop unless the muscle is innervated. Lubińska and Zelená (1967) found that when the sciatic nerve was cut in new-born or 10-day old rats, enzyme activity disappeared from the myotendinous junctions of soleus muscle and was still absent 4 months later. In contrast, if the lesion was made 20 or 30 days after birth, or in adult rats, AChE activity at the junctions was only slightly weaker than in control animals.

Brief reference must be made here to the effect of denervation on the activity of myosincholinesterase (ch. 1). Varga et al. (1957a) found that in the rabbit the specific activity of myosincholinesterase in the gastrocnemius muscle increased following section of the sciatic nerve. The significance of this change is unknown but it has been shown (Varga et al. 1957b; Szabolcs 1968) that during development the level of activity exceeds that at maturity hence the situation is somewhat analogous to that described earlier for AChE in Purkinje cells.

9.6.3 Substrate induction

The influence of innervation on the AChE activity at motor end-plates can be most easily explained in terms of the induction of enzyme synthesis by its substrate ACh but this has not yet been proved experimentally. Guth (1969b) made the point that although ACh seems to be essential for mediating the trophic effect of nerve on muscle it may be only one of a number of factors which normally ensure the proper synthesis of AChE. Hence, it may be difficult to establish unequivocally whether or not AChE appears in muscle as a result of substrate induction. The most promising approach seems to be the study of AChE activity developing in cultured cells. Useful data could come from a comparison of the response shown to added ACh by cells which are cholinergic or cholinoceptive, cells which are both cholinergic and cholinoceptive and cells which are neither. Cultured tissues in which the behaviour of cholinesterases has been examined include chick muscle (Engel 1961; Oh and Johnson 1972), chick brain (Burdick and Strittmatter 1965; Werner et al. 1971), chick spinal cord (Turbow and Burkhalter 1968; Kim et al. 1972), chick sympathetic ganglia (Kim and Munkacsi 1972), mouse cerebellum and sensory ganglia (Kim and Murray 1969), mouse neuroblastoma (Blume et al. 1970), rat cerebellum and brain stem (Hösli and Hösli 1970) and rat spinal cord (Hösli and Hösli 1971). Data from experiments on cultured cells need cautious interpretation. In their study of explants of chick lung Burkhalter et al. (1957a) showed that AChE (but not BuChE) activity increased when ACh was added to the medium. Since sodium chloride, choline and acetate were ineffective the authors concluded that ACh induced synthesis of AChE. This may be so but as Turbow and Burkhalter (1968) pointed out in connection with their findings on chick spinal cord, a beneficial effect of ACh on AChE levels is not necessarily evidence of substrate induction. What may be happening is that ACh, by increasing cell permeability improves the intracellular conditions for protein synthesis. Goodwin and Sizer (1965) looked at the effect of MeCh on the development of AChE in cultured chick muscle. In addition some muscles were cultured in media on which floated chambers sealed with millipore filters and containing embryonic chick cord. MeCh was found to have but little inductive effect while spinal cord depressed AChE levels. More recently Lentz (1971) showed that in cultures of newt muscle, cholinesterase activity is stimulated by the addition of homogenized nervous tissue and of pieces of spinal cord and liver. Perhaps the most significant finding was that the greatest effect was produced by the addition of whole spinal sensory ganglia. This bears out the idea that factors other than ACh may have an important influence on the synthesis of AChE. Among possible candidates is cyclic-adenosine monophosphate (cAMP), which has been shown to stimulate the synthesis of AChE (Furmanski et al. 1971), and also tyrosine hydroxylase (Waymire et al. 1972), in cultures of mouse neuroblastoma cells. Cyclic AMP has gained favour as a participant in synaptic transmission and is thought by some to promote the release of ACh (see Breckenridge et al. 1967; Goldberg and Singer 1969; Singer and Goldberg 1970). This idea immediately complicates the assessment of any effect of cAMP on AChE synthesis in cultures. The stimulus for synthesis could be cAMP itself or it could be ACh which is released by cAMP from some element in the culture. Although cAMP may prove to be important in the control of AChE synthesis in cell cultures extrapolation of these findings to the mechanisms operating in the whole animal is not warranted at this stage.

9.7 Summarizing conclusion

This chapter presents evidence that both acetyl- and pseudocholinesterase activity in brain and other tissues can be influenced by a multiplicity of factors. Those discussed include sex, age, genetic background, season and time of day; reproductive and other endocrinological factors, and the nutritional and pathological state of the animal are also considered. Because some of these variants will be operating in almost any quantitative study of cholinesterases it seems important to recognize them. If they are borne in mind when experiments are designed, their effects can be minimized. In addition to these factors which might be termed physiological variants, experimental procedures can alter cholinesterase activity. These procedures have been considered under the headings of surgical interference, X-irradiation, electroshock and training. Interpretation of data from experiments of these types is not easy. Effects of certain procedures e.g. lesions, may differ in different species or within different brain areas of the same species. More confusing still, outwardly similar experiments may yield opposing results in different laboratories. In such a confused field dogmatism is out of place but one assertion is justified: data must be interpreted with extreme caution. This injunction is also true for attempts to explain the mechanisms which determine the level of enzyme synthesis.

CHAPTER 10

Pseudocholinesterases

10.1 Introduction

Those neurophysiologists and neurochemists who are primarily interested in the role of AChE in transmitter mechanisms tend to regard 'pseudocholinesterase' as a single troublesome enzyme which, unless inhibited, may upset their results. On the other hand, many of those who work on blood consider 'serum cholinesterase' as the interesting component and ignore the AChE of the erythrocytes. As a consequence, pseudocholinesterases of the nervous system have been relatively neglected while those in the serum have been studied from various angles, although the question of function has received surprisingly little attention. Because of these circumstances it seems justifiable to review the subject of pseudocholinesterases rather broadly and not to confine discussion to those in nervous tissue.

10.2 Nomenclature and classification

The problem of naming the cholinesterases was mentioned in ch. 1 and I explained my reasons for retaining 'pseudocholinesterases' as the systematic name for the non-specific types of enzyme. Where appropriate, individual pseudocholinesterases will be distinguished according to their substrate preference.

The first question to settle is whether an enzyme which has a high affinity for choline esters should be classed as an acetylcholinesterase or as a pseudocholinesterase. As explained in ch. 1 pseudocholinesterases hydrolyse higher esters of choline, such as BuCh and PrCh more rapidly than they hydrolyse ACh. This is the reverse of the pattern shown by acetylcholinesterases. In the case of the cholinesterases from mammals it is usually reasonably easy to decide whether an enzyme meets the criteria for a particular class. Difficulties arise, however, when attempts are made to classify enzymes from non-mammalian species in the same way. It is sometimes hard to know whether an enzyme which falls short of certain criteria for AChE should, nevertheless, be regarded as a variant of the 'specific' type or whether it should be relegated to the 'pseudo' class (see Augustinsson 1959a, b, c, 1968). This is particularly true of the cholinesterases of the invertebrates (see ch. 5).

It will become clear from what follows that the pseudocholinesterases comprise a whole family of enzymes with properties that may vary from species to species. Since results obtained from one species are not applicable to another, it is imperative that the source of the enzyme is noted in any discussion of experimental findings.

10.3 Characteristics

10.3.1 Substrate specificity

One of the criteria that can be used in deciding how a cholinesterase should be classified is substrate specificity (§10.2). Because of various historical accidents the subject of substrate specificity and substrate preference is, at first sight, confusing. When Mendel and Rudney (1943a) initially investigated the cholinesterases from several species they happened to use benzoylcholine (BzCh) as one of their substrates. This proved 'specific' for what are now called pseudocholinesterases in that hydrolysis by AChE was only slight. On the other hand, it is not the preferred substrate for pseudoChE: in all species tested (for references see Kalow et al. 1956) it is hydrolysed much less rapidly than is ACh (see fig. 10.1). In ruminants it is split scarcely at all, a fact which gave rise to the belief that ruminants lack pseudocholinesterases (Mendel and Rudney 1943a; Gunter 1946). In 1949 Cohen et al. advocated the use of butyrylcholine as a specific substrate for pseudoChE. That this compound is hydrolysed faster than ACh by the cholinesterase of human serum was shown by Adams and Whittaker (1949a). Shortly after this Ord and Thompson (1951) discovered that the partially purified cholinesterase from the ventricle of rat

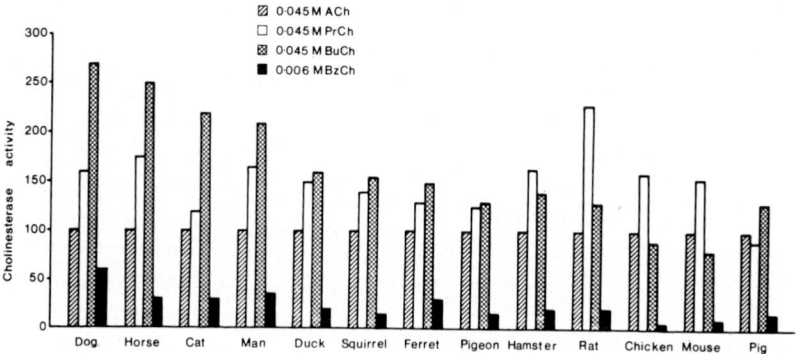

Fig. 10.1. Substrate specificity patterns of the pseudocholinesterases from sera of different species. Activity is expressed in relation to an arbitrary value of 100 for ACh. Most enzyme preparations were obtained from their respective sera by fractional precipitation between 0.50 and 0.85 saturation with $(NH_4)_2 SO_4$. (From Myers 1953.)

heart was able to hydrolyse propionylcholine faster than either ACh or BuCh. The rates relative to an arbitrary value of 100 for ACh were 239 for propionylcholine, 137 for butyrylcholine and 27 for benzoylcholine.

Using ACh, BuCh, PrCh and BzCh, Myers (1953) surveyed the substrate specificity patterns in sera of a number of species. His findings are illustrated in fig. 10.1. Of the 13 species examined eight, dog, horse, cat, man, duck, squirrel, ferret and pig showed a clear preference for BuCh. In the pigeon, activity towards BuCh was only slightly greater than that towards PrCh while in hamster, chicken, mouse and, especially, in the rat the rate of hydrolysis of PrCh exceeded that of BuCh. The BuChE of the pig was atypical in that PrCh was split more slowly than was ACh.

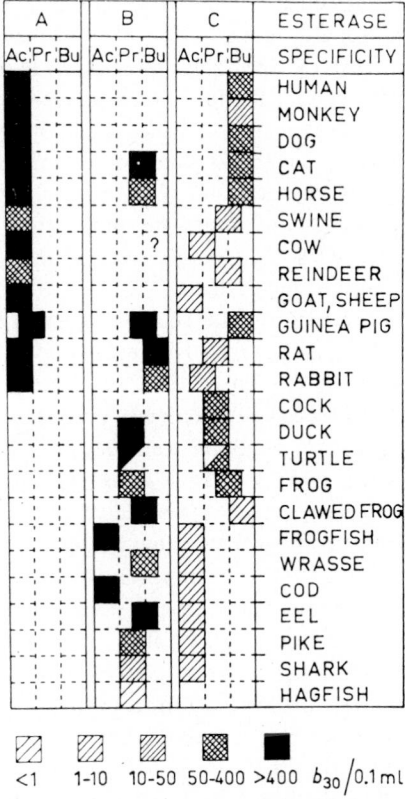

Fig. 10.2. Plasma-esterase patterns of vertebrates. Groups of esterases: A, arylesterase; B, aliesterase, lipase; C, cholinesterase. Relative esterase specificity based on the acyl radical of the substrate selectively hydrolysed by the enzyme in each group: Ac, acetyl; Pr, propionyl; Bu, butyryl. Key: activity values, $b_{30}/0.1$ ml plasma (total hydrolysis minus non-enzymic hydrolysis expressed as μl CO_2 evolved/30 min), refer to the hydrolysis rates observed with the substrate which is hydrolysed most rapidly by each esterase. (From Augustinsson 1959c.)

Subsequently, Augustinsson and Olsson (1959a) showed that this porcine enzyme was able to split phenylacetate almost as fast as it split BuCh.

Myers made a special study of avian sera. He confirmed the findings of Earl and Thompson (1952a, see also Blaber and Cuthbert 1962) that the cholinesterase in chicken serum (a propionylcholinesterase) shows considerable activity towards acetyl-β-methylcholine (MeCh). As mentioned in ch. 2, in the case of mammals this substrate is regarded as virtually specific for AChE. The butyrylcholinesterases of duck and pigeon sera were less active than the chicken enzyme towards MeCh but, even so, were considerably more active than mammalian sera.

To determine whether the same or different types of cholinesterase occur in the different organs of a particular species, Myers also examined a wide range of tissues from rat and a limited number from some other species. In all cases, the substrate specificity pattern was virtually identical for every tissue of a given animal. As a result, Myers concluded that 'the specificity pattern of the particular pseudocholinesterase concerned is species specific but not organ specific'. Although this seems generally true there may be exceptions. Davies et al. (1953) detected some slight hydrolysis of BzCh by sheep kidney cholinesterase while that in the plasma was inactive. More recently, Mittag et al. (1971b) working with guinea-pigs found differences between the BuChE of serum and gut with respect to the preferred concentration of substrate. This is probably evidence for the existence of different isozymes (§ 10.5.1) in the two preparations.

Further data on the diversity of pseudocholinesterases have been provided by Augustinsson in two wide-ranging studies (Augustinsson 1948, 1959a, b, c). In the latter series of experiments he examined the serum esterases (aryl- and carboxyl- as well as cholinesterases) in 27 species of mammals, birds, reptiles, amphibia and fish. Using electrophoretic separation and other techniques Augustinsson confirmed that in some mammals the serum cholinesterase is a butyrylcholinesterase and in some a propionylcholinesterase. The ruminants showed a variety of specificity patterns: in the cow the enzyme is an atypical propionylcholinesterase with some of the properties characteristic of AChE; the enzyme in the reindeer is classed as a 'butyryl (propionyl) cholinesterase' and that in sheep and goat is an acetylcholinesterase. Acetylcholinesterase also occurs in place of pseudoChE in the sera of the teleosts but in the turtle the single esterase present shows properties intermediate between those of a propionylcholinesterase and a carboxylesterase (fig. 10.2).

Augustinsson (1959a) established that the propionylcholinesterases (for example in rabbit, cow and rat serum) showed appreciable species differences in sensitivity towards inhibitors, in substrate specificity (particularly in their ability to hydrolyse non-choline esters) and in the relation between activity and substrate concentration. In contrast, all the butyrylcholinesterases (as in serum from man, dog and horse) were very similar in substrate specificity and in other properties although electrophoresis showed them to be different in molecular form. A recent study by Ecobichon and Comeau (1973) gave similar results.

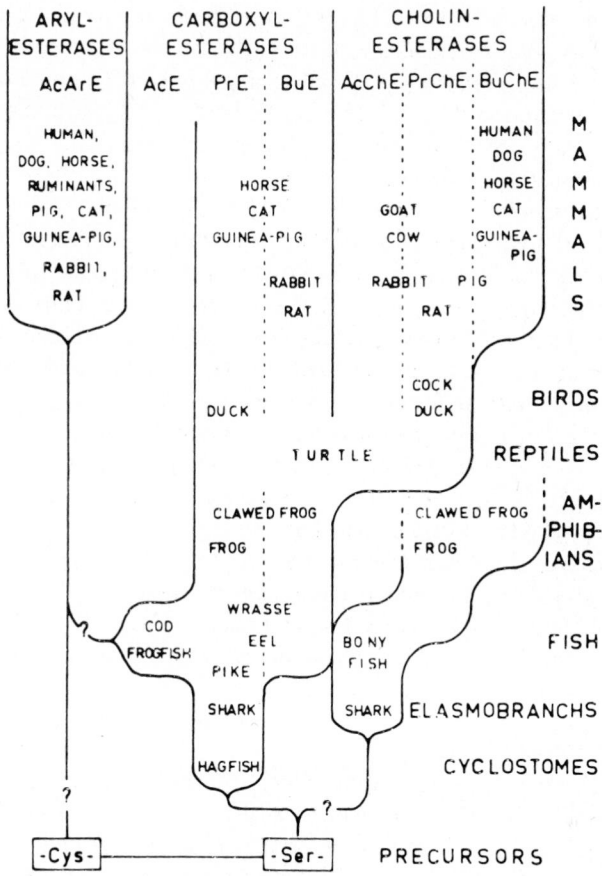

Fig. 10.3. Scheme showing the possible evolution of individual blood plasma esterases of vertebrates. Ac, acetyl; Pr, propionyl; Bu, butyryl; Cys, cystein; Ser, serine. (From Augustinsson 1968.)

Augustinsson (1959c, 1968) has considered the evolutionary implications of the different patterns of substrate specificity (fig. 10.3). He has proposed that the plasma cholinesterases and carboxylesterases were evolved from a common 'serine-enzyme' precursor, possibly a propionylesterase (see fig. 10.3). He suggests that during evolution the cholinesterases acquired an anionic site which conferred on them their special ability to bind cationic substrates. Augustinsson further suggests that mutational changes resulted in the development of differing specificity, the butyrylcholinesterases possibly being more specialized forms of the propionyl type. The sole esterase of the turtle serum, with properties both of a carboxylesterase and a cholinesterase, may represent a type of enzyme which appeared early in evolution. Serum acetylcholinesterases are also thought to have arisen early but Augustinsson considers they probably developed separately and are not precursors of the other types. In evolutionary terms the AChEs pose a problem because although commonest in lower vertebrates such as fish (see also Burnstock 1969 re AChE in place of tissue pseudocholinesterase in amphibia etc) they also occur in mammals (for example the rabbit). In hamster blood 23% of the total AChE activity is found in serum, while 35% of the pseudoChE is in the erythrocytes (Vincent et al. 1965).

The emphasis so far has been on activity towards choline esters. Another feature which differentiates pseudo- from acetylcholinesterases is the ability of the former to hydrolyse aliphatic and aromatic esters such as tributyrin and phenylbutyrate at an appreciable rate (see Augustinsson 1948, 1959a, b, c). It is sometimes assumed that AChE is totally inactive towards non-choline esters but, as emphasized in ch. 2, this is wrong. The distinction is relative but becomes more marked when butyrates rather than acetates are tested (Sturge and Whittaker 1950). As with choline esters, species differences are also apparent in the activity of pseudocholinesterases towards these non-choline substrates. The enzyme in rat heart (a propionyl ChE) was shown by Ord and Thompson (1951) to be inactive towards tributyrin but this substrate, and also methylbutyrate, is appreciably hydrolysed by the BuChE of dog, man and horse (Mendel and Rudney 1943a; Adams and Whittaker 1949b). Further data on differences of this type are given in the papers by Augustinsson cited above.

10.3.2 The effect of inhibitors and of excess substrate

The use of inhibitors as a further means of identifying the types of ChE present in a tissue was discussed in ch. 3 and will be dealt with in some detail in the next chapter. All that is needed here is a brief summary of the principles involved. An enzyme can be identified as a pseudocholinesterase either on the basis of its susceptibility to specific inhibitors of pseudoChE or on the basis of its resistance to inhibitors of AChE. DFP can be used as a differential inhibitor because in a particular species the pseudocholinesterase is more susceptible than the AChE. The degree of difference in susceptibility varies, however, from species to species. Ecobichon and Comeau (1973) measured I_{50} values for the inhibition of serum cholinesterases from 11 mammals and showed an 800-fold difference in the values for man and goat. Its toxicity and volatility makes the use of DFP particularly dangerous and iso-OMPA is both less hazardous and more specific. Of the 'specific' inhibitors (see ch. 11) ethopropazine is probably one of the most widely used but, again, it is necessary to be sure that the concentration is adequate for each species investigated. If an inhibitor is used which is specific for AChE (e.g. BW284C51) activity due to pseudocholinesterase should, in theory, be unaffected but the specificity is rarely absolute and the finding that an enzyme believed to be a pseudocholinesterase is slightly susceptible to inhibitors of AChE does not vitiate the identification.

Another feature which distinguishes pseudoChEs from AChEs is their behaviour towards different concentrations of ACh. It is generally true to say that excess of ACh will inhibit AChE but not pseudoChE. In addition, the K_Ms (ch. 2) of pseudocholinesterases exceed those of acetylcholinesterases for choline esters. This means that pseudoChEs are relatively less efficient than AChE at low concentrations of substrate, but more efficient at high concentrations. The lack of inhibition by high concentrations of ACh is a useful criterion for classifying an enzyme as a

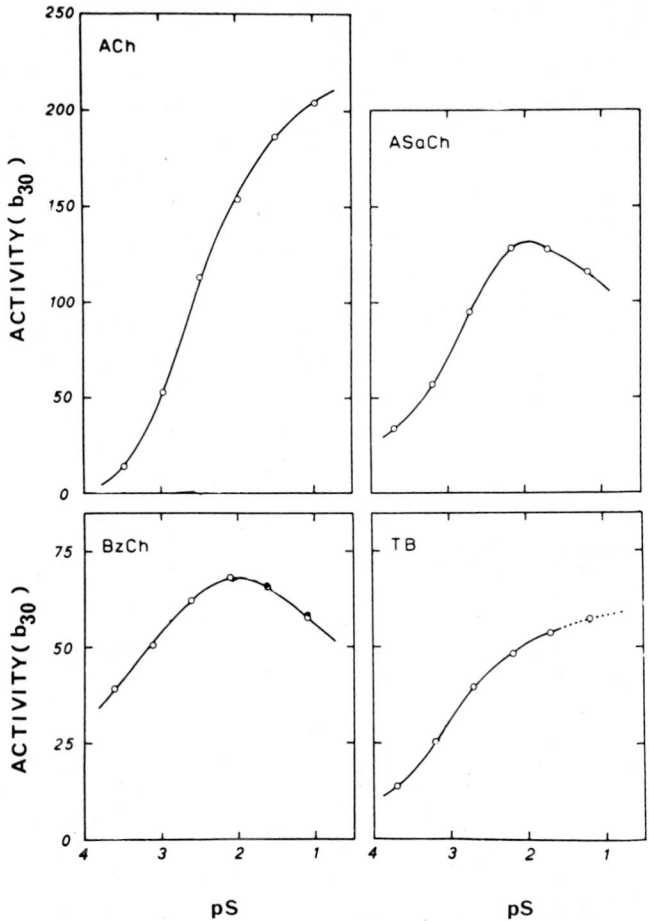

Fig. 10.4. Activity-*pS* curves for the hydrolysis of ACh, BzCh, ASaCh and TB by pseudocholinesterase purified from horse serum. Note sigmoid curve for ACh and TB (no substrate inhibition) but bell-shaped curve (indicative of inhibition) for ASaCh and BzCh. *pS*, −log molar concentration of substrate; b_{30}, enzyme activity (total hydrolysis minus non-enzymic hydrolysis) expressed as μl CO_2 evolved/30 min. (Modified from Augustinsson 1948.)

pseudoChE but this does not mean that they are altogether immune from substrate inhibition. Augustinsson (1948, 1963) demonstrated that although the BuChEs in dog, horse and human sera were not inhibited by an excess of BuCh or PrCh they were inhibited by an excess of certain aromatic esters such as BzCh and acetyl-salicylcholine (fig. 10.4). The propionylcholinesterase of cow serum is atypical and one of the features which distinguishes it from most other pseudocholinesterases is that it is inhibited by an excess of PrCh (Augustinsson 1959a).

10.4 Chemical nature and the problem of the active site

Most of the data about the chemical nature and the kinetics of pseudocholinesterases have come from experiments on the enzymes (butyrylcholinesterases) present in the plasma of horse and man but some generalizations hold to a greater or lesser extent for the enzymes in other tissues.

Pseudocholinesterases appear to be glycoproteins. The serum enzymes are highly soluble in water and although tissue cholinesterases are not quite so soluble they

TABLE 10.1

Amino acid analysis of butyrylcholinesterase from pig parotid gland. (From Tucci and Seifter 1969.)

Amino acid	Residues/1000	Residues/368,000*
Aspartic acid	115	340
Threonine	53	156
Serine	89	263
Glutamic acid	114	337
Proline	76	225
Glycine	105	310
Alanine	101	298
Valine	40	118
Methionine	10	30
Isoleucine	23	68
Leucine	76	224
Tyrosine	21	62
Phenylalanine	51	150
Lysine	53	156
Histidine	19	56
Arginine	44	130
Cysteic acid	9	27
Pre-aspartic acid	8**	
Prelysine	21**	

 * Molecular weight.
** Not included in residues per thousand with the other amino acids. Leucine equivalents. Ammonia, 0.83 μmoles. Total nitrogen, 4.84 μmoles.

are, nevertheless, more easily extractable than is AChE (see Surgenor and Ellis 1954; Das and Liddell 1970). Electrophoretic data suggest that the protein of the serum enzyme in man is a type of α_2-globulin (Surgenor et al. 1949). Svensmark (1965) found a cholinesterase with similar electrophoretic characteristics in human brain but he also found a second type which moved more slowly, resembling the γ-globulins (see below). This question of electrophoretic mobility and its relation to molecular weight will be considered again in §10.5.1. As yet, the carbohydrate moiety of the pseudocholinesterases has not been precisely defined but in both horse and man the enzyme is associated with sialic acid (N-acetylneuraminidic acid). If this is split off with neuraminidase the electrophoretic mobility is decreased and the enzyme becomes less stable, but its hydrolytic activity is unchanged. AChE, in contrast to pseudoChE, is not, apparently, a sialo-protein (Augustinsson 1971b).

Table 10.1 shows the amino acid analysis of a highly purified preparation of BuChE from the pig parotid gland (Tucci and Seifter 1969). Some information is available about the possible sequence of the amino acids in the vicinity of the active centre of BuChE from horse serum. Jansz et al. (1959) treated the enzyme with DFP and then used a degradation technique to examine the phosphorus-containing peptide. As with other esterases including AChE, the phosphorus was bound to a serine hydroxyl group which formed part of the following sequence.

$$\text{Phe}-\text{Gly}-\text{Glu}-\underset{\underset{\text{P}}{|}}{\text{Ser}}-\text{Ala}-\text{Gly}(\text{Ala}_2-\text{Gly})$$

(The order of the amino acids in brackets was not established)

The result does not provide unequivocal evidence that the serine is the catalytic amino acid because the binding of the phosphorus to the serine hydroxyl group could have occurred during processing rather than during the primary reaction between the inhibitor and enzyme (Augustinsson 1960; Svensmark 1965; see also ch. 2). Nevertheless, the peptide sequence is strikingly similar to that found in other enzymes where serine seems to have a part in the catalytic activity (see Watts 1968 for comparative data).

As discussed in ch. 2 a piece of evidence against serine hydroxyl itself being the active group of cholinesterases is that its *pK* does not accord with the data on the pH dependence of the enzyme. Bergmann et al. (1956) showed that pseudo-cholinesterase (from human serum) like AChE from *Torpedo* had a pH optimum of about 8 and they took this as evidence that the esteratic sites of both enzymes had a basic group of *pK* 6–7 and an acidic group of *pK* 9–10 (see also Augustinsson 1963; Svensmark 1965). Evidence that the imidazole group of histidine is a more likely candidate for the active group was presented in ch. 2 in relation to AChE; the same arguments hold for pseudoChE. As Augustinsson (1963) points out, it, alone of all of the groups in proteins, has a *pK* (6.95) which is in the range required of the basic group. Moreover, its heat of ionization is identical to that of human serum

ChE (Shukuya and Shinoda 1956). The mechanism by which imidazole might function has already been discussed (ch. 2). The nature of the acidic group in the esteratic site of cholinesterases is, to some extent, still an open question. Bergmann et al. (1956, 1958) suggested the phenolic hydroxyl of tyrosine but this has been neither established nor refuted. Another early idea, that -SH groups were functionally important has been discounted however (see Mounter and Whittaker 1953).

It is generally held that the differences in the behaviour of AChE and pseudo-ChEs (summarized in table 2.1) are mainly attributable to differences in or around their anionic sites although, as indicated below, minor differences between the esteratic sites may contribute. Zeller and Bissegger (1943; ch. 2) took the view that the high affinity of AChE for cationic substrates, and also the phenomenon of substrate inhibition, indicated that the active centre of the enzyme contained a negatively charged site in addition to the esteratic site. It was subsequently argued that if inhibition of AChE by excess ACh is to be attributed to the presence of an anionic site, then the failure of ACh to cause a similar inhibition of pseudocholinesterase would indicate that this enzyme has no such site (Adams and Whittaker 1950). This question, whether pseudocholinesterase has or has not an anionic site, has proved difficult to answer (see Augustinsson 1963). Some authors continued to favour the original idea, that it possesses only the esteratic site, while others believed that, like AChE, it has an anionic site as well (Župančič 1964a, b). Bergmann (1955, 1958, see also Bergmann and Wurzel 1954) suggested that the behavioural differences between the two types of cholinesterase could be explained if it were assumed that AChE had two anionic and pseudocholinesterase only one. The idea that the difference lay in the number of anionic sites has not been substantiated and it now seems likely that the difference is qualitative rather than quantitative. Augustinsson (1966) examined AChE from the electric organ of *Torpedo marmorata* and BuChE from human serum and compared their behaviour towards carbinol acetates of pyridine and N-methylpyridine. His main findings were these: the introduction of a charge on the pyridine nitrogen atom (quaternization) resulted in an increase in hydrolysis by BuChE but a decrease in hydrolysis by AChE; with ACh as substrate, pyridyl-2- and -3-carbinols inhibited BuChE but activated AChE. BuChE showed the same type of activity-pH relation for tertiary and for quaternary compounds (although the optimal pH was different for the two types), but AChE gave very differently shaped activity-pH curves for the two classes. Augustinsson interpreted this data as evidence for the existence in BuChE of a non-esteratic site in addition to an esteratic site. This second site, although 'anionic' in type, apparently differed from that in AChE with regard to the forces involved in enzyme-substrate binding. Whereas the anionic site in AChE exerts predominantly coulombic attractions, the dominant forces at the non-esteratic site of BuChE are Van der Waals' forces. This would explain the earlier findings (e.g. Adams and Whittaker 1950), which were confirmed in Augustinsson's experiments on pH dependence, that the charge on substrates and inhibitors is of less significance in the reactions of BuChE than of AChE. In these experiments Augustinsson's main object was to establish whether or not an anionic site was present in BuChE but some of his findings – notably the activating effect of pyridyl-2- and -3-carbinols on AChE – also led him to suggest that the esteratic sites were probably different in the two enzymes. Something that must not be overlooked in discussing these results is the question of species differences: AChE and BuChE may well have distinguishable esteratic sites but, on the other hand, the differences which Augustinsson detected could represent a general difference between the type of esteratic site characteristic of *Torpedo* cholinesterases and that characteristic of human cholinesterases.

Kabachnik et al. (1970) have put forward a different idea to account for the dissimilar behaviour of AChE (from bovine erythrocytes) and BuChE (from horse serum) towards various series of organophosphorus inhibitors. They have proposed that both the anionic and the esteratic site may be virtually identical in the two types but structures surrounding or adjacent to these active sites are different and it is these which determine the characteristic behaviour. According to this scheme a hydrophobic area surrounds the anionic site in both enzymes but it exerts a greater effect in BuChE than in AChE. In addition there is a second hydrophobic group near the anionic site which is thought to extend further in AChE than in BuChE. There are also hydrophobic regions in the vicinity of the esteratic site, AChE having one and BuChE, two. Other methods which have been used to study the behaviour of pseudocholinesterases include NMR (Kato 1968, 1969); radiochemistry (Goedde et al. 1968); substrate-specific electrodes (Baum and Ward 1971) and the analysis of free energy relationships (Zimmerman and Goyan 1971). Despite this ongoing work the precise features which are responsible for the differences between pseudoChE and AChE remain undetermined. Since certain enzymes may possess characteristics typical of AChE and of pseudoChE it seems unlikely that the chemical nature of the groups in the active centres will be very different in the various types. More probably the particular activity of each enzyme is a reflection of the specific structural arrangement of the active centre and this conformation may in turn be influenced by the nature of the groups in its vicinity.

10.5 Multiple forms of pseudocholinesterases

10.5.1 Isozymes

The early work of Surgenor and Ellis (1954) suggested that the pseudocholinester-ase activity of human serum was due to a single protein. As electrophoretic and chromatographic procedures became more refined several components with differ-ent mobilities were, however, revealed. Harris et al. (1962), who also give references to earlier work, used two-dimensional filter paper-starch gel electrophoresis in an attempt to define more precisely the various components which had been reported.

Four bands C_1, C_2, C_3 and C_4 were discernible in the region between the α_2- and β-globulins. On starch gel, C_1 moved fastest and C_4, the major component, slowest, but on paper the mobility of C_2 was appreciably greater than that of the other bands. Gel filtration on Sephadex G-200 (Harris and Robson 1963) showed C_4 to have the highest molecular weight, the weights of the other fractions decreasing in the order C_3, C_2 and C_1. Since the hydrolytic properties of the various bands are generally identical they would appear to represent isozymes of pseudocholinesterase (see ch. 2 for definition and general discussion of isozymes). As more species have been studied, it has become clear that isozymes are a characteristic of pseudocholinesterases. They have been described in the horse (Oki et al. 1964, 1965; Reiner et al. 1965) and in chicken, dog, rat and guinea-pig (Vincent et al. 1970b). Different workers have, however, reported different results and it would be premature to make dogmatic assertions about the normal complement of isozymes in a particular species. The number of forms present in an individual appears to be genetically determined and the occurrence of 'extra' isozymes (e.g. C_5) in certain people will be considered in more detail in § § 10.5.2 and 3. Some evidence suggests that the physiological state may also influence the form of pseudocholinesterase which is detectable at a particular moment. Augustinsson (1968) using starch gel found two bands of pseudocholinesterase activity in plasma from a rat during its pregnancy but only one band after parturition.

Technical differences undoubtedly account for some of the divergent results which have been reported. Thus Hess et al. (1963) used starch-gel electrophoresis on rabbit serum and obtained two bands with AcThCh as substrate but no bands at all with BuThCh. In contrast, Vincent et al. (1970b), using polyacrylamide gel, identified seven separate bands with BuThCh. La Motta et al. (1968, 1970) found evidence in human serum that the different isozymes could be interconverted under experimental conditions, and suggested they represented a polymerization sequence. If this is so, preparative methods will influence the number of bands obtained and it may not be easy to establish which bands represent genuine, naturally occurring isozymes and which are an experimental artifact (see Boutin and Brodeur 1971). Juul (1968), for example, demonstrated 12 bands in human serum whereas Das and Liddell (1970) using the same electrophoretic method, but a different staining technique, obtained only four.

Isozymes have been reported in tissues as well as in serum. Svensmark (1965) gives references to his earlier work in which he differentiated two enzymically identical but electrophoretically dissimilar components of pseudocholinesterase from human brain. One fraction behaved like the serum enzyme, migrating in the α_2-β-globulin range, but the other travelled more slowly with the γ-globulins and thus resembled the sialic acid-free moiety which results from treatment of the serum enzyme with neuraminidase (§ 10.4). Since brain tissue contains neuraminidase the difficulty is to decide whether the slower component is a normal constituent of intact tissue or whether it is an artifact produced during processing.

In human liver Svensmark (1963b) found components with three different mobilities. One was identical with the pseudocholinesterase of serum but the other two lacked sialic acid; Svensmark suggested these might be precursors of the serum enzyme. Holmes and Masters (1967) made an extensive survey of isozymes of a number of esterases in various tissues of the guinea-pig but, as they point out, the method used was probably suitable for showing only those components with a molecular weight of less than 300,000. Although they demonstrated 5 different bands of pseudocholinesterase from liver they obtained no activity at all from serum. This result supports, to some extent, Svensmark's suggestion that the low molecular weight components of liver cholinesterase are precursors of the heavier serum enzyme.

The existence of isozymes complicates any consideration of the exact molecular weight of serum or tissue pseudocholinesterases. In 1954 Surgenor and Ellis gave a figure of the order of 300,000 for human serum pseudocholinesterase, based on the sedimentation coefficient obtained in the analytical ultracentrifuge. From studies on the reaction between DFP and horse serum pseudoChE, Jansz and Cohen (1962) calculated that the molecular weight per active site could not exceed 84,000 but they pointed out that the sedimentation coefficient indicated a much larger molecule. This suggests (see Svensmark 1965) that there may be 2 to 4 active sites per molecule as proposed for AChE (ch. 2). Using a method which resulted in a 13,000-fold purification with a 54% yield, Das and Liddell (1970) obtained a single protein from human serum pseudocholinesterase. On the basis of polyacrylamide-gel electrophoresis this appeared to correspond to the C_4 component and had a molecular weight of 365,000. La Motta et al. (1970) determined the molecular weight of 5 isozymes of human serum cholinesterase and in contrast to Das and Liddell give a figure of only 200,000 for C_4. Values for the other isozymes were: C_1, 82,000; C_2, 110,000; C_3, 170,000; and C_5, 260,000. They point out that the difference in molecular weight between successive pairs approximately follows the progression n, $2n$, n, $2n$ where $n = 30,000$. Since they believe the forms to be interconvertible (La Motta et al. 1965) they suggest that the different components represent different degrees of aggregation of subunits. Boutin and Brodeur (1971) were similarly of the opinion that interconversions occur in vitro but recognized the possibility that they might not happen in vivo. They found one fraction in human serum pseudocholinesterase with a molecular weight of about 86,000 and another, possibly a tetramer, with a value of 348,000. Tucci and Seifter (1969) purified the BuChE from pig parotid gland. It behaved in the analytical centrifuge and on starch gel as if it were a single entity. The figure for the molecular weight, 368,000, is close to Das and Liddell's value of 365,000 for the C_4 component of human serum.

It is by no means generally accepted that the interconversion or aggregation of common subunits could explain the existence of multiple forms of pseudocholinesterases. For instance, Harris et al. (1963a, see §10.5.2) consider that the C_5 isozyme of human serum has no counterpart in those who lack it. This suggests that

C_5 is not simply a polymerized form of a common unit. Holmes and Masters (1967), working with tissue isozymes from guinea-pig, were similarly against the idea that polymerization could account for the multiplicity of forms found (§10.5.2).

10.5.2 Atypical variants

The different molecular forms of pseudoChE discussed so far are differentiated by their electrophoretic characteristics and their behaviour on columns, but not by their enzymic properties. Other variants exist which do differ in their activity towards substrates and inhibitors and an 'abnormal' form and the normal form can occur together in one individual (see Lehmann et al. 1963 for a summary of possible combinations). The so-called 'atypical' or 'Dibucaine-resistant' variant is unable to hydrolyse a number of substrates, including succinylcholine, at the normal rate and is unusually resistant to cholinesterase inhibitors. Kalow and Genest (1957) expressed this resistance in terms of the Dibucaine number which is the percentage inhibition of the enzyme produced by Dibucaine (a local anaesthetic) under certain standard conditions. Harris and Whittaker (1961) used sodium fluoride as an inhibitor in a similar way and showed that the Dibucaine-resistant variant had a low fluoride number. In addition, they distinguished another variant which had a low fluoride number but a Dibucaine number which was only slightly lower than normal; this they termed the 'fluoride-resistant' variant. Its activity towards succinylcholine was greater than that of the Dibucaine-resistant type but was still below normal. A third variant was reported by Liddell et al. (1962). They described a patient whose serum was apparently devoid of pseudocholinesterases and the gene responsible for the condition was therefore dubbed the 'silent gene'. It has subsequently been found that 'silent gene' serum does contain a protein which is very similar to normal pseudocholinesterase and under suitable experimental conditions enzymic activity can be detected (see Goedde et al. 1965). The four behaviourally different pseudocholinesterases which have been recognized are, 1) normal or usual, 2) atypical or Dibucaine-resistant, 3) fluoride-resistant, 4) silent-gene type. An automated method which simultaneously measures total pseudocholinesterase and Dibucaine and fluoride numbers (Boutin and Brodeur 1970) is valuable for detecting the different types.

The atypical variant occurs in combination with the normal type in about 4% of European and North American populations (see §10.3.3.2) and it has been studied in some detail. It is more sensitive than the normal form to inhibition by alkyl alcohols (Whittaker 1968) and by formaldehyde (Whittaker 1969; Whittaker and Hardisty 1969). There is some evidence that the variants may also differ in their electrophoretic and chromatographic characteristics (see Svensmark 1965) but the differences are probably very subtle and whether they are revealed may depend on the method used for their detection. Brody et al. (1965) noticed abnormalities in 'atypical' human serum run on starch gel whereas Svensmark (1963a) using paper electrophoresis found that both native and neuraminidase-treated preparations of atypical enzyme had mobilities identical with those of similar preparations of normal enzyme. Beckett et al. (1968) suggested that atypical enzyme arises as a result of changes in individual amino acids, or in their sequence, which distort the active centre.

10.5.3 Genetic studies

The existence of the 'atypical' variant of human serum pseudocholinesterase was first recognized when succinylcholine (suxamethonium) produced alarmingly prolonged apnoea in certain patients to whom it was given as a relaxant in surgery (for references see Svensmark 1965; Goedde et al. 1967). It was found that these patients had a low or undetectable level of serum pseudocholinesterase and this suggested that the liver might be diseased (§ 10.10). However, in those patients who were examined, the liver function tests were apparently normal. Although these results seemed to exonerate the liver as the cause of the condition, Forbat et al. (1953) thought it worth examining the brother of a succinylcholine-sensitive patient on the grounds that faulty liver function was sometimes due to inherited metabolic errors. The discovery that the serum pseudocholinesterase activity of the brother was even lower than that of the patient provided evidence of a genetic basis for the abnormality.

The genetics of pseudocholinesterases are outside the scope of this book but a few general points can be made. It was originally thought that the polymorphism of pseudocholinesterase, as indicated by the usual and atypical variants, could be accounted for on the basis of only two allelic genes (see glossary) but as more variants were recognized Lehmann et al. (1963) suggested that at least 4 alleles must be involved. These, they proposed were the 'normal' gene, the 'atypical' or 'Dibucaine-resistant' gene, the 'fluoride-resistant' gene and the 'silent' gene. Some more recent work (Rubenstein et al. 1970) indicates that there may be a family of genes of the silent type.

Many of the studies on the familial occurrence of the variants (for references see Goedde et al. 1967) indicated a Mendelian pattern of inheritance based, as suggested above, on 4 alleles belonging to one locus. However, since such a scheme with a single locus was inadequate to account for the inheritance of additional isozymes such as the C_5 isozyme, Harris et al. (1963b) proposed that a second locus is involved (see also Simpson 1966). The independent occurrence of the C_5 isozyme and the atypical, Dibucaine-resistant variant was well illustrated by Harris et al. (1963a) in their study on British and Tristan de Cunhan subjects. 4% of the British population examined possessed the Dibucaine-resistant variant but in the 5% who were C_5-positive the Dibucaine number was normal. Conversely, none of the Tristan de Cunhans had the atypical variant but 17% were C_5-positive.

The field of pseudocholinesterase inheritance has been widely explored — literally — by geneticists and ethnologists and the list of references to this work reads like some sort of travel brochure. Subjects investigated include Australian aborigines, Berbers, Britons, Canadians, Czechoslovakians, Eskimos, Germans, Greeks, Red Indians, Mexican Indians, Moroccan Jews, Pakistanis and Portuguese (for references see Kattamis et al. 1962; Horsfall et al. 1963; Goedde et al. 1967; Neumann and Walter 1968). In most cases the search has been for the atypical, Dibucaine-resistant type. Heterozygotes possessing both the normal and atypical variant can be detected on the basis of Dibucaine and fluoride numbers. The frequency with which such heterozygotes occur is surprisingly constant, at about 4%, in many of the Caucasian populations and in the Australian aborigines. The atypical variant was absent from the population of Tristan de Cunha when examined by Harris et al. (1963a) and is probably absent or rare in certain of the Eskimo and Red Indian tribes (Gutsch et al. 1967), in Icelanders (Neumann and Walter 1968), in Negroes, both American and African (Motulsky and Morrow 1968) and in Japanese and other oriental populations (Omoto and Goedde 1965; Motulsky and Morrow 1968).

Far fewer populations have been scanned for the possession of abnormal numbers of electrophoretically detectable isozymes. In addition to the usual complement of the bands C_1, C_2, C_3 and C_4, a fifth band designated C_5 was recognized by Harris et al. (1962), working in England, in 14 out of 300 randomly selected adults. The frequency with which this component occurs in related subjects is greater than in the population as a whole; this indicates a genetic factor. As mentioned earlier, C_5 has been reported in some of the inhabitants of Tristan de Cunha (Harris et al. 1963a) and from the pedigrees of those who were C_5-positive it is possible to pinpoint with fair certainty which pair of early settlers could have introduced the relevant gene. Ashton and Simpson (1966) found the C_5 isozyme in Brazilians and still more bands, C_6, C_{7a} and C_{7b} were present in serum from certain Africans (Van Ros and Druet 1966). Harris et al. (1963a) found that in C_5-positive subjects, the level of serum pseudoChE was 30% higher than in C_5-negative subjects and they suggested that the C_5 band represents an extra component which has no homologue in people in whom it is lacking.

Almost all the genetic studies have been done on serum pseudocholinesterase but Liddell et al. (1963) investigated tissues, including the brain, from a patient with atypical serum enzyme and found that the tissue enzyme was similarly atypical. Reference was made earlier to the work of Holmes and Masters (1967) on tissue cholinesterases in the guinea-pig. These authors believed only one gene to be responsible for the synthesis of the five isozymes that they detected. They suggested that the variation between forms is due to the attachment of small molecules and rejected the possibility (cf. La Motta et al. 1965, 1968, 1970) that polymerization could account for the experimentally demonstrable bands of activity.

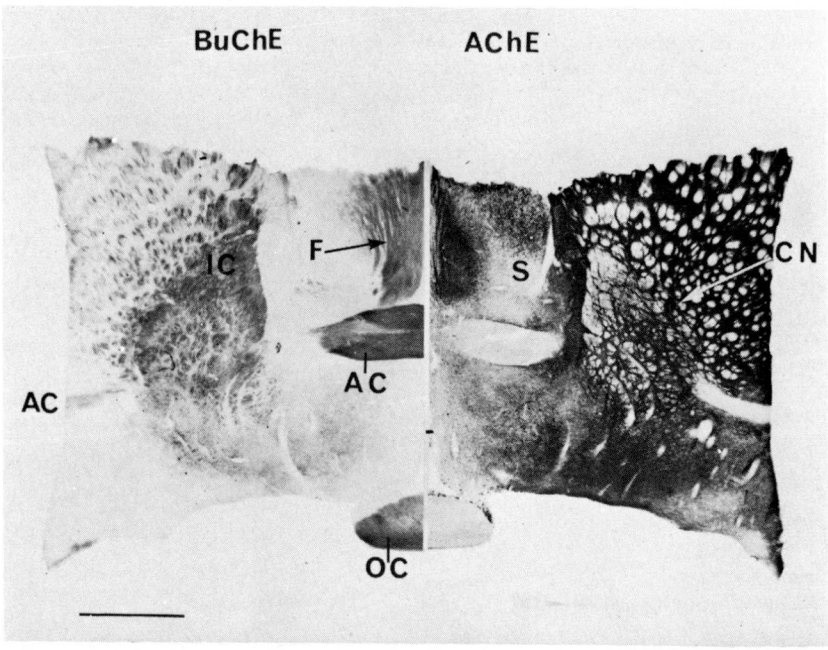

Plate 10.1. To demonstrate the presence of BuChE in brain tissue and to show how its distribution differs from that of AChE. Transverse sections of guinea-pig hypothalamus. AC, anterior commissure; CN, caudate nucleus; F, fornix; IC, internal capsule; OC, optic chiasma; S, septum. Scale bar: 1 mm.

10.6 Distribution in nervous tissue

10.6.1 CNS

At one time it was thought that central nervous tissue contained only acetylcholinesterase (Mendel and Rudney 1943a, b; Nachmansohn and Rothenberg 1944), but subsequent evidence showed that pseudocholinesterase was present in appreciable amounts (see plate 10.1). Burgen and Chipman (1951) measured the activity of AChE (ChEI) and pseudoChE (ChEII) in the nervous system of dog, using MeCh and BzCh as substrates, and found that in areas composed predominantly of white matter the level of pseudocholinesterase activity exceeded that of AChE (see table 6.9.1). The following year, Ord and Thompson (1952) reported a similar situation in the subcortical white matter of man and a number of other species. They used ACh, MeCh and BuCh as substrates and showed clearly that although the greatest amounts of pseudocholinesterase are generally to be found in white matter appreciable activity is present in grey matter as well. As table 10.2 indicates there are

TABLE 10.2

Distribution of cholinesterases in the grey and white matter of the cerebrum of different mammals. (After Ord and Thompson 1952.)

Enzyme source	Absolute activity (μl CO_2/g/hr)			Activity as percentage of ACh activity		
	ACh	BuCh	MeCh	BuCh	MeCh	Ratio: MeCh/BuCh
Cortical grey matter						
Human	476	204	260	43	55	1.3
Rabbit	8340	1981	6710	24	81	3.4
Rat	6100	1935	4555	32	75	2.4
Guinea-pig	4370	1009	2720	23	62	2.7
Cat	2340	117	2295	5	98	20
Dog	1799	1173	1063	65	59	0.9
Subcortical white matter						
Human						
Pre- and post-central gyri	275	594	82	216	30	0.13
Frontal lobe	523	1024	64	196	12	0.06
Occipital lobe	319	–	53	–	17	–
Rabbit	7790	9084	2387	117	31	0.3
Rat	6525	3257	2389	50	37	0.7
Guinea-pig	7145	18925	1564	266	22	0.1
Cat	2350	4550	549	194	23	0.1
Dog	328	595	151	181	46	0.3

wide species differences, however. In the grey matter the absolute value for pseudoChE is low in the cat and high in the dog. In the white matter the situation is reversed but even so, in agreement with the observations of Burgen and Chipman (1951), the value for pseudocholinesterase in dog white matter still exceeds the value for AChE. Ord and Thompson, speculating on the possible function of pseudoChE in white matter, suggested its involvement in processes occurring in myelin rather than in axons and pointed out that Sawyer (1946) had demonstrated the survival of pseudoChE in sectioned peripheral nerves in which the axons had degenerated.

With the development of histochemical techniques (see §10.6.3 re electron microscopy) the view that pseudocholinesterase was associated with supportive elements and not with axons became strengthened. The general consensus was that in the CNS the enzyme was localized in glia and associated structures such as the septa in the optic nerve (Hebb et al. 1953). Because structures which are stained for pseudoChE are seldom as sharply delineated as those stained for AChE, it has proved difficult to establish exactly which type of glial cell reacts for pseudoChE. Koelle (1954) identified the majority of stained cells in the rat brain as fibrous astrocytes. Brightman and Albers (1959) made a similar identification in cat and chicken but in sheep, monkey and man the staining was weak and could not be attributed to a particular structure. More recently Friede (1967) has claimed that the majority of stained cells are not astroglia but oligodendroglia. He based his identification on the shape and arrangement of the cells and on the fact that they were too numerous to be astrocytes. Friede has also suggested, contrary to orthodox theory, that in certain tracts pseudoChE activity occurs in axons and not in the associated glia. This will be discussed again below but certain points must be made here. First, most histochemists would agree that sections stained for pseudoChE rarely present the crisp, well-defined appearance of sections stained for AChE. In very many reports the stain is described as 'diffuse', 'mottled' or 'patchy'. The strikingly clear staining of individual glial cells, as demonstrated in the cat by Brightman and Albers (1959) or of glial processes (see Friede and Fleming 1964 for squirrel monkey, and Silver 1967 for sheep) is an exception rather than a general finding. Friede's results were obtained from tissues which were incubated for 18 hr at 38°C at a pH of 6.6. Under these conditions localization might be inaccurate, particularly since pseudoChE, unlike AChE, is a readily soluble enzyme. Friede's finding, that in tracts in which the axons were stained the glia were unstained, might reflect some diffusion phenomenon.

Despite these grounds for questioning some of Friede's interpretations, there is good evidence to support his assertion that pseudoChE is not confined to astroglia. Relatively high levels of pseudocholinesterase have been detected biochemically in gliomas and this is true of oligodendrogliomas as well as astrocytomas (Bülbring et al. 1953b; Cavanagh et al. 1954; Wollemann and Zoltan 1962). On the basis of histochemical experiments Roessmann and Friede (1966) proposed that the oligodendroglia are the primary site of pseudocholinesterase activity in the cat, only minimal activity being found in the astrocytes of normal tissue. Following injury the astrocytic staining increased; in contrast the reaction in the oligodendroglia disappeared. Because of this and because the activity in the astroglia was less susceptible to inhibition by iso-OMPA than was that in oligodendroglia, they postulated that the cholinesterase in the two types of cell could be different. It is relevant to mention that Koelle (1955) had earlier reported the presence, in the CNS and ganglia of various species, of a glial enzyme which hydrolysed AcThCh and BuThCh but which was not apparently a cholinesterase since it was resistant to selective inhibitors.

According to Roessmann and Friede (1966) in neither man nor cat do the glial cells of the cortex (in contrast to those in white matter) react for pseudoChE. This might explain why Rosenberg and Echlin (1965) did not detect a change in the BuChE activity of chronic, partially isolated slabs of cortex despite the marked gliosis which occurs in such slabs (see Krnjević et al. 1970). To some extent the histochemical findings in rat CNS appear confusing. Brightman and Albers (1959) did not obtain any staining of rat glial cells although their method (incubation at 38°C for 1–1½ hr, pH 6.0–6.4) revealed activity in the glia of cat and other species and, moreover, it produced a marked reaction in some elements, thought to be Schwann cells, in rat spinal roots. Navaratnam and Lewis (1970) who incubated their sections at pH 5.0 for 14 hr at room temperature were similarly unable to detect activity in either the white matter or the neuropil of the grey matter of rat spinal cord. On the other hand, Friede (1967) working at a very high pH (6.6), a temperature of 38°C and with an incubation of 18 hr, reported staining of glial cells in many areas of rat CNS including the alveus of Ammon's horn (cf. negative results of Storm-Mathisen and Blackstad 1964). The conditions used by Friede are not ideal for ensuring an accurate localization of end-product nor can one be certain that the staining seen after so long an incubation at so high a pH is a result of enzyme activity. However, the demonstration of a histochemical reaction is in keeping with biochemical findings. As table 10.2 shows, Ord and Thompson (1952) detected BuChE in both white and grey matter of the rat cerebrum. This suggests that negative histochemical findings could reflect the insensitivity of the method, a result, perhaps, of some diffusion barrier imposed by the glial cell membrane. There may, however, be another explanation for the biochemical findings. Since the endothelial cells of the blood vessels in rat CNS are rich in pseudocholinesterase (Brightman and Albers 1959; Joó and Várkonyi 1969; Navaratnam and Lewis 1970; Flumerfelt et al. 1973), these, rather than the glial cells, could be responsible for the greater part of the biochemically measurable activity in the cerebrum.

In addition to the enzyme in extraneuronal elements, be they astroglia, oligo-dendroglia or endothelial cells in blood vessels, BuChE is also to be found in certain neurones of various species. These are indicated in some of the tables in ch. 6. Shute and Lewis (1963) found a number of nuclear groups in rat brain which contained both AChE and BuChE. In most, AChE predominated but in the anterodorsal nucleus of the thalamus and also in the dorsal vagal motor nucleus the histochemical reaction of the cells for BuChE exceeded that for AChE (see also Lewis and Shute 1967). In the anteromedial nucleus of the thalamus BuChE staining is related to the neuropil and not to cells. Friede (1967) confirmed the staining in the rat dorsal vagal motor nucleus and also found scattered cells showing a strong reaction in the nucleus of the spinal trigeminal tract and in the reticular formation. In guinea-pig, we (Cottle and Silver 1970b) occasionally saw quite strongly stained cells close to the IIIrd ventricle at the level of the caudal end of the paraventricular nucleus. Cholinesterases in the reticular formation of cat and rabbit have been described by Papp and Bozsik (1966). In the rabbit, BuChE activity was weak and could not be related to particular structures but in the cat moderate staining occurred in the neuropil of most nuclear groups and, in some of these, BuChE was also present in neurones. Utley (1966) measured BuChE in the cat medial genicu-late body. Because the activity per mg tissue increased after cortical lesions or after lesions of the brachium of the inferior colliculus he concluded that the greater part of the activity was in non-neuronal elements. The increase in specific activity following degeneration was, however, lower than would be expected if the enzyme were entirely confined to such elements, hence Utley deduced that some activity was attributable to neurones. Abrahams (1963a; Abrahams and Edery 1964) also working on the cat, described BuChE-containing neurones in a number of nuclei in the brain stem. These include the lateral nucleus of the substantia nigra, the IIIrd nerve nucleus, red nucleus and various nuclei in the thalamus. Friede (1967) showed strong BuChE activity in cells of the inferior olive and the facial and lateral reticular nuclei but it is not always clear from his description of other nuclei whether activity is in cells or neuropil (this also applies to his account of man, monkey and rat). In cat spinal cord we (Silver and Wolstencroft 1971) found a few BuChE-positive cells in the medial part of the intermediate and dorsal horns (see plate 10.2). They seemed more numerous in kittens than in adult cats but it is possible that in the latter the stronger activity in the neuropil masked them. In rat spinal cord a large number of neurones in the sacral intermediolateral nucleus contain BuChE as well as AChE but at thoraco-lumbar levels the cells contain only AChE (Navaratnam and Lewis 1970). In addition, small cells ($15-25~\mu$ in diameter) present in the centre of the ventral horn react for BuChE at all levels. Navaratnam and Lewis very tentatively suggested that the latter cells might be Renshaw cells (see ch. 7 §3.6).

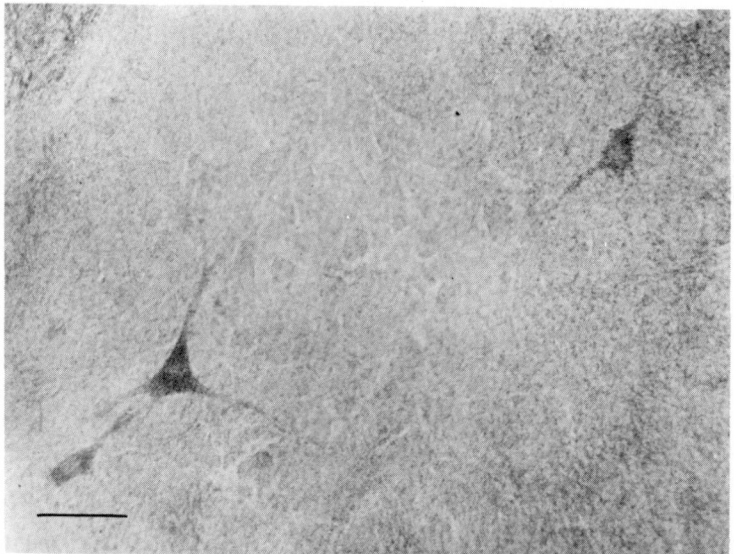

Plate 10.2. BuChE activity in cells in the dorsal horn of the spinal cord of the cat. Scale bar: $50~\mu$. (From Silver and Wolstencroft 1971.)

Klingman et al. (1968) measured AChE and BuChE activity in various sympathetic ganglia of the rat and found that about 30–40% of the total cholinesterase activity was due to BuChE; absolute values for total enzyme varied between different types of ganglia and also in the same type, from rat to rat (see table 6.27.2). A few comparative results suggested that levels of pseudoChE in the superior cervical ganglion and nerve trunk of the cat are lower than in the rat. Histochemical investigations have shown that the majority of neurones in peripheral ganglia do not contain detectable BuChE but Koelle (1954) found occasional stained cells in the rat superior cervical ganglion (see also Cauna et al. 1961) and in 1955 he reported strong BuChE activity in neurones in Auerbach's plexus of cat, rabbit and monkey. Similar reports of BuChE in certain other ganglion cells have appeared subsequently (e.g. Mann 1971). Activity in the satellite cells of the ganglia varies in different types (see tables in ch. 6). For instance, Naik and Cauna (1971) have shown that in man the satellite cells in sympathetic ganglia react less strongly than those in parasympathetic ganglia. In sensory ganglia BuChE activity is relatively uncommon; Cauna and Naik (1963) examined 5 species and found BuChE-positive satellite cells only in the cat. The nerve cells were unstained except in the guinea-pig and in the human foetus (tissue from adult man was negative); and only in the cat and the human foetus was there fibre staining.

10.6.2 Distribution in peripheral nerves

It has long been accepted that the pseudocholinesterase associated with peripheral nerve fibres is present in Schwann cells, the main evidence being that when a nerve is sectioned enzyme activity persists in the isolated segment despite degeneration of the axon (see Sawyer 1946). Histochemical findings support the view that where BuChE is present in a peripheral nerve it is associated with the Schwann cells rather than the axons but what is not clear is whether BuChE is part of the enzymic load of every Schwann cell. The various tables in ch. 6 which list cholinesterases in peripheral nerves suggest there are regional and species differences. Regional variation is well illustrated in the case of the rabbit. Hebb and Linzell (1970) concluded that in the mammary gland all types of fibre contained BuChE and they suggested that BuChE-histochemistry could be used in the rabbit as a general nerve stain. In contrast, Mann (1971) showed that in the bronchial muscle BuChE-containing fibres were sparse compared with fibres which stained for AChE. In the rat, Csillik and Koelle (1965) demonstrated BuChE in the Schwann cells of fibres in the iris but Eränkö et al. (1970) found that in the pineal the fine fibres within the gland reacted for BuChE only after nerve section.

The observation made by Eränkö et al. exemplifies the possibility, discussed in ch. 9 §6.2, that Schwann cells in histochemically negative nerves may be free of BuChE until synthesis is initiated by some process associated with nerve section. Alternatively, each Schwann cell might contain a small amount of enzyme which becomes histochemically detectable as the number of Schwann cells increases in the degenerating fibre. Eränkö and Teräväinen (1967) found that BuChE was demonstrable in the Schwann cells of the degenerating rat sciatic nerve on the 4th day after nerve section. The activity reached a maximum about the 10th day and remained elevated for at least 8 months (studies were not continued longer). The latency of 4 days is very similar to that for the proliferation of nuclei (the majority of these represent Schwann cells) in the degenerating tibial and peroneal nerves of the rabbit (Abercrombie and Johnson 1946): no increase occurred during the first 3 post-operative days but thereafter numbers rose in a sigmoid fashion reaching a peak at 25 days. At 225 days (i.e. nearly 8 months) the number of nuclei approximately equalled that in a nerve 12 days after section. Eränkö and Teräväinen (1967) noted that when the rat sciatic nerve was crushed rather than sectioned the Schwann cells in the degenerating part showed the same initial increase in BuChE activity but as the fibres regenerated the reaction declined and was eventually lost. The lack of demonstrable activity in Schwann cells of intact and regenerated sciatic nerves is a further example of regional differences in histochemically detectable BuChE since rat spinal roots stain particularly strongly (Brightman and Albers 1959). Using biochemical methods, Cavanagh et al. (1954; see also Cavanagh and Webster 1955) analysed pseudoChE activity in degenerating chicken sciatic nerve and reported somewhat different results. First, BuChE was detectable in intact nerves; secondly, the increase induced by a lesion started before there was any appreciable proliferation of cell nuclei and the peak was reached by 7–9 days; thirdly, normal values were re-established by 20 days. This time-course suggests that it is a transient stimulation of enzyme synthesis rather than a mere increase in the number of Schwann cells which produces the rise in activity in the chicken. The difference between these results for chicken and those of Eränkö and Teräväinen for rat probably reflects considerable species variation but the greater sensitivity of the biochemical as opposed to the histochemical method is also an important factor. That intact mammalian nerves do contain biochemically detectable pseudocholinesterase was demonstrated by Boell (1945) and by Sawyer (1946) in the sciatic nerves of rat and guinea-pig respectively. Sawyer's conclusion that the activity was unchanged in degenerating nerve was later questioned by Cavanagh et al. (1954). They pointed out that since no allowance was made for the change in water content of the sectioned nerve, any alteration in pseudocholinesterase activity might have been masked. Because of concomitant changes of this sort particular care is needed in the expression of results from work on degenerating nerve.

10.6.3 Electron microscope histochemistry of central and peripheral nervous tissue

The behaviour of BuChE in sectioned nerves is probably best studied by a combination of biochemistry with histochemistry, at both the optical and electron microscope level. So far this latter aspect has, however, been largely neglected not only in peripheral nerves but in the CNS as well. Many workers are interested only in 'cholinergic' mechanisms and in looking for AChE activity they necessarily inhibit BuChE. Another reason for the dearth of information is that much of the electron microscope histochemistry is done on rat, the CNS of which is not so well endowed with pseudoChE as that of some other species. Torack and Barrnett (1962) reported BuChE in astroglia in rat medulla but their thiolacetic acid method is now considered to lack specificity. Using a more specific technique, Shute and Lewis (1966a) failed to find any reaction for BuChE in glia of the rat hippocampus but positive reactions have been reported for some neurones, both central and peripheral. Among nerve cells in which the RER gives a reaction for BuChE are those of the intermediolateral nucleus of the spinal cord (Navaratnam and Lewis 1970) and those in the adrenal medulla (Palkama 1967). Schlaepfer (1968) found that in rat spinal ganglia the neurones lacked BuChE activity but the reticulum of the satellite cells was positive. A similar reaction in the reticulum, and also in the nuclear envelope, was observed in Schwann cells in the rat adrenal medulla by Lewis and Shute (1969). More recently Mazza et al. (1973) examined the trigeminal ganglion and showed that in addition to BuChE activity in the RER, mitochondria and

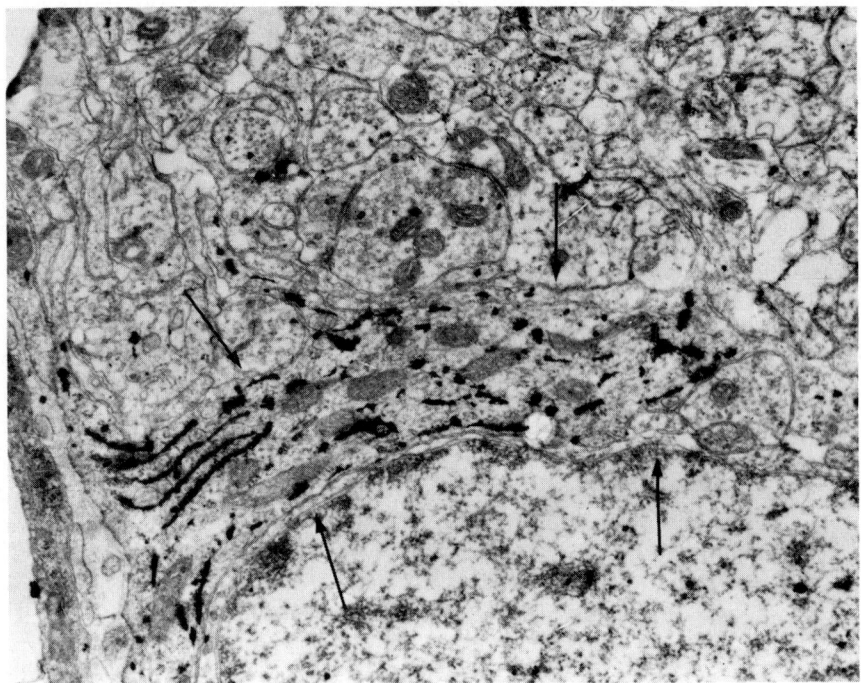

Plate 10.3. BuChE activity in the endoplasmic reticulum of glial cell (arrowed) in the molecular layer of the cat cerebellar cortex. (× 19,200) (P. Kása unpublished.)

nuclear envelope of the satellite cells, end-product was also associated with neuronal mitochondria (see ch. 4 §2.3). As indicated in ch. 6, ultrastructural studies of BuChE in species other than rat are rare. In the cat, Kása (personal communication) demonstrated BuChE activity in the ER of glia in the cerebellum (plate 10.3) while in guinea-pig cerebellum, Kása and Csillik (1966) found well-defined end-product between the lamellae of a myelinated fibre (plate 10.4). In both cases the lead-copper-thiocholine technique was used. With the Karnovsky method, Robinson (1969) obtained a slight reaction for BuChE on axonal and Schwann cell membranes of both adrenergic and cholinergic fibres innervating the longitudinal muscle of guinea-pig vas deferens.

10.7 Distribution in non-nervous tissue

Pseudocholinesterases can be found in almost all the major systems of the mammalian body but there are considerable species differences in the level of activity and in its precise distribution. No attempt will be made to give a detailed coverage of the subject, the main object being to provide a background against which the questions of synthesis and function can be examined. A more comprehensive account has been given by Gerebtzoff (1959).

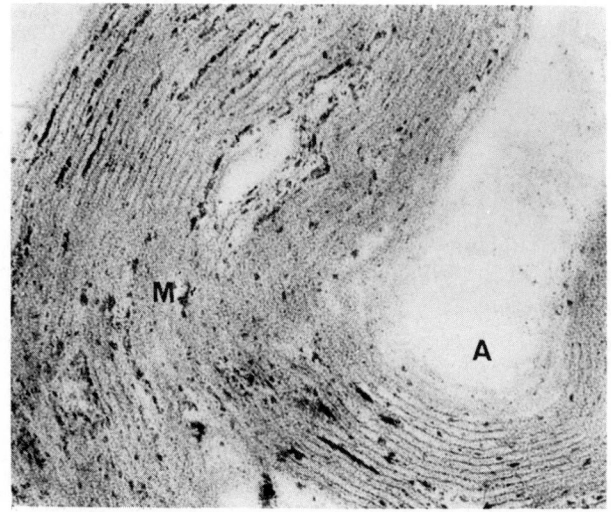

Plate 10.4. BuChE activity between the lamellae of a myelinated fibre in the white matter of the guinea-pig cerebellum. A, axon; M, myelin. (× 30,000) (From Kása and Csillik 1966.)

10.7.1 Vascular system

Pseudocholinesterase in the heart has been measured biochemically in a number of species and, in general, the concentration in the auricles is higher than that in the ventricles (Ord and Thompson 1951; Vlk and Tuček 1962a). Hegab and Ferrans (1966) confirmed these results histochemically in rat and noted particularly strong staining in the auriculoventricular bundle.

The carotid body contains considerably more pseudoChE than AChE (Hollinshead and Sawyer 1945) but the precise localization of both enzymes has been in dispute. Koelle (1951), in the cat, and Rogers (1965) in the rat, identified the histochemically-stained elements as chemoreceptor cells, but Biscoe and I concluded (Biscoe and Silver 1966) that, in the cat, the stain which appeared to be associated with the cells was attributable to nerve plexuses. In addition to fibre staining and staining in occasional ganglion cells, we observed marked activity in blood vessel walls. At all these sites AChE was present with BuChE.

The occurrence of cholinesterases in the endothelium and smooth muscle of blood vessels was discussed in ch. 8 and the point was made that one of the features which makes the problem of function particularly difficult to solve is the marked species difference both in the distribution of vessels which react for ChE and in the nature of the enzyme (see tables 8.4 and 5). Whereas vessels in the rat spinal cord stain strongly for BuChE (Navaratnam and Lewis 1970) those in the cat cord are totally unreactive (Silver and Wolstencroft 1971). Some of the vessels which contain BuChE contain AChE as well; examples include those in the goldfish CNS (Brightman and Albers 1959) and in the cat carotid body mentioned above. Yet other vessels stain for AChE but not for BuChE; for example, in rabbit cerebellum (Crook 1963).

10.7.2 Respiratory system

Using histochemistry Gerebtzoff (1959) studied ChE in the respiratory system of a variety of animals including cat, rat and rabbit. Staining patterns were similar in all species except the rabbit which showed no activity toward BuThCh; subsequently, however, Mann (1971) has found some BuChE in rabbit bronchial muscle. BuChE-containing structures reported by Gerebtzoff include mucous glands, some goblet cells and their secretions, and some of the smooth muscle of the bronchi and bronchioles. Mann (1971) found particularly marked activity in smooth muscle of the sheep bronchus and in associated ganglia.

10.7.3 Digestive system and associated structures

Augustinsson (1948) showed biochemically that salivary glands from various species were well endowed with BuChE although those of the cow and rabbit contained

AChE instead. Histochemical studies in the guinea-pig (Gerebtzoff 1959) indicate that the enzyme is present in both mucous and serous glands; in the latter it is associated mainly with zymogen granules. Snell (1959) noted appreciable hydrolysis of both AcThCh and BuThCh in the secretory cells of the rat parotid gland; in the submaxillary and sublingual glands, however, the cytoplasm was non-reactive (Snell 1958b).

In the alimentary tract there is a noticeable variation in enzyme activity from species to species and in similar structures at different levels of the tract in the same individual. Gerebtzoff (1959) gives a neat summary of the complex situation and this is quoted here with minor alterations. 'Beside localizations of cholinesterases in nerve fibres and in smooth muscle, the enzymes are present in epithelial and glandular cells: chief cells [peptic or zymogenic] of gastric glands of the guinea-pig, absorbing and goblet cells of small and large intestine of the rat, goblet cells of duodenum and a few goblet cells of large intestine of guinea-pig.

The enzyme is ChE [pseudocholinesterase] except in the rabbit (and the cat: Koelle et al. 1950). Species differences concern not only localization, repartition and type of enzyme, but its ultimate destiny: in guinea-pig, ChE passes into the mucus of goblet cells and is finally excreted; in the rat, mucus gives a negative reaction and the enzyme remains in the cytoplasm of goblet cells.

Two types of activity gradient are present in gastro-intestinal mucosa. The oro-anal gradient of ChE activity goes as follows in the guinea-pig:

stomach < duodenum > colon > rectum.

In the rat the series is:

duodenum > jejunum > ileum > colon ≥ rectum.'

Other studies on gut cholinesterases include those of Burn et al. (1952); Donhoffer (1959) and Ambache et al. (1971).

The liver is rich in pseudocholinesterase but, again, the exact cellular localization, whether in histiocytes, parenchymal cells or both, varies from species to species. Furthermore, in any one animal the distribution within the lobules may depend on whether it is well-fed or fasting (see Gerebtzoff 1959). It has been suggested that the pseudocholinesterase in the guinea-pig liver is a benzoylcholinesterase (and as such, is not contributed to the serum); in the rabbit the liver enzyme is predominantly AChE.

The pancreas is another organ with a particularly high content of pseudoChE, but only in certain species. In dog, Hebb and Hill (1955b) demonstrated the enzyme in the Islets of Langerhans and in the acinar cells and pancreatic secretion. In contrast, in rabbit, horse, sheep and goat the glandular tissue was inactive, while in the cat BuChE was confined to the Pacinian corpuscles (see also Hebb and Hill 1955a). Coupland (1958) also reported pseudoChE in cat Pacinian corpuscles and confirmed that the glandular tissue in this species (and also in rat) was inactive. In some other respects his results differed from those of Hebb and Hill, possibly because he used fixed tissue. In cat, rabbit and rat BuChE was present in smooth muscle cells of ducts and arteries and in the rabbit it occurred, together with AChE, in the Islets of Langerhans; acinar cells were negative. This pattern of distribution in the rabbit is the reverse of that in the guinea-pig: Islets are negative but the cyto-plasm and zymogen granules of the acinar cells are strongly positive for pseudoChE (see Gerebtzoff 1959). The cholinesterase activity is destroyed by enterokinase when the granules are secreted into the gut (Goutier and Goutier-Pirotte 1955). According to Augustinsson (1959c) the pancreas is the source of serum cholin-esterase in the pig; the enzyme has an atypical specificity pattern (§ 10.3.1).

10.7.4 Urogenital system

BuChE is present in parts of the kidney and this has been invoked in support of theories that the enzyme is concerned with ion transport (§ 10.9; see also ch. 8). For this reason some details of its distribution may be of interest. Pseudocholin-esterase activity is apparently affected by the degree of hydration of the animals and this, together with technical differences, may account for discrepancies in the literature. Marx and Carter (1963) examined mammalian kidneys manometrically, and also histochemically with the Koelle technique. Although they obtained bio-chemical evidence of pseudoChE in all the species they studied (rat, mouse, dog, cat, rabbit and guinea-pig), only in rat and dog was the enzyme demonstrated unequivocally by histochemistry. Staining of the glomeruli, nerves and vessels was inhibited by 10^{-5} M eserine but that in the tubules was not. Fourman (1966a) using Coupland and Holmes's histochemical method obtained rather different results. In kidneys from normally hydrated rats, rabbits, cats and mice, eserine-sensitive pseudocholinesterase staining was present in a short segment of tube in the medulla. Part of the vasa recta, but not the glomeruli also stained. In dehydrated rats, activity was more widespread in the thick ascending limb of Henle's loops and was especially strong in the tubules belonging to juxtamedullary glomeruli.

Koelle (1950) reported an almost total lack of pseudoChE in the cat bladder; the AChE activity which he described in muscle cells could be the result of a diffusion artifact since El-Badawi and Schenk (1966) have shown that these cells are sur-rounded by a rich network of AChE-containing fibres. Data about pseudoChE in bladders of other species are sparse but Gerebtzoff (1959) observed it in smooth muscle fibres of the detrusor muscle of guinea-pig. Sites of pseudoChE activity in the male reproductive tract of guinea-pig include the epithelium of the epididymis, glandular cells of the prostate gland and muscle fibres of the genital ducts (Gerebt-zoff 1959). In the guinea-pig, pseudoChE is also found in the spermatozoa (Gerebt-zoff 1959) but in most of the other species studied so far AChE is more common (see ch. 8 §4). In the ovary of the guinea-pig Gerebtzoff (1959) demonstrated BuChE in the theca and interstitial cells and also in atresic follicles and the corpora lutea of pregnancy. Arvy (1960) found that in the rabbit BuChE was present in interstitial cells but the corpora lutea, granulosa cells and follicular fluid contained

AChE. Bulmer (1965) examined ovaries of rat, rabbit, monkey and guinea-pig; his results in the latter differ from those of Gerebtzoff in that activity was marked in the granulosa cells but weak in interstitial cells.

10.7.5 Other sites

Pseudocholinesterase is present in some endocrine glands including the thyroid of guinea-pig (Gerebtzoff 1959) and rat (Welsch and Pearse 1969); in rat, but not in guinea-pig, it is also present in the parathyroid. In the adrenal glands the distribution of enzyme within cellular elements varies with species (Coupland and Holmes 1958; Eränkö et al. 1959) but it is probably true to say that activity is low or absent in chromaffin cells and much of the enzyme is associated with the Schwann cells of the innervating nerve fibres (for references to optical and electron microscopic studies, see Lewis and Shute 1969). Using histochemistry M. Cottle and I (unpublished) observed appreciable BuChE activity in the guinea-pig pituitary gland, staining being most marked in the ventral part of the anterior lobe. In the rabbit neurohypophysis, Lederis and Livingston (1969) found a ratio of AChE:BuChE of 10:1 as determined biochemically by the method of Ellman et al. (1961). As they point out, their results differ from those for bovine neurohypophysis obtained by La Bella and Shin (1968) using the colorimetric method of McOsker and Daniel (1959). In bovine tissue, the AChE activity was lower than in rabbit, the BuChE higher and the ratio 2:1. Whether this reflects a genuine species difference or a methodological difference is not clear.

Eränkö et al. (1970) found that the parenchymal cells of the rat pineal gland were devoid of BuChE, the only histochemically demonstrable activity was associated with nerve trunks in the capsule. Following denervation (see §10.6.2) the reaction persisted in these trunks and also appeared in fine fibres within the gland, suggesting the enzyme is localized in supportive elements. Other miscellaneous sites at which pseudocholinesterase is present include lymph nodes (Ballantyne and Burwell 1965), the salt gland of the duck (Fourman 1966b, 1969; Ash et al. 1969) and goose (R.W. Ash, J.W. Pearce and A. Silver, unpublished; J.R. McLean and M. Peaker, personal communication) and the Harderian gland of duck (Fourman and Ballantyne 1967). In dog and pig, but not in man or cat, pseudoChE is present in the colostrum and milk (McCance et al. 1949). The activity in dog colostrum exceeds that in pig colostrum but in milk the pattern is reversed (Hines and McCance 1953). In pups fed on maternal milk, the serum ChE rose rapidly during the first 4 days after birth and then fell. If pups received evaporated cows' milk, which lacks ChE, no such rise was seen but the pups throve nevertheless (McCance et al. 1949).

Enzymes with some cholinesterase-like properties have been found in certain plants. In pine-trees, both buds and seeds contain an enzyme with a preference for BuCh (Sovershaev 1968). Other reports of choline ester hydrolysis by enzymes from plants include those of Riov and Jaffe (1973) on the Mung bean (*Phaseolus aureus*) and of Tzagoloff (1963) on white mustard (*Brassica hirta*). In both cases the enzymes were active towards ACh but they were relatively resistant to inhibition by eserine. Because of this resistance and because of the substrate pattern, Tzagoloff regarded the enzyme from mustard as being different from other esterases which had been described to date.

10.8 Sites of synthesis

It is widely accepted that in most species plasma pseudocholinesterase is synthesised in the liver and this accounts for the profound effect of liver disease on blood levels (§10.10). Gerebtzoff (1959) pointed out that the guinea-pig may be exceptional because the preferred substrate of the serum pseudoChE is BuCh while the preferred substrate of the liver esterase seems to be BzCh. According to Augustinsson (1963) it is even doubtful whether the liver enzyme is a cholinesterase at all. It may, in fact, be an arylesterase. Goutier-Pirotte and Goutier (1956) have suggested that in the guinea-pig synthesis of the serum enzyme occurs in the intestine. The possibility that in pig serum ChE comes from the pancreas was mentioned earlier (§10.7.3).

   Studies specifically designed to determine the site of synthesis of the enzyme in the nervous system seem to be lacking. Svensmark (1965) made the tentative comment that since the same genetic type of pseudocholinesterase is present in all tissues its synthesis may be controlled by the same genes or may take place at one site. Indirect evidence which is against this latter possibility, and which suggests that neural pseudocholinesterases have a more local origin, within the CNS itself, has come from a number of developmental studies. Bonichon (see Gerebtzoff 1959) and Filogamo (1960) found that in chick optic lobes BuChE was present in the ependyma at hatching. In foetal cats we (Krnjević and Silver 1966) likewise found cholinesterase activity (which was more characteristic of BuChE than AChE) in the ependyma of the striatal eminence. Of course, these observations do not rule out the possibility that the enzyme is arriving via the blood from elsewhere, but this seems unlikely. A further indication that neural tissue is capable of synthesis of pseudocholinesterase is provided by tissue culture experiments. Hösli and Hösli (1970) found BuChE activity in glia and, less frequently, in neurones cultured from rat brain stem and cerebellum. Using a pure culture of rat glial cells Cotman et al. (1971) showed that the specific activity of BuChE was 0.011 μmoles BuThCh/mg protein/min, which is about ten times the figure for whole brains.

   Friede (1967) has suggested (§10.6) that in certain areas of white matter the BuChE is localized in axons and not, as had generally been assumed, in glia. If this idea is correct — at present the localization has not been confirmed — it raises the question of where axonal enzyme could be synthesised. The possibility that the glia contribute enzyme to the axons (a suggestion once mooted for AChE) does not seem likely since Friede found that in those tracts in which the axons contained BuChE the glia did not, and vice versa. Another factor which would seem to argue against the transfer of glial enzyme to axons is the apparent size of the molecule. As indicated in § 10.5.1 the molecular weight of the enzyme in serum possibly exceeds 300,000 but it might be argued that this figure represents an aggregate of much smaller subunits each of which might cross the interposed membranes individually. The most likely site of synthesis of axonal enzyme would seem to be the peri-karyon but, again, Friede's findings are not compatible with this idea. Some of the tracts (e.g. corticospinal) which have axonal enzyme appear to arise from cells with little or no BuChE activity. Similarly, all the motor cranial nerve roots showed pronounced axonal activity but in the rat, for example, Friede found little cellular activity except in the dorsal vagal nucleus; this latter result confirms an earlier observation of Shute and Lewis (1963). Electron microscopic data about the possible translocation of pseudoChE are sparse because, as mentioned above, most of the studies of cholinesterases in the CNS at the EM level have been concerned primarily with AChE. The few results that are available do, however, indicate that BuChE resembles AChE in being associated with RER and axonal membranes (see, for example, Palkama 1967). On the other hand, BuChE is much more soluble than AChE hence transport of the two enzymes may occur quite differently.

Evidence for the synthesis of BuChE by Schwann cells of degenerating peripher-al nerves was discussed earlier (§ 10.6.2). Another possible site of peripheral synthesis is the subneural apparatus of muscle, a structure known to produce AChE (ch. 9 §6.2). Eränkö and Teräväinen (1967) showed that although BuChE activity at the end-plate decreased immediately after denervation, some enzyme was still present 8 months later. They pointed out that this does not necessarily represent 'old' molecules of enzyme that have persisted but may mean that the subneural apparatus, like the Schwann cell, is capable of synthesising new enzyme.

10.9 What are the functions of pseudocholinesterases?

It seems extraordinary that a good 40 years after their recognition pseudocholin-esterases are still a mystery so far as their function and natural substrates are concerned. The sort of problem we are up against in trying to find an explanation for the existence of BuChE is nicely exemplified by the situation in the pancreas (see § 10.7.3). In the cat, rat and rabbit the acinar cells can apparently function without involving BuChE at all. In the guinea-pig, on the other hand, the acinar cells are particularly well endowed with BuChE but it is rapidly destroyed when secreted into the gut. One is left wondering why it was produced in the first place.

A number of possible roles for pseudocholinesterase have been mooted and for each there is some evidence but for none is this convincing and unequivocal.

Functions proposed for the enzyme outside the nervous system include its involvement in assimilation of food (Gerebtzoff 1959), in the destruction of any butyrylcholine which might be produced in fatty acid metabolism (Clitherow et al. 1963), in the regulation of choline metabolism (Zeller and Bissegger 1943) or choline plasma levels (Funnell and Oliver 1965), in the metabolism of lipid (Ballantyne 1967; Fourman and Ballantyne 1967), in the regulation of tissue growth (Ballantyne and Burwell 1965) and in membrane permeability with particular emphasis on ionic movements. This last possibility has already been discussed in ch. 8. The point was made that although cholinesterases, both AChE and BuChE, are frequently associated with structures involved in permeability changes, indisputable evidence to show that they play an essential role in these functions is still lacking. To suggest such a role has nevertheless proved tempting and BuChE-containing sites at which the hypothesis has been invoked include the following: pine-tree buds (Sovershaev 1968); frog skin (Koblick 1958); blood–brain barrier of frog (Greig and Holland 1949b) and of mouse (Greig and Mayberry 1951); rat kidney (Fourman 1966a, 1967); duck nasal gland (Wood and Ballantyne 1968) and duck salt-gland (Ballantyne and Fourman 1967; Fourman 1966b, 1969). Ash et al. (1969) using histochemical methods were unable to confirm the claim of the latter authors that BuChE activity in the salt-glands was increased by loading the bird with salt, and subsequently Smith et al. (1971a, b) were similarly unable to substantiate the finding biochemically. Conflicting results of this kind are not easily explained but conflicting interpretations can sometimes be resolved in the light of subsequent data. For example, Greig et al. (1950) found that procaine could anaesthetize the rabbit cornea only in the presence of eserine; they tentatively interpreted this as evidence that the inhibition of cholinesterases resulted in permeability changes which enabled the procaine to penetrate the corneal epithelium. Since there is evidence that procaine is hydrolysed by pseudocholinesterase in some species (Huc 1950; Hazard et al. 1967) it seems possible that the significant consequence of the inhibition of the enzyme by eserine is not the change in permeability but is the preservation of the procaine.

The occurrence of BuChE, in some cases together with AChE, in walls of certain blood vessels in the CNS has been taken as evidence that the enzyme could have some part in controlling the permeability of blood vessels (ch. 8 §2.3). It may be significant that in the rat no BuChE-containing vessels were found in structures such as the pineal gland, the pituitary gland and the choroid plexus which lie outside the blood–brain barrier (Joó and Csillik 1966). On the other hand, if the enzyme does have some important role in those vessels in which it is present, it is difficult to explain the marked species variation in its occurrence (see table 8.4). We (Cottle and Silver 1970a) found strange BuChE-containing globules lying between blood vessels which showed AChE activity, at the ventral surface of the external layer of the anterior median eminence of the female guinea-pig (plate 10.5). We tentatively suggested that the globules could be concerned in the transfer into the pituitary portal system of active material produced in the hypothalamus but as with most of the other proposals this is only speculation on a very shaky basis.

There is equal scope for speculation about the function of BuChE in the rest of the nervous system. As described in §10.6, the predominant site of activity is in the white matter where the enzyme is mainly associated with glia although, in some species, it is also found in certain neurones. Some years ago there was considerable support for the view that the pseudocholinesterase of white matter was involved in

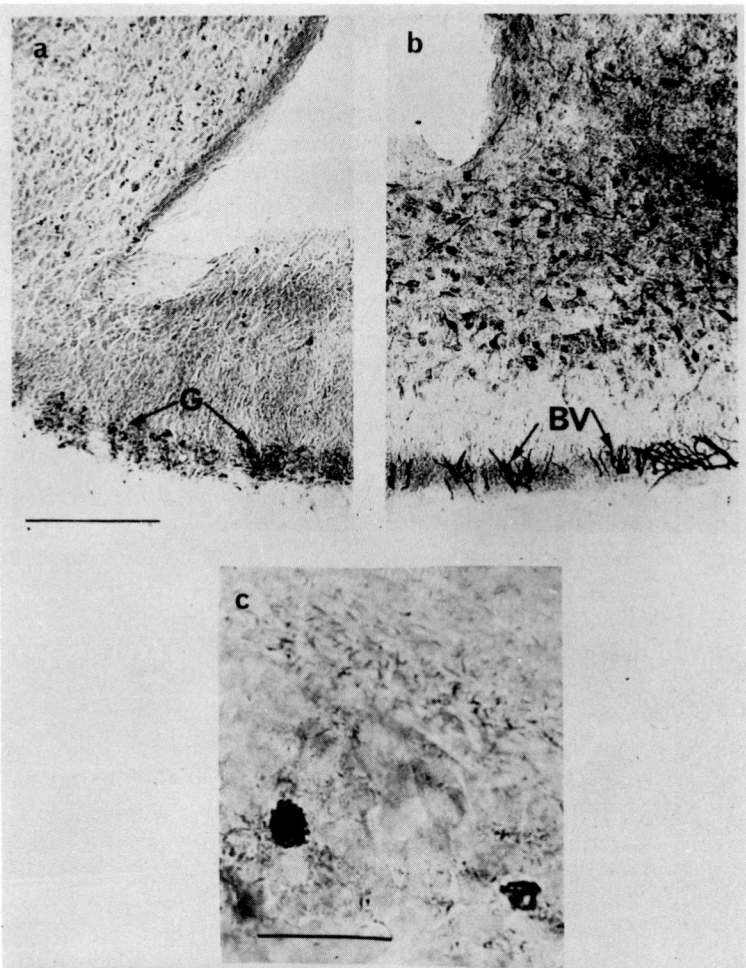

Plate 10.5. Median eminence of the guinea-pig. a) Section showing BuChE-containing globules (G). b) Section showing AChE-containing blood vessels (BV) in the same region. c) High-power view of BuChE-containing globules. Scale bar: 200 μ (a, b); 20 μ (c). (From Cottle and Silver 1970a.)

myelin maintenance. Much of the evidence for this idea came from experiments with those organophosphorus anticholinesterases (see ch. 11) which produce de-myelination and resultant paralysis. In 1941 Bloch had shown that TOCP and its meta-isomer were potent inhibitors of cholinesterase in horse serum and he suggest-ed that the paralysis was a consequence of the inhibition of ChE at the end-plate. Subsequently Earl and Thompson (1952a, b) showed this theory to be untenable since the cholinesterase at the end-plate, by then recognized to be AChE, was inhibited only very slightly. To replace Bloch's theory they suggested that pseudo-cholinesterase was necessary for maintaining the integrity of the myelin sheath and that demyelination was a direct result of its inhibition. Shortly afterwards this new hypothesis itself became untenable: as more and more organophosphorus com-pounds were tested it became clear that demyelination was not an invariable consequence of the depression of pseudocholinesterase but that its occurrence depended on the compound used and on the species tested (see Davison 1953b). Although these experiments failed to establish the initial premise, that pseudo-cholinesterase has a role in *normal* myelin maintenance, a theory was advanced by Davies et al. (1960) that pseudocholinesterase nevertheless did play a part in demyelination produced by the neurotoxic fluorine-containing alkyl organophos-phorus compounds. Their hypothesis was this: both neurotoxic and non-neurotoxic compounds react with pseudocholinesterase but, depending on the presence or absence of other groups in the drug which stabilise the P-F bond, ionic fluorine may or may not be released. If it is released, that is, if the bond is not stable, it causes a biochemical lesion resulting in demyelination. This theory has been criticized on a number of grounds by Aldridge et al. (1969). In particular, it makes no provision for the toxicity of compounds devoid of fluoride. The current idea, that an esterase other than pseudocholinesterase is the site of the attack by organophosphorus compounds, will be discussed in the next chapter (ch. 11 §6.2.2).

A theory that the glial enzyme was associated with neuronal function rather than with structure was advanced by Desmedt and La Grutta (1955, 1957). They showed that in the encéphale isolé preparation of the cat, acceleration and desyn-chronization of spontaneous electrical activity of the cerebral cortex could be produced by intracarotid injections of pseudocholinesterase inhibitors. They also compared the relative effectiveness of inhibitors of pseudoChE and of AChE in altering brain potentials and in potentiating the twitch in a nerve-muscle prepara-tion. The concentration of pseudocholinesterase inhibitor needed to produce an effect on the brain was lower than the concentration required for the AChE inhibitor but with the nerve-muscle preparation the situation was reversed. Desmedt and La Grutta did not think that differences in the lipid solubility of the inhibitors could account for the differential effects in the brain, and they tentatively inter-preted the results as evidence that the pseudocholinesterase of the glia has some role akin to that of a hormone and may be able to influence the activity of nearby neurones. Another suggestion that glial pseudoChE might hydrolyse γ-aminobutyrylcholine (GABuCh) has not found much support. As Holmstedt and Sjöqvist (1959b) pointed out when they made the proposal, the rate of hydrolysis of GABuCh by purified pseudoChE or by brain tissues is very slow.

The occurrence of BuChE in neurones has already been mentioned (§ 10.6.1 and ch. 6) and its role in these cells is as obscure as that anywhere else. Shute and Lewis (1963) found that in the rat almost all the neurones that contained BuChE contained AChE as well — the anterodorsal nucleus of the thalamus which contained BuChE and virtually no AChE being an exception. They suggested that the pseudocholinesterase could provide a safety mechanism, taking over the hydrolysis of ACh should AChE become inhibited by excess transmitter. Not only is BuChE immune from inhibition by high concentrations of ACh but it is also relatively inefficient against low concentrations (§ 10.3.2). Hebb and Krnjević (1962) pointed out that because of this, ACh can accumulate to a greater extent in pseudoChE-containing regions than in regions rich in AChE. This, they suggested, could be the virtue of BuChE in visceral structures where the response to ACh is relatively slow. Here again is the question posed in ch. 2. Is the type of response to ACh solely determined by the character of the receptor or does cholinesterase have some influence as well?

In his comparative study of rat, cat, man and monkey, Friede found that the location of BuChE-containing neurones varied widely with species. He considered that enzymes which showed such marked species differences (and these included alkaline phosphatases as well as BuChE) are probably not involved in basic metabolic or functional processes. Whether the apparently occult role of BuChE in neurones is the same as that in glia is another unsolved problem, and so, too, is the question of why BuChE is present in sensory structures such as the Pacinian corpuscle of cat pancreas (Hebb and Hill 1955a, b).

10.10 Variability of pseudocholinesterase levels and clinical implications

In man, serum pseudocholinesterase is produced in the liver hence the level of activity in the blood is, in general, a good index of liver function (see Szász 1968; Terazani et al. 1968). As discussed in the previous chapter, many other things besides liver dysfunction can, however, affect serum cholinesterase. Conditions in which serum levels are below normal include carcinoma (Vacarezza and Willson 1965b; Mustea 1969; see also ch. 9), eczema (Trotter and Fairburn 1966), rheumatic fever (Pisconti et al. 1964), typhus (Borisova 1966), tetanus (Nicholas et al. 1967), kwashiorkor (Begum and Prathapkumar 1969) and epilepsy (Koinov and Popov 1966). Cherchi et al. (1964) measured the blood levels of a number of enzymes in coalminers working below ground, and in surface workers and workers with other jobs. Pseudocholinesterase, alone of all the enzymes tested (and these included AChE), was below average in the underground workers.

During early pregnancy in women, serum pseudocholinesterase levels are below those found in late pregnancy or in non-pregnant women (Hazel and Monier 1971). In cases of toxaemia in the later stages of pregnancy cholinesterase activity in serum and in the erythrocytes and cerebrospinal fluid is depressed (Boczkowski 1966). Robertson (1967) found that serum pseudocholinesterase levels fell during the first month in which women were taking certain oral contraceptive pills but the levels after 3 months and after one year showed no further fall. From the point of view of pharmacological screening tests it should be mentioned here that Davies and Ojha (1950) reported species differences in the effect of stilboestrol on serum cholin-esterases in rats and cats. Findings of this type give practical significance to the otherwise academic interest in the diversity of pseudocholinesterases. Two other findings illustrate this same point. Foldes (1966) considered the relative toxicity to man of ester and amide types of local anaesthetic. In man, the level of pseudo-cholinesterase in serum is generally higher than that in most laboratory animals but, conversely, the level of activity of the liver microsomal enzymes is usually lower. Because of this, local anaesthetics of the ester type which can be hydrolysed by serum cholinesterases are less toxic to man than are the amide type metabolized by the liver but in other animals the situation may be the reverse and toxicity trials performed on these species could be misleading. Hobbiger and Peck (1970) investi-gated the role of serum pseudocholinesterases in the metabolism of suxamethonium (succinylcholine) and likewise concluded that studies on rat and cat were not relevant to man.

The majority of conditions which affect pseudoChE cause a fall in activity but kidney diseases tend to cause abnormally high serum levels. Raab (1969) studied this rise experimentally in rats injected with sodium tetrathionate, a substance which produces nephrosis-like changes. After 24 hours, serum pseudoChE was 40% above control levels but by 7 days after injection, levels were only slightly raised. It had been suggested previously (Szász et al. 1964) that the increased level might reflect a general increase of protein synthesis by the liver in response to protein loss resulting from renal failure. Raab postulated that in addition to an increase in synthesis, enzyme might escape from the damaged kidney cells (§10.7) and so add to the level in the serum. Whether any such contribution from damaged cells would be large enough to produce a measurable change in serum levels could probably be determined in experiments in which protein synthesis was prevented.

McKerracher et al. (1966) reported raised levels of serum pseudoChE in patients with mental abnormalities (see also ch. 9 §4). In their sample only 19% of control subjects gave values of more than 430 'units' whereas 65% of the mentally sub-normal and psychopathic patients had levels above this figure. Eleven of the patients were considered to be relatively stable, emotionally, and none of these had abnormal pseudoChE activity. According to Iordanova and Gotsev (1970) the stress associated with sitting university examinations can cause an increase in ChE levels in whole blood from healthy students.

In view of the apparent association of pseudoChE with glia it is not surprising that enzyme activity is high in many glial tumours (see §10.6.1). Robinson (1966) histochemically examined tissue from patients with Friedreich's ataxia; this is a demyelinating disease accompanied by gliosis and mainly affects the spinal cord. He found that in some areas of the brain the pseudocholinesterase in glia had increased but in the tracts of the spinal cord, levels were normal even in degenerating areas. Some anterior horn cells (like certain neurones in the medulla) stained less strongly than those in normal tissue.

10.11 Comment

Of the many questions about pseudocholinesterases that remain unanswered, perhaps the most puzzling is the question of function. Pseudocholinesterases are often treated with scant respect by neurophysiologists and neurochemists whose only concern is to ensure that they are adequately inhibited. Because of this, there is a slight tendency, particularly among histochemists, to regard the enzymes as second-class. But should this be so? According to Augustinsson's (1968) scheme for serum enzymes they evolved separately from AChE and were not, as one might think, a less efficient precursor. Why, if they are of no significance, have they been retained?

CHAPTER 11

Anticholinesterases

11.1 Introduction

Agents which inhibit cholinesterases must be counted in thousands rather than hundreds and, in consequence, they form a vast field of study which impinges on many disciplines. They have a relevance to biochemistry, pharmacology, physiology, toxicology and ecology and during the last 30 years have engendered a truly enormous literature. A number of monographs have been included in this spate of writing and these cover most aspects of the subject with a competence I could never match. With so much specialist literature already available it seems justifiable to treat the subject somewhat superficially and to select aspects which may be of general interest to those whose prime concern is with the cholinesterases themselves rather than with their inhibitors. Comprehensive lists of anticholinesterase agents are given by Saunders (1957), Holmstedt (1951, 1959), Tammelin (1958 a, b), Heath (1961) and O'Brien (1967), in numerous chapters in the Handbook on Cholinesterases and Anticholinesterase Agents edited by Koelle (1963) and in the relevant section by Wills (1970) in the International Encyclopedia of Pharmacology and Therapeutics. Papers on more recently developed compounds include those of Beckett et al. (1971); Beddoe et al. (1971) and Lüllmann et al. (1971). Many of these articles also deal with the mechanism and kinetics of enzyme-inhibitor interactions as do those of Cohen and Oosterbaan (1963), Belleau and Tani (1966), Kitz et al. (1967), Usdin (1970), Patočka and Bajgar (1971), Post (1971), Reiff et al. (1971) and Aldridge and Reiner (1972).

The compounds to be considered here are, in general, ones which are or which have been most commonly used as experimental tools in the physiological, biochemical or histochemical investigation of cholinergic mechanisms. Some of the myriad compounds which have been synthesised as pesticides were mentioned in ch. 5 and these will be discussed briefly in the context of pollution. The monographs of Metcalf (1955), Heath (1961) and O'Brien (1967) deal extensively with this type of anticholinesterase (see also numerous papers in Ann. N.Y. Acad. Sci. 160, 1969, on the biological effects of pesticides in mammalian systems). While the ability to inhibit cholinesterases is the main characteristic of the compounds under consideration, this single feature should not obscure the likelihood that these drugs

TABLE 11.1
Some anticholinesterases more commonly encountered in the literature.

No. Trivial name and synonyms	Chemical nomenclature and structure
1 Ambenonium chloride; Mytelase; Mysuran; WIN 8077	*N,N*′-bis(2-chlorobenzyldiethylammoniummethyl)oxamide dichloride Oxalbis(iminomethylene)bis(*o*-chlorobenzyl)diethylammonium chloride

2a   BW284C51   1:5-bis (4-allyldimethylammoniumphenyl)-pentan-3-one dibromide

2b   BW297C50 is the corresponding diiodide

3   BW62C47   1:5-bis(4-trimethylammoniumphenyl)-pentan-3-one diiodide

4   DFP; Dyflos; Diflupyl; Floropryl; isoflurophate; fluostigmine   Di*iso*propylphosphorofluoridate (see § 11.3.2 for variations)

TABLE 11.1 (continued)

Mol. wt.	Guide to concentration (see note) *	Comment
609	I.V. 0.0005–0.5 μmol/kg (Koelle 1957b)	'Selective' inhibitor of AChE. Low lipid solubility. Slowly reversible. Used in studies of extracellular AChE (Koelle 1957b). Used by Koelle & Gromadski (1966) for histochemistry with a gold-containing substrate (ch. 3)
566	In vitro 10^{-5} M (Koelle 1955) or 5 × 10^{-6} M (Holmstedt 1957a, b). 1–3 μg/kg I.V. (McIsaac & Koelle 1959)	'Selective' reversible inhibitor of AChE. Low lipid solubility. See Austin & Berry (1953); Fulton & Mogey (1954); and § 11.3.6
660		
608	In vitro 10^{-5} M (Bayliss & Todrick 1953a)	'Selective' reversible inhibitor of AChE. Low lipid solubility. See Fulton & Mogey (1954) and § 11.3.6
184	10^{-8} M–10^{-5} M in vitro depending on enzyme to be inhibited, and species (see ch. 3 § 1.1.2). In vivo, doses have ranged from 100μg– 8 mg/kg (see e.g. Holaday et al. 1954)	Differential inhibitor: at low concentrations inhibits BuChE, at higher concentrations inhibits AChE as well. Highly soluble in lipids. Volatile, causes demyelination. Use with extreme care. See § 11.3.2

* Note: The concentration depends on the purpose of the experiment and on the species.

TABLE 11.1 (continued)

No. Trivial name and synonyms	Chemical nomenclature and structure

5a Echothiophate; Phospholine iodide; 217-MI

O,O'-diethyl S-ethyltrimethylamine phosphorothiolate iodide; (2-mercaptoethyl) trimethylammonium iodide *O,O'*-diethyl phosphorothioate (see § 11.3.3 for other variations)

5b 217-AO

O,O'-diethyl S-ethyldimethylaminephosphorothiolate acid-oxalate;
(2-mercaptoethyl) dimethylamine acidoxalate *O,O'*-diethyl phosphorothioate

6 Edrophonium bromide; Tensilon; R02-3198

Ethyl (*m*-hydroxyphenyl) dimethylammonium bromide; (3-hydroxyphenyl) dimethylethylammonium bromide; dimethylethyl (3-hydroxyphenyl) ammonium bromide; 3-hydroxy-*N, N*-dimethyl-*N*-ethylanilinium bromide

7 Ethopropazine HCl; Lysivane; Dibutil; Parsidol; Rodipal; Parsitan; Parphezin etc.

10-(2-diethylaminopropyl)phenothiazine hydrochloride (see § 11.3.5 for variations)

TABLE 11.1 (continued)

Mol. wt.	Guide to concentration (see note) *	Comment
382	35–50 μg/kg I.V. (McIsaac & Koelle 1959). 2 × 10^{-5} M in vitro (Bourdois & Szerb 1972)	Non-selective cholinesterase inhibitor of low lipid solubility (cf. 217-AO). Binds to anionic and esteratic sites thereby differing from most other organophosphorus compounds. See § 11.3.3
331	106–300 μg/kg I.V. (McIsaac & Koelle 1959)	Tertiary analogue of Echothiophate. Lipid soluble, penetrates ganglia and CNS. Has been used in combination with Echothiophate in experiments in localization of intra- and extracellular enzyme (McIsaac & Koelle 1959), also as anticholinesterase in investigation of tremor (Lalley et al. 1970)
246	1–2 mg/kg I.V. (McIsaac & Koelle 1959)	Short acting inhibitor. Somewhat more effective against BuChE than AChE (see Long 1963). Binds only to anionic site (§ 11.2.2). Is used, as are similar compounds, decamethonium, tetraethyl ammonium (TEA) and tetrapropylammonium (TPA), to protect enzyme against other inhibitors particularly in the investigation of intra- and extracellular enzyme (e.g. Koelle 1955)
349	10^{-4} M for most biochemical and histochemical work	'Specific' inhibitor of BuChE. Enters CNS. Has some atropine-like effects. See § 11.3.5

TABLE 11.1 (continued)

No. Trivial name and synonyms	Chemical nomenclature and structure
8 iso-OMPA; DPDA	Tetramono*iso*propylpyrophosphortetramide; *N,N'*-di*iso*propylpyrophosphorodiamidic anhydride

| 9 Malathion; American
Cynamid 4049;
Malathon; Phos-
phothion | *O,O'*-dimethyl S-(1,2-dicarboxyethoxyethyl) phosphoro-
thiolothionate (or phosphorodithioate;
S-(1,2-dicarbethoxyethyl) *O,O'*-dimethyldithiophosphate;
O,O'-dimethyl S-(1,2-dicarbethoxyethyl) thiothiono-
phosphate |

| 10 Mipafox; Isopestox;
Pestox XV (Pest Con-
trol Ltd) | *N,N'*-di*iso*propylphosphorodiamidofluoridate;
N,N'-di*iso*propylphosphorodiamidic fluoride;
bis (*iso*propylamino) fluorophosphine oxide;
di(*iso*propylamido)phosphoryl fluoride |

| 11 Neostigmine bromide;
Prostigmin-, Proserine-,
Philostigmin-, Eustigmin-
bromide | (*m*-hydroxyphenyl) trimethylammonium bromide dimethyl-
carbamate;
(3-dimethylcarbamoxyphenyl) trimethylammonium
bromide |

TABLE 11.1 (continued)

Mol. wt.	Guide to concentration (see note) *	Comment
342	See Austin & Berry (1953) re wide species differences	'Selective' 'irreversible' inhibitor of BuChE. Water soluble (Aldridge 1953) also sufficiently lipid soluble to enter CNS (Davison 1953b). See § 11.3.4.
330		Used primarily as an insecticide but also in experiments with oximes. Lipid soluble. Oxidises to *Malaoxon* which is more toxic (see Heath 1961). Female rats more susceptible than males (Grob & Harvey 1958)
182	4×10^{-6} M recommended for histochemistry and biochemistry (Holmstedt 1957a, b)	Differential 'irreversible' inhibitor of BuChE in low concentrations. In higher concentrations inhibits both BuChE and AChE. Lipid soluble. Dangerous, causes demyelination
303 (methyl sulphate 334)	10^{-5} M to inhibit cholinesterase in bioassay of ACh. See Machne & Unna (1963) for in vivo doses	Stable synthetic compound structurally related to physostigmine. Is of equal or greater potency but effects confined to periphery since it is of low lipid solubility. Burgen & Hobbiger (1951) report slightly greater effect on human red cell AChE than human serum BuChE. Foldes et al. (1958) report the reverse

TABLE 11.1 (continued)

No. Trivial name and synonyms	Chemical nomenclature and structure

12 OMPA; Schradan

Octamethylpyrophosphortetramide

13 Parathion; E605; Thiophos 3422; DNTP; Niran; AAT; Etilon; DPP; Paraphos; Alkron etc.

*O,O'*diethyl *O''p*-nitrophenylphosphorothionate (or thioate)

14 Physostigmine; Eserine; Physotol

1'-methylpyrrolidino (2':3':2:3)1,3-dimethylindolin-5-yl *N*-methylcarbamate

15 Sarin; GB; T144

*Iso*propylmethylphosphonofluoridate; *Iso*propoxymethylphosphorylfluoride

16 Soman

Pinacolylmethylphosphonofluoridate; Pinacolyloxymethylphosphorylfluoride

TABLE 11.1 (continued)

Mol. wt.	Guide to concentration (see note) *	Comment
286		Insecticide, effective against only some orders but highly toxic to mammals; see O'Brien (1967) re possible metabolic basis of selective toxicity. Little activity in vitro. Low lipid solubility (Grob 1963a)
291.27	1–6 mg I.V. in cats (Erdmann & Schaefer 1954)	Primarily an insecticide but also used experimentally. More active against BuChE than AChE (man). Not very potent in vitro. In vivo metabolized to very toxic compound *Paraoxon* in which oxygen replaces the sulphur molecule. (*Paraoxon* itself is known as *Mintacol* or *E600*). Female rats more susceptible than males (Grob 1950). High lipid solubility
275 648 (sulphate)	10^{-5} M is adequate for total inhibition of cholinesterases in most species. Eccles et al. (1956) used 0.2–0.5 mg/kg for potentiation of 'cholinergic' effects in cat	Very potent but short acting inhibitor. BuChE slightly more susceptible than AChE (see Blaschko et al. 1949). Lipid soluble (cf. Neostigmine). See § 11.3.1
140.09		*Extremely toxic* volatile irreversible inhibitor. AChE slightly more susceptible than BuChE (see Grob 1963a). High lipid solubility. Developed as a nerve gas but used experimentally especially in investigations of oximes
182		Nerve gas of *even greater toxicity* than Sarin (Saunders 1957). Irreversible inhibition resistant to oxime therapy

TABLE 11.1 (continued)

No. Trivial name and synonyms	Chemical nomenclature and structure
17 Tabun; D7; Gelan; Trilon 83	Ethyl-*N*-dimethylphosphoramidocyanidate; Dimethylamidoethoxyphosphoryl cyanide

| 18 TEPP; Bladan; Nifos T;
Vapatone; Killax etc. | Tetraethylpyrophosphate |

| 19 TOCP
Tri-orthocresyl phosphate | Tri-*o*-cresyl phosphate |

TABLE 11.1 (continued)

Mol. wt.	Guide to concentration (see note) *	Comment
162		Nerve gas, less lipid soluble than Sarin and slightly less toxic. Equipotent against AChE and BuChE (see Grob 1963a)
290	I.V. 0.2–0.5 mg/kg for potentiation of 'cholinergic' effects in cat (Eccles et al. 1956)	Insecticide but also used in physiological experiments and in the investigation of oximes. Freely soluble in water but lipid solubility low. More active against BuChE than AChE. Aging of inhibited enzyme slower than with DFP (Grob & Harvey 1958). Although TEPP is about 10× more active than DFP as an anti-ChE in vitro it can be regarded as less dangerous because it is non-volatile, is not absorbed through the skin and does not cause demyelination
368	0.1–1.0 mg/kg in hens for demyelination studies (see Silver 1960)	Pure compound inactive in vitro. Inhibits BuChE in vivo. Demyelinating agent. See § 11.5.2.2. Commercial samples often impure and may contain many isomers of differing toxicity (see Silver 1960)

possess other pharmacologically significant properties. This point is emphasized very strongly in Karczmar's (1967) valuable article on anticholinesterases. Of the various effects which are independent of the inhibitory action on the enzymes, two deserve special mention. First, certain anticholinesterases act directly on the post-synaptic membrane. Some — eserine is a good example — act like curare and block the ACh receptor (Tauc and Gerschenfeld 1962; Levitan and Tauc 1972). Others, for example, neostigmine and edrophonium, may be excitatory (Holmstedt 1959). Since both eserine and neostigmine have been shown to excite neurones in cat brain stem which are unresponsive to ACh (Bradley et al. 1966) some of the actions of these compounds would appear to be independent of cholinergic mechanisms. Another feature of certain anticholinesterases is that they interfere with the mechanism by which ACh can be taken up from an incubation medium into slices of brain tissue. Polak (1969; see also Liang and Quastel 1969) showed that eserine has this effect but Soman and Tabun do not; neither, according to Heilbronn (1970a), does Sarin. Recognition of the fact that some anticholinesterases have this property and others do not, becomes important in experiments such as those of Bourdois and Szerb (1972) designed to study the phenomenon of 'surplus' ACh (ch. 4). In connection with this type of experiment one other point should be made. When AChE is inhibited the resulting accumulation of ACh may itself alter conditions and so complicate the interpretation of results. Szerb and Somogyi (1973) were able to show that when the AChE activity of rat cortical slices was inhibited by eserine, neostigmine or Echothiophate the release of ACh evoked by electrical stimulation was depressed. This effect appeared to be a consequence of the abnormally high level of extracellular ACh and was not a direct action of the anticholinesterases.

11.2 Types of anticholinesterase agents and their mode of inhibitory action

11.2.1 General introduction

The normal reaction between a cholinesterase and its substrate involves the attachment of the substrate to the enzyme with the production of an acylated form of enzyme which undergoes rapid hydrolysis to yield re-usable enzyme (ch. 2 §4). Anticholinesterases interfere with this orderly sequence, the point of attack depending on the nature of the inhibitor. They may block i) the anionic site, ii) the anionic and esteratic site, or iii) the esteratic site, but in almost all cases the reaction is to some extent similar to that between substrate and enzyme (see figs 2.3 and 2.4). Some inhibitors are almost equally active against AChE and BuChE, but others have a preferential action on one enzyme or the other. It is perhaps worth re-emphasizing the point, made in ch. 3, that none of the inhibitors is absolutely specific to one type of cholinesterase. The differential action is one of degree only and is reflected in the different concentrations at which the inhibitor affects the two types of

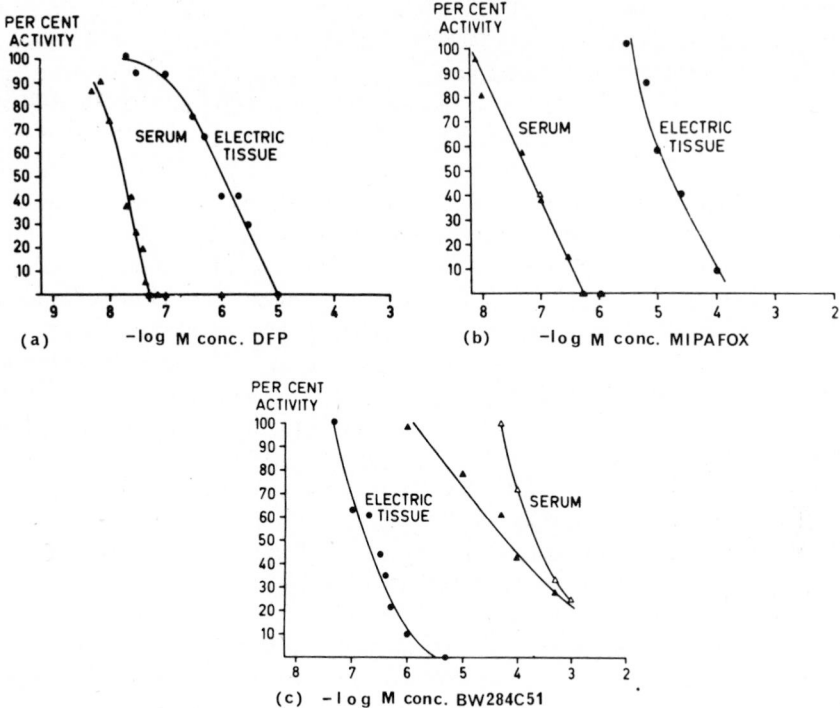

Fig. 11.1. To demonstrate the difference in inhibitory potency of DFP, Mipafox and BW284C5i towards AChE of eel electric tissue and pseudoChE of human serum. Substrates: AcThCh (black symbols); BuThCh (open symbols). a) DFP; note that the concentration of DFP which ensures complete inhibition of serum enzyme causes some inhibition of AChE. b) Mipafox; note that the curves do not overlap, which indicates a high level of selectivity towards serum enzyme. c) BW284C51; marked selectivity towards AChE. Substrate and inhibitor were added simultaneously and the protection of the serum enzyme afforded by BuThCh, the preferred substrate, is greater than that afforded by AcThCh hence the two curves. (From Holmstedt 1957a.)

enzyme (fig. 11.1). With some compounds, complete inhibition of one enzyme is achieved only by concentrations which cause, in certain species at least, significant inhibition of the other enzyme as well. With other compounds there may be little or no overlap in effective concentrations and it is these compounds which are termed 'selective'. This designation is justifiable so long as the breakdown in selectivity at higher concentrations is recognized. In addition to being classed as 'selective' or 'non-selective', anticholinesterases are also divided into those which are 'reversible' and those which are 'irreversible'. Again, this distinction is not absolute and it will be considered more fully below (§11.2.5).

The potency of an anticholinesterase in vivo depends on a number of factors which do not operate in vitro. If it is susceptible to detoxication by systems in the liver (see, for example, Neal and Dubois 1965; Murphy 1967) or other organs, it may prove less toxic than its performance in vitro would suggest. On the other hand some compounds are metabolized in such a way that a relatively weak inhibitor is rendered more potent. The metabolism of Malathion to Malaoxon and Parathion to Paraoxon (see table 11.1) exemplify just such an increase in toxicity. TOCP (triortho-cresyl phosphate, see table 11.1) provides an extreme case; while the pure compound has no anticholinesterase activity in vitro it is metabolized in vivo to a powerful inhibitor of pseudoChE. OMPA (see table 11.1) is also inactive in vitro but in mammals and a limited number of insects it produces marked inhibition of cholinesterases (see O'Brien 1967). Another factor which will influence the physio-logical effects produced by an anticholinesterase is its lipid solubility. If it is a polar (i.e. charged) molecule it will not readily enter the CNS and its actions will be largely restricted to the periphery. Bajgar (1972) has evidence that the rate of penetration of a charged inhibitor through the blood—brain barrier into the CNS may vary from area to area. He studied EDMM [*O*-ethyl S-(2-dimethylaminoethyl) methylphospho-nothioate] which at a physiological pH is mainly in the charged form and he found that the AChE in different parts of the rat medulla was inhibited at different rates. In vitro experiments fail to reveal factors of this type which can be very important to the understanding of physiological effects. In this context it is relevant to men-tion Lancaster's (1972) observation that the distribution within the tissue of qua-ternary compounds applied to brain slices in vitro is not the same as the distribu-tion resulting from in vivo administration. He found that in vitro, under the right conditions of inhibitor concentration and incubation time a compound of low lipid solubility ultimately reached *all* sites, intracellular as well as extracellular, of AChE activity. This means that the percentage inhibition of AChE observed in slices incubated with quaternary compounds is not a reliable guide to the amount of 'functional' (i.e. extracellular) enzyme present in the tissue.

11.2.2 Compounds which bind to the anionic site

Certain compounds interfere with the attachment of the substrate to the enzyme by occupying the anionic site. Examples of this type of inhibitor include some of the quaternary compounds such as decamethonium, tetraethylammonium (TEA), tetrapropylammonium (TPA) and edrophonium (Tensilon) (see table 11.1). The quaternary nitrogen attaches itself to the anionic site so rendering it unavailable to the substrate. This inhibitor-enzyme reaction differs from the substrate-enzyme reaction in that there is no direct reaction with the esteratic site. The orientation and dimensions of the compound may, however, be such that when the molecule is attached to the anionic site it extends far enough to mask the esteratic site as well (see fig. 5.1). In this way it can protect the enzyme from phosphorylation by other types of inhibitor (see §§11.2.3 and 4). BW284C51 (table 11.1; see §11.3.6) which is often used to inhibit AChE in the thiocholine histochemical technique belongs to a particularly interesting series of bis-quaternary compounds, the mono-quaternary analogues of which are generally inactive (see Long 1963). Attachment is at the anionic site but the stereochemical feature which demands two quaternary groups for efficient inhibition is not clear. This does not seem to be a case of additional masking of the esteratic site: Koelle and Gromadski (1966) found that BW284C51 failed to prevent the histochemical reaction when gold thiolacetic acid was used as the substrate in place of thiocholine. This indicates that the esteratic site, the only site involved in the binding of gold thiolacetic acid, is in no way shielded by the compound. The inability of BW284C51 to prevent the hydrolysis of gold thiolacetic acid brings out an important practical point about the choice of an inhibitor. An inhibitor which binds only to the anionic site can prevent the hydrolysis of a substrate such as acetylcholine or acetylthiocholine which must be bound to the anionic as well as to the esteratic site. To prevent the hydrolysis of a substrate which reacts only with the esteratic site, the inhibitor must bind to, or at least mask, the esteratic site (§§11.2.3 and 4).

11.2.3 Compounds which bind to both the anionic and esteratic site

The reaction between the enzyme and compounds of this type is qualitatively analogous to the reaction between enzyme and ACh. The inhibitory action arises because of the relative slowness with which the enzyme is regenerated by hydrolysis of the intermediate complex. Different inhibitors yield different intermediates and it is the rate of hydrolysis of these intermediates which determines the duration and reversibility of the block. An example of a relatively short-acting inhibitor is physostigmine (eserine). This is discussed more fully below (§11.3.1) but its action is briefly this. It binds to the anionic site by virtue of its amino group and at the same time the carbamyl group reacts at the esteratic site. The cationic part of the inhibitor is then split off leaving a carbamylated enzyme (Wilson et al. 1960; see O'Brien 1968, 1969; Aldridge and Reiner 1972). This is gradually hydrolysed but at a rate which is very much slower than that at which the acylated enzyme complex is broken down in the normal reaction (see Post 1971 for kinetics of inhibition). In some experiments on the release of ACh from the isolated rat diaphragm in vitro we (Mitchell and Silver 1963) found that ACh was preserved for 2 hr following eserinization of the tissue but thereafter the amount of ACh recovered from the bathing fluid fell very sharply, none being detectable 3½ hr after the experiment started. A further application of eserine restored the amount of recoverable ACh to its initial level. This showed that the decline in ACh in the bathing fluid had been due to its destruction by regenerated AChE and not to any diminution in its release by the tissue.

Certain organophosphorus compounds resemble physostigmine in that they bind at both the anionic and esteratic site but they yield a phosphorylated enzyme which is very much more stable than the carbamylated intermediate. An example of this type of inhibitor is Echothiophate (table 11.1), which is discussed below (§11.3.3).

11.2.4 Compounds which bind to the esteratic site

The majority of organophosphorus compounds react with the enzyme only at the esteratic site. As mentioned above, the phosphorylated enzyme is extremely resistant to hydrolysis but the exact degree of stability depends on the type and source of the cholinesterase as well as on the nature of the alkyl group attached to the phosphorus. Davison (1953a, 1955) has shown that in various rat tissues, pseudo-cholinesterase recovers from inhibition by Paraoxon faster than does AChE, and that brain pseudocholinesterase recovers more slowly than that of serum and heart. Welsch and Dettbarn (1972b) confirmed in principle Davison's findings for rat diaphragm but observed no spontaneous recovery of enzyme activity in lobster peripheral nerve following inhibition by Paraoxon. The degree of stability conferred by a particular alkyl group also depends on the enzyme involved. With rabbit erythrocyte AChE the rate of recovery from inhibition by compounds with different alkyl groups is in the order methyl > ethyl > *iso*propyl (Aldridge 1954). In contrast, rat pseudoChE recovers from diethylphosphorylation very much faster than from either dimethyl- or di*iso*propylphosphorylation. The half life, at pH 7.8 and 37°C, is only 5 hr for the diethylphosphorylated enzyme and more than 200 hr for the two latter intermediates (Davison 1955). Human serum enzyme shows yet another pattern: recovery from dimethyl- and diethylphosphorylation occurs at the same rate but there is scarcely any recovery of the di*iso*propylphosphorylated enzyme (Mengle and O'Brien 1960). Hobbiger (1963) gives a useful list of the types of phosphorylated enzyme which result from the reaction of AChE with different organophosphorus compounds.

11.2.5 Duration of action and aging

The literature is often confusing on whether a particular inhibitor is reversible or irreversible. Spontaneous hydrolysis of the carbamylated or phosphorylated enzyme occurs to some extent, at least, in virtually all cases (see Reiner 1971 for table of recovery rates) but although there is reversal of inhibition, the rate is often extremely slow and the physiological effect of any recovery achieved during an experiment lasting a few hours may be negligible. From the point of view of the experimental physiologist (as opposed to the enzymologist) the inhibition produced by something like DFP or Mipafox (table 11.1) appears 'irreversible'. This may be good or bad. If continuous inhibition is required an 'irreversible' inhibitor is more reliable than a short-acting compound such as physostigmine. On the other hand, if a series of observations is to be made, as in the iontophoretic application of a number of different drugs, prolonged inhibition may be inappropriate. In practice then, an inhibitory reaction may be termed 'irreversible' if the rate of spontaneous hydrolysis of the enzyme-intermediate is too slow to produce a significant effect either in vitro or in vivo. An enzyme which is irreversibly inhibited in this sense is nevertheless susceptible to reactivation by oximes (§ 11.5). This susceptibility is not permanent, however, because the phosphorylated enzyme undergoes an aging process (for a detailed description see Aldridge and Reiner 1972) during which it is converted to a form which cannot be dephosphorylated by any known means. The aging process involves a dealkylation reaction during which an ionized hydroxyl group is attached to the phosphorus atom of the phosphorylated enzyme. The result is that the inhibited enzyme will no longer yield to nucleophilic attack which is the basis of oxime reactivation. The rate of aging depends on the phosphorylating agent and also on temperature and pH; the lower the temperature the slower the reaction but the lower the pH the faster the reaction (Davies and Green 1956; see also Koenig and Koelle 1961; O'Brien 1967; Aldridge 1969). Other evidence indicates that the rate of aging also depends on the source of enzyme under test. Tabun-phosphorylated cholinesterase from horse plasma, for example, ages very much more slowly than that from human plasma (see Davies and Green 1956). Similarly, Andersen et al. (1972) found that with DFP the enzyme from frog brain aged 10 times more slowly than that from mouse, rat and chicken brains.

In experiments to measure the rate of spontaneous reactivation it is essential to establish that any return of hydrolytic activity following poisoning represents the disinhibition of pre-existing enzyme and not the synthesis of new enzyme. Welsch and Dettbarn (1972b) make this point in discussing their work on rat diaphragm and lobster walking-leg nerve (see also Dettbarn et al. 1970). If rat diaphragm was repeatedly washed following inhibition with Paraoxon appreciable enzyme activity returned in two hours. In lobster nerve, by contrast, recovery was extremely slow and virtually unaffected by washing. Although it seems probable that the differences are attributable to differences in the enzymic binding of inhibitor in the two tissues, the possibility that some additional synthesis is occurring in rat but not lobster, cannot be eliminated. There is indeed evidence that anticholinesterases may stimulate protein synthesis (Clouet and Waelsch 1963; Welsch and Dettbarn 1971). That synthesis could be rapid for certain types of AChE is suggested by the discovery in rat retina of an isozyme of AChE with a half-life of only 3 hr (Davis and Agranoff 1968).

11.3 Some compounds commonly used as anticholinesterase agents in the laboratory

11.3.1 Physostigmine (Eserine)

11.3.1.1 History Physostigmine is probably the best known of all anticholinesterases. It is an alkaloid containing a methyl carbamino group (see table 11.1) and is obtained from Calabar beans, the seeds of the West African vine *Physostigmina venosum* Balfouri. These seeds are also known as Ordeal beans and as Esére nuts, hence the use of eserine as an alternative term for physostigmine. Many people who use eserine pharmacologically or clinically may be unaware of its anthropological significance. The bean was used by certain African tribes in the trials of those suspected of witchcraft. According to Christison (1855) the natives were confident that innocent people vomited and were safe, while the guilty retained the poison and died. This misplaced confidence led to the death of many who were innocent of witchcraft but the belief could have had a physiological basis. It has been suggested (see Goodman and Gilman 1955) that an innocent person would eat the beans rapidly since he had nothing to fear. This caused gastric irritation and the resultant vomiting prevented poisoning. Witches, on the other hand, being more apprehensive would swallow the beans too slowly to induce vomiting and were thus poisoned. More recently Bhattacharya and Sanyal (1971) investigated the anticholinesterase activity of another substance used in tribal rites, Bufotenine. This is a hallucinogenic agent obtainable from a number of plants and is a structural analogue of physostigmine.

Calabar beans were brought to Britain in 1840 but were not examined immediately. In 1855 Sir Robert Christison, Professor of *Materia medica* in the University of Edinburgh reported the effects he had observed both on rabbits and on himself. He described how he swallowed one-eighth of a bean (6 grains) after a scanty supper and felt no obvious effects. Next morning, before breakfast, he ate a quarter of a bean and when he had noted his symptoms drank his shaving water as an emetic and survived to do further equally hair-raising experiments (Christison 1855; see also Christison 1885, 1886).

11.3.1.2 Uses Physostigmine is an extremely potent anticholinesterase and has two main uses as an experimental tool. In bioassay it will prevent the cholinesterase of the test-organ (e.g. dorsal muscle of leech) from hydrolysing any ACh in the solution under test. Similarly, in physiological investigations of cholinergic mechanisms it will preserve endogenous ACh, or ACh and other cholinomimetics which are injected into the animal locally or systemically (see Machne and Unna 1963; Megazzini et al. 1965; Brezenoff 1972). Without such an inhibitor, cholinesterase-susceptible drugs would be hydrolysed before they could exert their effects. When anticholinesterases are given to whole animals experiments must, however, be carefully controlled otherwise results may be misinterpreted. Eserine can, for example, produce a rise in blood pressure when injected systemically. This effect which contrasts with a reported hypotensive action of Soman, Sarin and DFP (see Preston and Heath 1972a, b) has been attributed to the central stimulation of the sympathetic system by accumulating ACh (Varagić and Krstić 1966). Generalized actions resulting from the widespread inhibition of cholinesterase will inevitably affect a number of physiological parameters. Mršulja et al. (1968), for instance, showed that in rat the brain glycogen is lowered by intravenous injections of physostigmine. Factors like this must be taken into account before anticholinesterase-induced effects (see for example Bradley and Nicholson 1962) in a particular organ can be interpreted in terms of a specific cholinergic mechanism. The second important laboratory use of physostigmine is in biochemical or histochemical studies where it is necessary to distinguish reactions which may be due to cholinesterases from those due to other, eserine-resistant, esterases. For most vertebrate species a concentration of 10^{-5} M is more than adequate to inhibit all cholinesterase activity in tissues in vitro. If a reaction still occurs despite the presence of 10^{-5} M physostigmine it is unlikely to be due to a cholinesterase. There are some exceptions, however; the AChE in frog brain, for example, is still only 94% inhibited by eserine at a concentration of 10^{-4} M (Hawkins and Mendel 1946).

Clinically, physostigmine is applied topically in the treatment of glaucoma but in most other instances where an anticholinesterase is required for medical use the more stable compound neostigmine (Prostigmin) is preferred. Neostigmine is chemically related to eserine being, as table 11.1 shows, a complex ester of methyl carbamic acid. Its main clinical use is in the treatment of myasthenia gravis but it is also used in cases of paralytic ileus (particulary post-operatively) and in the treatment of atony of the bladder and to cause expulsion of renal calculi. Because neostigmine is a quaternary compound it cannot easily cross the blood—brain barrier hence it does not produce significant effects on the central nervous system (cf. eserine).

Physostigmine salts and solutions are prone to oxidize on exposure to light and air and this must be borne in mind whenever these compounds are used. Containers should be airtight, light-resistant and of alkali-free glass. Any solution with a pinkish tinge is suspect and should be discarded (see Hemsworth and West 1970 re degradation products).

11.3.1.3 General effects   As mentioned above it must not be assumed that the effects which eserine or any other inhibitor produce are solely attributable, either directly or indirectly, to an action on cholinesterases. Evidence from various types of experiment indicates that eserine has a number of independent effects. Feldberg and Hebb (1948) showed that in a concentration of 1:10,000 it abolished the response of the cat superior cervical ganglion to ACh or to nerve stimulation and reduced the response to ATP, creatine phosphate and citrate ions. If the concentration was raised to 1:1000 these latter responses were abolished too. Eserine was effective on the denervated as well as on the innervated ganglion which added support to the view that its blocking action was independent of its anticholinesterase activity (see also Levitan and Tauc 1972). Koelle et al. (1971) have recently suggested that, contrary to previous opinion, much of the cholinesterase is associated with postganglionic neurones (see ch. 7 § 2.2); even if this proves to be the case it does not vitiate the evidence that eserine has other actions in addition to those resulting from the inhibition of cholinesterase. Krnjević et al. (1970) working on small isolated slabs of cat cerebral cortex from which all histochemically detectable AChE had disappeared found that iontophoretic application of eserine to cortical cells consistently depressed discharges evoked by a simultaneous application of glutamate (cf. Bradley et al. 1966 § 11.1). They pointed out that this apparently depressant effect of eserine itself must be taken into account when it is used to identify reputedly cholinergic inhibitory synapses (see Phillis and York 1968b). Bartolini et al. (1973) have recently examined the effect of the in vivo administration of eserine on ACh content and release from cat (and rat) brains. Their results indicate that eserine may increase the ACh available at subcortical structures, not only by inhibiting its hydrolysis but possibly by promoting its release. Another property of eserine, that of reducing the re-uptake of ACh into tissue slices was mentioned in §11.1.

11.3.2 Diisopropylphosphorofluoridate (DFP)

11.3.2.1 History   DFP (table 11.1) has a number of names including Dyflos, Diflupyl, Floropryl, isoflurophate and fluostigmine. With changes in chemical nomenclature over the years it has been variously designated di*iso*propoxyphosphoryl fluoride, di*iso*propyl fluorophosphate or -phosphonate, *iso*propylfluophosphate, and currently di*iso*propylphosphorofluoridate. DFP was synthesised in 1941 by Saunders but details were not published until after the war (see Saunders 1957). It was one of many organophosphorus compounds which were being screened as potential nerve gases. Saunders and his colleagues from the Ministry of Supply tested the effects of exposure to low concentrations on themselves. High concentrations were tested on laboratory animals and the inhibitory effect on cholinesterases was recognized (see Adrian et al. 1947). W. Feldberg (personal communication) has told how he was given a number of compounds to investigate. These, for security reasons, were unnamed. Needing to label them he used the notation L_1, L_2 and so on, the 'L' standing for Lohengrin with his very apt plea to Elsa 'Never ...... to question move thee, from whence to thee I came, or what my race and name' (Wagner 1848).

11.3.2.2 Uses   DFP has been, and still is, widely used as a laboratory tool despite some practical disadvantages, the major one being its extreme toxicity. Because it is volatile and also readily absorbed through the intact skin, all possible precautions must be taken against spillage, skin contamination and inhalation whenever DFP is in use (see appendix). Glassware should be decontaminated in sodium hydroxide of an adequate concentration. Saunders (1957) pointed out that 0.49 N NaOH at room temperature will hydrolyse only 16% of DFP in 30 min unless the solution is vigorously shaken. The concentration preferred in this laboratory is 10% (2.5 N).

Pure DFP can react with certain types of plastic and this can be dangerous. To avoid the risk of leakage neither plastic containers nor plastic syringes should be used.

Although the anhydrous compound is stable and may be kept in glass containers at room temperature, aqueous solutions hydrolyse in a matter of days (Saunders 1957). For laboratory work stock solutions (10^{-1} M) can be made up in anhydrous propylene glycol and aqueous solutions of the required concentration prepared immediately before use. Anhydrous peanut oil or sesame oil are the normal vehicles for DFP in medicinal preparations. Being lipid-soluble and 'irreversible', DFP is valuable in physiological experiments where prolonged inhibition of both central and peripheral cholinesterases is needed. Its use as a differential inhibitor in biochemical and histochemical techniques was mentioned in ch. 3. In all species examined so far, the concentration which affects pseudocholinesterase activity is lower than that which affects AChE but in many species the concentration necessary for complete inhibition of pseudoChE produces significant inhibition of AChE as well. DFP has been employed in many types of physiological investigation of possible cholinergic mechanisms (for example, Eccles et al. 1956; Dettbarn and Rosenberg 1962; Baker and Benedict 1968; Winson and Miller 1970; Katz et al. 1973). It has also been used in the experimental study of memory and learning. Work on rats by Deutsch et al. (1966) and by Deutsch and Leibowitz (1966) has shown that the effect of intracerebral injections of DFP on a learned response depends on the time interval between the training and the injection. At certain intervals it may cause amnesia but if given when the rat has almost forgotten the response it enhances recall. Subsequent work (Deutsch and Lutzky 1967) has indicated that if a task is only partially learned the recall achieved after injections of DFP is far better than that of a task which was well learned initially. The authors interpret their results in terms of synaptic activation. With a half-learned task only a few synapses are involved and the preservation by DFP of the small amount of ACh released may boost their activity. With a well-learned task, far more synapses are active and inhibition of AChE could cause an accumulation of ACh sufficient to block the response.

The use of [^{32}P]-labelled DFP in the autoradiography of end-plates (Salpeter 1969) has been mentioned (ch. 3). Radioactive DFP has also been used to tag red blood cells. On injection, the DFP binds to the erythrocytic AChE and the isotope provides a useful marker in studies of the life-span of the cells (I.C.S.H. Report 1971; Tucker 1974).

Clinically, DFP has been used for the treatment of glaucoma and myasthenia gravis. Its prolonged action is inappropriate in the latter disease since the condition of the patient tends to fluctuate and the dose of drug needed may change from day to day. Side effects with DFP, as with other organophosphorus compounds, are often troublesome and may include hallucinations and nightmares.

11.3.3 Echothiophate (or Echothiopate)

Echotiophate (table 11.1) is an 'irreversible' non-selective (Tammelin 1957) inhibitor of cholinesterases also known as pholine iodide or 217—MI. Chemically it is (2-mercaptoethyl) trimethylammonium iodide O,O'-diethyl phosphorothioate. Other designations in the literature include diethoxyphosphinylthiocholine iodide; 2-diethoxyphosphinylthiocholine iodide; 2-diethoxyphosphinylthiotrimethylammonium iodide; S-(2-trimethylammonium ethyl) phosphorothioate iodide. It is water-soluble and far more stable than DFP. Being a quaternary compound it does not penetrate the CNS following intravenous injection hence it has been used to inhibit AChE in a number of studies designed to differentiate peripheral and central cholinergic effects (see Koelle and Steiner 1956; Schaumann and Job 1958; McIsaac and Koelle 1959). In some experiments its actions have been compared with those of its lipid-soluble tertiary analogue 217—AO (see table 11.1); in others Echothiophate has been injected intracisternally to bypass the blood—brain barrier. The report by Bourdois and Szerb (1972) that Echothiophate, unlike eserine, does not interfere with the uptake of ACh from the medium into brain slices has been mentioned earlier (§11.1).

When using any anticholinesterase it is important to check the formulation of the preparation since those dispensed for therapeutic use may contain additives which could affect the experiment. Echothiophate serves as a good example. The most readily available form is that prepared (Ayerst Laboratories) for ophthalmological use and this contains potassium acetate. If such a preparation is to be used as the anticholinesterase in studies of choline acetyltransferase a possible activating effect of potassium must be considered.

Clinically, Echothiophate has been used for treating the same conditions as DFP. It is more convenient than DFP because preparations are more stable, they can be aqueous rather than unpleasantly oily and there is no risk of side effects involving the CNS.

11.3.4 Iso-OMPA (Tetramonoisopropyl pyrophosphortetramide)

Iso-OMPA (table 11.1) was originally called bis(di-*iso*propylamino) phosphonous anhydride. The abbreviation iso-OMPA was criticized by Austin and Berry (1953) on the grounds that it incorrectly implied that the compound was an isomer of OMPA (Schradan; see table 11.1). They suggested the initials DPDA as an abbreviation of yet another form of chemical nomenclature $N N'$-di*iso*propylphosphorodiamidic anhydride. DPDA is not, however, often used in the literature, iso-OMPA having been retained.

Iso-OMPA is employed as an 'irreversible' selective inhibitor of pseudocholinesterase but there are considerable species differences in the degree of selectivity exhibited. Austin and Berry (1953) showed, for example, that differential inhibition of plasma pseudocholinesterase relative to erythrocyte AChE was far greater for horse than for man. The compound is more stable than DFP and is water-soluble (Aldridge 1953) but for certain types of experiment it has been dissolved in propylene glycol (Davison 1953b; Silver 1960). Unlike DFP it does not cause demyelination (Davison 1953b) and this, together with the fact that it is virtually non-volatile and has relatively little effect on AChE, means that it is free from the hazards associated with DFP. It should be emphasized that this relative safety is not true of OMPA which is metabolized in mammals to a highly toxic compound. Despite the similarity in names the two drugs are very different in character and cannot be substituted for each other.

Table 11.1 includes a few other organophosphorus inhibitors commonly encountered in the literature. In general these compounds are, like DFP, highly toxic and should be avoided if other, less dangerous inhibitors can be used instead (e.g. § 11.3.5 and 6).

11.3.5 Ethopropazine

Chemically this is 10-(2-diethylaminopropyl) phenothiazine (table 11.1). For laboratory work the methosulphate or the hydrochloride are used. Proprietary names for the latter include Lysivane, Dibutil, Parsidol, Rodipal and a number of others. It is a valuable, safe, selective inhibitor of pseudocholinesterase widely used in measurements of AChE by the colorimetric method of Ellman et al. (1961), and in histochemistry (see ch. 3). Ethopropazine is classed as a reversible inhibitor and must be included in the incubation as well as the preincubation medium. In my experience it remains effective in histochemical incubations lasting many hours but Berry and Rutland (1971) found it to be unstable when used for biochemical assays on muscle preparations in NaOH-KCl. One practical point should be noted. According to the Merck Index (1968) 1 gm will dissolve in 20 ml H_2O at 40°C; the solubility of the commercially available product seems, however, to be very much lower. As mentioned in ch. 3, solutions tend to discolour on exposure to light but this does not affect the inhibitory action.

Ethopropazine enters the CNS and is effective in the treatment of Parkinsonian tremor and rigidity; it will also control the excessive salivation associated with the condition. This therapeutic action of ethopropazine is attributable to an atropine-like blocking of receptors and not to anti-pseudocholinesterase effects.

11.3.6 BW284C51, BW297C50 and BW62C47

These three compounds are selective inhibitors of AChE. As table 11.1 shows all are substituted bis-phenylketones and they belong to the series mentioned earlier (§11.2.2) in which the monoquaternary analogues are generally inactive against AChE. They are water-soluble but like other quaternary compounds their lipid

solubility is low. Chemically BW284C51 and 297C50 are, respectively, the dibromide and the diiodide of 1:5-bis(4-allyldimethylammoniumphenyl)-pentan-3-one.
BW62C47 is 1:5-bis(4-trimethylammoniumphenyl)-pentan-3-one diiodide. Information about other members of the series is given by Fulton and Mogey (1954) and by
Long (1963). A point to note about these agents is that because they are neither
carbamates nor organophosphorus compounds, oximes are ineffective in cases of
poisoning (§11.4).

BW28C51 in particular has been widely used in histochemistry. It has given very
satisfactory results when used in the thiocholine method, but, for reasons discussed
earlier (§11.2.2) it cannot inhibit activity towards substrates (e.g. gold thiolacetic
acid) which bind only to the esteratic site. Authors are not in general agreement about
the ideal concentration which should be used to ensure maximum inhibition of
AChE and minimum inhibition of BuChE. For cat enzymes Koelle (1955) suggested
3×10^{-5} M and Holmstedt (1957a) 5×10^{-6} M. From my own experience of
histochemistry on cat tissues I would suggest a concentration of not less than
10^{-5} M but this is unlikely to be appropriate for all species and methods (ch. 3).

BW284C51 has been used in physiological experiments on cortical arousal in cats
(Desmedt and La Grutta 1957). Despite its low lipid solubility intra-arterial injections of comparatively large doses (of the order of 225 µg) did produce some
central effects. The authors raised the possibility that the actions were not a reflection of the inhibition of AChE but were attributable to the quaternary ammonium
group acting directly on the receptors. McIsaac and Koelle (1959) have also demonstrated that high doses of BW284C51 enter nervous tissue while low doses do not.
In the same series of experiments McIsaac and Koelle used a low dose of
BW284C51 to inhibit only the external AChE. They showed that a subsequent
injection of lipid-soluble DFP was without effect on ganglionic function indicating
that intracellular AChE (which would have been inactivated by DFP) was functionally unimportant.

11.4 Miscellaneous compounds which possess some anticholinesterase activity

Long (1963, see also Augustinsson 1948) gives extensive lists of compounds known
to have some activity against cholinesterases. In most cases effective concentrations
are in the range of 10^{-2} to 10^{-4} M and Long points out that virtually all classes of
agent known to pharmacologists are represented in his tables. Although the anticholinesterase action may be insignificant compared with the primary effect, it is
nevertheless important to bear in mind that any drug used in an experiment could
be affecting cholinesterase levels. Types of compound known to have some inhibitory effect include alcohols and aldehydes; amino acids and amines; anaesthetics
(both local and general, including barbiturates) and other CNS depressants; anal-

gesics such as morphine and amidone (see also Lane et al. 1966; Weinstock 1971); antibiotics; anticholinergic drugs (e.g. atropine); anticoagulants; anticonvulsants; antihistamines; antimalarial drugs; antipyretics; bile salts; CNS stimulants (e.g. strychnine and picrotoxin; also LSD, see Nandy and Bourne 1964); hormones; inorganic ions; tranquilizers (e.g. chlorpromazine, see Maickel 1968); tubocurarine; vesicants; vitamins, and xanthine derivatives such as caffeine and theophylline.

This is a formidable list but in practice the concentration at which the drug achieves its specific purpose in an experiment will generally be below that at which it affects cholinesterase. In the case of barbiturate anaesthetics, for example, the concentrations shown to be somewhat inhibitory in the test-tube were of the order of 10^{-2} or 10^{-3} M (see Long 1963) but the concentration in the body following an anaesthetic dose (say 20 mg/kg) of sodium pentobarbitone would be of the order of 10^{-4} M. Where the inhibitory effect might become significant is in tissues from animals which have been killed with an unnecessarily large overdose. Although the anaesthetic Urethane (ethylcarbamate) is structurally related to physostigmine its inhibitory action is weak. Shukuya (1953b) found that a concentration of 2.1×10^{-1} M was necessary to match the degree of inhibition of human erythrocyte AChE that is produced by 10^{-7} M physostigmine.

Among the amines shown to be somewhat inhibitory are adrenaline (see also Zsigmond 1972), histamine and 5-hydroxytryptamine. Amino acids such as glycine and glutamic acid, which are widely used in iontophoretic studies will inhibit ChE but the concentration needed for 50% inactivation generally exceeds 10^{-1} M. Although the concentration used in multibarrel electrodes is of the order of $1-2$ M, the quantity of drug released into the tissue at any time is extremely small. One further point to keep in mind is that certain solvents used for dissolving drugs may be inhibitory. For instance, it has been shown by Sams and Carroll (1966) and by Gandiha and Marshall (1972) that DMSO (dimethyl sulfoxide) has some anticholinesterase activity. This is a compound which has been used not only as a solvent for drugs under physiological investigation but also as a flotation fluid in the production of ultrathin frozen sections (Bernhard and Viron 1971).

11.5 Reactivation of inhibited cholinesterase: Oximes

As already discussed (§ 11.2) the reaction between cholinesterase and an organophosphorus compound yields a stable phosphorylated enzyme complex in place of the very unstable acylated complex which is the normal intermediate in the enzymic reaction with a carboxylic ester. The phosphorylated complex breaks down only very slowly and this means that the enzyme may be inactivated for a long period. A number of agents have been developed which are capable of dephosphorylating the enzyme and so reversing the inhibition. Briefly these are nucleophilic compounds (see Wilson 1952) with a high affinity for phosphorus (table

TABLE 11.2
Oximes

1.   DAM   Diacetylmonoxime;
2-oximino-3-butanone

$$H_3C - \overset{\overset{\displaystyle O}{\|}}{C} - C\overset{\diagup CH_3}{\diagdown NOH}$$

2.   MINA   Mono*iso*nitrosoacetone

$$H_3C - \overset{\overset{\displaystyle O}{\|}}{C} - C\overset{\diagup H}{\diagdown NOH}$$

3.   PAD   Pyridine-2-aldoxime dodecyliodide

4.   2-PAM   Pyridine-2-aldoxime methiodide;
Pralidoxime iodide;
2-hydroxyiminomethyl-N-methyl pyridinium iodide;
1-methyl-2-formylpyridinium iodide oxime

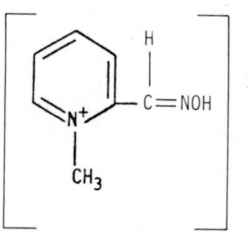

Pyridine-2-aldoxime   Pyridine-2-aldoxime methiodide

5.   P$_2$S   Pralidoxime methanesulphonate;
Protopam; Contrathion;
2-hydroxyiminomethyl-N-methyl pyridinium
methane sulphonate;
1-methyl-2-formylpyridinium methane sulphonate

TABLE 11.2 (continued)

6.   TMB-4; 4D3   1,3-bis(pyridinium-4-aldoxime) propane dibromide;
N'N-trimethylene bis(pyridinium-4-aldoxime) bromide;
1,1'-trimethylene bis(4-formylpyridinium bromide dioxime)

7.   Toxogonin;   1,3-bis (pyridinium-4-aldoxime)2-oxapropane dichloride;
       Lü6;   N'N-oxydimethylenebis (pyridinium-4-aldoxime) chloride;
       Obidoxime   bis 4-hydroxyiminomethylpyridinium-1-methyl ether dichloride.

11.2), the most efficient reactivators being those with a strongly ionizable oxime group (= NOH). Oximes are of some use in raising the LD_{50} of carbamate anticholinesterases but they are ineffective against those other non-phosphorylating inhibitors such as ambenonium (table 11.1) and BW284C51 which are not carbamates (see Hobbiger 1963). Although the protection afforded in vivo against carbamates may depend primarily on mechanisms other than the reactivation of the enzyme (see below), there is now evidence that oximes can effect the decarbamylation of inhibited enzyme (see, for example, Kuhnen 1972). This point was at one time in doubt (Hobbiger 1963). The need to develop antidotes to organophosphorus poisoning, made urgent by the widespread use of insecticides, has led to the synthesis and testing of a large number of possible candidates (Childs et al. 1955; Berry et al. 1959; Hobbiger and Sadler 1959; Grob 1963a; Hobbiger 1963; Kitz et al. 1965; Patočka et al. 1970; see also Benz 1970) and to studies on the interactions between enzyme, inhibitor and reactivator (Hobbiger 1955, 1963; Davies and Green 1956; Hobbiger and Vojvodić 1966; Wang and Braid 1967; Usdin 1970; Kuhnen 1972).

Among the points to emerge from this work is that a particular oxime is not equally effective against all organophosphorus compounds nor in all species. These findings are of obvious importance in considering therapeutic measures but they are also relevant to questions of enzyme kinetics and the configuration of the active site. In considering therapy it is vital to realize that although an oxime such as 2-PAM (see table 11.2) can effect life-saving reactivation following poisoning with certain compounds it may be virtually useless against others. Hobbiger (1957a, b) showed that about 20–30 times more 2-PAM is needed to reactivate an enzyme blocked with a di*iso*propylphosphoryl group (e.g. by DFP) than to reactivate one blocked with a diethylphosphoryl group (e.g. by Parathion, Paraoxon or TEPP; see table 11.1). Similarly, 2-PAM has little effect on enzymes poisoned with Tabun or OMPA (table 11.1) which have amidophosphoryl groups (Davies and Green 1959); Tabun-poisoned nerves can, however, be readily reactivated by TMB-4 (table 11.2, Heilbronn and Sundwall 1964). Enzymes poisoned with Soman (table 11.1) can be reactivated only slightly, even with TMB-4. The reason for this resistance is not fully understood. It is probably due mainly to the rapid aging of the inhibited enzyme, but steric effects and the toxicity of TMB-4 itself may also be involved (see Heilbronn and Tolagen 1965; Berry et al. 1966; Loomis and Johnson 1966; Fleischer et al. 1967; Bošković et al. 1968).

In some instances the use of two oximes in combination is recommended. O'Leary et al. (1961) showed that in rabbits poisoned with Sarin, atropine with 2-PAM plus TMB-4 was more effective in raising the LD_{50} than was atropine plus a single oxime. In dogs it was necessary to give the two oximes together, plus atropine, to combat Tabun poisoning but, in contrast to the findings in the rabbit, atropine with a single oxime was as efficient as atropine with two oximes in counteracting the effects of Sarin. This use of atropine as well as oximes is generally essential for efficient therapy (see, for example, Johnson and Stewart 1970; Natoff and Reiff 1970). It is not a question of synergism but stems from the fact that most oximes are only slightly lipid-soluble (DAM and MINA are exceptions) and do not readily enter the CNS (Loomis 1963; Hobbiger and Vojvodić 1966, 1967). Atropine, on the other hand, crosses the blood–brain barrier and, by blocking receptors, protects against the muscarinic effects of accumulating ACh.

Certain precautions are necessary in conducting experiments to test the ability of an oxime to reactivate phosphorylated cholinesterase. Karlog and Peterson (1963) examined the reactivation of inhibited ChE using the method of Ellman et al. (1961). They found that oximes in a concentration of 10^{-3} M or above would accelerate the non-enzymic hydrolysis of AcThCh and, unless this effect is allowed for, the value obtained for reactivation will be too high. A source of error in assessing the slight reactivation which can occur in the CNS in vivo was discussed by Hobbiger and Vojvodić (1966). They pointed out that significant reactivation can occur during the processing and storage of samples since oxime may be released from the tissue as it is homogenized. They also showed in rats that the amount of reactivation of phosphorylated brain enzyme which can be achieved by TMB-4 or by Toxogonin differs in different areas of the brain, hence homogenates of whole brain give misleading results. The authors suggested that the variation in degree of reactivation reflected regional differences in the amount of functional (i.e. more accessible, extracellular) enzyme and non-functional (i.e. less accessible, intracellular) enzyme. This thesis was extended in a subsequent paper (Hobbiger and Vojvodić 1967) in which it was reported that while TMB-4 and 2-PAM reactivate respectively 1/3rd and 1/5th of functional AChE in rat brain both are virtually without effect on non-functional enzyme. The authors suggested that homogenates of cerebellum rather than whole brain should be used in the assessment of in vivo reactivation of brain AChE. Although this work showed that compounds such as TMB-4 and 2-PAM which are only slightly lipid-soluble do have some central effects in rats and mice, Hobbiger and Vojvodić were of the opinion that in terms of therapy the peripheral effects were of far greater importance. They did, however, point out that the permeability of the blood—brain barrier is not uniform and it cannot be assumed that the degree of reactivation of functional AChE in one area, for example the cerebellum, gives an accurate indication of conditions at the all-important respiratory and cardiovascular centres (see also Bajgar et al. 1971; Filbert et al. 1972).

Oximes usually afford some protection against poisoning if taken prior to exposure to organophosphorus compounds (see e.g. Davies et al. 1959; Heilbronn and Sundwall 1964) but in clinical practice poisoning has generally occurred before oximes are given. In such cases speed is essential because, as explained above (§11.2.5), the phosphorylated enzyme, initially susceptible to dephosphorylation by the oxime gradually 'ages' to an acid form which is no longer able to react with the anionic oxime (see Davies and Green 1956). Another different type of protection has been demonstrated by Berry et al. (1971) in the case of Soman. These authors showed that if animals were treated with an oxime-responsive organophosphorus compound such as TEPP before the administration of Soman, and were given oximes and atropine afterwards, the LD_{50} of Soman was increased. Species differences were, however, evident, the regimen being far more effective in guinea-pig than rat.

Species differences in the antidotal effect of oximes against organophosphorus poisoning are important in two ways. First, there is the question common to any toxicity trial of how far results obtained on laboratory animals are applicable to man. Second, when organophosphorus compounds are used as sprays on crops, or on animals to control ticks and warblefly, or as general insect repellents in barns, hen-houses and other accommodation, a variety of animals are at risk (see for instance Anderson and Machin 1969; Schlinke 1970; Solly et al. 1971a, b; Khan 1973). A few examples will illustrate the complexities of species differences. Askew (1956) showed DAM was about 10 times more effective against Sarin poisoning in rat than in mouse, guinea-pig, rabbit and monkey. Similarly, Cohen and Wiersinga (1959) found the LD_{50} of Sarin in mice to be unchanged by 2-PAM, MINA and DAM although the two latter were effective in rats (see also Davies and Green 1959). In the case of Haloxon-poisoning of poultry, 2-PAM has a marked effect on the rate of enzyme reactivation in geese and ducks but does not increase the already rapid spontaneous recovery in hens (Lee and Pickering 1967).

Oximes are themselves toxic (Berry et al. 1959; Hopff and Waser 1970) and here again species differences are evident. With MINA the LD_{50} for mice is 150 mg/kg (Dultz et al. 1957) but in rats it is lower and, furthermore, shows a sex difference, males being the more susceptible (see Davies and Willey 1958). P_2S (Protopam) is about equally toxic to rat, mouse and guinea-pig but its toxicity to dog is much greater (Davies and Willey 1958). Since injections of atropine sulphate are an essential part of the treatment of organophosphorus poisoning (see §11.6.2.1) Davies and Willey examined the toxicity of P_2S in the presence as well as the absence of atropine. In rats the administration of atropine raised the LD_{50} of P_2S, in mice it had no effect, and in guinea-pigs it lowered it. The way in which oximes exert their toxic effect is only partly understood and it is likely that a number of factors are involved. The signs produced by an overdose of an oxime include hyperventilation followed by respiratory failure, convulsions and death, a pattern similar to that seen in cyanide poisoning. It was established by Askew et al. (1956) that lethal doses of MINA and DINA yield hydrogen cyanide in sufficient amounts to account for their toxicity, but in the case of DAM it is the compound itself rather than any metabolite which is toxic. Enander et al. (1961a, b) have shown that although P_2S forms thiocyanate and another toxic metabolite, N-methylpyridinium-2-nitrile the quantities produced are too small to explain the toxicity. Work by Jović and Bošković (1970) has indicated further complexities of oxime toxicity. They examined, in mice, the effectiveness of LüH6 (Toxogonin), PAM and TMB-4 against GT-45 (chemical formula O,O'-diethyl-S-[2-(N-phenylamino)ethyl] thiophosphonate methylsulfomethylate) and against two of its analogues, called Compounds I and II. With GT-45 and Compound I, all of the oximes when given with atropine were effective reactivators. With Compound II, PAM produced some reactivation, LüH6 had no effect and TMB-4 potentiated the toxicity. Tests on ChE, in vitro and in vivo seemed to eliminate the possibility that TMB-4 and Compound II reacted to form an even more potent anticholinesterase of the type which results when PAM reacts with Sarin (Hackley et al. 1959). Because atropine, but not d-tubocurarine nor hexamethonium, abolished the additional toxicity, the authors tentatively suggested that TMB-4 and Compound II form, not an anticholinesterase, but a potent cholinomimetic which acts at postganglionic parasympathetic sites. One or two other pieces of evidence have suggested that dephosphorylation of ChE is not the only way in which oximes can influence cholinergic mechanisms. Grob and Johns (1958) found in man that neuromuscular block produced by non-phosphorylating carbamate inhibitors such as neostigmine and bispyridostigmine was relieved by oximes just as effectively as that produced by the phosphorylating anticholinesterase, Sarin. The degree of reactivation of cholinesterases both in vivo and in vitro was, however, much lower for carbamylated than phosphorylated AChE. A possible explanation for the physiological effect is that oximes interfere with the reaction of ACh at receptors, so minimizing the effect of ACh accumulation. The difficulty about this idea is that, as mentioned above, oximes raise the LD_{50} of non-phosphorylating anticholinesterases only in the case of carbamates. They are without effect in cases of poisoning with BW284C51 for example. One could suggest that the enzyme-inhibitor complex formed with non-carbamates is so orientated as to prevent oxime, but not ACh, from reaching the receptor — but this is pure speculation. Quite recently Mayer and Michalek (1971) have shown that although obidoxime (Toxogonin) does not reactivate DFP-inhibited cholinesterase in rat brain it does, nevertheless, antagonize the rise in brain ACh. Among possible explanations, they suggest that ACh synthesis may be modified by the oxime.

Most of what has been mentioned so far concerns oximes in their therapeutic role but they are also of some use as an experimental tool. Baker and Benedict (1968), for example, investigating cholinergic mechanisms in the cat hippocampus gave 2-PAM by microinjection to reverse the effects produced by DFP. As mentioned above, oximes have also been employed in studies of the active site and of other features of cholinesterases. The efficiency with which a compound can reactivate a phosphorylated enzyme will be influenced to some extent by steric considerations (see Childs et al. 1955). In theory, the most efficient reactivators should be those with dimensions which ensure that when the cationic group is bound to the anionic site of the enzyme, the nucleophilic moiety is in a favorable position for attracting the phosphorus-containing group off the esteratic site of the enzyme. Because other factors such as *pK* and lipid solubility can have a marked effect on potency there is, however, no simple relation between biological action and chemical structure. Berry et al. (1959) have shown that considerable alterations can be made to the oxime molecule without reducing its potency as a reactivator. Despite these reservations some conclusions about the geometry of the active site of cholinesterases have been drawn from experiments with oximes. O'Brien (1967) pointed out that there is no obvious reason why TMB-4 is twenty times better than 2-PAM at reactivating phosphorylated enzyme and tentatively suggested that the finding revives the idea (see ch. 2) that AChE might have two anionic sites for each esteratic site. Kuhnen (1972) working on bovine erythrocyte AChE used carbamylcholine as a competitive inhibitor with Toxogonin as the reactivator and found evidence that a secondary binding site, in addition to the active centre, was involved in the mechanism of reactivation.

11.6 Environmental pollution and accidental poisoning

11.6.1 Pollution

The use of anticholinesterase agents as pesticides was discussed in ch. 5 and the point made that the control of reputedly undesirable creatures may itself prove a threat to the environment. Unless pesticides are highly selective they may destroy not only the creatures directly under attack but others as well. In addition, the breaking of food chains will jeopardize environmental well-being still further. In general the anticholinesterase type of pesticides are much less stable than the organochlorine type such as DDT. With some exceptions (see DuBois 1963; O'Brien 1967) their half-lives are measured in days or weeks rather than months or years. This means that they do not create a serious problem with regard to the persistence of toxic residues on crops. On the other hand, if an individual is repeatedly exposed to these compounds, for example while spraying crops, his blood cholinesterase may decrease progressively and reach a dangerously low level despite precautions to avoid obvious contamination. Another hazard is that some compounds are absorbed by the plants on to which they are sprayed and may be metabolized to compounds which are more toxic than the original (see Heath 1961). The likelihood that organophosphorus compounds may yield different metabolites in plants and in animals (see Iqbal and Menzer 1972) must be kept in mind in toxicity trials. The plants themselves may be adversely affected by anticholinesterases applied to 'protect' them. An example of this is given by Russian workers in the case of Phosphamidon (Dimecron). Akmal'khanov et al. (1969) found that when lucerne was treated with this pesticide, the nutrient value was 10 to 15% less than normal. Cows fed on the treated lucerne had lowered blood cholinesterases, they gave less milk than controls and the milk-fat and protein content was also below normal. Dimecron was detectable in the milk, and, if present in high concentration, was only partially destroyed by boiling.

11.6.2 Accidental poisoning

The risks inherent in the widespread use in agriculture and horticulture of compounds which act by virtue of their anticholinesterase properties are now well documented and, in consequence, their potential danger is lessened. A more insidious threat comes from substances which may be used in a number of processes and which may not be recognized as being, or as giving rise to anticholinesterases. The literature, particularly that from Russia, contains reports that workers in certain industries have subnormal levels of blood cholinesterase. These include nylon spinners (Antonova 1967), painters in shipyards (Bublewska et al. 1969) and men producing chloroprene rubber (Gasparyan 1964). Reduced ChE activity does not necessarily indicate exposure to an anticholinesterase: as discussed in chs. 9 and 10 other indirect factors could be responsible but until the suspected compounds have been tested against cholinesterases they must be regarded as potential inhibitors. While it is obviously important to acknowledge the existence of unrecognized anticholinesterase agents the problem should be kept in proportion. As mentioned earlier an enormous number of compounds possess some anticholinesterase activity but this may be of no physiological significance in terms of the concentrations likely to be encountered in daily life. Thus while Galzigna et al. (1969) showed that in vitro there was some inhibition of horse serum ChE by triethyl lead – the main metabolite of the extensively used petrol additive, tetraethyl lead – this particular effect may be unimportant in the context of atmospheric pollution. Similarly, although some acrylamides employed as flocculators in the purification of water in reservoirs cause damage to the CNS and a fall in ChE when added to the drinking water of rats, relatively high doses are needed for this effect (Strizhak 1967).

Some of the most historically interesting instances of accidental poisoning by anticholinesterases involve the aryl organophosphorus compound tri-orthocresyl phosphate (TOCP). This substance is heat-stable and has many applications in industry. Among other uses it has been employed as a plasticizer and as an additive in paints, printers ink, aviation spirit and other fuels, and in some types of lipstick. In 1930 Smith et al. established that TOCP was responsible for 20,000 or more cases of paralysis then occurring in the U.S.A. This was the era of prohibition and all the victims had been drinking extracts of Jamaica Ginger to which various oily substances, some containing TOCP, had been added to increase the potency. Subsequently numerous similar episodes have occurred, many of them during the Second World War when mineral oils produced for torpedoes and machine guns were appropriated for cooking. In 1959 over 2000 people in the Meknes district of Morocco developed paralysis after cooking with olive oil which had been diluted with aircraft lubricating oils. These oils had been synthesised specifically for certain types of turbo-jet engines but as the engines became obsolete the oils were of no further use and had come into the hands of a dealer who sold them for cooking (see Smith and Spalding 1959).

Man has not been the only victim in these episodes. In the Meknes outbreak a family who were suspicious of the oil fed it to their dog. Since it showed no ill-effects they used the oil for their own meals but later all, including the dog, developed signs of poisoning. Grossmann and Parnitzke (1952) described how tigers in a German zoo became paralysed following treatment of their cage with floor oil which contained TOCP.

11.6.2.1 Clinical course of poisoning   Anticholinesterase poisoning does not constitute a single condition but is manifest in different ways depending on the causative agent. Effects can be immediate (e.g. with eserine), delayed (e.g. with TOCP) or both (e.g. with DFP). The immediate effects are directly attributable to the excess of ACh which is able to accumulate in the absence of AChE. Most symptoms can be controlled by atropine which prevents the ACh acting at muscarinic sites. Effects include copious salivation, bronchoconstriction, bradycardia, stimulation of the gut, vomiting, pupillary constriction, muscular twitching, mental confusion, convulsions, coma and respiratory paralysis. This picture is seen in cases of acute poisoning following ingestion of, or exposure to, a single toxic dose; some of the signs may also be present to a greater or lesser degree in patients with subnormal cholinesterase levels produced by prolonged exposure to smaller amounts of anticholinesterases (see Metcalf and Holmes 1969 for psychological effects). These chronic 'cholinergic' effects must not be confused with the chronic neurological deficit which may be a long-term consequence of poisoning. Clinical signs of neurological damage appear about 7–15 days after contact with certain anticholinesterases and result from demyelinating lesions in various areas of the peripheral and central nervous system (see Davies 1963 for a description of events in man and other species; see also Cavanagh and Patangia 1965).

11.6.2.2 Mechanism of demyelination   The most notorious demyelinating agents are fluorine-containing alkyl organophosphorus compounds such as DFP and Mipafox and aryl inhibitors like TOCP. As discussed in the previous chapter the mechanism responsible for demyelination is not clear. Early evidence (see ch. 10 §10.9) suggested that the lesions were a direct consequence of the inhibition of pseudo-cholinesterase but this now seems unlikely. DFP, which is a potent demyelinating agent, causes only a transient fall in pseudocholinesterase whereas certain other compounds, for example iso-OMPA and TIPP (tetra*iso*propylpyrophosphate), produce prolonged depression of the enzyme and yet are without effect on the CNS (see Davison 1953b; Witter and Gaines 1963). The dramatic and easily measurable inhibition of cholinesterases tended, originally, to overshadow the possibility that anticholinesterases could be influencing a number of systems but evidence of multiple effects has gradually accumulated. Porcellati et al. (1961) found elevated proteinase activity in sciatic nerve and spinal cord (but not in brain) from hens poisoned with DFP and TOCP. Similarly, Clouet and Waelsch (1963) showed that the incorporation of $[^{14}C]$-lysine into rat brain was increased 2 days after the rats had been injected with the *non-demyelinating* inhibitor 217–AO. They tentatively interpreted this as a response to increased protein catabolism. Porcellati (1965) injected DFP into cats and found a significant decrease in protein synthesis in spinal cord and ganglia very shortly after injection; in spinal roots, however, there was a small increase. More recently, Jović et al. (1971) have shown that Soman and DFP cause a depression of tissue respiration and of the activity of Na^+/K^+-ATPase, aldolase and succinic dehydrogenase (but not hexokinase) in slices of rat brain. The effects were qualitatively similar whether the slices were merely exposed to the inhibitor in vitro or were obtained from animals which had received a lethal dose in vivo. Rather different results were found by Glow et al. (1972) in rats which had been treated with sublethal doses of DFP for up to 24 weeks. In these animals there was an increase in the level of Na^+/K^+-ATPase in microsomal fractions from brain and kidney. Stevens et al. (1972) have reported that a number of insecticides including Parathion and Malathion, but excluding OMPA, depressed the ability of microsomes from mouse liver to metabolize hexabarbital, ethylmorphine and aniline.

By far the most attention has been given to the question of lipid metabolism. In demyelination something has gone wrong with the phospholipids but it has proved extremely difficult to establish the precise nature of the biochemical lesion. Any theory of demyelination must account for the delay of some 7–15 days which occurs between poisoning and the manifestation of neurological damage. Experimental evidence suggests there is some metabolic failure, either in the cell body or in the sheath itself, which interferes with myelin maintenance. A direct chemical assault on a component of the sheath seems unlikely in view of the finding that phospholipid levels in nerves from hens poisoned with DFP remain close to normal for at least 25 days after the administration of the poison (Porcellati and Mastrantonio 1964). Papers on lipid metabolism include the following: Austin (1957); Majno and Karnovsky (1961); Montanini and Porcellati (1964); Berry and Cevallos (1966). Findings in the different studies are not always consistent and the discrepancies may indicate that the mechanisms involved are not identical for different agents nor for different species. Porcellati and his colleagues (see Porcellati 1967) working on peripheral nerves from hens poisoned with DFP and TOCP demonstrated a decreased synthesis of phosphatidylcholine, phosphatidylethanolamine and monophosphoinositide. The difference in results from experiments in vivo and in vitro indicates some interference in the conversion of phospholipid phosphoric esters to their cytidine diphosphate derivatives. Porcellati has, however, questioned whether these effects are involved in the production of neurological lesions since rather similar changes in phospholipid metabolism occur in the mouse (Nelson and Barnum 1960), a species which does not normally show demyelination. He suggested that it may be the effects on protein biosynthesis, rather than those on phospholipid metabolism which are responsible for initiating demyelination. This view had also been put forward by Ansell and Chojnacki (1967). Aldridge et al. (1969) favoured the idea that the neurotoxic effect involved the phosphorylation of some protein, probably a hydrolytic enzyme which, while not immediately essential for the life of the neurone was vital for the continued maintenance of the structural integrity of its processes. In a series of papers Johnson (1969a, b, c, 1970) described how, by removing many of the apparently irrelevant sites that bind organophosphorus compounds in nervous tissue, he was able to identify the 'neurotoxic' site as an esterase which he termed the 'neurotoxic esterase'. He found (Johnson 1969b) in hen tissue that non-neurotoxic organophosphorus compounds such as TEPP did not inhibit the enzyme. Subsequently it became clear that it is not inhibition of the enzyme per se which is essential for the neurotoxic effects since carbamates, sulphonates and phosphinates will inhibit the enzyme yet they do not produce lesions (M.K. Johnson 1970, 1973). It seems that neurotoxicity results from the formation of a phosphorylated or phosphonylated protein. Although Johnson's elegant work may pinpoint the site of the initial attack by the neurotoxic agents the sequence of events which eventually culminates in demyelination is still not clear. It remains to be established which of the other reported changes, particularly those involving protein and phospholipid synthesis, are directly involved in the demyelinating reaction, which are a consequence of it and which are purely independent effects of the poison. Johnson's basic theory has subsequently been applied by Flockart and Casida (1972) to the question of the teratogenic action of some organophosphorus compounds and eserine in developing hens' eggs. They suggested that a similar phosphorylation (or carbamylation) of a specific esterase or esterases in the yolk-sac membrane by the teratogenic type of compound may initiate the events which culminate in embryonic abnormalities.

Among the factors which complicate the study of demyelination are those of species and age differences in the susceptibility of animals to the various agents. Lesions are produced less easily in chicks than hens (see Barnes and Denz 1953; Johnson and Barnes 1970) and rabbits are far more vulnerable than rats (Barnes and Denz 1953; see also Aldridge et al. 1969). While this might indicate, as suggested above, that the biochemical lesion is not identical in all cases other factors, such as the efficiency of detoxication and the integrity of the blood—brain barrier, could be responsible for the variations. The different effect of different compounds in the same species may also be a reflection of the ease with which each crosses the blood—brain barrier. Reports of the many outbreaks of TOCP poisoning (§11.6.2) have shown that the clinical picture can differ from episode to episode. In some instances the degree of involvement of the CNS exceeds that of the periphery and in others the reverse is true. Since the TOCP responsible for these outbreaks is unlikely to be pure, the severity of lesions and their distribution may well depend on the extent to which the particular toxic constituents can enter the CNS. Work with TOCP in vitro is complicated by the fact (§11.2.1) that the pure compound (but not the commercial samples) has no anticholinesterase activity. In vivo it is converted into a potent inhibitor of pseudocholinesterase but since it does not affect AChE this explains why there is no acute phase in TOCP poisoning. This is an advantage for in vivo experiments on demyelination. It should perhaps be emphasized that if DFP or Mipafox are used in place of TOCP in long-term experiments on neurotoxicity, severe cholinergic poisoning will occur in the early stages and this must be controlled with atropine. Although both DFP and Mipafox are used by histochemists as inhibitors of pseudocholinesterase they are differential rather than selective in action (see §11.3.2 and table 11.1) and it is their potential as inhibitors of AChE, as well as their demyelinating action, that makes them so hazardous (see Bidstrup and Hunter 1952).

11.7 Concluding remarks

The agricultural importance of organophosphorus compounds — over 50,000 had been synthesised by 1965 — coupled with their threat to health has made them the subject of intensive study. Many facets are available for investigation — their reaction with enzymes, their toxicity, their antidotes — and each of these has opened up a separate field of specialization which has ramified to include other types of cholinesterase inhibitor. Faced with the mass of available data it has proved difficult to exercise ruthless selectivity in compiling this chapter. While it is very far from comprehensive some of the points most pertinent to physiologists may, nevertheless, have become submerged beneath speculation, contention, claim and counterclaim.

Set out below are some of the major factors which have a bearing on the choice of an anticholinesterase agent and on the interpretation of experimental results, both one's own and those in the literature.

1) Selectivity is not absolute and an inhibitor 'specific' for one type of cholinesterase (e.g. BuChE) will usually inhibit the other (i.e. AChE) if the concentration is increased. The degree of selectivity is assessed by the extent to which the minimum concentration required to give maximum inhibition of one enzyme will also inhibit the other.

2) Inhibitors are not equipotent in all species.

3) An inhibitor is not necessarily equipotent in vitro and in vivo nor against the hydrolysis of different types of substrate by the same enzyme.

4) Some inhibitors cross the blood—brain barrier, others do not. With the appropriate combination of lipid-soluble and -insoluble inhibitors, central or peripheral and intra- or extracellular effects may be distinguished.

5) Inhibitors are often termed 'reversible' or 'irreversible' but in certain cases the effects of 'irreversible' inhibitors may wear off, at least to some extent.

6) Some compounds produce chronic neurological lesions. Many are extremely hazardous and their lipid solubility and volatility influences the outcome in a case of exposure.

7) Oximes (plus atropine) form an essential part of therapy in poisoning with some types of anticholinesterase but note a) prompt treatment is essential because in certain instances the inhibited enzyme 'ages' rapidly and becomes completely irreversible; b) oximes are not equally efficient in all species nor against all inhibitors; c) oximes are themselves toxic.

8) A very large range of compounds which may be used in physiological experiments (as anaesthetics etc.) possess some activity against cholinesterases but their inhibitory effect is usually insignificant at the concentration required for their specific action.

9) Many compounds exert effects which are independent of their inhibitory action on cholinesterases and which may complicate interpretion of results. Side effects of particular importance include the blocking or stimulating of receptors.

11.8 Appendix

11.8.1 Safe handling of the highly toxic anticholinesterases

These precautions apply to the laboratory rather than the field. Where possible the

TABLE 11.3

Relative toxicity of various anticholinesterases in rat and man. (From Grob 1963a.)

Relative anti-ChE activity in vitro	Rat LD₅₀ (mg/kg)		Rat Dose (mg/kg) that produced 50% depression of RBC AChE		Man Dose (mg/kg) that produced moderate symptoms		Man Estimated lethal dose (mg/kg)	
	I.M.	oral	I.M. or I.A.	oral	I.M. or I.A.	oral	I.M.	oral
Sarin 1	0.17	0.6	0.003 (I.A.)	0.01	0.006 (I.A.)	0.028	0.03	0.14
Tabun 1/5	0.80	3.7						
TEPP 1/10	0.65	1.4	0.025 (I.M.)	0.10	0.083 (I.M.)	0.35	0.38	1.7
DFP 1/100	1.8	6	0.07 (I.M.)	0.28	0.083 (I.A.)	?0.32	0.48	2.1
Parathion 1/4,000	6	10						
Malathion		1375 (males) 1000 (females)				?150		?600

I.M., Intramuscular
I.A., Intra-arterial
RBC, Red blood corpuscle

use of volatile, lipid-soluble inhibitors such as DFP (and, in particular, Soman which is resistant to antidotes) should be avoided. If such compounds must be used the following safety precautions are recommended.

1) All glassware etc. to be used in the experiment should be assembled before the inhibitor is handled (see below).

2) The operator should wear an eye-shield and gloves; arms should be covered and there must be no gap between gloves and sleeves.

3) No one should handle toxic inhibitors unless someone competent to give or to fetch assistance in case of accident is also present.

4) All operations should be done in an efficient fume-cupboard in which there is a large bowl of 10% sodium hydroxide or freshly prepared Decon (10%). The cold tap should be kept running and there should be a good supply of paper tissues in case of spillage.

5) All glassware, gloves, tissues etc. must be put into sodium hydroxide or Decon immediately after use.

6) A clearly labelled First Aid box should be instantly available and should contain

Surgical gloves (to protect doctor from contamination)

Swabs

70% alcohol

10×5 ml sterile syringes with needles to fit

20×1 ml ampoules of atropine sulphate for injection (1.2 mg/ml)

Doses of up to 2–4 mg repeated every 10–20 min may be needed so several boxes of ampoules are necessary

Glass knife for opening ampoules

10×20 ml sterile syringes with wide-bore needles to fit

2×100 ml sterile water (pyrogen free). This becomes out of date and should be replaced as appropriate

5×1 g Pralidoxime (Protopam, P_2S)

Clear instructions for the administration of atropine and for the preparation and injection of an aqueous solution of Pralidoxime, together with the names and telephone numbers of medically qualified staff

7) Medically qualified staff should each be given a copy of the instructions and, if possible, be warned whenever a dangerous amount of inhibitor is to be handled. They should be advised to read Grob (1963a) on the treatment of anticholinesterase poisoning in man. The relative toxicities of a number of compounds are shown in table 11.3

11.8.2. Handling of anticholinesterases in general

The need for protection against accidental poisoning when handling very toxic materials such as Soman or DFP is so obvious that the type of safety precautions outlined above are easily enforced. With non-volatile and less potent inhibitors the risk to life is small and this may lead to careless technique. At all times the compounds should be handled in such a way that contamination of balance controls, balance pans and other surfaces is avoided. The accidental and unrecognized presence of an inhibitor can prove disastrous in an experimental series. To minimize the possibility of contamination it is advisable to treat all inhibitors as if they were highly toxic. In this laboratory all the glassware etc. to be used in the preparation of the stock solution of inhibitor and for the addition of inhibitor to the preincubation and incubation media is selected, labelled and put on a tray together with the solvent for the inhibitor and the required volumes of media. Prior assembly of the necessary equipment prevents the contamination which could occur if glassware etc. has to be fetched from cupboards and drawers after the inhibitor has been handled. Preparation or dilution of inhibitor and its addition to the media is done in the fume-cupboard and used glassware placed at once in Decon. On removal from the fume-cupboard, the flasks containing media plus inhibitor are stood on petri dishes and never directly on the bench. After the experiment is over, any tubes, staining jars etc. which have been in contact with inhibitor-containing media are soaked in Decon.

CHAPTER 12

Postscript

When I submitted the preliminary outline of this book to the Editors I was happily unaware of the magnitude of the subject. Tidily but naïvely I included a final chapter headed 'Summarizing comments'. Now that the moment has come to write the last chapter I realize that to provide a summary would be inappropriate as well as difficult. A book such as this, written by one author, must inevitably lack the accurate detail and definitude to be found when individual experts contribute to a multiauthor work. On the other hand, the single author is forced to take a very broad view and in so doing becomes aware not only of the well documented 'facts' but also of discrepancies and controversies. Herein lies the difficulty of summarizing: it seems that almost any statement of fact requires some sort of qualification about experimental conditions, tissues or species. What would seem more useful in place of a summary of what is known is some indication of areas where we are still largely in the dark.

In the context of AChE which is involved in transmitter mechanisms the remaining problems are mainly ones of detail. How, exactly, does AChE get from the perikaryon to the synapse? What, exactly, is its relation to the ACh receptors in the membrane? What, exactly, is the mechanism of hydrolysis and is there, or is there not, substrate inhibition in vivo? The two areas which are far more murky concern first, the nonsynaptic role of AChE in, for example, developing neurones, blood vessels and erythrocytes, and second, the action of pseudocholinesterase. The data assembled in the relevant chapters shed little light on the problem, being largely negative so far as function is concerned. Perhaps the question to ask is not 'What do the enzymes do?' but 'Why are they there?'. Barker et al. (1972) have suggested that transmitter-degrading enzymes may be constitutive gene products — in other words the cells happen to make them whether or not they need them. But if this is the case it raises another question — and one with which it is appropriate to end since it epitomizes a feature which occurs again and again in considering cholinesterases — 'Why are there such wide species differences in the occurrence, distribution and nature of non-neuronal cholinesterases?'

Abbreviations

AAc	Amylacetate
AAD	L-amino acid decarboxylase
ACh	Acetylcholine
AChE	Acetylcholinesterase
AcSeCh	Acetylselenocholine
ACTH	Adrenocorticotrophic hormone
AcThCh	Acetylthiocholine
ASaCh	Acetylsalicylcholine
ATP	Adenosine triphosphate
b_{30}	A measure of enzyme activity (see ch. 3)
BuCh	Butyrylcholine
BuChE	Butyrylcholinesterase
BuThCh	Butyrylthiocholine
BW284C51	An anticholinesterase (see ch. 11)
BzCh	Benzoylcholine
ChAc	Choline acetyltransferase (or acetylase)
ChE	Cholinesterase
CNS	Central nervous system
C-R	Crown-rump length (used to estimate the age of foetuses)
DEAE-	Diethylaminoethyl-
DFP	Di*iso*propylphosphorofluoridate (an anticholinesterase, see ch. 11)
DOPA	Dihydroxyphenylalanine; 3-hydroxytyrosine
EDTA	N,N'-ethylenediaminediacetic acid
EP	Ethopropazine (an anticholinesterase, see ch. 11)
ER	Endoplasmic reticulum
GABA	γ-aminobutyric acid
GABuCh	γ-aminobutyrylcholine
GTP	Guanosine triphosphate
5-HT	5-hydroxytryptamine (serotonin)
5-HTPD	5-hydroxytryptophan decarboxylase
I_{50}	A measure of concentration of inhibitor (see glossary)
K_M	Michaelis constant (see glossary)

LD_{50}	A measure of toxicity. The dose which is lethal to 50% of the animals tested
MAO	Monoamine oxidase
MB	Methylbutyrate
MeCh	Acetyl-β-methylcholine
NPA	O-nitrophenylacetate
Nu 1250	An anticholinesterase (also known as RO 2-1250)
PA	Phenylacetate
PAM (or 2-PAM)	An oxime (see ch. 11)
PB	Phenylbutyrate
pI_{50}	A measure of concentration of inhibitor (see glossary)
PrCh	Propionylcholine
PrThCh	Propionylthiocholine
pS	A measure of substrate concentration (see glossary)
RBC	Red blood corpuscle
RER	Rough endoplasmic reticulum
RNA	Ribonucleic acid
SCG	Superior cervical ganglion
SER	Smooth (agranular) endoplasmic reticulum
SuccCh	Succinylcholine
TA	Triacetin
TB	Tributyrin
TDF	p-(trimethylammonium) benzenediazonium fluoborate
TEA	Tetraethylammonium
TEPP	An anticholinesterase (see ch. 11)
TMB-4	An oxime (see ch. 11)
TOCP	Tri-O-cresylphosphate (a compound which inhibits cholinesterase, see ch. 11)
TPA	Tetrapropylammonium
TRIEG	Gallamine triethiodide (Flaxedil)

Glossary

This lists some biochemical, enzymological and other terms which are used in connection with cholinesterases

Acyl	Radical of an organic acid, e.g. acetyl-, butyryl-
Agonist	An agonist is a drug which combines with receptors and initiates the specific effects subserved by those receptors. Drugs which block these effects are called antagonists
Allele (allelomorph)	Any one of a series of two or more genes occupying the same position or locus on homologous chromosomes
Ammon unit	A unit of cholinesterase activity defined as the amount of enzyme which when added to 3 ml of 0.0092 M AChBr will hydrolyse 0.01 mg AChBr in 1 min
Azole	$-N<$, tertiary N atom in a ring structure
Carbonyl	$=CO$
Carboxyl	$>C=O$
Cholinergic	Originally this denoted a cell which released ACh as its transmitter and also applied to the type of transmission occurring at synapses of such cells. The term has become broader and is used to describe drugs which mimic the action of ACh and also to describe cells rich in AChE. This latter usage may be misleading, see ch. 1
Cholinoceptive	This denotes a cell which responds to ACh
Coulombic forces	Attractive or repulsive forces between charged groups on molecules
Curares	
i) Pachycurares	Stabilizing agents, e.g. curare. These block depolarization by interfering with the access of ACh to the receptor
ii) Leptocurares	Depolarizing agents, e.g. decamethonium. These mimic ACh by producing depolarization which is prolonged because these agents are not hydrolysed by AChE. The membrane is thus rendered insensitive to ACh
ΔE^*	This is the extra internal energy which the molecule requires from thermal fluctuations to overcome the barrier to denaturation

Extinction coefficient	This is a measure of the amount of light absorbed by a substance in solution
First order kinetics	See 'Zero order kinetics'
I_{50}	The molar concentration of an anticholinesterase which causes 50% inhibition
Imino	HN= , an NH group double-bonded to another atom, usually C
Isoelectric point	The pH at which the net charge on a protein is zero
K_M	See Michaelis constant
Ligand	Ligand is a word very commonly encountered in the literature on receptors but it is rarely defined in this context. The term comes from inorganic chemistry (see for instance Cotton and Wilkinson 1966) where a ligand may be defined as any atom, ion or molecule capable of functioning as the donor in one or more co-ordinate (i.e. non-covalent) bonds. The formation of co-ordinate compounds is particularly characteristic of transition elements such as iron, and the pyrrole groups bound to iron in haemoglobin are good examples of ligands in the strict chemical sense. In 'receptor' jargon a ligand means any small molecule which forms a reversible non-covalent bond with the macro-molecule (e.g. ACh receptor, AChE) under investigation
Lineweaver—Burk plot	A plot of the reciprocal of the initial rate of an enzyme reaction against the reciprocal of the molar substrate concentration used. The plot is commonly used as a graphical method for K_M and V_{max} determinations
Michaelis constant	K_M, the molar concentration at which the rate of enzyme reaction is half of the maximum velocity. The maximum velocity is termed V_{max}
Monomer, dimer etc	Terms denoting the number of units (chains, molecules etc) in a protein
Normal molar volume	22.4 l, the volume occupied by 1 mole of gas at standard temperature and pressure
Nucleophilic	A nucleophilic group is one which, having an excess of electrons, attacks an electron-deficient or partially positively charged group
Neuropil(e)	The neuroanatomical term for the interlaced dendrites, axon terminals and glial processes which surround the nerve cell bodies. The suffix -pil (or -pile) is derived from the Greek word 'pilos' meaning 'felt'
pK	The negative logarithm (to the base 10) of a dissociation constant (K)

pI_{50}	The negative logarithm (to the base 10) of the molar concentration of an anticholinesterase which causes 50% inhibition
pS	The negative logarithm (to the base 10) of the molar concentration of substrate
Q_{10}	The proportional increase in rate of reaction for a $10°C$ rise in temperature
Quaternary compound	A positively charged compound formed from an element belonging to group 5 of the periodic table
Second order kinetics	See 'Zero order kinetics'
Sedimentation coefficient	This is expressed in Svedberg units, S, and denotes the rate of sedimentation of a macromolecule in a given centrifugal force. Used in conjunction with data provided by various other techniques it gives information about molecular shape and weight. $s_{20,w}$ indicates that the sedimentation coefficient has been corrected to the value obtaining in water at $20°C$
Specific activity	This is properly defined as the number of μmoles of substrate consumed or of product formed per mg enzyme protein per min under given incubation conditions. In practice many authors express it per hr
Tertiary structure	The structure of a protein is arbitrarily divided into 4 levels a) primary — the amino acid sequence b) secondary — that part of the structure formed by hydrogen bonding interaction between the $C = O$ and NH groups of peptide bonds, giving rise to the α-helix and pleated sheet conformations c) tertiary — the overall shape of the protein molecule or part thereof d) quaternary — concerned with the number of protein chains in the molecule
Turnover number	The older definition was the number of molecules of substrate transformed per minute per molecule of enzyme. Recently 'per catalytic centre' or 'per active site' has replaced 'per molecule'
Van der Waals' forces	The attractive force between atoms or molecules which is generated when electrons in adjacent atoms or molecules move in sympathy with each other
V_{max}	See Michaelis constant
Zero order kinetics	When an enzyme is incubated with increasing concentration of substrate and the rate of reaction is plotted against substrate concentration a hyperbolic curve is usually obtained.

(cont.)

At low substrate concentrations the rate is directly proportional to these concentrations and is said to be first order with respect to substrate. At the other end of the substrate concentration range, when further increases do not have any effect on the rate (V_{max}), the kinetics are described as zero order with respect to substrate. In more complex situations, where the rate is proportional to the square of the substrate concentration or where two substrates are involved, the kinetics are described as second order

References

Abdel-Latif, A.A. (1966). Acetylcholine and the incorporation of [^{32}P] phosphate into phospholipids and phosphoproteins of nerve endings of developing rat brains. Nature 211, 530–531.

Abdel-Latif, A.A., Smith, J.P. and Ellington, E.P. (1970). Subcellular distribution of sodium-potassium adenosine triphosphatase, acetylcholine and acetylcholinesterase in developing rat brain. Brain Res. 18, 441–450.

Abercrombie, M. and Johnson, M.L. (1946). Quantitative histology of Wallerian degeneration. 1. Nuclear population in rabbit sciatic nerve. J. Anat. 80, 37–50.

Ábráham, A. (1960). Zur Kenntnis der Struktur der Netzhaut mit besonderer Berücksichtigung der Ganglienzellenschicht. Z. Zellforsch. Mikroskop. Anat. 52, 529–548.

Abrahams, V.C. (1963a). Histochemical localization of cholinesterases in some brain stem regions of the cat. J. Physiol. London 165, 55 P.

Abrahams, V.C. (1963b). Cholinesterases of the cat brain-stem. Biochem. Pharmacol. 12, Abst. 569, 161–162.

Abrahams, V.C. and Edery, H. (1964). Brain stem electrical activity and the release of acetylcholine. In: Progress in Brain Research, Vol. 6, W. Bargmann and J.P. Schadé, eds. (Elsevier, Amsterdam) pp. 26–36.

Abrahams, V.C., Koelle, G.B. and Smart, P. (1957). Histochemical demonstration of cholinesterases in the hypothalamus of the dog. J. Physiol. London 139, 137–144.

Adams, C.W.M., Grant, R.T. and Bayliss, O.B. (1967). Cholinesterases in peripheral nervous system. I. Mixed, motor and sensory trunks. Brain Res. 5, 366–376.

Adams, D.H. (1949). The specificity of the human erythrocyte cholinesterase. Biochim. Biophys. Acta 3, 1–14.

Adams, D.H. and Whittaker, V.P. (1948). The specificity of the human erythrocyte cholinesterase. Biochem. J. 43, xiv–xv.

Adams, D.H. and Whittaker, V.P. (1949a). The cholinesterases of human blood. I. The specificity of the plasma enzyme and its relation to the erythrocyte cholinesterase. Biochim. Biophys. Acta 3, 358–366.

Adams, D.H. and Whittaker, V.P. (1949b). The characterization of the esterases of human plasma. Biochem. J. 44, 62–70.

Adams, D.H. and Whittaker, V.P. (1950). The cholinesterases of human blood. II. The forces acting between enzyme and substrate. Biochim. Biophys. Acta 4, 543–558.

Adlard, B.P.F. and Dobbing, J. (1971a). Vulnerability of developing brain. III. Development of four enzymes in the brains of normal and undernourished rats. Brain Res. 28, 97–107.

Adlard, B.P.F. and Dobbing, J. (1971b). Elevated acetylcholinesterase activity in adult rat brain after undernutrition in early life. Brain Res. 30, 198–199.

Adlard, B.P.F. and Dobbing, J. (1972). Permanent changes in the activity and subcellular distribution of acetylcholinesterase and lactate dehydrogenase in adult rat cerebellum after X-irradiation in infancy. Exptl. Neurol. 35, 547–550.

Adner, M.M., Isé, C., Schwab, R., Sherman, J.D. and Dameshek, W. (1966). Immunologic studies of thymectomized and non-thymectomized patients with myasthenia gravis. Ann. N.Y. Acad. Sci. 135, 536–554.

Adrian, E.D., Feldberg, W. and Kilby, B.A. (1947). The cholinesterase inhibiting action of fluorophosphonates. Brit. J. Pharmacol. 2, 56–58.

Akester, A.R. and Mann, S.P. (1969). Adrenergic and cholinergic innervation of the renal portal valve in the domestic fowl. J. Anat. 104, 241–252.

Akmal'khanov, Sh.A., Mirkhidoyatov, M., Atabaev, Sh.T. and Sarkisova, L.G. (1969). Protecting milk from pesticide residues (in Russian). Zhivotnovodstvo 31, 55–57.

Albuquerque, E.X., Sokoll, M.D., Sonesson, B. and Thesleff, S. (1968). Studies on the nature of the cholinergic receptor. Eur. J. Pharmacol. 4, 40–46.

Aldridge, W.N. (1953). The differentiation of true and pseudo cholinesterase by organophosphorus compounds. Biochem. J. 53, 62–67.

Aldridge, W.N. (1954). Anticholinesterases. Inhibition of cholinesterase by organo-phosphorus compounds and reversal of this reaction. Mechanisms involved. Chemistry and Industry (London) 1954, 473–476.

Aldridge, W.N. (1969). Organophosphorus compounds and carbamates, and their reaction with esterases. Brit. Med. Bull. 25, 236–240.

Aldridge, W.N. and Johnson, M.K. (1959). Cholinesterase, succinic dehydrogenase, nucleic acids, esterase and glutathione reductase in sub-cellular fractions from rat brain. Biochem. J. 73, 270–276.

Aldridge, W.N. and Reiner, E. (1972). Enzyme Inhibitors as Substrates. (North-Holland Publishing Company, Amsterdam).

Aldridge, W.N., Barnes, J.M. and Johnson, M.K. (1969). Studies on delayed neurotoxicity produced by some organophosphorus compounds. Ann. N.Y. Acad. Sci. 160, 314–322.

Alid, G. and Orrego, F.J. (1972). Inhibition of mammalian acetylcholinesterase by phenylmethanesulfonyl fluoride. Experientia 28, 13–14.

Alles, G.A. and Hawes, R.C. (1940). Cholinesterases in the blood of man. J. Biol. Chem. 133, 375–390.

Altman, J. and Das, G.D. (1970). Post-natal changes in the concentration and distribution of cholinesterase in the cerebellar cortex of rats. Exptl. Neurol. 28, 11–34.

Ambache, N., Freeman, A.M. and Hobbiger, F. (1971). Distribution of acetylcholinesterase and butyrylcholinesterase in the myenteric plexus and longitudinal muscle of the guinea-pig intestine. Biochem. Pharmacol. 20, 1123–1132.

Ammon, R. (1935). Die Cholinesterase. Ergeb. Enzymforsch. 4, 102–110.

Andersen, P. and Curtis, D.R. (1964). The pharmacology of the synaptic and acetylcholine-induced excitation of ventrobasal thalamic neurones. Acta Physiol. Scand. 61, 100–120.

Andersen, R.A., Laake, K. and Fonnum, F. (1972). Reaction between alkyl phosphates and acetylcholinesterase from different species. Comp. Biochem. Physiol. 42B, 429–437.

Anderson, H.K. (1905). The paralysis of involuntary muscle. Part II. On paralysis of the sphincter of the pupil with special reference to paradoxical constriction and the functions of the ciliary ganglion. J. Physiol. London 33, 156–174.

Anderson, P.H. and Machin, A.F. (1969). Organophosphorus warble fly dressings: some aspects of their toxicity to cattle including antidote therapy. Vet. Record 85, 484–487.

Anfinsen, C.B. (1944). The distribution of cholinesterase in the bovine retina. J. Biol. Chem. 152, 267–278.

Ansell, G.B. and Chojnacki, T. (1967). The effect of diisopropylfluorophosphonate on enzymes concerned in brain phospholipid synthesis. Progr. Biochem. Pharmacol. 3, 189–195.

Antonova, A.N. (1967). Hygenic evaluation of the educational industrial system for students in a professional-technical school specializing in the twisting of acetate and capron fibres (in Russian). Mater. Nauch.-Prakt. Konf. Molodykh Gig. Sanit. Vrachei 11th, 279–281.

Aoki, T. (1964). Cholinesterase activities associated with the sweat glands in the toe pads of the dog. Nature 202, 1124–1126.

Aoki, T. (1966). Evidence for the discharge of cholinesterase into canine eccrine sweat. Nature 211, 886–887.

Applegate, A. and Nelson, L. (1962). Acetylcholinesterase in *Mytilus* spermatozoa. Biol. Bull. 123, 475.

Aprison, M.H. and Himwich, H.E. (1954). Relationship between age and cholinesterase activity in several rabbit brain areas. Am. J. Physiol. 179, 502–506.

Aprison, M.H. and Hingtgen, J.N. (1970). Neurochemical correlates of behavior. Intern. Rev. Neurobiol. 13, 325–341.

Aprison, M.H., Nathan, P. and Himwich, H.E. (1954). Brain acetylcholinesterase activities in rabbits exhibiting three behavioral patterns following the intracarotid injection of di-isopropyl fluorophosphate. Am. J. Physiol. 177, 175–178.

Aprison, M.H., Takahashi, R. and Folkerth, T.L. (1964). Biochemistry of the avian central nervous system. I. The 5-hydroxytryptophan decarboxylase-monoamine oxidase and choline-acetylcholinesterase systems in several discrete areas of the pigeon brain. J. Neurochem. 11, 341–350.

Aprison, M.H., Kariya, T., Hingtgen, J.N. and Toru, M. (1968). Neurochemical correlates of behaviour. Changes in acetylcholine, norepinephrine and 5-hydroxytryptamine concentration in several discrete brain areas of the rat during behavioural excitation. J. Neurochem. 15, 1131–1139.

Arvy, L. (1960). Contribution à l'histoenzymologie de l'ovaire. Z. Zellforsch. Mikroskop. Anat. 51, 406–420.

Arvy, L. (1961). Contribution à la connaissance de la glande pinéale de *Bos taurus* L., d'*Ovis aries* L. et de *Sus scrofa* L. C.R. Acad. Sci. Paris 253, 1361-1363.

Arvy, L. (1962). Histochemical demonstration of enzymatic activities in neurosecretory centres of some homoiothermic animals. Mem. Soc. Endocrinol. 12, 215–225.

Arvy, L. (1965). Enzymic activity histochemically demonstrable in the pineal gland of some artiodactyls. In: Progress in Brain Research, Vol. 10, J. Ariëns Kappers and J.P. Schadé, eds. (Elsevier, Amsterdam) pp. 473–475.

Ash, R.W., Pearce, J.W. and Silver, A. (1969). An investigation of the nerve supply to the salt gland of the duck. Quart. J. Exptl. Physiol. 54, 281–295.

Ashton, G.C. and Simpson, N.E. (1966). C_5 types of serum cholinesterase in a Brazilian population. Am. J. Human Genet. 18, 438–447.

Askew, B.M. (1956). Oximes and hydroxamic acids as antidotes in anticholinesterase poisoning. Brit. J. Pharmacol. 11, 417–423.

Askew, B.M., Davies, D.R., Green, A.L. and Holmes, R. (1956). The nature of the toxicity of 2-oxo-oximes. Brit. J. Pharmacol. 11, 424–427.

Aster, R.H. and Enright, S.E. (1969). A platelet and granulocyte membrane defect in paroxysmal nocturnal hemoglobinuria. Usefulness for the detection of platelet antibodies. J. Clin. Invest. 48, 1199–1210.

Atherton, G.W. (1963). An investigation of the specificity of cholinesterase in the developing brain of the chick. Histochemie 3, 214–221.

Atsumi, S. (1971). The histogenesis of motor neurons with special reference to the correlation of their endplate formation. I. The development of endplates in the chick embryo. Acta Anat. 80, 161–182.

Augustinsson, K.-B. (1946a). Choline esterases in some marine invertebrates. Acta Physiol. Scand. 11, 141–150.

Augustinsson, K.-B. (1946b). Studies on the specificity of choline esterase in *Helix pomatia*. Biochem. J. 40, 343–349.

Augustinsson, K.-B. (1948). Cholinesterases. A study in comparative enzymology. Acta Physiol. Scand. 15, Suppl. 52, 1–182.

Augustinsson, K.-B. (1950). Acetylcholine esterase and cholinesterase. In: The Enzymes, Vol 1, part 1, J.B. Sumner and K. Myrbäck, eds. (Academic Press, New York) pp. 443–472.

Augustinsson, K.-B. (1951). Comparison between the acetylcholinesterases of *Helix* blood and *Cobra* venom. II. The hydrolysis of certain choline and non-choline esters. Acta Chem. Scand. 5, 712–723.

Augustinsson, K.-B. (1955a). The normal variation of human blood cholinesterase activity. Acta Physiol. Scand. 35, 40–52.

Augustinsson, K.-B. (1955b). The electric organs and their cholinesterase activity. Pubbl. Staz. Zool. Napoli 27, 189–198.

Augustinsson, K.-B. (1957). Assay methods for cholinesterases. In: Methods of Biochemical Analysis, Vol. 5, D. Glick, ed. (Interscience Publishers, New York) pp. 1–63.

Augustinsson, K.-B. (1959a). Electrophoresis studies on blood plasma esterases I. Mammalian plasmata. Acta Chem. Scand. 13, 571–592.

Augustinsson, K.-B. (1959b). Electrophoresis studies on blood plasma esterases. II. Avian, reptilian, amphibian and piscine plasmata. Acta Chem. Scand. 13, 1081–1096.

Augustinsson, K.-B. (1959c). Electrophoresis studies on blood plasma esterases. III. Conclusions. Acta Chem. Scand. 13, 1097–1105.

Augustinsson, K.-B. (1960). Butyryl- and propionylcholinesterases and related types of eserine-sensitive esterases. In: The Enzymes, Vol. 4, 2nd ed., P.D. Boyer, H. Lardy and K. Myrbäck, eds. (Academic Press, New York) pp. 521–540.

Augustinsson, K.-B. (1963). Classification and comparative enzymology of the cholinesterases and methods for their determination. In: Cholinesterases and Anticholinesterase Agents, G.B. Koelle, ed. (Springer, Berlin) pp. 89–128.

Augustinsson, K.-B. (1966). The nature of an 'anionic' site in butyrylcholinesterase compared with that of a similar site in acetylcholinesterase. Biochim. Biophys. Acta 128, 351–362.

Augustinsson, K.-B. (1968). The evolution of esterases in vertebrates. In: Homologous Enzymes and Biochemical Evolution, N. van Thoai and J. Roche, eds. (Gordon and Breach, New York) pp. 299–311.

Augustinsson, K.-B. (1971a). Determination of activity of cholinesterases. In: Methods of Biochemical Analysis, supplemental volume, Analysis of Biogenic Amines and their Related Enzymes, D. Glick, ed. (Interscience Publishers, New York) pp. 217–273.

Augustinsson, K.-B. (1971b). Comparative aspects of the purification and properties of cholinesterases. Bull. World Health Organ. 44, 81–89.

Augustinsson, K.-B. and Gustafson, T. (1949). Cholinesterase in developing sea-urchin eggs. J. Cell. Comp. Physiol. 34, 311–321.

Augustinsson, K.-B. and Nachmansohn, D. (1949). Distinction between acetylcholine-esterase and other choline ester-splitting enzymes. Science 110, 98–99.

Augustinsson, K.-B. and Olsson, B. (1959a). Esterases in the milk and blood plasma of swine. I. Substrate specificity and electrophoresis studies. Biochem. J. 71, 477–484.

Augustinsson, K.-B. and Olsson, B. (1959b). Esterases in milk and blood plasma of swine. 2. Activities at different stages during lactation and suckling periods, and plasma arylesterase as a gene controlled enzyme. Biochem. J. 71, 484–492.

Aus der Mühlen, K. (1966). Der Hypothalamus des Meerschweinchens. (S. Karger, Basel).

Austin, L. (1957). Lipid biosynthesis and chemically induced paralysis in the chicken. Brit. J. Pharmacol. 12, 356–360.

Austin, L. and Berry, W.K. (1953). Two selective inhibitors of cholinesterase. Biochem. J. 54, 695–700.

Austin, L. and James, K.A.C. (1970). Rates of regeneration of acetylcholinesterase in rat brain subcellular fractions following DFP inhibition. J. Neurochem. 17, 705–707.

Austin, L. and Phillis, J.W. (1965). The distribution of cerebellar cholinesterases in several species. J. Neurochem. 12, 709–717.

Austin, L., Phillis, J.W. and Steele, R.P. (1964). The distribution of cholinesterase in cat cerebellar cortex. Experientia 20, 218–219.

Avanzino, G.L., Bradley, P.B. and Wolstencroft, J.H. (1966). Pharmacological properties of neurones of the paramedian reticular nucleus. Experientia 22, 410.

Axelsson, J. and Thesleff, S. (1959). A study of supersensitivity in denervated mammalian skeletal muscle. J. Physiol. London 147, 178–193.

Babers, F.H. and Pratt, J.J. (1950). Studies on the resistance of insects to insecticides. I. Cholinesterase in house flies (*Musca domestica* L.) resistant to DDT. Physiol. Zool. 23, 58–63.

Babers, F.H. and Pratt, J.J., Jr. (1951). A comparison of the cholinesterase in the heads of the house fly, the cockroach, and the honey bee. Physiol. Zool. 24, 127–131.

Bacq, Z.M. (1935). Recherches sur la physiologie et la pharmacologie du système nerveux autonome. XIX. La choline-estérase chez les invertébrés. L'insensibilité des crustacés à l'acétylcholine. Arch. Intern. Physiol. 42, 47–60.

Bacq, Z.M. (1947). L'acétylcholine et l'adrénaline chez les invertébrés. Biol. Rev. Cambridge Phil. Soc. 22, 73–91.

Bacq, Z.M. and Coppée, G. (1937). Réaction des vers et des mollusques à l'ésérine. Existence de nerfs cholinergiques chez les vers. Arch. Intern. Physiol. 45, 310–324.

Bacq, Z.M. and Nachmansohn, D. (1937). Cholinesterase in invertebrate muscle. J. Physiol. London 89, 368–371.

Bacq, Z.M. and Oury, A. (1937). Note sur la répartition de la choline-estérase chez les êtres vivant. Bull. Classe Sci. Acad. Roy. Belg. 23, 891–893.

Bajgar, J. (1972). Time course of acetylcholinesterase inhibition in the medulla oblongata of the rat by O-ethyl S-(2-dimethylaminoethyl) methylphosphonothioate in vivo. Brit. J. Pharmacol. 45, 368–371.

Bajgar, J. and Žižkovský, V. (1971). Partial characterization of soluble acetylcholinesterase isoenzymes of the rat brain. J. Neurochem. 18, 1609–1614.

Bajgar, J., Jakl, A. and Hrdina, V. (1971). Influence of trimedoxime and atropine on acetylcholinesterase activity in some parts of the brain of mice poisoned by isopropylmethyl phosphonofluoridate. Biochem. Pharmacol. 20, 3230–3233.

Bajgar, J., Gold, V., Petr, R., Žáková, Z. and Spaček, J. (1972). Development of acetylcholinesterase activity in offspring's brain of adrenalectomized female rats. Brain Res. 44, 688–691.

Baker, W.W. and Benedict, F. (1968). Analysis of local discharges induced by intrahippocampal microinjection of carbachol or diisopropylfluorophosphate (DFP). Intern. J. Neuropharmacol. 7, 135–147.

Baldwin, E. (1967). Dynamic Aspects of Biochemistry, 5th edn. (Cambridge Univ. Press, Cambridge).

Ballantyne, B. (1966a). The reticulo–endothelial localization of hepatic acetylcholinesterase. Experientia 22, 25–26.

Ballantyne, B. (1966b). Distribution of cholinesterase in mammalian liver. J. Anat. 100, 704–705.

Ballantyne, B. (1967). Esterase histochemistry of reticuloendothelial cells. In: The Reticuloendothelial System and Atherosclerosis. Advances in Experimental Medicine and Biology, Vol. 1, N.R. DiLuzio and R. Paoletti, eds. (Plenum Press, New York) pp. 121–132.

Ballantyne, B. (1968). Potentiometric pH-stat titration : importance of an inert atmosphere in reaction vessels when using alkali titrant. Experientia 24, 329–330.

Ballantyne, B. (1970). The suitability of ferric potassium ferrocyanide as a perfusate in cholinesterase histochemistry : ultrastructural and biochemical observations. Histochem. J. 2, 243–247.

Ballantyne, B. and Burwell, R.G. (1965). Distribution of cholinesterase in normal lymph nodes and its possible relation to the regulation of tissue size. Nature 206, 1123–1125.

Ballantyne, B. and Fourman, J. (1967). Cholinesterases and the secretory activity of the duck supraorbital gland. J. Physiol. London 188, 32−33*P*.

Ballard, K.J. and Jones, J.V. (1971). The fine structural localization of cholinesterases in the carotid body of the cat. J. Physiol. London 219, 747−753.

Bangham, D.R. (1960). The transmission of homologous serum proteins to the foetus and to the amniotic fluid in the rhesus monkey. J. Physiol. London 153, 265−289.

Banister, J., Whittaker, V.P. and Wijesundera, S. (1953). The occurrence of homologues of acetylcholine in ox spleen. J. Physiol. London 121, 55−71.

Barker, D. and Ip, M.C. (1963). A silver method for demonstrating the innervation of mammalian muscle in teased preparations. J. Physiol. London 169, 73−74*P*.

Barker, D.L., Herbert, E., Hildebrand, J.G. and Kravitz, E.A. (1972). Acetylcholine and lobster sensory neurones. J. Physiol. London 226, 205−229.

Barker, J.L. and Levitan, H. (1971). Salicylate: effect on membrane permeability of molluscan neurons. Science 172, 1245−1247.

Barlow, J.J. (1971). The distribution of acetylcholinesterase and catecholamines in the vertical lobe of *Octopus vulgaris*. Brain Res. 35, 304−307.

Barnard, E.A. and Rogers, A.W. (1967). Determination of the number, distribution and some in situ properties of cholinesterase molecules in the motor end plate, using labeled inhibitor methods. Ann. N.Y. Acad. Sci. 144, 584−610.

Barnard, E.A., Rymaszewska, T. and Wieckowski, J. (1971a). Cholinesterases at individual neuromuscular junctions. In: Cholinergic Ligand Interactions, D.J. Triggle, J.F. Moran and E.A. Barnard, eds. (Academic Press, New York) pp. 175−200.

Barnard, E.A., Wieckowski, J. and Chiu, T.H. (1971b). Cholinergic receptor molecules and cholinesterase molecules at mouse skeletal muscle junctions. Nature 234, 207−209.

Barnes, E.W. and Irvine, W.J. (1973). Clinical syndromes associated with thymic disorders. Proc. Roy. Soc. Med. 66, 151−154.

Barnes, J.M. and Denz, F.A. (1953). Experimental demyelination with organo-phosphorus compounds. J. Pathol. Bacteriol. 65, 597−605.

Barrnett, R.J. (1962). The fine structural localization of acetylcholinesterase at the myoneural junction. J.Cell Biol. 12, 247−262.

Barrnett, R.J. and Palade, G.E. (1958). Applications of histochemistry to electron microscopy. J. Histochem. Cytochem. 6, 1−12.

Barrnett, R.J. and Seligman, A.M. (1951). Histochemical demonstration of esterases by production of indigo. Science 114, 579−582.

Barron, K.D. and Bernsohn, J. (1965). Brain esterases and phosphatases in multiple sclerosis. Ann. N.Y. Acad. Sci. 122, 369−399.

Barron, K.D. and Bernsohn, J. (1968). Esterases of developing human brain. J. Neurochem. 15, 273−284.

Barron, K.D., Bernsohn, J. and Hess, A.R. (1963). Separation and properties of human brain esterases. J. Histochem. Cytochem. 11, 139−156.

Bartels, E. (1968). Reactions of acetylcholine receptor and esterase studied on the electroplax. Biochem. Pharmacol. 17, 945−966.

Bartels, E., Brzin, M. and Dettbarn, W.-D. (1969). Action of acetylcholine in the presence of organophosphates on single axons of the lobster. Biochem. Pharmacol. 18, 2590−2595.

Bartolini, A., Bartolini, R. and Domino, E.F. (1973). Effects of physostigmine on brain acetylcholine content and release. Neuropharmacol. 12, 15−25.

Bastide, P., Laumin, A.M. and Dastugue, G. (1967). Activités enzymatiques dans les organes et humeurs de *Vipera aspis*. Bull. Soc. Pharm. Marseilles 16, 193−199.

Baum, G. and Ward, F.B. (1971). General enzyme studies with a substrate-selective electrode. Characterization of cholinesterases. Anal. Biochem. 42, 487−493.

Bauman, A., Benda, P. and Rieger, F. (1972). Identification des espèces acétylcholinestérasiques de gymnote après fractionnement d'organe électrique et électrophorèse en gel de polyacrylamide. Brain Res. 45 183–192.

Bayer, G. and Wense, T. (1936). Über des Nachweis von Hormonen in einzelligen Tieren. I Mitteilung. Cholin und Acetylcholin im Paramecium. Pflügers Arch. 237, 417–422.

Bayliss, B.J. and Todrick, A. (1953a). The use of specific inhibitors in the estimation of the pseudo cholinesterase in nervous tissue. Biochem. J. 54, xxix.

Bayliss, B.J. and Todrick, A. (1953b). The development of cholinesterases in the brain and spinal cord of the young rat. Biochem. J. 54, xxix.

Bayliss, B.J. and Todrick, A. (1956). The use of a selective acetylcholinesterase inhibitor in the estimation of pseudocholinesterase activity in rat brain. Biochem. J. 62, 62–67.

Beani, L., Bianchi, C., Santinoceto, L. and Marchetti, P. (1968). The cerebral acetylcholine release in conscious rabbits with semi-permanently implanted epidural cups. Intern. J. Neuropharmacol. 7, 469–481.

Beckett, A.H., Mitchard, M. and Clitherow, J.W. (1968). The importance of steric and stereochemical features in serum cholinesterase substrates. Biochem. Pharmacol. 17, 1601–1607.

Beckett, A.H., Lan, N.T. and Khokhar, A.Q. (1971). Anti-cholinesterase activity of some stereoisomeric aminobornanes. J. Pharm. Pharmacol. 23, 528–533.

Beckmann, A.L. and Eisenman, J.S. (1970). Microelectrophoresis of biogenic amines on hypothalamic thermosensitive cells. Science 170, 334–336.

Beddoe, F. and Smith, H.J. (1971). Inhibition of acetylcholinesterase by dibenamine and dibenzyline. J. Pharm. Pharmacol. 23, 37–49.

Beddoe, F., Nicholls, P. and Smith, H.J. (1971). Potential irreversible inhibitors of acetylcholinesterase: N-trimethyl-N'-iodoacetyl diaminoalkane iodides. J. Pharm. Pharmacol. 23, 865–867.

Beesley, P., Emson, P.C. and Kerkut, G.A. (1972). Change in K_M of insect ChE after behavioural training. J. Physiol. London 221, 26–27P.

Begum, A. and Prathapkumar, J. (1969). Effect of vitamin A on cholinesterase activity in normal children and in children with protein-calorie malnutrition. Clin. Chim. Acta 26, 343–349.

Bell, C. (1966). Use of the direct-coloring thiocholine technique for demonstration of intracellular neuronal cholinesterases. J. Histochem. Cytochem. 14, 567–570.

Bell, C. (1968). Dual vasoconstrictor and vasodilator innervation of the uterine arterial supply in the guinea pig. Circulation Res. 23, 279–289.

Bell, C. (1969). Fine structural localization of acetylcholinesterase at a cholinergic vasodilator nerve-arterial smooth muscle synapse. Circulation Res. 24, 61–70.

Bell, C. (1971). Distribution of cholinergic vasomotor nerves to the parametrial arteries of some laboratory and domestic animals. J. Reprod. Fertility 27, 53–58.

Bell, C. (1972). Autonomic nervous control of reproduction: circulatory and other factors. Pharmacol. Rev. 24, 657–736.

Bell, C. and Burnstock, G. (1965). Cholinesterases in the bladder of the toad (*Bufo marinus*). Biochem. Pharmacol. 14, 79–89.

Bell, C. and McLean, J.R. (1967). Localization of norepinephrine and acetylcholinesterase in separate neurons supplying the guinea-pig vas deferens. J. Pharmacol. Exptl. Therap. 157, 69–73.

Bell, C. and McLean, J.R. (1970). The distribution of cholinergic and adrenergic nerve fibres in the retractor penis and vas deferens of the dog. Z. Zellforsch. Mikroskop. Anat. 106, 516–522.

Belleau, B. (1970). The modification of acetylcholinesterase and the acetylcholine receptor by quaternary salts. In: Fundamental Concepts in Drug-Receptor Interactions, J.F. Danielli, J.F. Moran and D.J. Triggle, eds. (Academic Press, New York) pp. 121–131.

Belleau, B. and DiTullio, V. (1970). The anionic sites of acetylcholinesterase versus the acetyl-choline receptors. In: Drugs and Cholinergic Mechanisms in the CNS, E. Heilbronn and A. Winter, eds. (Research Institute of National Defence, Stockholm) pp. 441–453.

Belleau, B. and Tani, H. (1966). A novel irreversible inhibitor of acetylcholinesterase specifical-ly directed at the anionic binding site: structure-activity relationships. Mol. Pharmacol. 2, 411–422.

Benda, P., Tsuji, S., Daussant, J. and Changeux, J.-P. (1970). Localization of acetylcholinester-ase by immunofluorescence in eel electroplax. Nature 225, 1149–1150.

Bennett, E.L., Rosenzweig, M.R., Krech, D., Karlsson, H., Dye, N. and Ohlander, A. (1958a). Invidual strain and age differences in cholinesterase activity of the rat brain. J. Neurochem. 3, 144–152.

Bennett, E.L., Krech, D., Rosenzweig, M.R., Karlsson, H., Dye, N. and Ohlander, A. (1958b). Cholinesterase and lactic dehydrogenase activity in the rat brain. J. Neurochem. 3, 153–160.

Bennett, E.L., Diamond, M.C., Morimoto, H. and Hebert, M. (1966). Acetylcholinesterase activity and weight measures in fifteen brain areas from six lines of rats. J. Neurochem. 13, 563–572.

Bennett, E.L., Rosenzweig, M.R. and Diamond, M.C. (1970). Time courses of effects of differ-ential experience on brain measures and behavior of rats. In: Molecular Approaches to Learning and Memory, W.L. Byrne, ed. (Academic Press, New York) pp. 55–89.

Benson, W.M. (1948). Inhibition of cholinesterase by adrenaline. Proc. Soc. Exptl. Biol. Med. 68, 598–601.

Benz, F.W. (1970). Stereo- and bio-chemical evaluation of a series of oxime reactivators of diethylphosphorylacetylcholinesterase. Ph.D. Thesis: University of Iowa. pp. 1–124.

Berg, G.R. and Klein, D.C. (1972). Norepinephrine increases the [^{32}P] labelling of a specific phopholipid fraction of postsynaptic pineal membranes. J. Neurochem. 19, 2519–2532.

Bergmann, F. (1955). Fine structure of the active surface of cholinesterases and the mechanism of enzymatic ester hydrolysis. Discussions Faraday Soc. 20, 126–134.

Bergmann, F. (1958). The structure of the active surface of cholinesterases and the mechanism of their catalytic action in ester hydrolysis. Advan. Catalysis 10, 131–164.

Bergmann, F. and Wurzel, M. (1954). The structure of the active surface of serum cholinester-ase. Biochim. Biophys. Acta 13, 251–259.

Bergmann, F., Wilson, I.B. and Nachmansohn, D. (1950). The inhibitory effect of stilbamidine, curare and related compounds and its relationship to the active group of acetylcholine esterase. Action of stilbamidine upon nerve impulse conduction. Biochim. Biophys. Acta 6, 217–224.

Bergmann, F., Segal, R., Shimoni, A. and Wurzel, M. (1956). The pH-dependence of enzymic ester hydrolysis. Biochem. J. 63, 684–690.

Bergmann, F., Rimon, S. and Segal, R. (1958). Effect of pH on the activity of eel esterase towards different substrates. Biochem. J. 68, 493–499.

Bergner, A.D. and Bayliss, M.W. (1952). Histochemical detection of fatal anticholinesterase poisoning. U.S. Armed Forces Med. J. 3, 1637–1644.

Bergner, A.D. and Durlacher, S.H. (1951). Histochemical detection of fatal anticholinesterase poisoning. Am. J. Pathol. 27, 1011–1021.

Berman, J.D. and Young, M. (1971). Rapid and complete purification of acetylcholinesterases of electric eel and erythrocyte by affinity chromatography. Proc. Natl. Acad. Sci. U.S. 68, 395–398.

Bernhard, W. and Viron, A. (1971). Improved techniques for the preparation of ultrathin frozen sections. J.Cell Biol. 49, 731–746.

Bernheim, F. and Bernheim, M.L.C. (1936). The action of drugs on the choline esterase of the brain. J. Pharmacol. Exptl. Therap. 57, 427–436.

Bernsohn, J., Barron, K.D. and Hess, A.R. (1962). Multiple nature of acetylcholinesterase in nerve tissue. Nature 195, 285–286.

Bernsohn, J., Barron, K.D. and Hedrick, M.T. (1963). Some properties of isozymes of brain acetylcholinesterase. Biochem. Pharmacol. 12, 761–763.

Berry, J.F. and Cevallos, W.H. (1966). Lipid class and fatty acid composition of peripheral nerve from normal and organophosphorus-poisoned chickens. J. Neurochem. 13, 117–124.

Berry, W.K. (1951). The turnover number of cholinesterase. Biochem. J. 49, 615–620.

Berry, W.K. and Rutland, J.P. (1971). Choline ester hydrolases in diaphragm muscle. Biochem. Pharmacol. 20, 669–682.

Berry, W.K., Davies, D.R. and Green, A.L. (1959). Oximes of $\alpha\omega$-diquaternary alkane salts as antidotes to organophosphate anticholinesterases. Brit. J. Pharmacol. 14, 186–191.

Berry, W.K., Davies, D.R. and Rutland, J.P. (1966). Problems in the treatment with oximes and atropine of rats poisoned by organophosphates. Biochem. Pharmacol. 15, 1259–1266.

Berry, W.K., Davies, D.R. and Gordon, J.J. (1971). Protection of animals against Soman (1,2,2-trimethylpropyl methylphosphonofluoridate) by pretreatment with some other organophosphorus compounds, followed by oxime and atropine. Biochem. Pharmacol. 20, 125–134.

Bevan, S., Miledi, R. and Grampp, W. (1973). Induced transmitter release from Schwann cells and its suppression by Actinomycin D. Nature New Biol. 241, 85–86.

Bidstrup, P.L. and Hunter, D. (1952). Toxic chemical substances used in agriculture. Lancet i, 262–263.

Bieth, J., Wassermann, N., Vratsanos, S.M. and Erlanger, B.F. (1970). Photoregulation of biological activity by photochromic reagents. IV. A model for diurnal variation of enzymic activity. Proc. Natl. Acad. Sci. U.S. 66, 850–854.

Bingham, E. and Buckley, G.A. (1973). The occurrence and origin of cholinesterase in rat tears. J. Physiol. London 231, 50–51P.

Biological effects of pesticides in mammalian systems (1969). Ann. N.Y. Acad. Sci. 160, 1–422.

Birks, R.I. and Brown, L.M. (1960). A method for locating the cholinesterase of a mammalian myoneural junction by electron microscopy. J. Physiol. London 152, 5–7P.

Birks, R.I. and MacIntosh, F.C. (1961). Acetylcholine metabolism of a sympathetic ganglion. Can. J. Biochem. Physiol. 39, 787–827.

Birks, R., Katz, B. and Miledi, R. (1960). Physiological and structural changes at the amphibian myoneural junction, in the course of nerve degeneration. J. Physiol. London 150, 145–168.

Bischoff, C., Grab, W. and Kapfhammer, J. (1931). Acetylcholine in beef blood III. Z. Physiol. Chem. 200, 153–165.

Biscoe, T.J. (1971). Carotid body: structure and function. Physiol. Rev. 51, 437–495.

Biscoe, T.J. and Silver, A. (1966). The distribution of cholinesterases in the cat carotid body. J. Physiol. London 183, 501–512.

Biscoe, T.J. and Straughan, D.W. (1966). Micro-electrophoretic studies of neurones in the cat hippocampus. J. Physiol. London 183, 341–359.

Bhattacharya, S.K. and Sanyal, A.K. (1971). Anticholinesterase activity of bufotenine. Indian J. Physiol. Pharmacol. 15, 133–134.

Blaber, L.C. and Cuthbert, A.W. (1962). Cholinesterases in the domestic fowl and the specificity of some reversible inhibitors. Biochem. Pharmacol. 11, 113–124.

Blaschko, H., Bülbring, E. and Chou, T.C. (1949). Tubocurarine antagonism and inhibition of cholinesterases. Brit. J. Pharmacol. 4, 29–32.

Bligh, J. (1973). Temperature Regulation in Mammals and Other Vertebrates (North-Holland Publishing Co., Amsterdam).

Bligh, J., Cottle, W.H. and Maskrey, M. (1971). Influence of ambient temperature on the thermoregulatory responses to 5-hydroxytryptamine, noradrenaline and acetylcholine injected into the lateral cerebral ventricles of sheep, goats and rabbits. J. Physiol. London 212, 377–392.

Blinzinger, K. and Kreutzberg, G. (1968). Displacement of synaptic terminals from regenerating motoneurons by microglial cells. Z. Zellforsch. Mikroskop. Anat. 85, 145–157.

Bloch, H. (1941). Der Einfluss von Trikresylphosphat auf die Aktivität der cholinesterase. Helv. Med. Acta 8 Suppl. 7, 15–17.

Bloom, F.E., Costa, E. and Salmoiraghi, G.C. (1964). Analysis of individual rabbit olfactory bulb neuron responses to the microelectrophoresis of acetylcholine, norepinephrine and serotonin synergists and antagonists. J. Pharmacol. Exptl. Therap. 146, 16–23.

Bloom, F.E., Costa, E. and Salmoiraghi, G.C. (1965). Anesthesia and the responsiveness of individual neurons of the caudate nucleus of the cat to acetylcholine, norepinephrine and dopamine administered by microelectrophoresis. J. Pharmacol. Exptl. Therap. 150, 244–252.

Bloom, F., Iversen, L.L. and Schmitt, F.O. (1970). Macromolecules in synaptic function. Neurosciences Res. Prog. Bull. 8, 325–455.

Blume, A., Gilbert, F., Wilson, S., Farber, J., Rosenberg, R. and Nirenberg, M. (1970). Regulation of acetylcholinesterase in neuroblastoma cells. Proc. Natl. Acad. Sci. U.S. 67, 786–792.

Bobbin, R.P. and Konishi, T. (1971). Acetylcholine mimics crossed olivocochlear bundle stimulation. Nature New Biol. 231, 222–223.

Boczkowski, Z. (1966). Cholinesterase activity in the blood and cerebrospinal fluid of women in late intoxication of pregnancy (in Polish). Przegl. Lek. 22, 495–497.

Bodian, D. (1970). An electron microscopic characterization of classes of synaptic vesicles by means of controlled aldehyde fixation. J. Cell Biol. 44, 115–124.

Boell, E.J. (1945). Cholinesterase activity of peripheral nerves. J. Cell. Comp. Physiol. 25, 75–84.

Boell, E.J. and Nachmansohn, D. (1940). Localization of choline esterase in nerve fibers. Science 92, 513–514.

Boell, E.J. and Shen, S.C. (1944). Functional differentiation in embryonic development. I. Cholinesterase activity of induced neural structures in *Amblystoma punctatum*. J. Exptl. Zool. 97, 21–41.

Boell, E.J. and Shen, S.C. (1950). Development of cholinesterase in the central nervous system of *Amblystoma punctatum*. J. Exptl. Zool. 113, 583–599.

Boell, E.J., Greenfield, P. and Shen, S.C. (1955). Development of cholinesterase in the optic lobes of the frog (*Rana pipiens*). J. Exptl. Zool. 129, 415–451.

Bogart, B.I. (1970). Fine structural localization of cholinesterase activity in the rat submandibular gland. J. Histochem. Cytochem. 18, 730–739.

Bogart, B.I. (1971). The fine structural localization of acetylcholinesterase activity in the rat parotid and sublingual glands. Am J. Anat. 132, 259–266.

Bondy, S.C. (1972). Axonal migration of various ribonucleic acid species along the optic tract of the chick. J. Neurochem. 19, 1769–1776.

Bonichon, A. (1957). Evolution de l'activité de l'acétylcholine-estérase au niveau des lobes optiques chez l'embryon de poulet. Compt. Rend. Acad. Sci. Paris 245, 1345–1347.

Bonichon, A. (1958). L'acétylcholinestérase dans la cellule et la fibre nerveuse au cours du développement. I. Différenciation biochimique précoce du neuroblaste. Ann. Histochim. 3, 85–93.

Bonichon, A. (1961). L'acétylcholinestérase dans la cellule et la fibre nerveuse au cours du développement. Bibliotheca Anat. 2, 62–72.

Bonichon, A. (1962). Recherches histochimiques et biochimiques sur le développement des lobes optiques chez l'embryon de poulet. Thèse Sc. phys., Nancy.

Bonting, S.L. and Featherstone, R.M. (1956). Ultramicroassay of the cholinesterases. Arch. Biochem. Biophys. 61, 89–98.

Booth, G.M. and Metcalf, R.L. (1970). Phenylthioacetate : a useful substrate for the histo-chemical and colorimetric detection of cholinesterase. Science 170, 455—457.

Borisova, M.A. (1966). Activity of blood serum cholinesterase and catalase as an index of the course of typhus abdominalis (in Russian). Vrach. Delo 10, 110—112.

Bošković, B., Maksimović, M. and Minić, D. (1968). Ageing and reactivation of acetylcholines-terase inhibited with Soman and its thiocholine-like analogue. Biochem. Pharmacol. 17, 1738—1741.

Bosmann, H.B. and Hemsworth, B.A. (1970). Intraneural mitochondria. Incorporation of amino acids and monosaccharides into macromolecules by isolated synaptosomes and synap-tosomal mitochondria. J. Biol. Chem. 245, 363—371.

Bourdois, P.S. and Szerb, J.C. (1972). The absence of 'surplus' acetylcholine formation in prisms prepared from rat cerebral cortex. J. Neurochem. 19, 1189—1193.

Boutin, D. and Brodeur, J. (1970). An automated method for simultaneous determination of serum pseudocholinesterase activity, dibucaine number and fluoride number. Clin. Biochem. 3, 245—254.

Boutin, D. and Brodeur, J. (1971). Human serum pseudocholinesterases: molecular weight estimation of a subunit structure. Can. J. Physiol. Pharmacol. 49, 777—779.

Bradley, P.B. and Dray, A. (1972). Short-latency excitation of brain stem neurones in the rat by acetylcholine. Brit. J. Pharmacol. 45, 372—374.

Bradley, P.B. and Nicholson, A.N. (1962). The effect of some drugs on hippocampal arousal. Electroencephalog. Clin. Neurophysiol. 14, 824—834.

Bradley, P.B., Dhawan, B.N. and Wolstencroft, J.H. (1966). Pharmacological properties of cholinoceptive neurones in the medulla and pons of the cat. J. Physiol. London 183, 658—674.

Bradley, W.G., Murchison, D. and Day, M.J. (1971). The range of velocities of axoplasmic flow. A new approach, and its application to mice with genetically inherited spinal muscular atrophy. Brain Res. 35, 185—197.

Breckenridge, B.McL., Burn, J.H. and Matschinsky, F.M. (1967). Theophylline, epinephrine and neostigmine facilitation of neuromuscular transmission. Proc. Natl. Acad. Sci. U.S. 57, 1893—1897.

Brestkin, A.P. and Rozengart, E.V. (1965). Cholinesterase catalysis. Nature 205, 388—389.

Brestkin, A.P., Ivanova, L.A. and Svechnikova, V.V. (1965). Inhibition of the rate of hydrolysis of acetylcholine under the action of bovine erythrocyte acetylcholinesterase in the presence of high concentrations of substrate. Biokhimiya 30, 1154 (Eng. trans. 991—995).

Brestkin, A.P., Ivanova, L.A. and Svechnikova, V.V. (1966). On the influence of choline on the rate of hydrolysis of acetylcholine under the action of bovine erythrocyte acetylcholinester-ase. Biokhimiya 31, 416—423 (Eng. trans. 361—366).

Brezenoff, H.E. (1972). Cardiovascular responses to intrahypothalamic injections of carbachol and certain cholinesterase inhibitors. Neuropharmacol. 11, 637—644.

Brightman, M.W. and Albers, R.W. (1959). Species differences in the distribution of extraneuro-nal cholinesterases within the vertebrate central nervous system. J. Neurochem. 4, 244—250.

Brody, I.A., Resnick, J.S. and Engel, W.K. (1965). Detection of atypical cholinesterase by electrophoresis. Arch. Neurol. 13, 126—129.

Brooks, C.McC., Ishikawa, T., Koizumi, K. and Lu, H.-H. (1966). Activity of neurones in the paraventricular nucleus of the hypothalamus and its control. J. Physiol. London 182, 217—231.

Brown, G.L., Dale, H.H. and Feldberg, W. (1936). Reactions of the normal mammalian muscle to acetylcholine and to eserine. J. Physiol. London 87, 394—424.

Brown, J.C. and Howlett, B. (1968). The facial outflow and the superior salivatory nucleus. An histochemical study in the rat. J. Comp. Neurol. 134, 175—192.

Brown, J.C. and Howlett, B. (1970). The nucleus supragenualis, presumptive superior salivatory nucleus. Acta Anat. 76, 35—46.

Brown, J.C. and Howlett, B. (1971). Rat hindbrain thiocholine reactions relevant to the localization of the inferior salivatory nucleus. Acta Anat. 79, 333–359.

Brown, J.O. (1943). Pigmentation of substantia nigra and locus coeruleus in certain carnivores. J. Comp. Neurol. 79, 393–405.

Brown, L.M. (1961). A thiocholine method for locating cholinesterase activity by electron microscopy. Bibliotheca Anat. 2, 21–33.

Bryant, S.H. and Brzin, M. (1966). Cholinesterase activity of isolated giant synapses. J. Cell. Physiol. 68, 107–108.

Brzin, M. and Zeuthen, E. (1964). Notes on the possible use of the magnetic diver for respiration measurements (error 10^{-7} μl/hour). Compt. Rend. Trav. Lab. Carlsberg 34, 427–431.

Brzin, M., Kovič, M. and Oman, S. (1964). The magnetic diver balance. Compt. Rend. Trav. Lab. Carlsberg 34, 407–426.

Brzin, M., Dettbarn, W.-D., Rosenberg, P. and Nachmansohn, D. (1965). Cholinesterase activity per unit surface area of conducting membranes. J. Cell Biol. 26, 353–364.

Brzin, M., Tennyson, V.M. and Duffy, P.E. (1966). Acetylcholinesterase in frog sympathetic and dorsal root ganglia. A study by electron microscope cytochemistry and microgasometric analysis with the magnetic diver. J. Cell Biol . 31, 215–242.

Bublewska, A., Uselis, J. and Krnicki, A. (1969). Activity of some enzymes in blood serum investigated in a selected group of shipyard workers. Bull. Inst. Mar. Med. Gdansk 20, 175–181.

Buckley, G.A. and Heaton, J. (1968). A quantitative study of cholinesterase in myoneural junctions from rat and guinea-pig extraocular muscles. J. Physiol. London 199, 743–749.

Buckley, G.A. and Heaton, J. (1970). Cholinesterase in endplates of rat and chick, relationship of activity to log-dose response curves and the effects of some inhibitors. Brit. J. Pharmacol. 38, 434P.

Buckley, G., Consolo, S., Giacobini, E. and Sjöqvist, F. (1967). Cholinacetylase in innervated and denervated sympathetic ganglia and ganglion cells of the cat. Acta Physiol. Scand. 71, 348–356.

Bueding, E. (1952). Acetylcholinesterase activity of *Schistosoma mansoni.* Brit. J. Pharmacol. 7, 563–566.

Bülbring, E., Lourie, E.M. and Pardoe, U. (1949). The presence of acetylcholine in *Trypanosoma rhodesiense* and its absence from *Plasmodium gallinaceum.* Brit. J. Pharmacol. 4, 290–294.

Bülbring, E., Burn, J.H. and Shelley, H.J. (1953a). Acetylcholine and ciliary movement in the gill plates of *Mytilus edulis.* Proc. Roy. Soc. London Ser. B. 141, 445–466.

Bülbring, E., Philpot, F.J. and Bosanquet, F.D. (1953b). Amine oxidase, pressor amines, and cholinesterase in brain tumours. Lancet i, 865–866.

Bull, G., Hebb, C. and Ratković, D. (1963). Estimation of choline acetyltransferase in small samples of nervous tissue. Biochim. Biophys. Acta 67, 138–140.

Bull, G., Hebb, C. and Ratković, D. (1970). Choline acetyltransferase activity of human brain tissue during development and at maturity. J. Neurochem. 17, 1505–1516.

Bullock, T.H. and Nachmansohn, D. (1942). Choline esterase in primitive nervous systems. J. Cell. Comp. Physiol. 20, 239–242.

Bullock, T.H., Nachmansohn, D. and Rothenberg, M.A. (1946a). Effects of inhibitors of choline esterase on the nerve action potential. J. Neurophysiol. 9, 9–22.

Bullock, T.H., Grundfest, H., Nachmansohn, D., Rothenberg, M.A. and Sterling, K. (1946b). Effect of di-isopropyl fluorphosphate (DFP) on action potential and cholinesterase of nerve. J. Neurophysiol. 9, 253–260.

Bullock, T.H., Grundfest, H., Nachmansohn, D. and Rothenberg, M.A. (1947a). Generality of the role of acetylcholine in nerve and muscle conduction. J. Neurophysiol. 10, 11–21.

Bullock, T.H., Grundfest, H., Nachmansohn, D. and Rothenberg, M.A. (1947b). Effect of di-isopropyl fluorphosphate (DFP) on action potential and cholinesterase of nerve, II. J. Neurophysiol. 10, 63–78.

Bulmer, D. (1965). A histochemical study of ovarian cholinesterases. Acta Anat. 62, 254–265.

Burdick, C.J. and Strittmatter, C.F. (1965). Appearance of biochemical components related to acetylcholine metabolism during the embryonic development of chick brain. Arch. Biochem. Biophys. 109, 293–301.

Burgen, A.S.V. and Chipman, L.M. (1951). Cholinesterase and succinic dehydrogenase in the central nervous system of the dog. J. Physiol. London 114, 296–305.

Burgen, A.S.V. and Hobbiger, F. (1951). The inhibition of cholinesterases by alkylphosphates and alkylphenolphosphates. Brit. J. Pharmacol. 6, 593–605.

Burkhalter, A., Jones, M. and Featherstone, R.M. (1957a). Acetylcholine-cholinesterase relationship in embryonic chick lung cultivated in vitro. Proc. Soc. Exptl. Biol. Med. 96, 747–750.

Burkhalter, A., Featherstone, R.M., Schueler, F.W. and Jones, M. (1957b). The effects of some acetylcholine derivatives on the cholinesterases of chick embryo intestine cultured in vitro. J. Pharmacol. Exptl. Therap. 120, 285–290.

Burn, J.H. (1971). Release of noradrenaline from sympathetic endings. Nature 231, 237–240.

Burn, J.H. and Rand, M.J. (1959). Sympathetic postganglionic mechanism. Nature 184, 163–165.

Burn, J.H. and Rand, M.J. (1965). Acetylcholine in adrenergic transmission. Ann. Rev. Pharmacol. 5, 163–182.

Burn, J.H., Kordik, P. and Mole, R.H. (1952). The effect of X-irradiation on the response of the intestine to acetylcholine and on its content of 'pseudo'-cholinesterase. Brit. J. Pharmacol. 7, 58–66.

Burnasheva, S.A. and Efremenko, M.V. (1962). The role of adenosine triphosphate in the function of movement of the infusorial species *Tetrahymena pyriformis*. Biokhimiya 27, 167–172 (Eng. trans. 138–142).

Burnasheva, S.A. and Jurzina, G.A. (1968). Cited by Schuster and Hershenov (1969) q.v.

Burnstock, G. (1969). Evolution of the autonomic innervation of visceral and cardiovascular systems in vertebrates. Pharmacol. Rev. 21, 247–324.

Burnstock, G. and Iwayama, T. (1971). Fine-structural identification of autonomic nerves and their relation to smooth muscle. In: Progress in Brain Research, Vol. 34, O. Eränkö, ed. (Elsevier, Amsterdam) pp. 389–404.

Burt, A.M. (1968). Acetylcholinesterase and choline acetyltransferase activity in the developing chick spinal cord. J. Exptl. Zool. 169, 107–112.

Burt, A.M. (1973). Choline acetyltransferase and neuronal maturation. In: Progress in Brain Research, Vol. 40, D.H. Ford, ed. (Elsevier, Amsterdam) pp. 245–252.

Burt, A.M. and Dettbarn, W.-D. (1972). A histochemical study of the distribution of choline acetyltransferase and acetylcholinesterase activity in sensory ganglia and nerve roots of the bullfrog. Histochem. J. 4, 401–411.

Burt, A.M. and Narayanan, C.H. (1970). Effect of extrinsic neuronal connections on development of acetylcholinesterase and choline acetyltransferase activity in the ventral half of the chick spinal cord. Exptl. Neurol. 29, 201–210.

Burt, A.M. Brzin, M. and Davies, J. (1970). The allantoic membrane of the rabbit: evidence of a cholinergic electrical potential. Anat. Record 168, 453–456.

Cajal, S.R. (1911). Histologie du Système Nerveux de l'Homme et des Vertébrés, Vol. II. (Maloine, Paris) pp. 273–274.

Calabro, Q. (1933). Sulla regolazione neuro-umorale cardiaca. Riv. Biol. 15, 299–320.

Callahan, J.F. and Kruckenberg, S.M. (1967). Erythrocyte cholinesterase activity of domestic and laboratory animals: normal levels for nine species. Am. J. Vet. Res. 28, 1509–1512.

Campbell, G. (1970). Autonomic nervous supply to effector tissues. In: Smooth Muscle, E. Bülbring, A.F. Brading, A.W. Jones and T. Tomita, eds. (Arnold, London) pp. 451–495.

Capurro, S., Zaccheo, D. and Viale, G. (1958). Ricerche di istochimica ensimatica sulla retina. Riv. Istochim. Norm. Patol. 4, 327.

Capurro, S., Zaccheo, D. and Viale, G. (1959a). La distribuzione delle colinesterasi nella retina dei pesci. Riv. Istochim. Norm. Patol. 5, 539–540.

Capurro, S., Zaccheo, D. and Viale, G. (1959b). Richerche istochimica sulla attivita colinesterasica nella retina di rettili. Boll. Soc. Ital. Biol. Sper. 35, 609–610.

Capurro, S., Zaccheo, D. and Viale, G. (1959c). Studio dell'attiviti colinesterasica nella retina in rapporto al riposo visivo e agli stimoli luminosi. Boll. Soc. Ital. Biol. Sper. 35, 611–613.

Capurro, S., Zaccheo, D. and Viale, G. (1960). Recherches histochimiques sur l'activité cholinestérasique de l'écorce cérébelleuse. Compt. Rend. Anat. 46, 146–151.

Carbonell, L.M. (1956). Esterases of the conductive system of the heart. J. Histochem. Cytochem. 4, 87–95.

Casida, J.E. (1955a). Comparative enzymology of certain insect acetylesterases in relation to poisoning by organophosphate insecticides. J. Physiol. London 127, 20–21P.

Casida, J.E. (1955b). Comparative enzymology of certain insect acetylesterases in relation to poisoning by organophosphorus insecticides. Biochem. J. 60, 487–496.

Cauna, N. (1960). The distribution of cholinesterase in the cutaneous receptor organs, especially touch corpuscles of the human finger. J. Histochem. Cytochem. 8, 367–375.

Cauna, N. (1961). Cholinesterase activity in cutaneous receptors of man and of some quadrupeds. Bibliotheca Anat. 2, 128–138.

Cauna, N. and Alberti, P. (1961). Nerve supply and distribution of cholinesterase activity in the external nose of the mole. Z. Zellforsch. Mikroskop. Anat. 54, 158–166.

Cauna, N. and Naik, N.T. (1963). The distribution of cholinesterases in the sensory ganglia of man and of some mammals. J. Histochem. Cytochem. 11, 129–138.

Cauna, N., Naik, N.T., Leaming, D.B. and Alberti, P. (1961). The distribution of cholinesterases in the autonomic ganglia of man and of some mammals. Bibliotheca Anat. 2, 90–96.

Cauna, N., Cauna, D. and Hinderer, K.H. (1972). Innervation of human nasal glands. J. Neurocytol. 1, 49–60.

Cavallito, C.J. (1970). Formal discussion, Skokloster Conference. In: Drugs and Cholinergic Mechanisms in the CNS, E. Heilbronn and A. Winter, eds. (Research Institute of National Defence, Stockholm) p. 576.

Cavanagh, J.B. and Holland P. (1961a). Cholinesterase in the chicken nervous system. Nature 190, 735–736.

Cavanagh, J.B. and Holland, P. (1961b). Localization of cholinesterases in the chicken nervous system and the problem of the selective neurotoxicity of organophosphorus compounds. Brit. J. Pharmacol. 16, 218–230.

Cavanagh, J.B. and Patangia, G.N. (1965). Changes in the central nervous system in the cat as the result of tri-o-cresyl phosphate poisoning. Brain 88, 165–180.

Cavanagh, J.B. and Webster, G.R. (1955). On the changes in ali-esterase and pseudocholinesterase activity of chicken sciatic nerve during Wallerian degeneration and their correlation with cellular proliferation. Quart. J. Exptl. Physiol. 40, 12–23.

Cavanagh, J.B., Thompson, R.H.S. and Webster, G.R. (1954). The localization of pseudocholinesterase activity in nervous tissue. Quart. J. Exptl. Physiol. 39, 185–197.

Chacko, L.W. and Cerf, J.A. (1960). Histochemical localization of cholinesterase in the amphibian spinal cord and alterations following ventral root section. J. Anat. 94, 74–81.

Chadwick, L.E. (1963). Actions on insects and other invertebrates. In: Cholinesterases and Anticholinesterase Agents, G.B. Koelle, ed. (Springer, Berlin) pp. 741–798.

Chadwick, L.E. and Lovell, J.B. (1958). The effect of temperature on the activity of flyhead cholinesterase. Proc. 10th Intern. Congr. Entomol. 2, 19–27.

Chadwick, L.E., Lovell, J.B. and Egner, V.E. (1954). The relationship between pH and the activity of cholinesterase from flies. Biol. Bull. 106, 139–148.

Chagas, C., Penna-Franca, E., Nishie, K. and Garcia, E.J. (1958). A study of the specificity of the complex formed by gallamine triethiodide with a macromolecular constituent of the electric organ. Arch. Biochem. Biophys. 75, 251–259.

Chan, S.L., Shirachi, D.Y. and Trevor, A.J. (1972a). Purification and properties of brain acetylcholinesterase (EC 3.1.1.7). J. Neurochem. 19, 437–447.

Chan, S.L., Shirachi, D.Y., Bhargava, H.N., Gardner, E. and Trevor, A.J. (1972b). Purification and properties of multiple forms of brain acetylcholinesterase (EC 3.1.1.7). J. Neurochem. 19, 2747–2758.

Chang, C.C. and Lee, C.Y. (1963). Isolation of neurotoxins from the venom of *Bungarus multicinctus* and their modes of neuromuscular blocking action. Arch. Intern. Pharmacodyn. 144, 241–257.

Chang, P.-L., Bhagat, B. and Taylor, J.J. (1973). Effect of chronic administration of nicotine on acetylcholinesterase activity in the hypothalamus and medulla oblongata of the rat brain. An ultrastructural study. Brain Res. 54, 75–84.

Changeux, J.-P. (1966). Responses of acetylcholinesterase from *Torpedo marmorata* to salts and curarizing drugs. Mol. Pharmacol. 2, 369–392.

Changeux, J.-P. (1969). Remarks on the symmetry and cooperative properties of biological membranes. In: Symmetry and Function of Biological Systems at the Macromolecular Level, A. Engström and B. Strandberg, eds. (Almqvist and Wiksell, Stockholm) pp. 235–256.

Changeux, J.-P., Podleski, T. and Meunier, J.-C. (1969). On some structural analogies between acetylcholinesterase and the macromolecular receptor of acetylcholine. J. Gen. Physiol. 54, 225s–244s.

Changeux, J.-P., Kasai, M., Huchet, M. and Meunier, J.-C. (1970a). Extraction à partir du tissu électrique de gymnote d'une protéine presentant plusieurs propriétés caractéristiques du récepteur physiologique de l'acétylcholine. Compt. Rend. Acad. Sci. Paris. 270, 2864–2867.

Changeux, J.-P., Kasai, M. and Lee, C.-Y. (1970b). Use of a snake venom toxin to characterize the cholinergic receptor protein. Proc. Natl. Acad. Sci. 67, 1241–1247.

Changeux, J.-P., Kasai, M., Huchet, M. and Meunier, J.-C. (1971). In vitro studies with the cholinergic receptor of the eel electroplax. In: Cholinergic Ligand Interactions, D.J. Triggle, J.F. Moran and E.A. Barnard, eds. (Academic Press, New York) pp. 33–47.

Chapman, J.B. and McCance, I. (1967). Acetylcholine sensitive cells in the intracerebellar nuclei of the cat. Brain Res. 5, 535–538.

Chaudhary, K.D., Srivastava, V. and Lemonde, A. (1966). Acetylcholinesterase in *Tribolium confusum* Duval. Arch. Intern. Physiol. Biochim. 74, 416–428.

Cheng, K. (1963). Cholinesterase activity in human extraocular muscles. Japan. J. Ophthalmol. 7, 174–183.

Cheng, K. (1964a). Localization of cholinesterase activity in mammalian extraocular muscles (in Japanese). Acta Ophthal. Jap. 68, Suppl. 1, 1–16.

Cheng, K. (1964b). Localization of non-specific cholinesterase activity in extraocular muscles (in Japanese). Acta Ophthal. Jap. 68, Suppl. 17, 17–22.

Cherchi, P., Casula, D., Spinazzola, A., Nissardi, G.P., Sanna, R., Andaccio, F., Melis, L., Zedda, S. and Casciu, G. (1964). Effects of the gases present in the work environment on the biohumoral picture of coal miners (Sulcis basin). III. Behavior of some enzymic activities (in Italian). Lavoro Umano 16, 667–691.

Childs, A.F., Davies, D.R., Green, A.L. and Rutland, J.P. (1955). The reactivation by oximes and hydroxamic acids of cholinesterase inhibited by organophosphorus compounds. Brit. J. Pharmacol. 10, 462–465.

Chokroverty, S., Parameswar, K.S. and Co, C. (1971). Nonspecific esterases in the myoneural junction of human striated muscle. J. Histochem. Cytochem. 19, 798–800.

Chothia, C. (1970). Interaction of acetylcholine with different cholinergic nerve receptors. Nature 225, 36–38.

Chothia, C. and Pauling, P. (1969). Conformation of cholinergic molecules relevant to acetylcholinesterase. Nature 223, 919–921.

Chothia, C. and Pauling, P. (1970). Absolute configuration of cholinergic molecules; the crystal structure of (+)-*trans*- 2 - acetoxy cyclopropyl trimethylammonium iodide. Nature 226, 541–542.

Christensen, H.N. and Riggs, T.R. (1951). Physostigmine uptake by cells and its effect on potassium exchange. J. Biol. Chem. 193, 621–626.

Christison, R. (1855). The properties of the ordeal-bean of old Calabar, Western Africa. Monthly J. Med. XX 193–204; also Pharmaceutical J. XIV, 470–476.

Christison (1885, 1886). The Life of Sir Robert Christison, Bart, Vols I & II, edited by his sons (Wm Blackwood and Sons, Edinburgh).

Chu, N-S., Rutledge, L.T. and Sellinger, O.Z. (1971a). The effect of cortical undercutting and long-term electrical stimulation on synaptic acetylcholinesterase. Brain Res. 29, 323–330.

Chu, N-S., Rutledge, L.T. and Sellinger, O.Z. (1971b). Tubocurarine binding in undercut cerebral cortex and the effect of long-term electrical stimulation. Brain Res. 29, 331–337.

Churchill, J.A., Schuknecht, H.F. and Doran, R. (1956). Acetylcholinesterase activity in the cochlea. Laryngoscope 66, 1–15.

Cier, A., Cuisinaud, G., Legheand, J. and Solal, M. (1970). Influence de quelques cations divalents sur l'activité des cholinestérases. Bull. Trav. Soc. Pharm. Lyon 14, 19–30.

Ciliv, G. and Özand, P.T. (1972). Human erythrocyte acetylcholinesterase purification, properties and kinetic behavior. Biochim. Biophys. Acta 284, 136–156.

Clegg, M.T. and Doyle, L.L. (1967). Role in reproductive physiology of afferent impulses from the genitalia and other regions. In: Neuroendocrinology, Vol. II, L. Martini and W.F. Ganong, eds. (Academic Press, New York) pp. 1–17.

Clitherow, J.W., Mitchard, M. and Harper, N.J. (1963). The possible biological function of pseudocholinesterase. Nature 199, 1000–1001.

Clos, F. and Serfaty, A. (1958). Specificité zoologique de cholinestérases chez les poissons dulcaquicoles. Bull. Soc. Hist. Natl. Toulouse 93, 30–34.

Clos, F., Serfaty, A. and Cathala, M. (1957). Activités cholinestérasiques chez les poissons dulcaquicoles. Bull. Soc. Hist. Natl. Toulouse 92, 205–217.

Clouet, D.H. and Waelsch, H. (1961). Amino acid and protein metabolism of the brain. VIII. The recovery of cholinesterase in the nervous system of the frog after inhibition. J. Neurochem. 8, 201–215.

Clouet, D.H. and Waelsch, H. (1963). Amino acid and protein metabolism of the brain IX. The effect of an organophosphorus inhibitor on the incorporation of [^{14}C]lysine into the proteins of the brain. J. Neurochem. 10, 51–63.

Coërs, C. (1953). La détection histochimique de la cholinestérase au niveau de la jonction neuro-musculaire. Rev. Belge Pathol. Med. Exptl. 22, 306–315.

Coggeshall, R.E. (1965). A fine structural analysis of the ventral nerve cord and associated sheath of *Lumbricus terrestris* L. J. Comp. Neurol. 125, 393–437.

Coggeshall, R.E. and Fawcett, D.W. (1964). The fine structure of the central nervous system of the leech, *Hirudo medicinalis*. J. Neurophysiol. 27, 229–289.

Cohen, E.M. and Wiersinga, H. (1959). Oximes in the treatment of nerve gas poisoning. I. Acta Physiol. Pharmacol. Neerl. 8, 40–51.

Cohen, J.A. and Oosterbaan, R.A. (1963). The active site of acetylcholinesterase and related esterases and its reactivity towards substrates and inhibitors. In: Cholinesterases and Anti-cholinesterase Agents, G.B. Koelle, ed. (Springer, Berlin) pp. 299–373.

Cohen, J.A. and Warringa, M.G.P.J. (1953). Methods to estimate the turnover number of preparations of ox red cell cholinesterase. Biochim. Biophys. Acta 11, 52–58.

Cohen, J.A., Kalsbeek, F. and Warringa, M.G.P.J. (1949). The significance of butyrylcholine in the testing of cholinesterase-containing preparations. Acta Brevia Neerl. Physiol. Pharmacol. Microbiol. 17, 32–36.

Cohen, J.A., Oosterbaan, R.A. and Warringa, M.G.P.J. (1955). The turnover number of ali-esterase, pseudo- and true cholinesterase and the combination of these enzymes with diiso-propylfluorophosphonate. Biochim. Biophys. Acta 18, 228–235.

Colhoun, E.H. (1959). Physiological events in organophosphorus poisoning. Can. J. Biochem. Physiol. 37, 1127–1134.

Collier, B. and Katz, H.S. (1970). The release of acetylcholine by acetylcholine in the cat's superior cervical ganglion. Brit. J. Pharmacol. 39, 428–438.

Collier, B. and Katz, H.S. (1971). The synthesis, turnover and release of surplus acetylcholine in a sympathetic ganglion. J. Physiol. London 214, 537–552.

Collier, B. and Mitchell, J.F. (1966). The central release of acetylcholine during stimulation of the visual pathway. J. Physiol. London 184, 239–254.

Collier, B. and Mitchell, J.F. (1967). The central release of acetylcholine during consciousness and after brain lesions. J. Physiol. London 188, 83–98.

Collier, B., Poon, P. and Salehmoghaddam, S. (1972). The formation of choline and of acetyl-choline by brain in vitro. J. Neurochem. 19, 51–60.

Comis, S.D. and Davies, W.E. (1969). Acetylcholine as a transmitter in the cat auditory system. J. Neurochem. 16, 423–429.

Comis, S.D. and Guth, P.S. (1974). The release of acetylcholine from the cochlear nucleus upon stimulation of the crossed olivo-cochlear burdle. Neuropharmacol. (in press).

Comis, S.D. and Whitfield, I.C. (1968). Influence of centrifugal pathways on unit activity in the cochlear nucleus. J. Neurophysiol. 31, 62–68.

Costall, B. and Olley, J.E. (1971). Cholinergic- and neuroleptic-induced catalepsy : modification by lesions in the caudate-putamen. Neuropharmacol. 10, 297–306.

Costall, B., Naylor, R.J. and Olley, J.E. (1972). Catalepsy and circling behaviour after intracere-bral injections of neuroleptic, cholinergic and anticholinergic agents into the caudate–putamen, globus pallidus and substantia nigra of rat brain. Neuropharmacol. 11, 645–663.

Cotman, C.W. and Taylor, D.A. (1971). Autoradiographic analysis of protein synthesis in synaptosomal fractions. Brain Res. 29, 366–372.

Cotman, C., Herschman, H. and Taylor, D. (1971). Subcellular fractionation of cultured glial cells. J. Neurobiol. 2, 169–180.

Cottle, M.K.W. and Silver, A. (1970a). Fluorescent granules in the guinea-pig hypothalamus and their possible relation to neurosecretory substance. Z. Zellforsch. Mikroskop. Anat. 103, 559–569.

Cottle, M.K.W. and Silver, A. (1970b). Histochemical demonstration of acetylcholinesterase in the hypothalamus of the female guinea-pig. Z. Zellforsch. Mikroskop. Anat. 103, 570–588.

Cotton, F.A. and Wilkinson, G. (1966). Advanced Inorganic Chemistry, 2nd ed (Interscience Publishers, New York) p.139.

Cottrell, G.A. and Laverack, M.S. (1968). Invertebrate pharmacology. Ann. Rev. Pharmacol. 8, 273–298.

Cottrell, G.A., Powell, B. and Stanton, M. (1970). A simple method for measuring a picogram of acetylcholine using the clam (*Mya arenaria*) heart. Brit. J. Pharmacol. 40, 866–870.

Couceiro, A., Almeida, D.F. and Freire, J.R.C. (1953). Localisation histochimique de l'acétyl-cholinestérase dans le tissue électrique d'*Electrophorus electricus* (L). Anais Acad. Brasil. Cienc. 25, 205–214.

Couch, J.R. (1970). Responses of neurons in the raphe nuclei to serotonin, norepinephrine and acetylcholine and their correlation with an excitatory synaptic input. Brain Res. 19, 137–150.

Coupland, R.E. (1958). The innervation of pancreas of the rat, cat and rabbit as revealed by the cholinesterase technique. J. Anat. 92, 143–149.

Coupland, R.E. and Holmes, R.L. (1957). The use of cholinesterase techniques for the demonstration of peripheral nervous structures. Quart. J. Microscop. Sci. 98, 327–330.

Coupland, R.E. and Holmes, R.L. (1958). The distribution of cholinesterase in the adrenal glands of the rat, cat and rabbit. J. Physiol. London 141, 97–106.

Couteaux, R. (1951). Remarques sur les méthodes actuelles de détection histochimique des activités cholinestérasiques. Arch. Intern. Physiol. 59, 526–537.

Couteaux, R. and Nachmansohn, D. (1938). Cholinesterase at the end-plates of voluntary muscle after nerve degeneration. Nature 142, 481.

Couteaux, R. and Szabo, T. (1959). Siège de la jonction nerf-électroplaque dans les organes électriques à électroplaques pédiculées. Compt. Rend. Acad. Sci. Paris 248, 457–460.

Couteaux, R. and Taxi, J. (1952). Recherches histochimiques sur la distribution des activités cholinestérasiques au niveau de la synapse myoneurale. Arch. Anat. Microscop. Morphol. Exp. 41, 352–392.

Cowan, W.M. and Powell, T.P.S. (1962). Centrifugal fibres to the retina in the pigeon. Nature 194, 487.

Cowan, W.M., Adamson, L. and Powell, T.P.S. (1961). An experimental study of the avian visual system. J. Anat. 95, 545–563.

Cranmer, M.F. and Peoples, A. (1971). Sensitive gas chromatographic method for human cholinesterase determination. J. Chromatog. 57, 365–371.

Crawford, J.M., Curtis, D.R., Voorhoeve, P.E. and Wilson, V.J. (1963). Excitation of cerebellar neurones by acetylcholine. Nature 200, 579–580.

Crawford, J.M., Curtis, D.R., Voorhoeve, P.E. and Wilson, V.J. (1966). Acetylcholine sensitivity of cerebellar neurones in the cat. J. Physiol. London 186, 139–165.

Crescitelli, F., Koelle, G.B. and Gilman, A. (1946). Transmission of impulses in peripheral nerves treated with diisopropyl fluorophosphate (DFP). J. Neurophysiol. 9, 241–252.

Crevier, M. and Bélanger, L.F. (1955). Simple method for histochemical detection of esterase activity. Science 122, 556.

Crone, H.D. (1971). The dissociation of rat brain membranes bearing acetylcholinesterase by the non-ionic detergent Triton X-100 and an examination of the product. J. Neurochem. 18, 489–497.

Crook, J.C. (1963). Acetylcholinesterase activity of capillary blood vessels in the central nervous system of the rabbit. Nature 199, 41–43.

Csillik, B. and Koelle, G.B. (1965). Histochemistry of the adrenergic and the cholinergic autonomic innervation apparatus as represented by the rat iris. Acta Histochem. 22, 350–363.

Csillik, B. and Sávay, Gy. (1958). Die Regeneration subneuralen Apparate der motorischen End-platten. Acta Neuroveget. Vienna 19, 41–52.

Csillik, B. and Tóth, L. (1972). Histochemical identification of Renshaw elements. J. Histochem. Cytochem. 20, 385–387.

Csillik, B., Joó, F. and Kása, P. (1963). Cholinesterase activity of archicerebellar mossy fibre apparatuses. J. Histochem. Cytochem. 11, 113–114.

Csillik, B., Joó, F., Kása, P., Tomity, I. and Kalman, Gy. (1964). Development of acetylcholinesterase-active structures in the rat archicerebellar cortex. Acta Biol. Acad. Sci. Hung. 15, 11–17.

Csillik, B., Knyihár, E. and Halász, N. (1968). Ultrastructural localization of acetylcholinesterase in autonomic postganglionic axons. J. Neuro-visc. Rel. 31, 3–10.

Cuénod, M. and Schonbach, J. (1971). Synaptic proteins and axonal flow in the pigeon visual pathway. J. Neurochem. 18, 809–816.

Cunningham, L.W. (1957). Proposed mechanism of action of hydrolytic enzymes. Science 125, 1145–1146.

Curtis, D.R. (1963). Acetylcholine as a central transmitter. Can. J. Biochem. Physiol. 41, 2611–2618.

Curtis, D.R. and Crawford, J.M. (1965). Acetylcholine sensitivity of cerebellar neurones. Nature 206, 516–517.

Curtis, D.R. and Koizumi, K. (1961). Chemical transmitter substances in brain stem of cat. J. Neurophysiol. 24, 80–90.

Curtis, D.R. and Ryall, R.W. (1964). Nicotinic and muscarinic receptors of Renshaw cells. Nature 203, 652–653.

Curtis, D.R. and Ryall, R.W. (1966a). The excitation of Renshaw cells by cholinomimetics. Exptl. Brain Res. 2, 49–65.

Curtis, D.R. and Ryall, R.W. (1966b). The acetylcholine receptors of Renshaw cells. Exptl. Brain Res. 2, 66–80.

Curtis, D.R., Phillis, J.W. and Watkins, J.C. (1961). Cholinergic and non-cholinergic transmission in the mammalian spinal cord. J. Physiol. London 158, 296–323.

Curtis, D.R., Ryall, R.W. and Watkins, J.C. (1966). The action of cholinomimetics on spinal interneurones. Exptl. Brain Res. 2, 97–106.

Curtis, D.R., Duggan, A.W. and Johnston, G.A.R. (1971). The specificity of strychnine as a glycine antagonist in the mammalian spinal cord. Exptl. Brain Res. 12, 547–565.

Cuthbert, A.W. (1963). An acetylcholine-like substance and cholinesterase in the smooth muscle of the chick amnion. J. Physiol. London 166, 284–295.

D'Agostini, N. and Rossatti, B. (1959). The histochemical localization of specific cholinesterase in the lymphatic tissue of mammals. J. Anat. 93, 354–360.

D'Agostini, N. and Rossatti, B. (1961). Histochemical features of acetylcholinesterase activity in the lymphatic tissue of man and other mammals. Bibliotheca Anat. 2, 236–242.

Dale, H.H. (1914). The action of certain esters and ethers of choline and their relation to muscarine. J. Pharmacol. Exptl. Therap. 6, 147–190.

Dale, H.H. (1934). Nomenclature of fibres in the autonomic system and their effects. J. Physiol. London 80, 10–11*P.*

Dale, H.H. (1935). Pharmacology and nerve endings. Proc. Roy. Soc. Med. 28, 319–332.

Dale, H.H. (1938). The William Henry Welch Lectures (1937). Acetylcholine as a chemical transmitter. I. History of the ideas and evidence. II. Chemical transmission of ganglionic synapses and voluntary nerve endings. J. Mt. Sinai Hosp. 4, 401–429.

Dale, H.H. (1948). Transmission of effects from the endings of nerve fibres. British Association lecture. Nature 162, 558–560.

Dale, H.H. (1953). Adventures in Physiology (Pergamon Press, London).

Dale, H.H. (1954). The beginnings and the prospects of neurohumoral transmission. Pharmacol. Rev. 6, 7–13.

Dale, H.H. (1955). Junctional transmission of nervous effects by chemical agents. Proc. Mayo Clin. 30, 5–20.

Dalton, A.J. (1955). A chrome-osmium fixative for electron microscopy. Anat. Record 121, 281.

Danilova, O.A. (1971). Cholinesterase activity of the hypothalamo-hypophysial neurosecretory system in rats during the ontogenetic development. Histochemie 28, 255–264.

Das, P.K. and Liddell, J. (1970). Purification and properties of human serum cholinesterase. Biochem. J. 116, 875–881.

Dauterman, W.C. and Mehrotra, K.N. (1963). The *N*-alkyl group specificity of cholinesterase from the housefly, *Musca domestica* L., and the two-spotted spider mite, *Tetranychus telarius* L. J. Insect Physiol. 9, 257–263.

Dauterman, W.C., Talens, A. and van Asperen, K. (1962). Partial purification and properties of flyhead cholinesterase. J. Insect Physiol. 8, 1–14.

Davies, D.R. (1954). Cholinesterases and the mode of action of some anticholinesterases. J. Pharm. Pharmacol. 6, 1–26.

Davies, D.R. (1963). Neurotoxicity of organophosphorus compounds. In: Cholinesterases and Anticholinesterase Agents, G.B. Koelle, ed. (Springer, Berlin) pp. 860–882.

Davies, D.R. and Green, A.L. (1956). The kinetics of reactivation, by oximes, of cholinesterase inhibited by organophosphorus compounds. Biochem. J. 63, 529–535.

Davies, D.R. and Green, A.L. (1959). The chemotherapy of poisoning by organophosphate anticholinesterases. Brit. J. Indust. Med. 16, 128–134.

Davies, D.R. and Nicholls, J.D. (1955). A field test for the assay of human whole-blood cholinesterase. Brit. Med. J. 1, 1373–1375.

Davies, D.R. and Willey, G.L. (1958). The toxicity of 2-hydroxyiminomethyl-N-methylpyridinium methanesulphonate (P2S). Brit. J. Pharmacol. 13, 202–206.

Davies, D.R., Risley, J.E. and Rutland, J.P. (1953). The hydrolysis of choline esters by tissues of the ruminant. Biochem. J. 53, XV.

Davies, D.R., Green, A.L. and Willey, G.L. (1959). 2-hydroxyiminomethyl-N-methylpyridinium methanesulphonate and atropine in the treatment of severe organophosphate poisoning. Brit. J. Pharmacol. 14, 5–8.

Davies, D.R., Holland, P. and Rumens, M.J. (1960). The relationship between the chemical structure and neurotoxicity of alkyl organophosphorus compounds. Brit. J. Pharmacol. 15, 271–278.

Davies, R.E. and Ojha, K.N. (1950). Specific and non-specific serum cholinesterase in cats before and after treatment with stilboestrol. Brit. J. Pharmacol. 5, 395–397.

Davis, D.A., Wasserkrug, H.L., Heyman, I.A., Padmanabhan, K.C., Seligman, G.A., Plapinger, R.E. and Seligman, A.M. (1972). Comparison of ultrastructural cholinesterase demonstration in the motor end plate with α-acetylthiol-m-toluenediazonium ion and 3-acetoxy-5-indolediazonium ion. J. Histochem. Cytochem. 20, 161–172.

Davis, G.A. and Agranoff, B.W. (1968). Metabolic behaviour of isozymes of acetylcholinesterase. Nature 220, 277–280.

Davis, J.E. (1960). Erythropoietic stimulating action of acetylcholinesterase. Proc. Soc. Exptl. Biol. Med. 104, 698–699.

Davis, R. and Koelle, G.B. (1967). Electron microscopic localization of acetylcholinesterase and non-specific cholinesterase at the neuromuscular junction by the gold-thiocholine and gold-thiolacetic acid methods. J. Cell Biol. 34, 157–171.

Davis, R. and Vaughan, P.C. (1969). Pharmacological properties of feline red nucleus. Intern. J. Neuropharmacol. 8, 475–488.

Davison, A.N. (1953a). Return of cholinesterase activity in the rat after inhibition by organophosphorus compounds. 1. Diethyl p-nitrophenyl phosphate (E 600, Paraoxon). Biochem. J. 54, 583–590.

Davison, A.N. (1953b). Some observations on the cholinesterases of the central nervous system after the administration of organo-phosphorus compounds. Brit. J. Pharmacol. 8, 212–216.

Davison, A.N. (1955). Return of cholinesterase activity in the rat after inhibition by organophosphorus compounds. 2. A comparative study of true and pseudo cholinesterase. Biochem. J. 60, 339–346.

Dawson, R.M. and Crone, H.D. (1973). Inorganic ion effects on the kinetic parameters of acetylcholinesterase. J. Neurochem. 21, 247–249.

DeCastro, F. (1926). Sur la structure de l'innervation de la glande intercarotidienne (glomus caroticum) de l'homme et des mammifères, et sur un nouveau système d'innervation autonome du nerf glossopharyngien. Trab. Lab. Invest. Biol. Univ. Madrid 24, 365–432.

Deffenu, G., Bertaccini, G. and Pepeu, G. (1967). Acetylcholine and 5-hydroxytryptamine levels of the lateral geniculate bodies and superior colliculus of cats after visual deafferentation. Exptl. Neurol. 17, 203–209.

Dekirmenjian, H., Brunngraber, E.G., Lemkey-Johnston, N. and Larramendi, L.M.H. (1969). Distribution of gangliosides, glycoprotein-NANA and acetylcholinesterase in axonal and synaptosomal fractions of cat cerebellum. Exptl. Brain Res. 8, 97–104.

Delaunois, A.L. (1962). Automatized micromethod for the potentiometric determination of cholinesterase activity. Arch. Intern. Pharmacodyn. 140, 351–357.

De Lorenzo, A.J.D., Dettbarn, W.-D. and Brzin, M. (1969). Fine structural localization of acetylcholinesterase in single axons. J. Ultrastruct. Res. 28, 27–40.

Denz, F.A. (1953). The histochemistry of the myoneural junction. Brit. J. Exptl. Pathol. 34, 329–339.

De Robertis, E. (1971). Molecular biology of synaptic receptors. Science 171, 963–971.

De Robertis, E., Pellegrino De Iraldi, A., Rodríguez de Lores Arnaiz, G. and Salganicoff, L. (1961). Electron microscope observations on nerve endings isolated from rat brain. Anat. Record 139, 220–221.

De Robertis, E., Pellegrino de Iraldi, A., Rodríguez de Lores Arnaiz, G. and Salganicoff, L. (1962). Cholinergic and non-cholinergic nerve endings in rat brain. I. Isolation and subcellular distribution of acetylcholine and acetylcholinesterase. J. Neurochem. 9, 23–35

De Robertis, E., Rodríguez de Lores Arnaiz, G., Salganicoff, L., Pellegrino de Iraldi, A., and Zieher, L.M. (1963). Isolation of synaptic vesicles and structural organization of the acetylcholine system within brain nerve endings. J. Neurochem. 10, 225–235.

De Robertis, E., Azcurra, J.M. and Fiszer, S. (1967). Ultrastructure and cholinergic binding capacity of junctional complexes isolated from rat brain. Brain Res. 5, 45–56.

De Robertis, E., Fiszer, S., La Torre, J.L. and Lunt, G.S. (1970). Proteolipid cholinergic receptors isolated from the central nervous system and electric tissue. In: Drugs and Cholinergic Mechanisms in the CNS, E. Heilbronn and A. Winter, eds. (Research Institute of National Defence, Stockholm) pp. 505–517.

Desmedt, J.E. and La Grutta, G. (1955). Control of brain potentials by pseudo-cholinesterase. J. Physiol. London 129, 46–47P.

Desmedt, J.E. and La Grutta, G. (1957). The effect of selective inhibition of pseudocholinesterase on the spontaneous and evoked activity of the cat's cerebral cortex. J. Physiol. London 136, 20–40.

Dettbarn, W.-D. (1963). Hydrolysis of choline esters by invertebrate nerve fibers. Biochim. Biophys. Acta 77, 430–435.

Dettbarn, W.-D. (1967). The acetylcholine system in peripheral nerve. Ann. N.Y. Acad. Sci. 144, 483–503.

Dettbarn, W.-D. and Bartels, E. (1968). The action of acetylcholine and cholinesterase inhibitors on single axons of the lobster. Biochem. Pharmacol. 17, 1833–1844.

Dettbarn, W.-D. and Rosenberg, P. (1962). Sources of error in relating electrical and acetylcholinesterase activity. Biochem. Pharmacol. 11, 1025–1030.

Dettbarn, W.-D., Bartels, E., Hoskin, F.C.G. and Welsch, F. (1970). Spontaneous reactivation of organophosphorus-inhibited electroplax cholinesterase in relation to acetylcholine-induced depolarization. Biochem. Pharmacol. 19, 2949-2955.

Deutsch, J.A. and Leibowitz, S.F. (1966). Amnesia or reversal of forgetting by anticholinesterase, depending simply on time of injection. Science 153, 1017–1018.

Deutsch, J.A. and Lutzky, H. (1967). Memory enhancement by anticholinesterase as a function of initial learning. Nature 213, 742.

Deutsch, J.A., Hamburg, M.D. and Dahl, A. (1966). Anticholinesterase-induced amnesia and its temporal aspects. Science 151, 221–223.

Dewhurst, S.A., McCaman, R.E. and Kaplan, W.D. (1970). The time course of development of acetylcholinesterase and choline acetyltransferase in *Drosophila melanogaster*. Biochem. Genet. 4, 499–508.

Dickson, D.H., Flumerfelt, B.A., Hollenberg, M.J. and Gwyn, D.G. (1971). Ultrastructural localization of cholinesterase activity in the outer plexiform layer of the newt retina. Brain Res. 35, 299–303.

Dikshit, B.B. and Mahal, H.S. (1937). Choline esterase in toxaemia. Quart. J. Exptl. Physiol. 27, 41–48.

Dixon, J.S. and Gosling, J.A. (1971). Histochemical and electron microscopic observations on the innervation of the upper segment of the mammalian ureter. J. Anat. 110, 57–66.

Dixon, M. and Webb, E.C. (1964). Enzymes, 2nd ed. (Academic Press, New York) pp. 466–467.

Domino, E.F. and Bartolini, A. (1972). Effects of various psychotomimetic agents on the EEG and acetylcholine release from the cerebral cortex of brainstem transected cats. Neuropharmacol. 11, 703–713.

Donhoffer, A. (1959). Localization of cholinesterases in the nervous plexuses of the small intestine of the rat. Acta Morphol. Acad. Sci. Hung. 8, 375–379.

Douglas, W.W. and Paton, W.D.M. (1951). The mode of action of tetraethylpyrophosphate at the cat's neuromuscular junction. J. Physiol. London 115, 71–72P.

Dow, R.S. (1936). The fiber connections of the posterior parts of the cerebellum in the rat and cat. J. Comp. Neurol. 63, 527–548.

Drachman, D.B. (1967). Is acetylcholine the trophic neuromuscular transmitter? Arch. Neurol. 17, 206–218.

Droz, B. and Koenig, H.L. (1970). The turnover of proteins in axons and nerve endings. In: Cellular Dynamics of the Neuron, S.H. Barondes, ed. (Academic Press, New York). pp. 35–50.

Drukker, J. and Schadé, J.P. (1965a). Degeneration patterns in the optic lobe of cephalopods. In: Progress in Brain Research, Vol. 14, M. Singer and J.P. Schadé, eds. (Elsevier, Amsterdam) pp. 122–142.

Drukker, J. and Schadé, J.P. (1965b). Enzyme histochemistry of the developing cerebral cortex of the rabbit. Acta Physiol. Pharmacol. Neerl. 13, 219–220.

Dubois, K.P. (1963). Toxicological evaluation of the anticholinesterase agents. In: Cholinesterases and Anticholinesterase Agents, G.B. Koelle, ed. (Springer, Berlin) pp. 833–859.

Dubois, K.P., Doull, J. and Coon, J.M. (1950). Studies on the toxicity and pharmacological action of octamethyl pyrophosphoramide (OMPA; Pestox III). J. Pharmacol. Exptl. Therap. 99, 376–393.

Duckett, S. and Pearse, A.G.E. (1967). Histoenzymology of the developing human basal ganglia. Histochemie 8, 334–341.

Duckett, S. and Pearse, A.G.E. (1968). The cells of Cajal-Retzius in the developing human brain. J. Anat. 102, 183–187.

Duckett, S. and Pearse, A.G.E. (1969). Histoenzymology of the developing human spinal cord. Anat. Rec. 163, 59–66.

Dudley, H.W. (1933). The alleged occurrence of acetylcholine in ox blood. J. Physiol. London 79, 249–254.

Duggan, A.W. and Curtis, D.R. (1972). Morphine and the synaptic activation of Renshaw cells. Neuropharmacol. 11, 189–196.

Duffy, P.E., Tennyson, V.M. and Brzin, M. (1967). Cholinesterase in adult and embryonic hypothalamus. Arch. Neurol. 16, 385–403.

Dultz, L., Epstein, M.A., Freeman, G., Gray, E.H. and Weil, W.B. (1957). Studies on a group of oximes as therapeutic compounds in Sarin poisoning. J. Pharmacol. Exptl. Therap. 119, 522–531.

Duncan, C.J. (1967). The Molecular Properties and Evolution of Excitable Cells (Pergamon Press, New York) pp. 112–124.

Duncan, J.A., Rutledge, L.T. and Domino, E.F. (1968). Acetylcholinesterase activity in partially isolated cerebral cortex after prolonged intermittent stimulation. Exptl. Neurol. 20, 268–274.

Dupé, M. (1967). Relation between cholinesterase activity and telencephalic electric activity in *Protopterus annectens*. Compt. Rend. Acad. Sci. Paris, Ser. D. 265, 1063–1066.

Dupé, M. and Bocklée-Morvan, M.L. (1968). Mise en évidence d'une cholinestérase spécifique au niveau du système nerveux central et du système circulatoire chez un Dipneuste (*Protopterus annectens*). Compt. Rend. Soc. Biol. 162, 823–829.

Durell, J., Garland, J.T. and Friedel, R.O. (1969). Acetylcholine action: biochemical aspects. Science 165, 862–866.

Dyball, R.E.J. and Koizumi, K. (1969). Electrical activity in the supraoptic and paraventricular nuclei associated with neurohypophysial hormone release. J. Physiol. London 201, 711–722.

Earl, C.J. and Thompson, R.H.S. (1952a). The inhibitory action of tri-ortho-cresyl phosphate on cholinesterases. Brit. J. Pharmacol. 7, 261–269.

Earl, C.J. and Thompson, R.H.S. (1952b). Cholinesterase levels in the nervous system in tri-ortho-cresyl phosphate poisoning. Brit. J. Pharmacol. 7, 685–694.

Easson, L.H. and Stedman, E. (1937). The specificity of choline-esterase. Biochem. J. 31, 1723–1729.

Ebel, A., Hermetet, J.C. and Mandel, P. (1973). Comparative study of acetylcholinesterase and choline acetyltransferase enzyme activity in brain of DBA and C_{57} mice. Nature New Biol. 242, 56–57.

Eccles, J.C., Eccles, R.M. and Fatt, P. (1956). Pharmacological investigations on a central synapse operated by acetylcholine. J. Physiol. London 131, 154–169.

Eccles, J.C., Fatt, P. and Koketsu, K. (1954). Cholinergic and inhibitory synapses in a pathway from motor-axon collaterals to motoneurones. J. Physiol. London 126, 524–562.

Ecobichon, D.J. and Comeau, A.M. (1973). Pseudocholinesterases of mammalian plasma: physicochemical properties and organophosphate inhibition in eleven species. Toxicol. Appl. Pharmacol. 24, 92–100.

Ecobichon, D.J. and Israël, Y. (1967). Characterization of the esterases from electric tissue of *Electrophorus* by starch-gel electrophoresis. Can. J. Biochem. 45, 1099–1105.

Edery, H. and Levinger, I.M. (1971). Acetylcholine release into the perfused intermeningeal spaces of the cat spinal cord. Neuropharmacol. 10, 239–246.

Edström, A. (1964). The ribonucleic acid in the Mauthner neuron of the goldfish. J. Neurochem. 11, 309–314.

Edwards, A.J., Burt, J.S. and Ogilvie, B.M. (1971). The effects of immunity upon some enzymes of the parasitic nematode, *Nippostrongylus brasiliensis*. Parasitol. 62, 339–347.

Ehinger, B. (1967). Double innervation of the feline iris dilator. Arch. Ophthalmol. 77, 541–545.

Ehinger, B., Falck, B., Persson, H., Rosengren, A.-M. and Rosengren, E. (1966). Choline acetylase activity in the normal and denervated cat iris. Life Sci. Oxford 5, 481–483.

Ehinger, B., Falck, B., Persson, H., Rosengren, A.-M. and Sporrong, B. (1970). Acetylcholine in adrenergic terminals of the cat iris. J. Physiol. London 209, 557–565.

Ehrenpreis, S. (1959). Interaction of curare and related substances with acetylcholine receptor-like protein. Science 129, 1613–1614.

Ehrenpreis, S. (1967). Possible nature of the cholinergic receptor. Ann. N.Y. Acad. Sci. 144, 720–734.

Ehrenpreis, S., Chiesa, A., Bigo-Gullino, M. and Patrick, P. (1967). Correlation between acetyl-choline (ACh) potentiation and cholinesterase (ChE) inhibition by diisopropylfluorophos-phate (DFP) using a simplified radiometric assay for the enzyme. Federation Proc. 26, 296.

Ehrenpreis, S., Mittag, T.W. and Patrick, P. (1970). Radiometric assay of cholinesterases in intact tissues in the nanomolar concentration range of acetylcholine. Biochem. Pharmacol. 19, 2165–2169.

Ehrenpreis, S., Hehir, R.M. and Mittag, T.W. (1971). Assay and properties of essential (junc-tional) cholinesterases of the rat diaphragm. In: Cholinergic Ligand Interactions, D.J. Trig-gle, J.F. Moran and E.A. Barnard, eds. (Academic Press, New York) pp. 67–81.

Eichberg, J., Shein, H.M. and Hauser, G. (1973). Effect of neurotransmitters and other pharma-cological agents on the metabolism of phospholipids in pineal-gland cultures and cloned neuronal and glial cells. Biochem. Soc. Trans. 1, 352–359.

Eichner, D. (1955a). Zur Frage der Fermentlokalisation in der Netzhaut des Rindes. Z. Zell-forsch. Mikroskop. Anat. Abt. Histochem. 41, 493–508.

Eichner, D. (1955b). Zur Topochemie der Netzhaut. In: Auge und Zwischenhirn, H. Becher et al., eds. (Enke, Stuttgart) pp. 29–35.

Eichner, D. (1956). Zur der Fermentlokalisation in der Netzhaut des Menschen. Z. Zellforsch. Mikroskop. Anat. Abt. Histochem. 44, 339–344.

Eichner, D. (1957). Über Histologie und Topochemie der Sehschicht in der Netzhaut des Men-schen. Z. Mikroskop. anat. Forsch. Leipzig 63, 82–93.

Eichner, D. (1958). Zur Histologie und Topochemie der Netzhaut des Menschen. Z. Zellforsch. Mikroskop. Anat. Abt. Histochem. 48, 137–186.

Eichner, D. (1959). Zum Esterasennachweis in den Geweben des vorderens Augenabschnittes. Z. Mikroskop. Anat. Forsch. Leipzig 66, 37–44.

El-Badawi, A. and Schenk, E.A. (1966). Dual innervation of the mammalian urinary bladder. A histochemical study of the distribution of cholinergic and adrenergic nerves. Am. J. Anat. 119, 405–427.

El-Badawi, A. and Schenk, E.A. (1967a). The distribution of cholinergic and adrenergic nerves in the mammalian epididymis. A comparative histochemical study. Am. J. Anat. 121, 1–14.

El-Badawi, A. and Schenk, E.A. (1967b). Histochemical methods for separate, consecutive and simultaneous demonstration of acetylcholinesterase and norepinephrine in cryostat sections. J. Histochem. Cytochem. 15, 580–588.

El-Bermani, A.W.I. (1973). Innervation of the rat lung. Acetylcholinesterase-containing nerves of the bronchial tree. Am. J. Anat. 137, 19–29.

El-Bermani, A.-W., McNary, W.F. and Bradley, D.E. (1970). The distribution of acetylcholines-terase and catecholamine containing nerves in the rat lung. Anat. Record 167, 205–212.

Eldefrawi, M.E., Tripathi, R.K. and O'Brien, R.D. (1970). Acetylcholinesterase isozymes from the housefly brain. Biochim. Biophys. Acta 212, 308–314.

Elkes, J. and Todrick, A. (1955). On the development of cholinesterases in the rat brain. In: Biochemistry of the Developing Nervous System, H. Waelsch, ed. (Academic Press, New York) pp. 309–314.

Elliott, K.A.C. (1968). Formal discussion. In: Progress in Brain Research, Vol. 29, A. Lajtha and D.H. Ford, eds. (Elsevier, Amsterdam) p. 61.

Ellison, J.P. (1971). The nerves of the umbilical cord in man and the rat. Am. J. Anat. 132, 53–60.

Ellison, J.P. and Olander, K.W. (1972). Simultaneous demonstration of catecholamines and acetylcholinesterase in peripheral autonomic nerves. Am. J. Anat. 135, 23–32.

Ellman, G.L. (1959). Tissue sulfhydryl groups. Arch. Biochem. Biophys. 82, 70–77.

Ellman, G.L., Courtney, D.K., Andres, V. and Featherstone, R.M. (1961). A new and rapid colorimetric determination of acetylcholinesterase activity. Biochem. Pharmacol. 7, 88–95.

El-Rakhawy, M.T. and Bourne, G.H. (1961). Cholinesterases in the human tongue. Bibliotheca Anat. 2, 243–255.

Emmelin, N. and MacIntosh, F.C. (1956). The release of acetylcholine from perfused sympathetic ganglia and skeletal muscles. J. Physiol. London 131, 477–496.

Emson, P.C. and Kerkut, G.A. (1971). Acetylcholinesterase in snail brain. Comp. Biochem. Physiol. 39B, 879–889.

Enander, I., Sundwall, A. and Sörbo, B. (1961a). Metabolic studies on N-methylpyridinium-2-aldoxime. I. The conversion to thiocyanate. Biochem. Pharmacol. 7, 226–231.

Enander, I., Sundwall, A. and Sörbo, B. (1961b). Metabolic studies on N-methylpyridinium-2-aldoxime. II. The conversion to N-methyl-pyridinium-2-nitrile. Biochem. Pharmacol. 7, 232-236.

Engel, W.K. (1961). Cytological localization of cholinesterase in cultured skeletal muscle cells. J. Histochem. Cytochem. 9, 66–72.

Engelhard, N., Prchal, K. and Nenner, M. (1967). Acetylcholinesterase. Angew Chem. (Intern. ed.) 6, 615–626. (German ed. 79, 604–615).

Enzyme Commission (1965). Enzyme Nomenclature. Recommendations (1964) of the International Union of Biochemistry on the nomenclature and classification of enzymes together with their units and the symbols of enzyme kinetics. (Elsevier, Amsterdam).

Eränkö, L. (1972). Biochemical and histochemical observations on the postnatal development of cholinesterases in the sympathetic ganglion of the rat. Histochem. J. 4, 545–559.

Eränkö, O. (1967). Histochemistry of nervous tissues: catecholamines and cholinesterases. Ann. Rev. Pharmacol. 7, 203–222.

Eränkö, O. and Eränkö, L. (1971). Loss of histochemically demonstrable catecholamines and acetylcholinesterase from sympathetic nerve fibres of the pineal body of the rat after chemical sympathectomy with 6-hydroxydopamine. Histochem. J. 3, 357–363.

Eränkö, O. and Härkönen, M. (1964). Noradrenaline and acetylcholinesterase in sympathetic ganglion cells of the rat. Acta Physiol. Scand. 61, 299–300.

Eränkö, O. and Härkönen, M. (1965). Effect of axon division on the distribution of noradrenaline and acetylcholinesterase in sympathetic neurons of the rat. Acta Physiol. Scand. 63, 411–412.

Eränkö, O. and Räisänen, L. (1965). Fibres containing both noradrenaline and acetylcholinesterase in the nerve net of the rat iris. Acta Physiol. Scand. 63, 505–506.

Eränkö, O. and Räisänen, L. (1966). Demonstration of catecholamines in adrenergic nerve fibres by fixation in aqueous formaldehyde solution and fluorescence microscopy. J. Histochem. Cytochem. 14, 690–691.

Eränkö, O. and Teräväinen, H. (1967). Cholinesterases and eserine-resistant carboxylic esterases in degenerating and regenerating motor end plates of the rat. J. Neurochem. 14, 947–954.

Eränkö, O., Hopsu, V. and Räisänen, L. (1959). Effect of denervation on histochemically demonstrable acetylcholinesterase and non-specific cholinesterase in the rat adrenal. J. Neurochem. 4, 264–267.

Eränkö, O., Niemi, M. and Merenmies, E. (1961). Histochemical observations on esterases and oxidative enzymes of the retina. In: The Structure of the Eye, G.K. Smelser, ed. (Academic Press, New York) pp. 159–171.

Eränkö, O., Härkönen, M., Kokko, A. and Räisänen, L. (1964). Histochemical and starch gel electrophoretic characterization of desmo- and lyo-esterases in the sympathetic and spinal ganglia of the rat. J. Histochem. Cytochem. 12, 570–581.

Eränkö, O., Rechardt, L. and Hänninen, L. (1967a). Electron microscopic demonstration of cholinesterases in nervous tissue. Histochemie 8, 369–376.

Eränkö, O., Koelle, G.B. and Räisänen, L. (1967b). A thiocholine - lead ferrocyanide method for acetylcholinesterase. J. Histochem. Cytochem. 15, 674–679.

Eränkö, O., Kouvalainen, K., Mattila, M. and Takki, S. (1968). Histochemical and biochemical observations on cholinesterases of cat's tapeworm *Taenia taeniaformis*. Acta Physiol. Scand. 73, 226–233.

Eränkö, O., Rechardt, L., Eränkö, L. and Cunningham, A. (1970). Light and electronmicroscopic histochemical observations on cholinesterase-containing sympathetic nerve fibres in the pineal body of the rat. Histochem. J. 2, 479–489.

Erulkar, S.D., Nichols, C.W., Popp, M.B. and Koelle, G.B. (1968). Renshaw elements : localization and acetylcholinesterase content. J. Histochem. Cytochem. 16, 128–135.

Esila, R. (1963). Histochemical and electrophoretic properties of cholinesterases and nonspecific esterases in the retina of some mammals including man. Acta Ophthalmol. Kbh. Suppl. 77, 1–113.

Esterhuizen, A.C., Graham, J.D.P., Lever, J.D. and Spriggs, T.L.B. (1968). Catecholamine and acetylcholinesterase distribution in relation to noradrenaline release. An enzyme histochemical and autoradiographic study on the innervation of the cat nictitating muscle. Brit. J. Pharmacol. 32, 46–56.

Evans, F.T., Gray, P.W.S., Lehmann, H. and Silk, E. (1952). Sensitivity to succinylcholine in relation to serum-cholinesterase. Lancet i, 1229–1230.

Everett, J.W. and Sawyer, C.H. (1946). Effects of castration and treatment with sex steroids on the synthesis of serum cholinesterase in the rat. Endocrinology 39, 323–343.

Faeder, I.R., O'Brien, R.D. and Salpeter, M.M. (1970). A re-investigation of evidence for cholinergic neuromuscular transmission in insects. J. Exptl. Zool. 173, 187–202.

Falck, B. (1962). Observations on the possibilities of the cellular localization of monoamines by a fluorescence method. Acta Physiol. Scand. 56, Suppl. 197, 1–25.

Fehér, O. and Bokri, E. (1960). Beiträge zur Cholinesterase kinetik in vivo. II Vergleich der Azetylcholin- und Azetyl-β-methylcholinhydrolyse am oberen Haloganglion der Katze in vivo und in vitro. Acta Physiol. Acad. Sci. Hung. 18, 11–17.

Feldberg, W. (1945). Present views on the mode of action of acetylcholine in the central nervous system. Physiol. Rev. 25, 596–642.

Feldberg, W. (1957). Acetylcholine. In: Metabolism of the Nervous System, D. Richter, ed. (Pergamon Press, London) pp. 493–510.

Feldberg, W. and Hebb, C. (1948). The stimulating action of phosphate compounds on the perfused superior cervical ganglion of the cat. J. Physiol. London 107, 210–221.

Feldberg, W. and Vogt, M. (1948). Acetylcholine synthesis in different regions of the central nervous system. J. Physiol. London 107, 372–381.

Feldberg, W., Harris, G.W. and Lin, R.C.Y. (1951). Observations on the presence of cholinergic and non-cholinergic neurones in the central nervous system. J. Physiol. London 112, 400–404.

Ferry, C.B. (1966). Cholinergic link hypothesis in adrenergic neuroeffector transmission. Physiol. Rev. 46, 420–456.

Filbert, M.G., Fleisher, J.H. and Lochner, M.A. (1972). Failure of Toxogonin to reactivate Soman-inhibited brain acetylcholinesterase in monkeys and regeneration of the enzyme. Biochim. Biophys. Acta 284, 164–174.

Fillenz, M. (1970). The innervation of the cat spleen. Proc. Roy. Soc. London Ser. B. 174, 459–468.

Fillenz, M. and Wood, R.I. (1966). Some observations on the rabbit carotid body. J. Physiol. London 186, 39–40P.

Filogamo, G. (1960). Recherches experimentales sur l'activité des cholinestérases spécifique et non-spécifique dans le développement du lobe optique du poulet. Arch. Biol. Liège 71, 159–198.

Filogamo, G. and Candiollo, L. (1962). Observations on the behavior of acetylcholinesterase

(AChE) in the nerve cells of the spinal reflex arc, after section of the peripheral nerves (experimental investigations in *Lepus cuniculus* L.) Acta Anat. 51, 273–291.

Filogamo, G. and Gabella, G. (1966). Cholinesterase behavior in the denervated and reinnervated muscles. Acta Anat. 63, 199–214.

Filogamo, G. and Gabella, G. (1967). The development of neuro-muscular correlations in vertebrates. Arch. Biol. Liège 78, 9–60.

Filogamo, G. and Marchisio, P.C. (1971). Acetylcholine system and neural development. Neurosciences Res. 4, 29–64.

Filogamo, G. and Robecchi, M.G. (1969). Neuroblasts in the olfactory pits in mammals. Acta Anat. Suppl. 56, 182–187.

Finch, L., Haeusler, G., Kuhn, H. and Thoenen, H. (1973). Rapid recovery of vascular adrenergic nerves in the rat after chemical sympathectomy with 6-hydroxydopamine. Brit. J. Pharmacol. 48, 59–72.

Finlay, M. (1972). A study of the development of the intrinsic innervation of the rat heart. B.Sc. Thesis. Manchester University.

Firer, L.D. and Khamitov, Kh.S. (1957). On the kinetics of the cholinesterase of frog smooth muscle organs after removal of the pancreas. Bull. Exptl. Biol. Med. U.S.S.R. 44, 1298–1301.

Fitzgerald, G.G. and Cooper, J.R. (1971). Acetylcholine as a possible sensory mediator in rabbit corneal epithelium. Biochem. Pharmacol. 20, 2741–2748.

Flacke, W. and Yeoh, T.S. (1968a). The action of some cholinergic agonists and anticholinesterase agents on the dorsal muscle of the leech. Brit. J. Pharmacol. 33, 145–153.

Flacke, W. and Yeoh, T.S. (1968b). Differentiation of acetylcholine and succinylcholine receptors in leech muscle. Brit. J. Pharmacol. 33, 154–161.

Fleisher, J.H., Harris, L.W. and Murtha, E.F. (1967). Reactivation by pyridinium aldoxime methochloride (PAM) of inhibited cholinesterase activity in dogs after poisoning with pinacolyl methylphosphonofluoridate (SOMAN). J. Pharmacol. Exptl. Therap. 156, 345–351.

Flockhart, I.R. and Casida, J.E. (1972). Relationship of the acylation of membrane esterases and proteins to the teratogenic action of organophosphorus insecticides and eserine in developing hen eggs. Biochem. Pharmacol. 21, 2591–2603.

Flood, P.R., Fonnum, F. and Storm-Mathisen, J. (1970). Choline acetyltransferase and acetylcholinesterase in myo-tendinous and neuromuscular junctions of mouse skeletal muscle. Experientia 26, 964–965.

Flood, S. and Jansen, J. (1966). The efferent fibres of the cerebellar nuclei and their distribution on the cerebellar peduncles in the cat. Acta Anat. 63, 137–166.

Florey, E. (1962). Comparative neurochemistry : inorganic ions, amino acids and possible transmitter substances of invertebrates. In: Neurochemistry, K.A.C. Elliott, I.H. Page and J.H. Quastel, eds. (Thomas, Springfield Illinois) pp. 673–693.

Florey, E. (1963). Acetylcholine in invertebrate nervous systems. Can. J. Biochem. Physiol. 41, 2619–2626.

Florey, E. and Biederman, M.A. (1960). Studies on the distribution of Factor I and acetylcholine in crustacean peripheral nerve. J. Gen. Physiol. 43, 509–522.

Florey, E. and Winesdorfer, J. (1968). Cholinergic nerve endings in octopus brain. J. Neurochem. 15, 169–177.

Florey, E. and Woodcock, B. (1968). Presynaptic excitatory action of glutamate applied to crab nerve-muscle preparations. Comp. Biochem. Physiol. 26, 651–661.

Flumerfelt, B.A., Lewis, P.R. and Gwyn, D.G. (1973). Cholinesterase activity of capillaries in the rat brain. A light and electron microscopic study. Histochem. J. 5, 67–77.

Foldes, F.F. (1966). The influence of metabolic transformation on the toxicity of local anesthetic agents in man. Acta Anaesthesiol. Scand. Suppl. 23, 591–597.

Foldes, F.F., Van Hees, G., Davis, D.L. and Shanor, S.P. (1958). The structure-action relationship of urethane type cholinesterase inhibitors. J. Pharmacol. Exptl. Therap. 122, 457–464.

Foldes, F.F., Zsigmond, E.K., Foldes, V.M. and Erdös, E.G. (1962). The distribution of acetylcholinesterase and butyrylcholinesterase in the human brain. J. Neurochem. 9, 559–572.

Follett, B.K., Kobayashi, H. and Farner, D.S. (1966). The distribution of monoamine oxidase and acetylcholinesterase in the hypothalamus and its relation to the hypothalamo-hypophysial neurosecretetory system in the white-crowned sparrow, *Zonotrichia leucophrys gambelii.* Z. Zellforsch. Mikroskop. Anat. 75, 57–65.

Fonnum, F. (1966). Is choline acetyltransferase present in synaptic vesicles? Biochem. Pharmacol. 15, 1641–1643.

Fonnum, F. (1967). The 'compartmentation' of choline acetyltransferase within the synaptosome. Biochem. J. 103, 262–270.

Fonnum, F. (1969). Radiochemical micro assays for the determination of choline acetyltransferase and acetylcholinesterase activities. Biochem. J. 115, 465–472.

Fonnum, F. (1970). Topographical and subcellular localization of choline acetyltransferase in rat hippocampal region. J. Neurochem. 17, 1029–1037.

Fonnum, F. (1973). Recent developments in biochemical investigations of cholinergic transmission. Brain Res. 62, 497–507.

Forbat, A., Lehmann, H. and Silk, E. (1953). Prolonged apnoea following injection of succinyldicholine. Lancet ii, 1067–1068.

Fourman, J. (1966a). Cholinesterase in the mammalian kidney. Nature 209, 812–813.

Fourman, J. (1966b). Cholinesterase and sodium transport in the supra-orbital gland of the duck. J. Anat. 100, 693.

Fourman, J. (1967). The distribution and variations of cholinesterase activity in the nephron and in other tissues concerned with sodium transport. J. Physiol. London 191, 52–53P.

Fourman, J. (1969). Cholinesterase activity in the supra-orbital salt secreting gland of the duck. J. Anat. 104, 233–239.

Fourman, J. and Ballantyne, B. (1967). Cholinesterase activity in the Harderian gland of *Anas domesticus.* Anat. Record 159, 17–28.

Frady, C.H. and Knapp, S.E. (1967). A radioisotopic assay of acetylcholinesterase in *Fasciola hepatica.* J. Parasitol. 53, 298–302.

Francis, C.M. (1953). Cholinesterase in the retina. J. Physiol. London 120, 435–439.

Fray, R. (Jr.) and Leaders, F.R. (1967). Demonstration of separate adrenergic and cholinergic fibres to the vessels of the rear quarters of the rat by hemicholinium and a proposed role in peripheral vascular regulation. Brit. J. Pharmacol. 30, 265–273.

Fredricsson, B. and Sjöqvist, F. (1962). A cytomorphological study of cholinesterase in sympathetic ganglia of the cat. Acta Morphol. Neerl. Scand. 5, 140–166.

Friede, R.L. (1967). A comparative histochemical mapping of the distribution of butyryl cholinesterase in the brains of four species of mammals including man. Acta Anat. 66, 161–177.

Friede, R.L. and Fleming, L.M. (1964). A comparison of cholinesterase distribution in the cerebellum of several species. J. Neurochem. 11, 1–7.

Friedenberg, R.M. and Seligman, A.M. (1972). Acetylcholinesterase at the myoneural junction: cytochemical ultrastructure and some biochemical considerations. J. Histochem. Cytochem. 20, 771–792.

Fripp, P.J. (1967a). Histochemical localization of esterase activity in schistosomes. Exptl. Parasitol. 21, 380–390.

Fripp, P.J. (1967b). The sites 1-^{14}C glucose assimilation in *Schistosoma haematobium*. Comp. Biochem. Physiol. 23, 893–898.

Frizell, M., Hasselgren, P.O. and Sjöstrand, J. (1970). Axoplasmic transport of acetylcholinesterase and choline acetyltransferase in the vagus and hypoglossal nerve of the rabbit. Exptl. Brain Res. 10, 526–531.

Froede, H.C. and Wilson, I.B. (1970). Subunit structure of acetylcholinesterase. Israel J. Med. Sci. 6, 170–184.

Frontali, N. (1968). Histochemical localization of catecholamines in the brain of normal and drug-treated cockroaches. J. Insect Physiol. 14, 881–886.

Fukuda, T. and Koelle, G.B. (1959). The cytological localization of intracellular neuronal acetylcholinesterase. J. Biophys. Biochem. Cytol. 5, 433–440.

Fulton, J.F. and Nachmansohn, D. (1943). Acetylcholine and the physiology of the nervous system. Science 97, 569–571.

Fulton, M.P. and Mogey, G.A. (1954). Some selective inhibitors of true cholinesterase. Brit. J. Pharmacol. 9, 138–144.

Funnell, H.S. and Oliver, W.T. (1965). Proposed physiological function for plasma cholinesterase. Nature 208, 689–690.

Furmanski, P., Silverman, D.J. and Lubin, M. (1971). Expression of differential functions in mouse neuroblastoma mediated by dibutyrylcyclic adenosine monophosphate. Nature 233, 413–415.

Furness, J.B. (1973). Arrangement of blood vessels and their relation with adrenergic nerves in the rat mesentery. J. Anat. 115, 347–364.

Furness, J.B. and Iwayama, T. (1972). The arrangement and identification of axons innervating the vas deferens of the guinea-pig. J. Anat. 113, 179–196.

Gaddum, J.H. (1961). Push-pull cannulae. J. Physiol. London 155, 1–2P.

Gahery, Y. and Boistel, J. (1965). Study of some pharmacological substances which modify the electrical activity of the sixth abdominal ganglion of the cockroach, *Periplaneta americana*. In: The Physiology of the Insect Central Nervous System, J.E. Treherne and J.W.L. Beament, eds. (Academic Press, London) pp. 73–78.

Galindo, A., Krnjević, K. and Schwartz, S. (1967). Micro-iontophoretic studies on neurones in the cuneate nucleus. J. Physiol. London 192, 359–377.

Galindo, A., Krnjević, K. and Schwartz, S. (1968). Patterns of firing in cuneate neurones and some effects of Flaxedil. Exptl. Brain Res. 5, 87–101.

Galley, N., Klinke, R., Pause, M. and Storch, W.-H. (1971). The effect of Flaxedil (gallamine triethiodide) on the efferent endings in the cochlea. Pflügers Arch. 330, 1–4.

Galzigna, L., Corsi, G.C., Saia, B. and Rizzoli, A.A. (1969). Inhibitory effect of triethyl lead on serum cholinesterase in vitro. Clin. Chim. Acta 26, 391–393.

Gandiha, A. and Marshall, I.G. (1972). Some actions of dimethylsulphoxide at the neuromuscular junction. J. Pharm. Pharmacol. 24, 417–419.

Garland, J.T. and Durell, J. (1970). Chemical mechanisms of transmitter-receptor interaction. Intern. Rev. Neurobiol. 13, 159–180.

Garrett, J.R. (1966a). The innervation of salivary glands. I. Cholinesterase-positive nerves in normal glands of the cat. J. Roy. Microsc. Soc. 85, 135–148.

Garrett, J.R. (1966b). The innervation of salivary glands. III. The effects of certain experimental procedures on cholinesterase-positive nerves in glands of the cat. J. Roy. Microsc. Soc. 86, 1–13.

Garrett, J.R., Howard, E.R. and Lansdale, J.M. (1972). Myenteric nerves in the hind gut of the cat. J. Physiol. London 226, 103–104P.

Gaskell, J.F. (1914). The chromaffin system of annelids and the relation of this sytem to the contractile vascular system in the leech *Hirudo medicinalis*. Phil. Trans. Roy. Soc. London, Ser. B. 205, 153–211.

Gasparyan, E.I. (1964). On the content of acetylcholine and cholinesterase activity in the blood of workers employed in the production of chloroprene rubber (in Russian). Zh. Eksp. Klin. Med. 4, 39–45.

Geel, S.E. and Timiras, P.S. (1967). Influence of neonatal hypothyroidism and of thyroxine on the acetylcholinesterase and cholinesterase activities in the developing central nervous system of the rat. Endocrinology 80, 1069–1074.

Geneser-Jensen, F.A. (1972a). Distribution of acetyl cholinesterase in the hippocampal region of the guinea pig. II. Subiculum and hippocampus. Z. Zellforsch. Mikroskop. Anat. 124, 546–560.

Geneser-Jensen, F.A. (1972b). Distribution of acetyl cholinesterase in the hippocampal region of the guinea pig. III. The dentate area. Z. Zellforsch. Mikroskop. Anat. 131, 481–495.

Geneser-Jensen, F.A. and Blackstad, T.W. (1971). Distribution of acetyl cholinesterase in the hippocampal region of the guinea pig. I. Entorhinal area, parasubiculum and presubiculum. Z. Zellforsch. Mikroskop. Anat. 114, 460–481.

Gentinetta, R. and Brodbeck, U. (1972). Subunit molecular weights of acetylcholinesterases. Experientia 28, 735–736.

Gerebtzoff, M.A. (1953). Recherches histochimiques sur les acétylcholine et choline estérases. i. Introduction et technique. Acta Anat. 19, 366–379.

Gerebtzoff, M.A. (1959). Cholinesterases. A Histochemical Contribution to the Solution of Some Functional Problems (Pergamon press, London).

Gerebtzoff, M.A. (1970). Recherches histochimiques et histoenzymologiques sur la synergie métabolique entre neurones et névroglie dans la châine nerveuse ventrale de la sangsue 'Hirudo medicinalis' Bull. Acad. Roy. Med. Belg. VIII, ser. X. 337–359.

Gerlini, G., Ottaviano, S., Sbraccia, C., Carapella, E. and Bonanni, V. (1968). The behavior of erythrocytic acetylcholinesterase activity in the hemolytic disease of the newborn due to ABO incompatability in relation to the erythrocytic age (in Italian). Riv. Clin. Pediat. 81, 742–745.

Giacobini, E. (1956a). Histochemical demonstration of AChE in isolated nerve cells. Acta Physiol. Scand. 36, 276–290.

Giacobini, E. (1956b). Quantitative determination of cholinesterase in individual sympathetic cells. J. Neurochem. 1, 234–244.

Giacobini, E. (1959). Quantitative determination of cholinesterase in individual spinal ganglion cells. Acta Physiol. Scand. 45, 238–254.

Giacobini, E. (1969). Value and limitations of quantitative chemical studies in individual cells. J. Histochem. Cytochem. 17, 139–155.

Giacobini, E. and Holmstedt, B. (1958). Cholinesterase content of certain regions of the spinal cord as judged by histochemical and Cartesian diver techniques. Acta Physiol. Scand. 42, 12–27.

Giacobini, E., Hökfelt, T., Kerpel-Fronius, S., Koslow, S.H., Mitchard, M. and Noré, B. (1971). A micro-scale procedure for the preparation of subcellular fractions from individual autonomic ganglia. J. Neurochem. 18, 223–231.

Giacobini, G. (1972). Embryonic and post-natal development of choline acetyltransferase activity in muscles and sciatic nerve of the chick. J. Neurochem. 19, 1401–1403.

Giacobini, G., Marchisio, P.C., Giacobini, E. and Koslow, S.H. (1970). Developmental changes of cholinesterases and monoamine oxidase in chick embryo spinal and sympathetic ganglia. J. Neurochem. 17, 1177–1185.

Giberman, E., Silman, I. and Edery, H. (1973). Effect of cholinesterase inhibitors on active potassium influx in monkey erythrocytes. Biochem. Pharmacol. 22, 271–273.

Gilboa-Garber, N. and Mizrahi, L. (1972). Effect of acetylcholine on the osmotic fragility of papain-treated and untreated human red blood cells. Experientia, 28, 78–79.

Giller, E. Jr. and Schwartz, J.H. (1971a). Choline acetyltransferase in identified neurons of abdominal ganglion of *Aplysia californica*. J. Neurophysiol. 34, 93–107.

Giller, E. Jr. and Schwartz, J.H. (1971b) Acetylcholinesterase in identified neurons of abdominal ganglion of *Aplysia californica*. J. Neurophysiol. 34, 108–115.

Ginsborg, B.L. and Mackay, B. (1961). A histochemical demonstration of two types of motor innervation in avian skeletal muscle. Bibliotheca Anat. 2, 174–181.

Girgis, M. (1967). Distribution of cholinesterase in the basal rhinencephalic structures of the coypu (*Myocastor coypus*). J. Comp. Neurol. 129, 85–96.

Girgis, M. (1968). Distribution of cholinesterase in the basal rhinencephalic structures of the Grivet monkey (*Cercopithecus aethiops aethiops*). Acta Anat. 70, 568–576.

Girgis, M. (1969). Distribution of cholinesterase in the basal rhinencephalic structures of the Senegal bush baby (*Galago senegalensis senegalensis*). Acta Anat. 72, 94–100.

Glick, D. (1938). Studies in enzymatic histochemistry. XXV. A micro method for the determination of choline esterase and the activity-pH relationship of this enzyme. J. Gen. Physiol. 21, 289–295.

Glick, D. (1945). The controversy on cholinesterases. Science 102, 100–101.

Glick, D. (1967). Usage of 'histochemical', 'staining' and 'biochemical' in histochemical literature. J. Histochem. Cytochem. 15, 299.

Glow, P.H., Rose, S., and Richardson, A. (1966). The effect of acute and chronic treatment with diisopropylfluorophosphate on cholinesterase activities of some tissues of the rat. Aust. J. Exptl. Biol. Med. Sci. 44, 73–86.

Glow, P.H., Opit, L.J. and Charnock, J.S. (1972). Elevated sodium plus potassium activated adenosinetriphosphatase in rats after chronic diisopropylfluorophosphate poisoning. Arch. Intern. Pharmacodyn. 198, 22–28.

Goedde, H.W., Gehring, D. and Hofmann, R.A. (1965). On the problem of a 'silent gene' in pseudocholinesterase polymorphism. Biochim. Biophys. Acta 107, 391–393.

Goedde, H.W. Doenicke, A. and Altland, K. (1967). Pseudocholinesterasen. Pharmakogenetik, Biochemie, Klinik. (Springer, Berlin)

Goedde, H.W., Held, K.R. and Altland, K. (1968). Hydrolysis of succinyldicholine and succinylmonocholine in human serum. Molec. Pharmacol. 4, 274–287.

Goldberg, A.L. and Singer, J.J. (1969). Evidence for a role of cyclic AMP in neuromuscular transmission. Proc. Natl. Acad. Sci. U.S. 64, 134–141.

Goldberg, A.M. and McCaman, R.E. (1967). A quantitative microchemical study of choline acetyltransferase and acetylcholinesterase in the cerebellum of several species. Life Sci. Oxford 6, 1493–1500.

Goldberg, A.M., Pollock, J.J., Hartman, E.R. and Craig, C.R. (1972). Alterations in cholinergic enzymes during the development of cobalt-induced epilepsy in the rat. Neuropharmacol. 11, 253–259.

Goldberg, M.A. (1971). Protein synthesis in isolated rat brain mitochondria and nerve endings. Brain Res. 27, 319–328.

Goldberg, M.A. (1972). Inhibition of synaptosomal protein synthesis by neurotransmitter substances. Brain Res. 39, 171–179.

Gomori, G. (1948). Histochemical demonstration of sites of choline esterase activity. Proc. Soc. Exptl. Biol. Med. 68, 354–358.

Gomori, G. (1952). Microscopic Histochemistry. Principles and Practice (Univ. Chicago Press, Chicago) p. 210.

Goodman, L.S. and Gilman, A. (1955). The Pharmacological Basis of Therapeutics, 2nd ed. (The MacMillan Co., New York) p. 444.

Goodwin, B.C. and Sizer, I.W. (1965). Effects of spinal cord and substrate on acetylcholinesterase in chick embryonic skeletal muscle. Develop. Biol. 11, 136–153.

Gordon, M.W. and Deanin, G.G. (1968). Protein synthesis by isolated rat brain mitochondria and synaptosomes. J. Biol. Chem 243, 4222–4226.

Gosbee, J.L. and Lederis, K. (1972). In vivo release of antidiuretic hormone by direct application of acetylcholine or carbachol to the rat neurohypophysis. Can. J. Physiol. Pharmacol. 50, 618–620.

Goutier, R. and Goutier-Pirotte, M. (1955). Localisation intracellulaire des cholinestérases. II. Pancréas et suc pancréatique de chien. Biochim. Biophys. Acta 16, 558–565.

Goutier-Pirotte, M. and Gerebtzoff, M.A. (1955). L'acétylcholinestérase dans le placenta du cobaye. Premiers résultats de recherches histochimiques et biochimiques. Arch. Intern. Physiol. Biochim. 63, 445–457.

Goutier-Pirotte, M. and Goutier, R. (1956). Les cholinestérases de l'intestin grêle et du foie du cobaye adulte et foetal. Arch. Intern. Physiol. Biochim. 64, 20–33.

Goyer, G.R. (1968). Evaluation par méthode manométrique des cholinestérases des muscle diaphragme et tibial antérieur chez le rat. Rev. Can. Biol. 27, 209–216.

Graff, D.J. and Read, C.P. (1967). Specific acetylcholinesterase in *Hymenolepis diminuta*. J. Parasitol. 53, 1030–1031.

Grafius, M.A. and Millar, D.B. (1967). Reversible aggregation of acetylcholinesterase. II. Interdependence of pH and ionic strength. Biochemistry 6, 1034–1046.

Grafstein, B. (1969). Axonal transport: communication between soma and synapse. In: Advances in Biochemical Pharmacology, Vol. I., E. Costa and P. Greengard, eds. (Raven Press, New York) pp. 11–25.

Grafstein, B., Forman, D.S. and McEwen, B.S. (1972). Effects of temperature on axonal transport and turnover of protein in gold fish optic system. Exptl. Neurol. 34, 158–170.

Graham, J.D.P., Lever, J.D. and Spriggs, T.L.B. (1968). An examination of adrenergic axons around pancreatic arterioles of the cat for the presence of acetylcholinesterase by high resolution autoradiographic and histochemical methods. Brit. J. Pharmacol. 33, 15–20.

Grant, R.T., Bayliss, O.B. and Adams, C.W.M. (1967). Cholinesterases in peripheral nervous system. II. The motor, sensory and sympathetic nerves in the rabbit ear perichondrium and cat cremaster muscle. Brain Res. 6, 457–474.

Gray, E.G. (1959). Electron microscopy of synaptic contacts on dendrite spines of the cerebral cortex. Nature 183, 1592–1593.

Gray, E.G. (1966). Problems of interpreting the fine structure of vertebrate and invertebrate synapses. Intern. Rev. Gen. Exptl. Zool. 2, 139–170.

Gray, E.G. and Whittaker, V.P. (1962). The isolation of nerve endings from brain: an electron-microscopic study of cell fragments, derived by homogenization and centrifugation. J. Anat. 96, 79–88.

Gray, J.A.B. (1959). Initiation of impulses at receptors. In: Handbook of Physiology. Neurophysiology, Vol. I, H.W. Magoun, ed. (Amer. Physiol. Soc., Washinton D.C.) pp. 123–145.

Green, J.R., Halpern, L.M. and Van Niel, S. (1970). Alterations in the activity of selected enzymes in the chronic isolated cerebral cortex of cat. Brain 93, 57–64.

Greig, M.E. and Holland, W.C. (1949a). Studies on the permeability of erythrocytes. I. The relationship between cholinesterase activity and permeability of dog erythrocytes. Arch. Biochem. Biophys. 23, 370–384.

Greig, M.E. and Holland, W.C. (1949b). Increased permeability of the hemoencephalic barrier produced by physostigmine and acetylcholine. Science 110, 237.

Greig, M.E. and Mayberry, T.C. (1951). The relationship between cholinesterase activity and brain permeability. J. Pharmacol. Exptl. Therap. 102, 1–4.

Greig, M.E., Holland, W.C. and Lindvig, P.E. (1950). The anaesthetization of the rabbit's cornea by non-surface anaesthetics. Brit. J. Pharmacol. 5, 461–464.

Greig, M.E., Faulkner, J.S. and Mayberry, T.C. (1953). Studies on permeability. IX. Replacement of potassium in erythrocytes during cholinesterase activity. Arch. Biochem. Biophys. 43, 39–47.

Gridelet, J., Foidart, J.-M. and Wins, P. (1970). Contribution à l'étude des propriétés allostériques de l'acétylcholinestérase. Arch. Intern. Physiol. Biochim. 78, 259–264.

Grieten, J. (1956). Apparition de cholinestérase au cours de la maturation des spermatozoïdes du cobaye. Compt. Rend. Soc. Biol. 150, 1015–1016.

Grob, D. (1950). The anticholinesterase activity in vitro of the insecticide parathion (p-nitrophenyl diethyl thionophosphate) Bull. Johns Hopkins Hosp. 87, 95–105.

Grob, D. (1963a). Anticholinesterase intoxication in man and its treatment. In: Cholinesterases and Anticholinesterase Agents, G.B. Koelle, ed. (Springer, Berlin) pp. 989–1027.

Grob, D. (1963b). Therapy of myasthenia gravis. In: Cholinesterases and Anticholinesterase Agents, G.B. Koelle, ed. (Springer, Berlin) pp. 1028–1050.

Grob, D. and Harvey, J.C. (1958). Effects in man of the anticholinesterase compound Sarin (isopropyl methyl phosphonofluoridate). J. Clin. Invest. 37, 350–368.

Grob, D. and Johns, R.J. (1958a). Use of oximes in the treatment of intoxication by anticholinesterase compounds in normal subjects. Am. J. Med. 24, 497–511.

Grob, D. and Johns, R.J. (1958b). Use of oximes in the treatment of intoxication by anticholinesterase compounds in patients with myasthenia gravis. Am. J. Med. 24, 512–518.

Gromadzki, C. G. and Koelle, G.B. (1965). The effect of axotomy on the acetylcholinesterase of the superior cervical ganglion of the cat. Biochem. Pharmacol. 14, 1745–1754.

Grossmann, H. and Parnitzke, K.H. (1952). Beobachtungen über die Orthotrikresylphosphatvergiftung bei Tigern. Psychiat. Neurol. Med. Psychol. 4, 91–94.

Gruber, H. and Zenker, W. (1973). Acetylcholinesterase: histochemical differentiation between motor and sensory nerve fibres. Brain Res. 51, 207–214.

Guilbault, G.G., Kuan, S.S., Tully, J. and Hackney, D. (1970a). New procedure for rapid and sensitive detection of cholinesterase separated by polyacrylamide gel electrophoresis. Anal. Biochem. 36, 72–77.

Guilbault, G.G., Kuan, S.S. and Sadar, M.H. (1970b). Purification and properties of cholinesterases from honey bees - *Apis mellifera* Linnaeus and boll weevils - *Anthonomus grandis* Boheman. J. Agr. Food Chem. 18, 692–697.

Gunter, J.M. (1946). Absence of pseudocholinesterase from the tissues of ruminants. Nature 157, 369.

Gurdon, J.B. (1968). Transplanted nuclei and cell differentiation. Sci. Am. 219, pt. 6, 24–36.

Gurdon, J.B. (1971). Gene activity during embryogenesis. Triangle 10, 23–28.

Gustafson, T. and Toneby, M. (1970). On the role of serotonin and acetylcholine in sea urchin morphogenesis. Exptl. Cell Res. 62, 102–117.

Guth, L. (1968). 'Trophic' influences of nerve on muscle. Physiol. Rev. 48, 645–687.

Guth, L. (1969a). Effect of immobilization on sole-plate and background cholinesterase of rat skeletal muscle. Exptl. Neurol. 24, 508–513.

Guth, L. (1969b). 'Trophic' effects of vertebrate neurons. Neurosciences Res. Prog. Bull. 7, no.1., L. Guth, ed, pp. 30–36.

Guth, L. and Zalewski, A.A. (1963). Disposition of cholinesterase following implantation of nerve into innervated and denervated muscle. Exptl. Neurol. 7, 316–326.

Guth, L., Albers, R.W, and Brown, W.C. (1964). Quantitative changes in cholinesterase activity of denervated muscle fibers and sole plates. Exptl. Neurol. 10, 236–250.

Guth, L., Zalewski, A.A. and Brown, W.C. (1966). Quantative changes in cholinesterase activity of denervated sole plates following implantation of nerve into muscle. Exptl. Neurol. 16, 136–147.

Guth, L., Brown, W.C. and Watson, P.K. (1967). Studies on the role of nerve impulses and acetylcholine release in the regulation of cholinesterase activity of muscle. Exptl. Neurol. 18, 443–452.

Guth, P.S. and Amaro, J. (1969). A possible cholinergic link in olivo-cochlear inhibition. Intern. J. Neuropharmacol. 8, 49–53.

Guth, P.S. and Bobbin, R.P. (1971). The pharmacology of peripheral auditory processes; cochlear pharmacology. Adv. Pharmacol. Chemother. 9, 93–130.

Gutsche, B.B., Scott, E.M. and Wright, R.C. (1967). Hereditary deficiency of pseudocholinesterase in Eskimos. Nature 215, 322–323.

Gwyn, D.G. (1971a). Histochemical evidence for a somatotopic organisation of the rubrospinal projection in the rat. Experientia 27, 819–821.

Gwyn, D.G. (1971b). Acetylcholinesterase activity in the red nucleus of the rat. Effects of rubrospinal tractotomy. Brain Res. 35, 447–461.

Gwyn, D.G. and Flumerfelt, B.A. (1971). Acetylcholinesterase in non-cholinergic neurones: a histochemical study of dorsal root ganglion cells in the rat. Brain Res. 34, 193–198.

Gwyn, D.G. and Heardman, V. (1965). A cholinesterase-Bielschowsky method for mammalian motor end-plates. Stain Technol. 40, 15–18.

Gwyn, D.G. and Waldron, H.A. (1968). A nucleus in the dorsolateral funiculus of the spinal cord of the rat. Brain Res. 10, 342–351.

Gwyn, D.G. and Wolstencroft, J.H. (1966). Ascending and descending cholinergic fibers in cat spinal cord: histochemical evidence. Science 153, 1543–1544.

Gwyn, D.G. and Wolstencroft, J.H. (1967). Cholinesterases in a vascular structure in the floor of the fourth ventricle of the cat. Nature 214, 831–832.

Gwyn, D.G. and Wolstencroft, J.H. (1968). Cholinesterases in the area subpostrema. A region adjacent to the area postrema in the cat. J. Comp. Neurol. 133, 289–308.

Gwyn, D.G., Silver A. and Wolstencroft, J.H. (1969). Evidence for ascending and descending cholinergic fibres in cat spinal cord. J. Physiol. London 201, 23–24P.

Gwyn, D.G., Wolstencroft, J.H. and Silver, A. (1972). The effect of a hemisection on the distribution of acetylcholinesterase and choline acetyltransferase in the spinal cord of the cat. Brain Res. 47, 289–301.

Ha, H. and Liu, C.N. (1968). Cell origin of the ventral spinocerebellar tract. J. Comp. Neurol. 133, 185–206.

Hackley, B.E. Jr., Steinberg, G.M. and Lamb, J.C. (1959). Formation of potent inhibitors of AChE by reaction of pyridinaldoximes with isopropyl methylphosphonofluoridate (GB). Arch. Biochem. Biophys. 80, 211–214.

Häggqvist , G. (1960). Cholinesterases and innervation of skeletal muscles. Acta Physiol. Scand. 48, 63–70.

Hagopian, M., Tennyson, V.M. and Spiro, D. (1970). Cytochemical localization of cholinesterase in embryonic rabbit cardiac muscle. J. Histochem. Cytochem. 18, 38–43.

Haites, N., Don, M. and Masters, C.J. (1972). Heterogenicity and molecular weight interrelationships of the esterase isoenzymes of several invertebrate species. Comp. Biochem. Physiol. 42B, 303–322

Hajós, F., Priymak, E.Kh. and Kerpel-Fronius, S. (1970). The electron microscopic demonstration of acetylcholinesterase activity in some cholinergic and non-cholinergic synapses of the rat brain. Acta Histochem. 35, 114–122.

Hall, E. and Geneser-Jensen, F.A. (1971). Distribution of acetylcholinesterase and monoamine oxidase in the amygdala of the guinea pig. Z. Zellforsch. Mikroskop. Anat. 120, 204–221.

Hall, G.H. and Myers, R.D. (1972). Temperature changes produced by nicotine injected into the hypothalamus of the conscious monkey. Brain Res. 37, 241–251.

Halton, D.W. (1967). Histochemical studies of carboxylic esterase activity in *Fasciola hepatica*. J. Parasitol. 53, 1210–1216.

Hamberger, B., Norberg, K.-A. and Sjöqvist, F. (1965). Correlated studies of monoamines and acetylcholinesterase in sympathetic ganglia illustrating the distribution of adrenergic and cholinergic neurons. In: Pharmacology of Cholinergic and Adrenergic Transmission, G.B. Ķoelle, W.W. Douglas and A. Carlsson, eds. (Pergamon Press, Oxford) pp. 41–54.

Hamburger, V., Balaban, M., Oppenheim, R. and Wenger, E. (1965). Periodic motility of normal and spinal chick embryos between 8 and 17 days of incubation. J. Exptl. Zool. 159, 1–13.

Hamburgh, M. and Flexner, L.B. (1957). Biochemical and physiological differentiation during morphogenesis. XXI. Effect of hypothyroidism and hormone therapy on enzyme activities of the developing cerebral cortex of the rat. J. Neurochem. 1, 279–288.

Hanin, I., Massarelli, R. and Costa, E. (1970). Acetylcholine concentrations in rat brain: diurnal oscillation. Science 170, 341–342.

Hansen, B. (1957). The preparation of thiocholine esters. Acta Chem. Scand. 11, 537–540.

Hansson, H.-A. (1966). The distribution of acetylcholine esterase and other hydrolytic enzymes in retinal cultures. Acta Physiol. Scand. 70, Suppl. 288, 1–30.

Hanzon, V. and Toschi, G. (1959). Electron microscopy on microsomal fractions from rat brain. Exptl. Cell Res. 16, 256–271.

Hard, W.L. and Peterson, A.C. (1950). The distribution of choline esterase in nerve tissue of the dog. Anat. Record 108, 57–69.

Hardwick, D.C. and Hebb, C. (1956). Pseudo-cholinesterase in the central nervous system of the frog. Nature 177, 667.

Hardwick, D.C. and Palmer, A.C. (1961). Effect of formalin fixation on cholinesterase activity in sheep brain. Quart. J. Exptl. Physiol. 46, 350–352.

Hargreaves, A.B. (1961). Purification, enzyme determination and some physico-chemical characteristics of the acetylcholinesterase of the *Electrophorus electricus* (L.). In: Bioelectrogenesis, C. Chagas and A. Paes de Carvalho, eds. (Elsevier, Amsterdam) pp. 397–405.

Harrex, W.K. (1971). Observations of the distribution of cholinesterases in the carotoid body of the brush tailed possum, *Trichosurus vulpecula*. J. Anat. 110, 499.

Harris, H. and Robson, E.B. (1963). Fractionation of human serum cholinesterase components by gel filtration. Biochim. Biophys. Acta 73, 649–652.

Harris, H. and Whittaker, M. (1961). Differential inhibition of human serum cholinesterase with fluoride: recognition of two new phenotypes. Nature 191, 496–498.

Harris, H., Hopkinson, D.A. and Robson, E.B. (1962). Two-dimensional electrophoresis of pseudocholinesterase components in normal human serum. Nature 196, 1296–1298.

Harris, H., Hopkinson, D.A., Robson, E.B. and Whittaker, M. (1963a). Genetical studies on a new variant of serum cholinesterase detected by electrophoresis. Ann. Human Genet. London 26, 359–382.

Harris, H., Robson, E.B., Glen-Bott, A.M. and Thornton, J.A. (1963b) Evidence for non-allelism between genes affecting human serum cholinesterase. Nature 200, 1185–1187.

Hassón-Voloch, A. (1968). Curare and acetylcholine receptor substance. Nature 218, 330–333.

Hawkins, R.D. and Mendel, B. (1946). True cholinesterases with pronounced resistance to eserine. J. Cell. Comp. Physiol. 27, 69–85.

Hazard, R., Uriel, J. and Larno, S. (1967). Apparente identité de la cholinestérase et de la procaïnestérase sériques d'origine humaine. J. Physiol. Paris. 59, 5–8.

Hazel, B. and Monier, D. (1971). Human serum cholinesterase : variations during pregnancy and postpartum. Can. Anaesthesiol. Soc. J. 18, 272–277.

Heath, D.F. (1961). Organophosphorus Poisons. Anticholinesterases and Related Compounds (Pergamon Press, Oxford).

Hebb, C.O. (1954). Acetylcholine metabolism of nervous tissue. Pharmac. Rev. 6, 39–43.

Hebb, C.O. (1955). Choline acetylase in mammalian and avian sensory systems. Quart. J. Exptl. Physiol. 40, 176–186.

Hebb, C.O. (1957a). Biochemical evidence for the neural function of acetylcholine. Physiol. Rev. 37, 196–220.

Hebb, C.O. (1957b). The problem of identifying cholinergic neurones in the retina. Acta Physiol. Pharmacol. Neerl. 6, 621–631.

Hebb, C.O. (1968). Formal discussion. In: Proceedings of the Wates Symposium on Arterial Chemoreceptors 1966, R.W. Torrance, ed. (Blackwell Scientific Publications, Oxford) pp. 138–139.

Hebb, C.O. (1970). CNS at the cellular level: identity of transmitter agents. Ann. Rev. Physiol. 32, 165–192.

Hebb, C.O. and Hill, K.J. (1955a) Pseudocholinesterase in Pacinian corpuscles. Nature 175, 597.

Hebb, C. and Hill, K.J. (1955b). Distribution of cholinesterases in the mammalian pancreas. Quart. J. Exptl. Physiol. 40, 168–175.

Hebb, C.O. and Krnjević, K. (1962). The physiological significance of acetylcholine. In: Neurochemistry, K.A.C. Elliott, I.H. Page and J.H. Quastel, eds. 2nd edn. (Thomas, Springfield Illinois) pp. 452–521.

Hebb, C. and Linzell, J.L. (1970). Innervation of the mammary gland. A histochemical study in the rabbit. Histochem. J. 2, 491–505.

Hebb, C.O. and Ratković, D. (1962). Choline acetylase in the placenta of man and of other species. J. Physiol. London 163, 307–313.

Hebb, C.O. and Silver, A. (1956). Choline acetylase in the central nervous system of man and some other mammals. J. Physiol. London 134, 718–728.

Hebb, C.O. and Silver, A. (1961). Gradient of choline acetylase activity. Nature 189, 123–125.

Hebb, C.O. and Silver, A. (1970). Biochemical parameters of central cholinergic nerves, mapping the distribution of acetylcholine and the enzymes which control its metabolism in the brain. In: Neuropathology: Methods and Diagnosis, C.G. Tedeschi, ed. (Little, Brown and Co., Boston) pp. 665–690.

Hebb, C.O. and Waites, G.M.H. (1956). Choline acetylase in antero- and retrograde degeneration of a cholinergic nerve. J. Physiol. London 132, 667–671.

Hebb, C.O., Silver, A., Swan, A.A.B. and Walsh, E.G. (1953). A histochemical study of cholinesterases of rabbit retina and optic nerve. Quart. J. Exptl. Physiol. 38, 185–191.

Hebb, C.O., Krnjević, K. and Silver, A. (1963). Effect of undercutting on the acetylcholinesterase and choline acetyltransferase activity in the cat's cerebral cortex. Nature 198, 692.

Hebb, C., Mann, S. and Perkins, D. (1966). Histochemical demonstration of catecholamines and of cholinesterases on adjacent sections of tissue. J. Physiol. London 184, 12–13P.

Hegab, El-S.H.H. and Ferrans, V.J. (1966). A histochemical study of the esterases of the rat heart. Am. J. Anat. 119, 235–262.

Heilbronn, E. (1954). pH dependence of choline esterase activity at various substrate and inhibitor concentrations. Acta Chem. Scand. 8, 1368–1372.

Heilbronn, E. (1970a). Further experiments on the uptake of acetylcholine and atropine and the release of acetylcholine from mouse brain cortex slices after treatment with phospholipases. J. Neurochem. 17, 381–389.

Heilbronn, E. (1970b). Formal discussion, Skokloster Conference. In: Drugs and Cholinergic Mechanisms in the CNS, E. Heilbronn and A. Winter, eds. (Research Institute of National Defence, Stockholm) p. 575.

Heilbronn, E. and Sundwall, P. (1964). Studies on reactivation and ageing of blood cholinesterases of Tabun intoxicated dogs. Biochem. Pharmacol. 13, 59–67.

Heilbronn, E. and Tolagen, B. (1965). Toxogonin in Sarin, Soman and Tabun poisoning. Biochem. Pharmacol. 14, 73–77.

Heilbronn, E., Hause, S. and Lundgren, G. (1971). Chemical identification of acetylcholine in squid-head ganglion. Brain Res. 33, 431–437.

Hellenbrand, K. and Krupka, R.M. (1970). Kinetic studies on the mechanism of insect acetylcholinesterase. Biochemistry 9, 4665–4672.

Heller, H. and Ginsburg, M. (1966). Secretion, metabolism and fate of the posterior pituitary hormones. In: The Pituitary Gland. Vol. 3, G.W. Harris and B.T. Donovan, eds. (Butterworth, London) pp. 330–373.

Heller, M. and Hanahan, D.J. (1972). Human erythrocyte membrane bound enzyme: acetylcholinesterase. Biochim. Biophys. Acta 255, 251–272.

Hemsworth, B.A. and Mitchell, J.F. (1969). The characteristics of acetylcholine release mechanisms in the auditory cortex. Brit. J. Pharmacol. 36, 161–170.

Hemsworth, B.A. and West, G.B. (1970). Anticholinesterase activity of some degradation products of physostigmine. J. Pharm. Sci. 59, 118–120.

Herz, A. and Zieglgänsberger, W. (1968). The influence of microelectrophoretically applied biogenic amines, cholinomimetics and procaine on synaptic excitation in the corpus striatum. Intern. J. Neuropharmacol. 7, 221–230.

Herz, F., Kaplan, E. and Gleiman, E.J. (1967). Acetylcholinesterase and glucose-6-phosphate dehydrogenase activities in erythrocytes of fetal, newborn and adult sheep. Proc. Soc. Exptl. Biol. Med. 124, 1185–1187.

Hess, A. (1960). The effects of eye removal on the development of cholinesterase in the superior colliculus. J. Exptl. Zool. 144, 11–24.

Hess, A. (1961a). The effect of unilateral eye removal on the cholinesterase activity of the optic tectum of adult frogs. Anat. Record 140, 295–306.

Hess, A. (1961b). Structural differences of fast and slow extrafusal muscle fibres and their nerve endings in chickens. J. Physiol. London 157, 221–231.

Hess, A. (1962). Further morphological observations of 'en plaque' and 'en grappe' nerve endings on mammalian extrafusal muscle fibers with the cholinesterase technique. Rev. Can. Biol. 21, 241–248.

Hess, A. (1972). Histochemical localizaiton of cholinesterase in the brain of the cockroach (*Periplaneta americana*). Brain Res. 46, 287–295.

Hess, A.R., Angel, R.W., Barron, K.D. and Bernsohn, J. (1963). Proteins and isozymes of esterases and cholinesterases from sera of different species. Clin. Chim. Acta 8, 656–667.

Hestrin, S. (1949). The reaction of acetylcholine and other carboxylic acid derivatives with hydroxylamine, and its analytical application J. Biol. Chem. 180, 249–261.

Hillman, G.R. and Mautner, H.G. (1970). Hydrolysis of electronically and sterically defined substrates of acetylcholinesterase. Biochemistry 9, 2633–2638.

Himwich, H.E. and Aprison, M.H. (1955). The effect of age on cholinesterase activity of rabbit brain. In: Biochemistry of the Developing Nervous System, H. Waelsch, ed. (Academic Press, New York) pp. 301–307.

Himwich, W.A., Sullivan, W.T., Kelly, B., Benaron, H.B.W. and Tucker, B.E. (1955). Chemical constituents of the brain. J. Nervous Mental Disease 122, 441–447.

Hines, B.E. and McCance, R.A. (1953). Pseudo-cholinesterase activity in secretions and organs of piglets and pigs. J. Physiol. London 122, 188–192.

Ho, I.K. and Ellman, G.L. (1969). Triton solubilized acetylcholinesterase of brain. J. Neurochem. 16, 1505–1513.

Ho, A.K.S., Paddle, B.M. and Freeman, S.E. (1965). The estimation of the activity of acetylcholinesterase and other esterases in the rat brain by an amperometric method. Biochem. Pharmacol. 14, 151–157.

Hobbiger, F. (1955). Effect of nicotinhydroxamic acid methiodide on human plasma cholines-terase inhibited by organophosphates containing a dialkylphosphate group. Brit. J. Pharma-col. 10, 356–362.

Hobbiger, F. (1957a). Protection against the lethal effects of organophosphates by pyridine-2-aldoxime methiodide. Brit. J. Pharmacol. 12, 438–446.

Hobbiger, F. (1957b). Reactivation of phosphorylated acetocholinesterase by pyridine-2-aldoxime methiodide. Biochim. Biophys. Acta 25, 652–654.

Hobbiger, F. (1963). Reactivation of phosphorylated acetylcholinesterase. In: Cholinesterases and Anticholinesterase Agents, G.B. Koelle, ed. (Springer, Berlin) pp. 921–988.

Hobbiger, F. and Lancaster, R. (1971). The determination of acetylcholinesterase activity of brain slices and its significance in studies of extracellular acetylcholinesterase. J. Neuro-chem. 18, 1741–1749.

Hobbiger, F. and Peck, A.W. (1970). The relationship between the level of cholinesterase in plasma and the action of suxamethonium in animals. Brit. J. Pharmacol. 40, 775–789.

Hobbiger, F. and Sadler, P.W. (1959). Protection against lethal organophosphate poisoning by quaternary pyridine aldoximes. Brit. J. Pharmacol. 14, 192–201.

Hobbiger, F. and Vojvodić, V. (1966). The reactivating and antidotal actions of N, N'-trimeth-ylenebis(pyridinium-4-aldoxime) (TMB-4) and N, N'-oxydimethylenebis(pyridinium-4-aldoxime) (Toxogonin), with particular reference to their effect on phosphorylated acetyl-cholinesterase in the brain. Biochem. Pharmacol. 15, 1677–1690.

Hobbiger, F. and Vojvodić, V. (1967). The reactivation by pyridinium aldoximes of phos-phorylated acetylcholinesterase in the central nervous system. Biochem. Pharmacol. 16, 455–462.

Hoekman, T.B. and Dettbarn, W.-D. (1971). Acetylcholine – a possible mechanism for the depolarization response in giant axons of the lobster circumesophageal connective. Biochem. Pharmacol. 20, 1713–1717.

Hokin, L.E. (1965). Autoradiographic localization of the acetylcholine-stimulated synthesis of phosphatidylinositol in the superior cervical ganglion. Proc. Natl. Acad. Sci. U.S. 53, 1369–1376.

Hokin, L.E. (1969). Functional activity in glands and synaptic tissue and the turnover of phosphatidylinositol. Ann. N.Y. Acad. Sci. U.S. 165, 695–709.

Hokin, L.E. and Hokin, M.R. (1958a). Acetylcholine and the exchange of inositol and phos-phate in brain phosphoinositide. J. Biol. Chem. 233, 818–821.

Hokin, L.E. and Hokin, M.R. (1958b). Acetylcholine and the exchange of phosphate in phos-phatidic acid in brain microsomes. J. Biol. Chem. 233, 822–826.

Hokin, L.E. and Hokin, M.R. (1959). The mechanism of phosphate exchange in phosphatidic acid in response to acetylcholine. J. Biol. Chem. 234, 1387–1390.

Hokin, L.E. and Yoda, A. (1964). Inhibition by diisopropylphosphorofluoridate of a kidney transport adenosine triphosphatase by phosphorylation of a serine residue. Proc. Natl. Acad. Sci. U.S. 52, 454–461.

Hokin, M.R. (1969). Effect of norepinephrine on ^{32}P incorporation into individual phospha-tides in slices from different areas of the guinea pig brain. J. Neurochem. 16, 127–134.

Hokin, M.R. (1970). Effects of dopamine, gamma-aminobutyric acid and 5-hydroxytryptamine on incorporation of ^{32}P into phosphatides in slices from guinea pig brain. J. Neurochem. 17, 357–364.

Holaday, D.A., Kamijo, K. and Koelle, G.B. (1954). Facilitation of ganglionic transmission following inhibition of cholinesterase by DFP. J. Pharmacol. Exptl. Therap. 111, 241–254.

Holland, W.C. and Greig, M.E. (1952). The synthesis of acetylcholine by human erythrocytes. Arch. Biochem. Biophys. 39, 77–79.

Hollinshead, W.H. and Sawyer, C.H. (1945). Mechanisms of carotoid-body stimulation. Am. J. Physiol. 144, 79–86.

Hollunger, E.G. and Niklasson, B.H. (1973). The release and molecular state of mammalian brain acetylcholinesterase. J. Neurochem. 20, 821–836.

Holmes, R.L. (1961a). Phosphatase and cholinesterase in the hypothalamo-hypophysial system of the monkey. J. Endocrinol. 23, 63–67.

Holmes, R.L. (1961b). Esterases of the hypothalamo-hypophysial system. In: Cytology of nervous tissue. Proc. Anat. Soc. Gt. Brit. and Ireland (Taylor and Francis Ltd, London). pp. 1–4.

Holmes, R.L. and Wolstencroft, J.H. (1964). Cholinesterase in the medulla and pons of the cat. J. Physiol. London 175, 55–56P.

Holmes, R.S. and Masters, C.J. (1967). The developmental multiplicity and isoenzyme status of cavian esterases. Biochim. Biophys. Acta 132, 379–399.

Holmstedt, B. (1951). Synthesis and pharmacology of dimethylamidoethoxyphosphoryl cyanide (Tabun) together with a description of some allied anticholinesterase compounds containing the N-P bond. Acta Physiol. Scand. 25, Suppl. 90, 7–120.

Holmstedt, B. (1957a). A modification of the thiocholine method for the determination of cholinesterase. I. Biochemical evaluation of selective inhibitors. Acta Physiol. Scand. 40, 322–330.

Holmstedt, B. (1957b). A modification of the thiocholine method for the determination of cholinesterase. II. Histochemical application. Acta Physiol. Scand. 40, 331–337.

Holmstedt, B. (1959). Pharmacology of organophosphorus cholinesterase inhibitors. Pharmacol. Rev. 11. 567–688.

Holmstedt, B. and Sjöqvist, F. (1959a). Distribution of acetocholinesterase in the ganglion cells of various sympathetic ganglia. Acta Physiol. Scand. 47, 284–296.

Holmstedt, B. and Sjöqvist, F. (1959b). Pharmacological properties of γ-aminobutyrylcholine a supposed inhibitory neutrotransmitter. Biochem. Pharmacol. 3, 297–304.

Holmstedt, B. and Toschi, G. (1959). Enzymic properties of cholinesterases in subcellular fractions from rat brain. Acta Physiol. Scand. 47, 280–283.

Holmstedt, B., Lundgren, G. and Sjöqvist, F. (1963). Determination of acetylcholinesterase activity in normal and denervated sympathetic ganglia of the cat. A biochemical and histochemical comparison. Acta Physiol. Scand. 57, 235–247.

Holt, S.J. (1958). Indogogenic staining methods for esterases. In: General Cytochemical Methods, Vol. 1, J.F. Danielli, ed. (Academic Press, New York) pp. 375–398.

Holt, S.J. and O'Sullivan, D.G. (1958). Studies in enzyme cytochemistry. I. Principles of cytochemical staining methods. Proc. Roy. Soc. London. Ser. B. 148, 456–480.

Holt, S.J. and Withers, R.F.J. (1952). Cytochemical localization of esterases using indoxyl derivatives. Nature 170, 1012–1014.

Honjin, R. (1956). The innervation of the pancreas of the mouse with special reference to the structure of the peripheral extension of the vegetative nervous system. J. Comp. Neurol. 104, 331–371.

Hopff, W.H. and Waser, P.G. (1970). Warum können Reaktivatoren schädlich sein? Abgehandelt am Beispiel der Reaktivierung der blockierten Acetylcholinesterase. Pharm. Acta Helv. 45, 414–423.

Horsfall, W.R., Lehmann, H. and Davies, D. (1963). Incidence of pseudocholinesterase variants in Australian aborigines. Nature 199, 1115.

Hosein, E.A., Kato, A., Vine, E. and Hill, A.M. (1970). The identification of acetyl-1-carnityl choline in rat brain extracts and the comparison of its cholinomimetic properties with acetylcholine. Can. J. Physiol. Pharmacol. 48, 709–722.

Hoskin, F.C.G. (1971). Diisopropylphosphorofluoridate and Tabun: enzymatic hydrolysis and nerve function. Science 172, 1243–1245.

Hoskin, F.C.G., Kremzner, L.T. and Rosenberg, P. (1969). Effects of some cholinesterase inhibitors on the squid giant axon – their permeability, detoxication and effects on conduction and anticholinesterase activity. Biochem. Pharmacol. 18, 1727–1737.

Hösli, E. and Hösli, L. (1970). The presence of acetylcholinesterase in cultures of cerebellum and brain stem. Brain Res. 19, 494–496.

Hösli, E. and Hösli, L. (1971). Acetylcholinesterase in cultured rat spinal cord. Brain Res. 30, 193–197.

Hoyle, G. (1953). Potassium ions and insect nerve muscle. J. Exptl. Biol. 30, 121–135.

Huc, M. (1950). Recherches sur la détermination et la signification de l'activité procaïnestérasique du sérum humain. Thèse de Médicine, Lille, 1949-50. No. 37. (Lille: Douriez-Bataille).

Huikuri, K.T. (1966). Histochemistry of the ciliary ganglion of the rat and the effect of pre- and post-ganglionic nerve division. Acta Physiol. Scand. 69, Suppl. 286, 1–83.

Hyyppä, M. (1969). Histochemically demonstrable esterase activity in the hypothalamus of the developing rat. Histochemie 20, 29–39.

I.C.S.H. report (1971). Recommended methods for radioisotope red cell survival studies. A report by the I.C.S.H. panel on diagnostic application of radioisotopes in haematology. Brit. J. Haematol. 21, 241–250.

Iijima, K. and Bourne, G.H. (1968). Histochemical studies on the distribution of esterases, monoamine oxidase, and dephosphorylating enzymes in the area postrema of the squirrel monkey. Acta Histochem. 29, 349–362.

Iijima, K. Shantha, T.R. and Bourne, G.H. (1967). Enzyme histochemical studies on the hypothalamus with special reference to the supraoptic and paraventricular nuclei of squirrel monkey (*Saimiri sciureus*). Z. Zellforsch. Mikroskop. Anat. 79, 76–91.

Imagawa, N. (1969). The ciliary muscle and the autonomic nervous system. Report 1. The activity of acetylcholine esterase in the cat ciliary muscle in electron microscope preparations. Acta Soc. Ophthal. Jap. 73, 2151–2159.

Iordanova, L. and Gotsev, T. (1970). The influence of examination stress in students on the activity of certain enzymes. Nauch. Tr. Vissh. Med. Inst. Sofia 49, 7–12.

Ip, M.C. (1967). A combined method for demonstrating the cholinesterase activity and the nervous structure of mammalian peripheral motor endings in teased preparations. J. Physiol. London 192, 801–803.

Iqbal, Z.M. and Menzer, R.E. (1972). Metabolism of O-ethyl S, S-dipropylphosphorodithioate in rats and liver microsomal systems. Biochem. Pharmacol. 21, 1569–1584.

Iqbal, Z. and Talwar, G.P. (1971). Acetylcholinesterase in developing chick embryo brain. J. Neurochem. 18, 1261–1267.

Ishii, T. and Friede, R.L. (1967). Comparative histochemical mapping of the distribution of acetylcholinesterase and nicotinamide adenine dinucleotide-diaphorase activities in the human brain. Intern. Rev. Neurobiol. 10, 231–275.

Ishii, Y. (1957a). The histochemical studies of cholinesterase in the central nervous system. I. Normal distribution in rodents (in Japanese). Arch. Histol., Okoyama 12, 587–611.

Ishii, Y. (1957b). The histochemical studies of cholinesterase in the central nervous system. II. Histochemical alteration of cholinesterase of the brain of rats from late fetal life to adults (in Japanese). Arch. Histol. Okoyama 12, 613–637.

Ishiwata, K., Furukawa, M. and Takahashi, H. (1956). J. Jap. Biochem. Soc 28, 209–213, cited by T. Mann (1964). In: The Biochemistry of Semen and the Male Reproductive Tract. (Methuen, London) p. 208.

Israël, M. and Whittaker, V.P. (1965). The isolation of mossy fibre endings from the granular layer of the cerebellar cortex. Experientia 21, 325–326.

Ito, M., Yoshida, M. and Obata, K. (1964). Monosynaptic inhibition of the intracerebellar nuclei induced from the cerebellar cortex. Experientia 20, 575–576.

Iurato, S., Luciano, L., Pannese, E. and Reale, E. (1971). Histochemical localization of acetyl-cholinesterase (AChE) activity in the inner ear. Acta Oto-laryngol. Suppl. 279, 1–50.

Ivens, C., Mottram, D.R., Lever, J.D., Presley, R. and Howells, G. (1973). Studies on the acetylcholinesterase (AChE) -positive and -negative autonomic axons supplying smooth muscle in the normal and 6-hydroxydopamine (6-OHDA) treated rat iris. Z. Zellforsch. Mikroskop. Anat. 138, 211–222.

Iwata, N., Sakai, Y. and Deguchi, T. (1971). Effects of physostigmine on the inhibition to trigeminal motoneurones by cutaneous impulses in the cat. Exptl. Brain. Res. 3, 519–532.

Jackson, R.L. and Aprison, M.H. (1966a). Mammalian brain acetylcholinesterase. Purification and properties. J. Neurochem. 13, 1351–1365.

Jackson, R.L. and Aprison, M.H. (1966b). Mammalian brain acetylcholinesterase. Effects of surface active agents. J. Neurochem. 13, 1367–1371.

Jacob, F. and Monod, J. (1961). Genetic regulatory mechanisms in the synthesis of proteins. J. Mol. Biol. 3, 318–356.

Jacobowitz, D. and Koelle, G.B. (1963). Demonstration of both acetylcholinesterase (AChE) and catecholamines in the same nerve trunk. Pharmacologist 5, 270.

Jacobowitz, D. and Koelle, G.B. (1965). Histochemical correlations of acetylcholinesterase and catecholamines in postganglionic autonomic nerves of the cat, rabbit and guinea pig. J. Pharmacol. Exptl. Therap. 148, 225–237.

Jacobsen, C.F., Léonis, J., Linderstrøm-Lang, K. and Ottesen, M. (1957). The pH-stat and its use in biochemistry. In: Methods of Biochemical Analysis, Vol. IV, D. Glick, ed. (Interscience Publishers, New York) pp. 171–210.

Jaju, B.P. (1969). Burn and Rand's hypothesis. Indian J. Physiol. Pharmacol. 13, 1–27.

James, T.N. and Spence, C.A. (1966). Distribution of cholinesterase within the sinus node and AV node of the human heart. Anat. Record 155, 151–162.

Jankowska, E., Lubińska, L. and Niemierko, S. (1969). Translocation of AChE-containing particles in the axoplasm during nerve activity. Comp. Biochem. Physiol. 28, 907–913.

Jansen, J. and Brodal, A. (1954). Aspects of Cerebellar Anatomy. (Tanum, Oslo).

Jansz, H.S. and Cohen, J.A. (1962). Pseudocholinesterase from horse serum. I. Purification and properties of the enzyme. Biochim. Biophys. Acta 56, 531–537.

Jansz, H.S., Brons, D. and Warringa, M.G.P.J. (1959). Chemical nature of the DFP-binding site of pseudocholinesterase. Biochim. Biophys. Acta 34, 573–575.

Jarlstedt, J. and Karlsson, J.-O. (1973). Evidence for axonal transport of RNA in mammalian neurons. Exptl. Brain Res. 16, 501–506.

Jasser, A. and Guth, P.S. (1973). The synthesis of acetylcholine by the olivo-cochlear bundle. J. Neurochem. 20, 45–53.

Jell, R.M. (1973). Responses of hypothalamic neurones to local temperature and to acetylcholine, noradrenaline and 5-hydroxytryptamine. Brain Res. 55, 123–134.

Jenkinson, D. McE. and Maloiy, G.M.O. (1969). The distribution of nerves, monoamine oxidase and cholinesterases in the skin of the red deer (*Cervus elaphus*). Res. Vet. Sci. 10, 448–452.

Jensen-Holm, J., Lausen, H.H., Milthers, K. and Møller, K.O. (1959). Determination of the cholinesterase activity in blood and organs by automatic titration. With some observations on serious errors of the method and remarks of the photometric determination. Acta Pharmacol. Toxicol. 15, 384–394.

Johns, R.J. (1962). Familial reduction in red-cell cholinesterase. New England J. Med. 267, 1344–1348.

Johnson, D.D. and Stewart, W.C. (1970). The effects of atropine, pralidoxime and lidocaine on nerve-muscle and respiratory function in organophosphate-treated rabbits. Can. J. Physiol. Pharmacol. 48, 625–630.

Johnson, J.L. (1970). Changes in acetylcholinesterase, acid phosphatase and beta glucuronidase proximal to a nerve crush. Brain Res. 18, 427–440.

Johnson, M.K. (1969a). A phosphorylation site in brain and the delayed neurotoxic effect of some organophosphorus compounds. Biochem. J. 111, 487–495.

Johnson, M.K. (1969b). The delayed neurotoxic effect of some organophosphorus compounds. Identification of the phosphorylation site as an esterase. Biochem. J. 114, 711–717.

Johnson, M.K. (1969c). Delayed neurotoxic action of some organophosphorus compounds. Brit. Med. Bull. 25, 231–235.

Johnson, M.K. (1970). Organophosphorus and other inhibitors of brain neurotoxic esterase and the development of delayed neurotoxicity in hens. Biochem. J. 120, 523–531.

Johnson, M.K. (1973). Brain 'neurotoxic esterase'. Z. Physiol. Chem. 354, 6–7.

Johnson, M.K. and Barnes, J.M. (1970). Age and the sensitivity of chicks to the delayed neurotoxic effects on some organophosphorus compounds. Biochem. Pharmacol. 19, 3045–3047.

Jones, J.V. and Ballard, K.J. (1971). Cholinesterases in the carotid body of the cat as seen with the electron microscope. Nature New Biol. 233, 146–147.

Jones, M., Featherstone, R.M. and Bonting, S.L. (1956). The effect of acetylcholine on the cholinesterases of chick embryo intestine cultured in vitro. J. Pharmacol. Exptl. Therap. 116, 114–118.

Jones, V.E. and Ogilvie, B.M. (1972). Protective immunity to *Nippostrongylus brasiliensis* in the rat. III. Modulation of worm acetylcholinesterase by antibodies. Immunology 22, 119–129.

Joó, F. (1969). Electron histochemical structure of capillaries in the rat brain. Acta Biolog., Szeged 15, 79–88.

Joó, F. and Csillik, B. (1966). Topographical correlation between the hemato-encephalic barrier and the cholinesterase activity of brain capillaries. Exptl. Brain Res. 1, 147–151.

Joó, F. and Várkonyi, T. (1969). Correlation between the cholinesterase activity of capillaries and the blood brain barrier in the rat. Acta Biol. Acad. Sci. Hung. 20, 359–372.

Joó, F., Kása, P., Kálmán, Gy. and Csillik, B. (1963). Ontogenetical differentiation of cholinesterase-active structures in archicerebellar cortex. Folia Histochem. Cytochem. 1, Suppl. 1, 118.

Joó, F., Lever, J.D., Ivens, C., Mottram, D.R. and Presley, R. (1971). A fine structural and electron histochemical study of axon terminals in the rat superior cervical ganglion after acute and chronic preganglionic denervation. J. Anat. 110, 181–189.

Jordan, L.M. and Phillis, J.W. (1972). Acetylcholine inhibition in the intact and chronically isolated cerebral cortex. Brit. J. Pharmacol. 45, 584–595.

Jordan, S.M. (1971). Autonomic innervation of the ovary. J. Anat. 110, 500.

Jović, R. and Bošković, B. (1970). Antidotal action of pyridinium oximes in poisoning by O, O-diethyl-S-[2-(N-methyl-N-phenylamino)ethyl] thiophosphonate methylsulfomethylate (GT-45) and its two new analogs. Toxicol. Appl. Pharmacol. 16, 194–200.

Jović, R., Bachelard, H.S., Clark, A.G. and Nicholas, P.C. (1971). Effects of Soman and DFP in vivo and in vitro on cerebral metabolism in the rat. Biochem. Pharmacol. 20, 519–527.

Jung, M.J. and Belleau, B. (1972). Purification and fractionation of acetylcholinesterase into subspecies by affinity chromatography on a d-tubocurarine-sepharose column. Mol. Pharmacol. 8, 589–593.

Jurchenko, O.P., Vulfius, C.A. and Zeimal, E.V. (1973a). Cholinesterase activity in ganglia of gastropoda, *Lymnaea stagnalis* and *Planorbarius corneus*. I. Effect of anticholinesterase agents on giant neurone depolarization by acetylcholine and its analogues. Comp. Biochem. Physiol. 45A, 45–60.

Jurchenko, O.P., Kultas, K.N. and Vulfius, C.A. (1973b). Cholinesterase activity in ganglia of gastropoda, *Lymnaea stagnalis* and *Planorbarius corneus.* II. Histochemical investigation. Comp. Biochem. Physiol. 45A, 61–68.

Juul, P. (1968). Human plasma cholinesterase isoenzymes. Clin. Chim. Acta 19, 205–213.

Kabachnik, M.I., Brestkin, A.P., Godovikov, N.N., Michelson, M.J., Rozengart, E.V. and Rozengart, V.I. (1970). Hydrophobic areas on the active surface of cholinesterases. Pharmacol. Rev. 22, 355–388.

Kahlson, G. and Renvall, S. (1956). The distribution of cholinesterases in cats and changes caused by hypophysectomy, adrenalectomy, undernutrition and DOCA. Acta Physiol. Scand. 37, 159–176.

Kaita, A.A. and Goldberg, A.M. (1969). Control of acetylcholine synthesis – the inhibition of choline acetyltransferase by acetylcholine. J. Neurochem. 16, 1185–1191.

Kalina, M. and Bubis, J.J. (1969). Ultrastructural localization of acetylcholine esterase in neurones of rat trigeminal ganglia. Experientia 25, 388–389.

Kalow, W. and Genest, K. (1957). A method for the detection of atypical forms of human serum cholinesterase. Determination of Dibucaine numbers. Can. J. Biochem. Physiol. 35, 339–346.

Kalow, W., Genest, K. and Staron, N. (1956). Kinetic studies on the hydrolysis of benzoylcholine by human serum cholinesterase. Can. J. Biochem. Physiol. 34, 637–653.

Kamemoto, F.I. (1957). Cholinesterase in the nemertean *Prostoma rubrum.* Science 125, 351–352.

Kamemoto, F.I. (1961). The effects of eserine on sodium regulation in crayfish. Comp. Biochem. Physiol. 3, 297–303.

Kanagasuntheram, R., Wong, W.C. and Chan, H.L. (1969). Some observations on the innervation of the human nasopharynx. J. Anat. 104, 361–376.

Kanai, T. and Szerb, J.C. (1965). Mesencephalic reticular activating system and cortical acetylcholine output. Nature 205, 80–82.

Kaplan, E., Herz, F., and Hsu, K.S. (1964). Erythrocyte acetylcholinesterase activity in ABO hemolytic disease of the newborn. Pediatrics 33, 205–211.

Kaplay, S.S. and Jagannathan, V. (1970). Purification and properties of ox brain acetylcholinesterase. Arch. Biochem. Biophys. 138, 48–57.

Karassik, V.M. (1946). Pharmacologic characterization of cholinergic and adrenergic structures of the organism (in Russian). Usp. Sovrem. Biol. 21, 1 (cited Župančič 1967).

Karczmar, A.G. (1963). Ontogenesis of cholinesterases. In: Cholinesterases and Anticholinesterase Agents, G.B. Koelle, ed. (Springer, Berlin) pp. 129–186.

Karczmar, A.G. (1967). Pharmacologic, toxicologic and therapeutic properties of anticholinesterase agents. In: Physiological Pharmacology, Vol. III, The Nervous System - Part C, W.S. Root and F.G. Hofmann, eds. (Academic Press, New York) pp. 163–322.

Karczmar, A.G. (1969). Is the central cholinergic nervous system overexploited? Federation Proc. 28, 147–157.

Karlin, A. (1965). The association of acetylcholinesterase and membrane in subcellular fractions of the electric tissue of *Electrophorus.* J. Cell Biol. 25, 159–169.

Karlin, A. (1967). Chemical distinctions between acetylcholinesterase and the acetylcholine receptor. Biochim. Biophys. Acta 139, 358–362.

Karlin, A. (1969). Chemical modification of the active site of the acetylcholine receptor. J. Gen. Physiol. 54, 245s–264s.

Karlin, A. (1970). Investigations of the chemical nature of the acetylcholine receptor. In: Drugs and Cholinergic Mechanisms in the CNS, E. Heilbronn and A. Winter, eds. (Institute of National Defence, Stockholm) pp. 489–503.

Karlog, O. and Petersen, H.E.H. (1963). The influence of oximes on the acetylthiocholine hydrolysis rate. Biochem. Pharmacol. 12, 590–591.

Karlsson, J.-O. and Sjöstrand, J. (1971). Synthesis, migration and turnover of protein in retinal ganglion cells. J. Neurochem. 18, 749–767.

Karnovsky, M.J. (1964). The localization of cholinesterase activity in rat cardiac muscle by electron microscopy. J. Cell Biol. 23, 217–232.

Karnovsky, M.J. and Roots, L. (1964). A 'direct-coloring' thiocholine method for cholinesterases. J. Histochem. Cytochem. 12, 219–221.

Kása, P. (1968a). Ultrastructural localization of acetylcholinesterase in the cerebellar cortex with special reference to the intersynaptic organelles. Histochemie 14, 161–167.

Kása, P. (1968b). Acetylcholinesterase transport in the central and peripheral nervous tissue: the role of tubules in the enzyme transport. Nature 218, 1265–1267.

Kása, P. (1969). Electron histochemical evidence of different types of mossy fibre endings in the cerebellar cortex. Experientia 25, 740–741.

Kása, P. (1971). Ultrastructural localization of choline acetyltransferase and acetylcholinesterase in central and peripheral nervous tissue. In: Progress in Brain Research, Vol. 34, O. Eränkö, ed. (Elsevier, Amsterdam) pp. 337–344.

Kása, P. and Csillik, B. (1964). Histochemical studies on the effect of nerve degeneration in the cerebellar cortex. II Intern. Congr. of Histo. and Cytochemistry (Springer, Berlin) pp. 195–196.

Kása, P. and Csillik, B. (1965a). Cholinergic excitation and inhibition in the cerebellar cortex. Nature 208, 695–696.

Kása, P. and Csillik, B. (1965b). Comparative histochemistry of the cerebellum during development. Prelim. Abstr. Neurochem. Congr. Oxford. p. 54.

Kása, P. and Csillik, B. (1966). Electron microscopic localization of cholinesterase by a copper-lead-thiocholine technique. J. Neurochem. 13, 1345–1349.

Kása, P. and Csillik, B. (1968). AChE synthesis in cholinergic neurons: electron histochemistry of enzyme translocation. Histochemie 12, 175–183.

Kása, P. and Silver, A. (1969). The correlation between choline acetyltransferase and acetylcholinesterase activity in different areas of the cerebellum of rat and guinea pig. J. Neurochem. 16, 389–396.

Kása, P., Joó, F. and Csillik, B. (1965). Histochemical localization of acetylcholinesterase in the cat cerebellar cortex. J. Neurochem. 12, 31–35.

Kása, P., Csillik, B., Joó, F. and Knyihár, E. (1966). Histochemical and ultrastructural alterations in the isolated archicerebellum of the rat. J. Neurochem. 13, 173–178.

Kása, P., Mann, S.P., Karcsu, S., Tóth, L. and Jordan, S. (1973). Transport of choline acetyltransferase and acetylcholinesterase in the rat sciatic nerve: a biochemical and electron histochemical study. J. Neurochem. 21, 431–436.

Kasai, M. and Changeux, J.-P. (1971). In vitro excitation of purified membrane fragments by cholinergic agonists. III. Comparison of the dose response curves to decamethonium with the corresponding binding curves of decamethonium to the cholinergic receptor. J. Membr. Biol. 6, 58–80.

Kaswin, A. and Serfaty, A. (1946). L'activité cholinestérasique du sperme de Roussette (*Scyliorhinus canicula*). Compt. Rend. Soc. Biol. 140, 78–79.

Katchalski, E., Silman, I. and Goldman, R. (1971). Effect of the microenvironment on the mode of action of immobilized enzymes. In: Advances in Enzymology, Vol. 34, F.F. Nord, ed. (Interscience Publishers, New York) pp. 445–536.

Kato, G. (1968). Studies on serum cholinesterase kinetics by nuclear magnetic resonance spectroscopy. Mol. Pharmacol. 4, 640–644.

Kato, G. (1969). Nuclear magnetic resonance study of the interaction between acetylcholine and horse serum cholinesterase. Mol. Pharmacol. 5, 148–155.

Kato, G. (1972a). Acetylcholinesterase. I. A study by nuclear magnetic resonance of the binding of inhibitors to the enzyme. Mol. Pharmacol. 8, 575–581.

Kato, G. (1972b). Acetylcholinesterase. II. A study by nuclear magnetic resonance of the acceleration of acetylcholinesterase by atropine and inhibition by eserine. Mol. Pharmacol. 8, 582–588.

Kato, G. and Yung, J. (1971). The use of nuclear magnetic resonance to describe the binding of atropine analogues to acetylcholinesterase. Mol. Pharmacol. 7, 33–39.

Kato, G., Yung, J., and Ihnat, M. (1970). Nuclear magnetic resonance studies on acetylcholinesterase. The use of atropine and eserine to probe binding sites. Mol. Pharmacol. 6, 588–596.

Kato, G., Tan, E. and Yung, J. (1972). Acetylcholinesterase. Kinetic studies on the mechanism of atropine inhibition. J. Biol. Chem. 247, 3186–3189.

Kato, J. and Minaguchi, H. (1964). Cholinergic and adrenergic mechanisms in the female rat hypothalamus with special reference to reproductive functions. Fluctuation in choline acetylase and monoamine oxidase activities. Gunma Symp. Endocr. 1, 269–281.

Kattamis, C., Zannos-Mariolea, L., Franco, A.P., Liddell, J., Lehmann, H. and Davies, D. (1962). Frequency of atypical pseudocholinesterase in British and Mediterranean populations. Nature 196, 599–600.

Katz, B. and Miledi, R. (1973). The binding of acetylcholine to receptors and its removal from the synaptic cleft. J. Physiol. London 231, 549–574.

Katz, H.S., Salehmoghaddam, S. and Collier, B. (1973). The accumulation of radioactive acetylcholine by a sympathetic ganglion and by brain: failure to label endogenous stores. J. Neurochem. 20, 569–579.

Kavaler, F. and Kimel, V.M. (1952). Biochemical and physiological differentiation during morphogenesis. XV. Acetylcholinesterase activity of the motor cortex of the fetal guinea pig. J. Comp. Neurol. 96, 113–119.

Kerkut, G.A. (1967). Biochemical aspects of invertebrate nerve cells. In: Invertebrate Nervous Systems, C.A.G. Wiersma, ed. (Univ. Chicago Press, Chicago) pp. 5–37.

Kerkut, G.A. and Cottrell, G. (1963). Acetylcholine and 5-hydroxytryptamine in the snail brain. Comp. Biochem. Physiol. 8, 53–63.

Kerkut, G.A. and Walker, R.J. (1962). The specific chemical sensitivity of *Helix* nerve cells. Comp. Biochem. Physiol. 7, 277–288.

Kerkut, G.A. and Walker, R.J. (1967). The action of acetylcholine, dopamine and 5-hydroxytryptamine on the spontaneous activity of the cells of Retzius of the leech, *Hirudo medicinalis*. Brit. J. Pharmacol. 30, 644–654.

Kerkut, G.A., Shapira, A. and Walker, R.J. (1967). The transport of ^{14}C-labelled material from CNS ⇌ muscle along a nerve trunk. Comp. Biochem. Physiol. 23, 729–748.

Kerkut, G.A., Pitman, R.M. and Walker, R.J. (1969). Sensitivity of neurons of the insect central nervous system to iontophoretically applied acetylcholine or GABA (γ-amino butyric acid). Nature 222, 1075–1076.

Kerkut, G.A., Oliver, G., Rick, J.T. and Walker, R.J. (1970a). Biochemical changes during learning in an insect ganglion. Nature 227, 722–723.

Kerkut, G.A., Oliver, G.W.O., Rick, J.T. and Walker, R.J. (1970b). The effects of drugs on learning in a simple preparation. Comp. Gen. Pharmacol. 1, 437–483.

Kerkut, G.A., Beesley, P., Emson, P., Oliver, G. and Walker, R.J. (1971). Reduction in ChE during avoidance learning in the cockroach CNS. Comp. Biochem. Physiol. 39B, 423–424.

Kerkut, G.A., Emson, P.C. and Beesley, P.W. (1972). Effect of leg-raising learning on protein synthesis and ChE activity in the cockroach CNS. Comp. Biochem. Physiol. 41B, 635–645.

Kernan, R.P. (1968). Membrane potential and chemical transmitter in active transport of ions by rat skeletal muscle. J. Gen. Physiol. 51, 204s–210s.

Keyl, M.J., Michaelson, I.A. and Whittaker, V.P. (1957). Physiologically active choline esters in certain marine gastropods and other invertebrates. J. Physiol. London 139, 434–454.

Keynes, R.D. (1969). From frog skin to sheep rumen: a survey of transport of salts and water across multicellular structures. Quart. Rev. Biophys. 2, 177–281.

Khan, M.A. (1973). Toxicity of systemic insecticides: toxicological considerations in using organophosphorus insecticides. Vet. Record 92, 411–419.

Khera, K.S. and Laham, Q.N. (1965). Cholinesterases and motor end-plates in developing duck skeletal muscle. J. Histochem. Cytochem. 13, 559–565.

Kienhuis, H. (1964). The possible significance of the amino dicarboxylic acid next to the reactive serine residue in esterases. Abst. 1st Meeting Federation Europ. Biochem. Soc. p.4.

Kim, S.U. and Munkacsi, I. (1972). Cytochemical demonstration of catecholamines and acetylcholinesterase in cultures of chick sympathetic ganglia. Experientia 28, 824–825.

Kim, S.U. and Murray, M.R. (1969). Histochemical demonstration of acetylcholinesterase in organized cultures of mouse cerebellum and sensory ganglia. Anat. Record 163, 310.

Kim, S.U., Oh, T.H. and Johnson, D.D. (1972). Developmental changes of acetylcholinesterase and pseudocholinesterase in organotypic cultures of spinal cord. Exptl. Neurol. 35, 274–281.

Kingsley, R.E. and Barnes, C.D (1973). An electrophysiological study. The origin of the crossed olivocochlear bundle. Exptl. Neurol. 39, 323–335.

Kiraly, J.K. and Phillis, J.W. (1961). Action of some drugs on the dorsal root potentials of the isolated toad spinal cord. Brit. J. Pharmacol. 17, 224–231.

Kirkpatrick, J.B., Bray, J.J. and Palmer, S.M. (1972). Visualization of axoplasmic flow in vitro by Nomarski microscopy. Comparison to rapid flow of radioactive proteins. Brain Res. 43, 1–10.

Kirkpatrick, W.E. and Lomax, P. (1970). Temperature changes following iontophoretic injection of acetylcholine into the rostral hypothalamus of the rat. Neuropharmacol. 9, 195–202.

Kirschner, L.B. (1953). Effect of cholinesterase inhibitors and atropine on active sodium transport across frog skin. Nature 172, 348–349.

Kitz, R.J. and Kremzner, L.T. (1968). Conformational changes of acetylcholinesterase. Mol. Pharmacol. 4, 104–107.

Kitz, R.J., Ginsburg, S. and Wilson, I.B. (1965). Activity-structure relationships in the reactivation of diethylphosphoryl acetylcholinesterase by phenyl-1-methyl pyridinium ketoximes. Biochem. Pharmacol. 14, 1471–1477.

Kitz, R.J., Ginsburg, S. and Wilson, I.B. (1967). The reaction of acetylcholinesterase with diethylphosphoryl esters of quaternary and tertiary aminophenols. Mol. Pharmacol. 3, 225–232.

Kitz, R.J., Braswell, L.M. and Ginsburg, S. (1970). On the question: Is acetylcholinesterase an allosteric protein? Mol. Pharmacol. 6, 108–121.

Kivalo, E., Rinne, U.K. and Mäkelä, S. (1958). Acetylcholinesterase, acid phosphatase, and succinic dehydrogenase after chlorpromazine administration. Experientia 14, 293–294.

Klemm, W.R. (1972). Ascending and descending excitatory influences in the brain stem reticulum. A re-examination. Brain Res. 36, 444–452.

Kling, A., Finer, S. and Nair, V. (1965). Effects of early handling and light stimulation on the acetylcholinesterase activity of the developing rat brain. Intern. J. Neuropharmacol. 4, 353–357.

Kling, A., Finer, S. and Gilmour, J. (1969). Regional development of acetylcholinesterase activity in the maternally reared and maternally deprived cat. Intern. J. Neuropharmacol. 8, 25–31.

Kling, A., Tucker, T. and Finer, S. (1970). Effects of dorsolateral frontal cortex lesions and discrimination-alteration training on regional acetylcholinesterase activity in monkey. Brain Res. 20, 401–408.

Klinge, E. and Pohto, P. (1971). Innervation of bull retractor penis muscle and peripheral autonomic mechanism of erection. In: Progress in Brain Research, Vol. 34, O. Eränkö, ed. (Elsevier, Amsterdam) pp. 415–421.

Klinge, E., Pohto, P. and Solatunturi, E. (1970). Adrenergic innervation and structure of the bull retractor penis muscle. Acta Physiol. Scand. 78, 110–116.

Klingman, G.I., Klingman, J.D. and Poliszczuk, A. (1968). Acetyl- and pseudocholinesterase activities in sympathetic ganglia of rats. J. Neurochem. 15, 1121–1130.

Knox, G.V., Campbell, C. and Lomax, P. (1973). The effects of acetylcholine and nicotine on unit activity in the hypothalamic thermoregulatory centers of the rat. Brain Res. 51, 215–224.

Kobayashi, H. and Farner, D.S. (1964). Cholinesterases in the hypothalamo-hypophysial neuro-secretory system of the white crowned sparrow, *Zonotrichia leucophrys gambelii*. Z. Zell-forsch. Mikroskop. Anat. 63, 965–973.

Kobayashi, T., Kobayashi, T., Kato, J. and Minaguchi, H. (1966). Cholinergic and adrenergic mechanisms in the female rat hypothalamus with special reference to feedback of ovarian steroid hormones. In: Steroid Dynamics, G. Pincus, T. Nakao and J.F. Tait, eds. (Academic Press, New York) pp. 303–339.

Koblick, D.C. (1958). The characterization and localization of frog skin cholinesterase. J. Gen. Physiol. 41, 1129–1134.

Koblick, D.C. (1959). An enzymatic ion exchange model for active sodium transport. J. Gen. Physiol. 42, 635–645.

Koblick, D.C., Goldman, M.H. and Pace, N. (1962). Cholinesterase and active sodium transport in frog skin. Am. J. Physiol. 203, 901–902.

Koch, H.J. (1954). Cholinesterase and active transport of sodium chloride through the isolated gills of the crab, *Eriocheir sinensis*. Recent Devel. Cell Physiol. Colston Symp. 7, 15–27.

Koelle, G.B. (1950). The histochemical differentiation of types of cholinesterases and their localizations in tissues of the cat. J. Pharmacol. Exptl. Therap. 100, 158–179.

Koelle, G.B. (1951). The elimination of enzymatic diffusion artifacts in the histochemical localization of cholinesterases and a survey of their cellular distributions. J. Pharmacol. Exptl. Therap. 103, 153–171.

Koelle, G.B. (1954). The histochemical localization of cholinesterases in the central nervous system of the rat. J. Comp. Neurol. 100, 211–235.

Koelle, G.B. (1955). The histochemical identification of acetylcholinesterase in cholinergic, adrenergic and sensory neurons. J. Pharmacol. Exptl. Therap. 114, 167–184.

Koelle, G.B. (1957a). Histochemical demonstration of reactivation of acetylcholinesterase in vivo. Science 125, 1195–1196.

Koelle, G.B. (1957b). Histochemical demonstration of reversible anticholinesterase action at selective cellular sites in vivo. J. Pharmacol. Exptl. Therap. 120, 488–503.

Koelle, G.B. (1961). A proposed dual neurohumoral role of acetylcholine: its functions at pre- and post- synaptic sites. Nature 190, 208–211.

Koelle, G.B. (1962). A new general concept of the neurohumoral functions of acetylcholine and acetylcholinesterase. J. Pharm. Pharmacol 14, 65–90.

Koelle, G.B. (1963). Cytological distributions and physiological functions of cholinesterases. In: Cholinesterases and Anticholinesterase Agents, G.B. Koelle, ed. (Springer, Berlin) pp. 187–298.

Koelle, G.B. (1969). Significance of acetylcholinesterase in central synaptic transmission. Federation Proc. 28, 95–100.

Koelle, G.B. (1970). Improvement in the accuracy of histochemical localization of acetylcho-linesterase. Facts and artifacts. In: Drugs and Cholinergic Mechanisms in the CNS, E. Heilbronn and A. Winter, eds. (Research Institute of National Defence, Stockholm) pp. 431–439.

Koelle, G.B. (1971). Current concepts of synaptic structure and function. Ann. N.Y. Acad. Sci. U.S. 183, 5–20.

Koelle, G.B. and Friedenwald, J.S. (1949). A histochemical method for localizing cholinesterase activity. Proc. Soc. Exptl. Biol. Med. 70, 617–622.

Koelle, G.B. and Friedenwald, J.S. (1950). Histochemical localization of cholinesterase in ocular tissue. Am. J. Ophthalmol. 33, 253–256.

Koelle, G.B. and Geesey, C.N. (1961). Localization of acetylcholinesterase in the neurohypophysis and its functional implications. Proc. Soc. Exptl. Biol. Med. 106, 625–628.

Koelle, G.B. and Gromadzki, C.G. (1966). Comparison of the gold-thiocholine and gold-thiolacetic acid methods for the histochemical localization of acetylcholinesterase and cholinesterases. J. Histochem. Cytochem. 14, 443–454.

Koelle, G.B. and Horn, R.S. (1968). Acetyl disulphide, $(CH_3COS)_2$ a major active component in the thiolacetic acid histochemical method for acetylcholinesterase. J. Histochem. Cytochem. 16, 743–753.

Koelle, G.B. and Steiner, E.C. (1956). The cerebral distributions of a tertiary and a quaternary anticholinesterase agent following intravenous and intraventricular injection. J. Pharmacol. Exptl. Therap. 118, 420–434.

Koelle, G.B., Koelle, E.S. and Friedenwald, J.S. (1950). The effect of inhibition of specific and non-specific cholinesterase on the motility of the isolated ileum. J. Pharmacol. Exptl. Therap. 100, 180–191.

Koelle, G.B., Wolfand, L., Friedenwald, J.S. and Allen, R.A. (1952). Localization of specific cholinesterase in ocular tissues of the cat. Am. J. Ophthalmol. 35, 1580–1584.

Koelle, G.B., Davis, R. and Devlin, M. (1968). Acetyl disulphide $(CH_3COS)_2$ and bis-(thioacetoxy) aurate (I) complex, Au $(CH_3COS)_2^-$ histochemical substrates of unusual properties with acetylcholinesterase J. Histochem. Cytochem. 16, 754–764.

Koelle, G.B., Davis, R. and Smyrl, E.G. (1971). New findings concerning the localization by electron microscopy of acetylcholinesterase in autonomic ganglia. In: Progress in Brain Research, Vol. 34, O. Eränkö, ed. (Elsevier, Amsterdam) pp. 371–375.

Koelle, G.B., Davis, R., Smyrl, E.G. and Fine, A.V. (1974). Refinement of the bis-(thioacetoxy) aurate (I) method for the electron microscopic localization of acetylcholinesterase (AChE) and non-specific cholinesterase (ChE). J. Histochem. Cytochem. 22, 252–259.

Koelle, W.A. and Koelle, G.B. (1959). The localization of external or functional acetylcholinesterase at the synapses of autonomic ganglia. J. Pharmacol. Exptl. Therap. 126, 1–8.

Koelle, W.A., Hossaini, K.S., Akbarzadeh, P. and Koelle, G.B. (1970). Histochemical evidence and consequences of the occurrence of isoenzymes of acetylcholinesterase. J. Histochem. Cytochem. 18, 812–819.

Koenig, E. (1965a). Synthetic mechanisms in the axon - I. Local axonal synthesis of acetylcholinesterase. J. Neurochem. 12, 343–355.

Koenig, E. (1965b). Synthetic mechanisms in the axon - II. RNA in myelin free axons of the cat. J. Neurochem. 12, 357–361.

Koenig, E. (1967). Synthetic mechanisms in the axon - IV. In vitro incorporation of [^{3}H] precursors into axonal protein and RNA. J. Neurochem. 14, 437–446.

Koenig, E. and Koelle, G.B. (1960). Acetylcholinesterase regeneration in peripheral nerve after irreversible inactivation. Science 132, 1249–1250.

Koenig, E. and Koelle, G.B. (1961). Mode of regeneration of acetylcholinesterase in cholinergic neurons following irreversible inactivation. J. Neurochem. 8, 169–188.

Koinov, R. and Popov, A. (1966). Importance of epinephrine-like substances and cholinesterase activity in the pathogenesis of the epileptic syndrome (in Bulgarian). Nevrolog. Psikhiatr. Nevrokhirung 5, 186–192.

Koizumi, K., Ishikawa, T. and Brooks, C.McC. (1964). Control of activity of neurons in the supraoptic nucleus. J. Neurophysiol. 27, 878–892.

Koketsu, K., Karczmar, A.G. and Kitamura, R. (1969). Acetylcholine depolarization of the dorsal root nerve terminals in the amphibian spinal cord. Intern. J. Neuropharmacol. 8, 329–336.

Kokko, A., Mautner, H.G. and Barrnett, R.J. (1969). Fine structural localization of acetylcholinesterase using acetyl-β-methylthiocholine and acetylselenocholine as substrates. J. Histochem. Cytochem. 17, 625–640.

Konishi, T. (1972). Action of tubocurarine and atropine on the crossed olivocochlear bundles. Acta Oto-laryngol. 74, 252–264.

Kooistra, G. (1950). Action of acetyl choline in intestine of *Periplaneta americana* L'. Physiol. Comparata et Oecol. 2, 75–79.

Korn, E.D. (1966). Structure of biological membranes. Science 153, 1491–1498.

Korn, E. (1969). Choline esterase activity in the tissues of the snail *Helix aspera*. Comp. Biochem. Physiol. 28, 923–929.

Koshland, D.E. (1963). Correlation of structure and function in enzyme action. Science 142, 1533–1541.

Koslow, S.H. and Giacobini, E. (1969). An isotopic micromethod for the measurement of cholinesterase activity in individual cells. J. Neurochem. 16, 1523–1528.

Kostowski, W. (1971). Effects of some cholinergic and anticholinergic drugs injected intracerebrally to the midline pontine area. Neuropharmacol. 10, 595–605.

Köver, A. and Kovács, T. (1957). On the specificity of myosincholinesterase. Acta Physiol. Acad. Sci. Hung. 11, 259–265.

Köver, A., Kovács, T. and König, T. (1957). On the properties of myosincholinesterase. Acta Physiol. Acad. Sci. Hung, 11, 253–258.

Kremzner, L.T. and Wilson, I.B. (1963). A chromatographic procedure for the purification of acetylcholinesterase. J. Biol. Chem. 238, 1714–1717.

Kremzner, L.T. and Wilson, I.B. (1964). A partial characterization of acetylcholinesterase. Biochemistry 3, 1902–1905.

Kreutzberg, G.W. (1969). Neuronal dynamics and axonal flow, IV. Blockage of intra-axonal enzyme transport by colchicine. Proc. Natl. Acad. Sci. U.S. 62, 722–728.

Kreutzberg, G.W., Schubert, P., Tóth, L. and Rieske, E. (1973). Intradendritic transport to postsynaptic sites. Brain Res. 62, 399–404.

Kreutzberg, G.W., Tóth, L., Weikert, M. and Schubert, P. (1974). Changes in perineuronal capillaries accompanying chromatolysis of motoneurons. In: The Pathology of Microcirculation, J. Cervos-Navorra ed. (De Gruyter, Berlin) pp. 282–288.

Krishnan, N. and Singer, M. (1973). Penetration of peroxidase into peripheral nerve fibers. Am. J. Anat. 136, 1–14.

Kristensson, K., Olsson, Y. and Sjöstrand, J. (1971). Axonal uptake and retrograde transport of exogenous proteins in the hypoglossal nerve. Brain Res. 32, 399–406.

Křivánek, J. and Burešová, O. (1972). Cortical acetylcholinesterase and 'handedness' in rats. Experientia 28, 291–292.

Krnjević, K. (1969). Central cholinergic pathways. Federation Proc. 28, 113–120.

Krnjević, K. and Mitchell, J.F. (1961). The release of acetylcholine in the isolated rat diaphragm. J. Physiol. London 155, 246–262.

Krnjević, K. and Phillis, J.W. (1963a). Acetylcholine-sensitive cells in the cerebral cortex. J. Physiol. London 166, 296–327.

Krnjević, K. and Phillis, J.W. (1963b). Pharmacological properties of acetylcholine-sensitive cells in the cerebral cortex. J. Physiol. London 166, 328–350.

Krnjević, K. and Silver, A. (1963). Cholinesterase staining in the cerebral cortex. J. Physiol. London 165, 3–4P.

Krnjević, K. and Silver, A. (1965). A histochemical study of cholinergic fibres in the cerebral cortex. J. Anat. 99, 711–759.

Krnjević, K. and Silver, A. (1966). Acetylcholinesterase in the developing forebrain. J. Anat. 100, 63–89.

Krnjević, K., Reiffenstein, R.J. and Silver, A. (1970). Chemical sensitivity of neurons in long-isolated slabs of cat cerebral cortex. Electroencephalog. Clin. Neurophysiol. 29, 269–282.

Krnjević, K., Pumain, R. and Renaud, L. (1971). The mechanism of excitation by acetylcholine in the cerebral cortex. J. Physiol. London 215, 247–268.

Krvavica, S., Lui, A. and Bečejac, S. (1967). Acetylcholinesterase and butyrylcholinesterase in the liver fluke (*Fasciola hepatica*). Exptl. Parasitol. 21, 240–248.

Krupka, R.M. (1963). The mechanism of action of acetylcholinesterase: substrate inhibition and the binding of inhibitors. Biochemistry 2, 76–82.

Krupka, R.M. (1964). Acetylcholinesterase: trimethylammonium-ion inhibition of deacetylation. Biochemistry 3, 1749–1754.

Krupka, R.M. (1966a). Hydrolysis of neutral substrates by acetylcholinesterase. Biochemistry 5, 1983–1988.

Krupka, R.M. (1966b). Chemical structure and function of the active center of acetylcholinesterase. Biochemistry 5, 1988–1998.

Krupka, R.M. (1967). Evidence for an intermediate in the acetylation reaction of acetylcholinesterase. Biochemistry 6, 1183–1190.

Krupka, R.M. and Laidler, K.J. (1960). The influence of pH on the rates of enzyme reactions. 6. Reactions catalyzed by acetylcholinesterase. Trans. Faraday Soc. 56, 1477–1480.

Krupka, R.M. and Laidler, K.J. (1961). Molecular mechanisms for hydrolytic enzyme action. IV. Structure of the active center and the reaction mechanism. J. Am. Chem. Soc. 83, 1458–1460.

Kuenzle, C.C., Pelloni, R.R. and Kistler, G.S. (1972). Zonal centrifugation of neuronal peri-karya and isolation of neuronal membranes rich in acetylcholinesterase. J. Neurochem. 19, 2333–2339.

Kuhnen, H. (1972). Influence of acetyl-β-methylcholine, carbamoylcholine, and bis-pyridinium compounds on the activity of acetylcholinesterase. Biochem. Pharmacol. 21, 1187–1196.

Kuno, M. and Rudomin, P. (1966). The release of acetylcholine from the spinal cord of the cat by antidromic stimulation of motor nerves. J. Physiol. London 187, 177–193.

Kunstling, T.R. and Rosse, W.F. (1969). Erythrocyte acetylcholinesterase deficiency in paroxysmal nocturnal hemoglobinuria (PNH). A comparison of the complement-sensitive and insensitive populations. Blood 33, 607–616.

Kupfer, C. (1951). Histochemistry of muscle cholinesterase after motor nerve section. J. Cell. Comp. Physiol. 38, 469–473.

Kupfer, C. and Palmer, P. (1964). Lateral geniculate nucleus: histological and cytochemical changes following afferent denervation and visual deprivation. Exptl. Neurol. 9, 400–409.

Kuwashima, S. (1957). Experimental studies on the quantitative fluctuations of acetylcholine-like substances in brain during sexual cycles. J. Jap. Obst. Gyn. Soc. 9, 135–143. (English ed. 1958, Vol. 5, 173.)

La Bella, F.S. (1968). Storage and secretion of neurohypophyseal hormones. Can. J. Physiol. Pharmacol. 46, 335–345.

La Bella, F.S. and Shin, S. (1968). Estimation of cholinesterase and choline acetyltransferase in bovine anterior pituitary, posterior pituitary, and pineal body. J. Neurochem. 15, 335–342.

Laidler, K.J. (1955a). The influence of pH on the rates of enzyme reactions. Part I, General theory. Trans. Faraday Soc. 51, 528–539.

Laidler, K.J. (1955b). The influence of pH on the rates of enzyme reactions. Part 2, The nature of the enzyme-substrate interaction. Trans. Faraday Soc. 51, 540–549.

Laidler, K.J. (1955c). The influence of the pH on the rates of enzyme reactions. Part 3, Analysis of experimental results for various enzyme systems. Trans. Faraday Soc. 51, 550–561.

Laing, A.C., Miller, H.R. and Bricknell, K.S. (1967). Purification and properties of the inducible cholinesterase of *Pseudomonas fluorescens* (Goldstein). Can. J. Biochem. 45, 1711–1724.

Lalley, P.M., Rossi, G.V. and Baker, W.W. (1970). Analysis of local cholinergic tremor mechanisms following selective neurochemical lesions. Exptl. Neurol. 27, 258–275.

La Motta, R.V., McComb, R.B. and Wetstone, H.J. (1965). Isozymes of serum cholinesterase: a new polymerization sequence. Can. J. Physiol. Pharmacol. 43, 313–318.

La Motta, R.V., McComb, R.B., Noll, C.R., Wetstone, H.J. and Reinfrank, R.F. (1968). Multiple forms of serum cholinesterase. Arch. Biochem. Biophys. 124, 299–305.

La Motta, R.V., Woronick, C.L. and Reinfrank, R.F. (1970). Multiple forms of serum cholinesterase: molecular weights of the isoenzymes. Arch. Biochem. Biophys. 136, 448–451.

Lancaster, R. (1971). Measurement of the rate of acetylcholine diffusion through a brain slice and its significance in studies of cellular distribution of acetylcholinesterase. J. Neurochem. 18, 2329–2334.

Lancaster, R. (1972). Inhibition of acetylcholinesterase in the brain and diaphragm of rats by a tertiary organophosphorus anticholinesterase and its quaternary analogue; in vivo and in vitro studies. J. Neurochem. 19, 2587–2597.

Landmesser, L. (1972). Pharmacological properties, cholinesterase activity and anatomy of nerve-muscular junctions in vagus-innervated frog sartorius. J. Physiol. London 220, 243–256.

Landolt, A.M. and Sandri, C. (1966). Cholinergische Synapsen in Oberschlundganglion der Waldameise (*Formica lugubris* Zett.). Z. Zellforsch. Mikroskop. Anat. 69, 246–259.

Lane, A.C., MacFarlane, I.R. and McCoubrey, A. (1966). Inhibition of cholinesterases by complex derivatives of morphine. Biochem. Pharmacol. 15, 122–123.

Langley, J.N. and Anderson, H.K. (1892). On the mechanism of the movements of the iris. J. Physiol. London 13, 554–597.

Lapetina, E.G., Soto, E.F. and De Robertis, E. (1967). Gangliosides and acetylcholinesterase in isolated membranes of the rat-brain cortex. Biochim. Biophys. Acta 135, 33–43.

Larrabree, M. (1967a). Phospholipid responses to physiological activity in isolated ganglia and nerve trunks. Neurosci. Res. Prog. Bull. 5, 37–40.

Larrabree, M. (1967b). Formal discussion, Neurosci. Res. Prog. Bull. 5, 43.

Lasek, R.J. (1970). Protein transport in neurons. Intern. Rev. Neurobiol. 13, 289–324.

La Torre, J.L., Lunt, G.S. and De Robertis, E. (1970). Isolation of a cholinergic proteolipid receptor from electric tissue. Proc. Natl. Acad. Sci. U.S. 65, 716–720.

Lawler, H.C. (1961). Turnover time of acetylcholine esterase. J. Biol. Chem. 236, 2296–2301.

Lawler, H.C. (1963). Purification and properties of an acetylcholinesterase polymer. J. Biol. Chem. 238, 132–137.

Lawler, H.C. (1964). The preparation of a soluble acetylcholinesterase from brain. Biochim. Biophys. Acta 81, 280–288.

Lederis, K. and Livingston, A. (1969). Acetylcholine and related enzymes in the neural lobe and anterior hypothalamus of the rabbit. J. Physiol. London 201, 695–709.

Lee, D.L. (1962). The distribution of esterase enzymes in *Ascaris lumbricoides*. Parasitol. 52, 241–260.

Lee, D.L. (1970). The fine structure of the excretory system in adult *Nippostrongylus brasiliensis* (Nematoda) and a suggested function for the 'excretory glands'. Tissue Cell 2, 225–231.

Lee, D.L., Rothman, A.H. and Senturia, J.B. (1963). Esterases in *Hymenolepis* and in *Hydatigera*. Exptl. Parasitol. 14, 285–295.

Lee, R.M. and Hodsden, M.R. (1963). Cholinesterase activity in *Haemonchus contortus* and its inhibition by organophosphorus anthelmintics. Biochem. Pharmacol. 12, 1241–1252.

Lee, R.M. and Pickering, W.R. (1967). The toxicity of Haloxon to geese, ducks and hens and its relationship to the stability of the di-(2-chloroethyl) phosphoryl cholinesterase derivatives. Biochem. Pharmacol. 16, 941–948.

Leela, K., Kanagasuntheram, R. and Ahmed, M.M. (1971). Innervation of the nasopharynx in *Macaca fascicularis*. J. Anat. 110, 49–56.

Lehrer, G.M. (1966). The localization of cholinesterase in peripheral synapses. In: Biochemistry and Pharmacology of the Basal Ganglia, E. Costa, L.J. Côté and M.D. Yahr, eds. (Raven Press, New York) pp. 89–99.

Lehrer, G.M. and Ornstein, L. (1959). A diazo coupling method for the electron microscopic localization of cholinesterase. J. Biophys. Biochem. Cytol. 6, 399–406.

Lehmann, H. and Ryan, E. (1956). The familial incidence of low pseudocholinesterase level. Lancet ii, 124.

Lehmann, H., Liddell, J., Blackwell, B., O'Connor, D.C. and Daws, A.V. (1963). Two further serum pseudocholinesterase phenotypes as causes of suxamethonium apnoea. Brit. Med. J. 1, 1116–1118.

Lentz, T.L. (1966). Histochemical localization of neurohumors in a sponge. J. Exptl. Zool. 162, 171–180.

Lentz, T.L. (1967). Fine structure of nerve cells in a planarian. J. Morphol. 121, 323-338.

Lentz, T.L. (1968). Histochemical localization of acetylcholinesterase activity in a planarian. Comp. Biochem. Physiol. 27, 715–718.

Lentz, T.L. (1971). Nerve trophic function: in vitro assay of effects of nerve tissue on muscle cholinesterase activity. Science 171, 187–189.

Lentz, T.L. and Barrnett, R.J. (1961). Enzyme histochemistry of hydra. J. Exptl. Zool. 147, 125–149.

Leonardelli, J. (1966). Recherches histoenzymologiques au niveau de l'hypothalamus du cobaye. Compt. Rend. Assoc. Anat. 135, 594–598.

Leplat, G. and Gerebtzoff, M.A. (1956). Localization de l'acétylcholinestérase, et des médiateurs diphénoliques dans la rétine. Ann. Oculist Paris 189, 121–128.

Leuzinger, W. (1969). Structure and function of acetylcholinesterase. In: Progress in Brain Research, Vol. 31, K. Akert and P.G. Waser, eds. (Elsevier, Amsterdam) pp. 241–245.

Leuzinger, W. (1971a). Studies on the subunits of acetylcholinesterase. In: Cholinergic Ligand Interactions, D.J. Triggle, J.F. Moran and E.A. Barnard, eds. (Academic Press, London) pp. 19–31.

Leuzinger, W. (1971b). The number of catalytic sites in acetylcholinesterase. Biochem. J. 123, 139–141.

Leuzinger, W. and Baker, A.L. (1967). Acetylcholinesterase, I. Large-scale purification, homogeneity, and amino acid analysis. Proc. Natl. Acad. Sci. U.S. 57, 446–451.

Leuzinger, W. and Schneider, M. (1972). Acetylcholine-induced excitation on bilayers. Experientia 28, 256–257.

Leuzinger, W., Baker, A.L. and Cauvin, E. (1968). Acetylcholinesterase, II. Crystallization, absorption spectra, isoionic point. Proc. Natl. Acad. Sci. U.S. 59, 620–623.

Leuzinger, W., Goldberg, M. and Cauvin, E. (1969). Molecular properties of acetylcholinesterase. J. Mol. Biol. 40, 217–225.

Levi-Montalcini, R. and Levi, G. (1943). Recherches quantitatives sur la marche du processus de différenciation des neurones dans les ganglions spinaux de l'embryon de poulet. Arch. Biol. Liège 54, 189–206.

Levinson, S.R. and Keynes, R.D. (1972). Isolation of acetylcholine receptors by chloroformmethanol extraction: artifacts arising in use of Sephadex LH-20 columns. Biochim. Biophys. Acta 288, 241–247.

Levitan, H. and Tauc, L. (1972). Acetylcholine receptors: topographic distribution and pharmacological properties of two receptor types on a single molluscan neurone. J. Physiol. London 222, 537–558.

Lewis, P. R. (1958). A simultaneous coupling azodye technique suitable for whole mounts. Quart. J. Microsc. Sci. 99, 67–71.

Lewis, P.R. (1961). The effect of varying the conditions in the Koelle technique. Bibliotheca Anat. 2, 11–20.

Lewis, P.R. and Flumerfelt, B.A. (1970). The pseudocholinesterase-containing neurones of the rat hypoglossal nucleus. J. Anat. 106, 189.

Lewis, P.R. and Hughes, A.F.W. (1957). The cholinesterase of developing neurones in *Xenopus laevis*. In: Metabolism of the Nervous System, D. Richter, ed. (Pergamon Press, London) pp. 511–514.

Lewis, P.R. and Shute, C.C.D. (1963). Alginate gel; an embedding medium for facilitating the cutting and handling of frozen sections. Stain Technol. 38, 307–310.

Lewis, P.R. and Shute, C.C.D. (1964). Demonstration of cholinesterase activity with the electron microscope. J. Physiol. London 175, 5–7P.

Lewis, P.R. and Shute, C.C.D. (1965a). Changes in cholinesterase activity following axotomy. J. Physiol. London 178, 25–26P.

Lewis, P.R. and Shute, C.C.D. (1965b). Fine localization of acetylcholinesterase in the optic nerve and retina of the rat. J. Physiol. London 180, 8–10P.

Lewis, P.R. and Shute, C.C.D. (1966). The distribution of cholinesterase in cholinergic neurons demonstrated with the electron microscope. J. Cell Sci. 1, 381–390.

Lewis, P.R. and Shute, C.C.D. (1967). The cholinergic limbic system. Projections to hippocampal formation, medial cortex, nuclei of the ascending cholinergic reticular system, and the subfornical organ and supra-optic crest. Brain 90, 521–540.

Lewis, P.R. and Shute, C.C.D. (1969). An electron-microscopic study of cholinesterase distribution in the rat adrenal medulla. J. Microsc. Sci. 89, 181–193.

Lewis, P.R., Shute, C.C.D. and Silver, A. (1964). Confirmation from choline acetylase analyses of a massive cholinergic innervation to the hippocampus. J. Physiol. London 172, 9–10P.

Lewis, P.R., Shute, C.C.D. and Silver, A. (1967). Confirmation from choline acetylase analyses of a massive cholinergic innervation to the rat hippocampus. J. Physiol. London 191, 215–224.

Lewis, P.R., Scott, J.A. and Navaratnam, V. (1970). Localization in the dorsal motor nucleus of the vagus in the rat. J. Anat. 107, 197–208.

Lewis, P.R., Flumerfelt, B.A. and Shute, C.C.D. (1971). The use of cholinesterase techniques to study topographical localization in the hypoglossal nucleus of the rat. J. Anat. 110, 203–213.

Liang, C.C. and Quastel, J.H. (1969). Effects of drugs on the uptake of acetylcholine in rat brain cortex slices. Biochem. Pharmacol. 18, 1187–1194.

Liddell, J., Lehmann, H. and Silk, E. (1962). A 'silent' pseudocholinesterase gene. Nature 193, 561–562.

Liddell, J., Newman, G.E. and Brown, D.F. (1963). A pseudocholinesterase variant in human tissues. Nature 198, 1090–1091.

Lim, R., Davis, G.A. and Agranoff, B.W. (1971). Electrophoretic studies on solubilized proteins of goldfish brain. Brain Res. 25, 121–131.

Limperos, G. and Ranta, K.E. (1953). A rapid screening test for the determination of the approximate cholinesterase activity of human blood. Science 117, 453–455.

Linderstrøm-Lang, K. and Glick, D. (1938). Micromethod for determination of cholinesterase activity. C.R. Trav. Lab. Carlsberg, Sér. Chim. 22, 300–306.

Lindvig, P.E., Greig, M.E. and Peterson, S.W. (1951). Studies on permeability. V. The effects of acetylcholine and physostigmine on the permeability of human erythrocytes to sodium and potassium. Arch. Biochem. Biophys. 30, 241–250.

Lineweaver, H. and Burk, D. (1934). The determination of enzyme dissociation constants. J. Am. Chem. Soc. 56, 658–666.

Ling, A.S.C. (1970). The influence of the thyroid gland on brain cholinesterase activity of mature rats. Brain Res. 22, 73–80.

Liu, H.-C. and Maneely, R.B. (1968). The development of motor end-plates in the embryonic and regenerative tail of *Hemidactylus bowringi* (Gray). Acta Anat. 71, 249–267.

Loe, P.R. and Florey, E. (1966). The distribution of acetylcholine and cholinesterase in the nervous system and in innervated organs of *Octopus dofleini*. Comp. Biochem. Physiol. 17, 509–522.

Loewi, O. and Navratil, E. (1926). Über humorale Übertragbarkeit der Herznervenwirkung. XI. Über den Mechanismus der Vaguswirkung von Physostigmin und Ergotamin. Pflügers Arch. 214, 689–696.

Lomax, P. and Jenden, D.J. (1966). Hypothermia following systematic and intracerebral injection of oxotremorine in the rat. Intern. J. Neuropharmacol. 5, 353–359.

Lomax, P., Foster, R.S. and Kirkpatrick, W.E. (1969). Cholinergic and adrenergic interactions in the thermoregulatory centers of the rat. Brain Res. 15, 431–438.

Long, J.P. (1963). Structure-activity relationships of the reversible anticholinesterase agents. In: Cholinesterases and Anticholinesterase Agents, G.B. Koelle, ed. (Springer, Berlin) pp. 374–427.

Loomis, T.A. (1963). Distribution and excretion of pyridine-2-aldoxime methiodide (PAM); atropine and PAM in Sarin poisoning. Toxicol. Appl. Pharmacol. 5, 489–499.

Loomis, T.A. and Johnson, D.D. (1966). Aging and reversal of Soman-induced effects on neuromuscular function with oximes in the presence of dimethyl sulfoxide. Toxicol. Appl. Pharmacol. 8, 533–539.

Lord, K.A. and Potter, C. (1953). Hydrolysis of esters by extracts of insects. Nature 172, 679–681.

Lorente de Nó, R. (1944). Effects of choline and acetylcholine chloride upon peripheral nerve fibers. J. Cell. Comp. Physiol. 24, 85–97.

Lorez, H.P., Kuhn, H. and Tranzer, J.P. (1973). The adrenergic innervation of the renal artery and vein of the rat. A fluorescence histochemical and electron microscopical study. Z. Zellforsch. Mikroskop. Anat. 138, 261–272.

Lubińska, L. (1961a). Demyelination and remyelination in the proximal parts of regenerating nerve fibers. J. Comp. Neurol. 117, 275–289.

Lubińska, L. (1961b). Sedentary and migratory states of Schwann cells. Exptl. Cell Res. Suppl. 8, 74–90.

Lubińska, L. (1964). Axoplasmic streaming in regenerating and in normal nerve fibres. In: Progress in Brain Research, Vol. 13, M. Singer and J.P. Schadé, eds. (Elsevier, Amsterdam) pp. 1–71.

Lubińska, L. (1974). On axoplasmic flow. Intern. Rev. Neurobiol. (in press).

Lubińska, L. and Niemierko, S. (1971). Velocity and intensity of bidirectional migration of acetylcholinesterase in transected nerves. Brain Res. 27, 329–342.

Lubińska, L. and Zelená, J. (1967). Acetylcholinesterase at muscle-tendon junctions during postnatal development in rats. J. Anat. 101, 295–308.

Lubińska, L., Niemierko, S. and Oderfeld, B. (1961). Gradient of cholinesterase activity and of choline acetylase activity in nerve fibres. Nature 189, 122–123.

Lubińska, L., Niemierko, S., Oderfeld, B. and Szwarc, L. (1963). The distribution of acetylcholinesterase in peripheral nerves. J. Neurochem. 10, 25–41.

Lubińska, L., Niemierko, S., Oderfeld-Nowak, B. and Szwarc, L. (1964). Behaviour of acetylcholinesterase in isolated nerve segments. J. Neurochem. 11, 493–503.

Lüdtke, A.H. and Ohnesorge, F.K. (1966). Charakterisierung der Cholinesterasen in verschiedenen Geweben der Schleie (*Tinca vulgaris*) und des Kaninchens. Z. Vergleich. Physiol. 52, 260–275.

Lüllmann, H., Ohnesorge, F.K., Tonner, H.D., Wassermann, O. and Ziegler, A. (1971). Influence of alkane-bis-onium compounds upon the activity of the AChE and upon its inhibition by DFP. Biochem. Pharmacol. 20, 2579–2586.

Lundin, S.J. (1962). Comparative studies of cholinesterases in body muscles of fishes. J. Cell. Comp. Physiol. 59, 93–105.

Lundin, S.J. (1968). Studies on cholinesterases in fish muscle with special regard to a cholinesterase from plaice (*Pleuronectes platessa*). FOA Report, 2, 3, 1–22.

Lundin, J. and Hellström, B. (1968). The ultrastructural localization of a cholinesterase in the body muscle of plaice (*Pleuronectes platessa*). Z. Zellforsch. Mikroskop. Anat. 85, 264–270.

Lunt, G.G., Canessa, O.M. and De Robertis, E. (1971). Association of the acetylcholine-phosphatidyl inositol effect with a 'receptor' proteolipid from cerebral cortex. Nature New Biol. 230, 187–189.

Luppa, H. and Feustel, G. (1966). Histoenzymologische Untersuchungen am caudalen neurosekretorischen System von *Cyprinus carpio* L. Acta Histochem. 25, 159–182.

Luppa, H., Weiss, J. and Feustel, G. (1968). Histochemische Untersuchungen zur Lokalisation von Acetylcholinesterase, Monoaminooxydase and Monoaminen im Kaudalen neurosekretorischen System von *Cyprinus carpio*. Z. Zellforsch. Mikroskop. Anat. 89, 499–508.

Lwoff, A. (1969). Death and transfiguration of a problem. Bacteriol. Rev. 33, 390–403.

Lynch, G.S., Lucas, P.A. and Deadwyler, S.A. (1972). The demonstration of acetylcholinesterase containing neurones within the caudate nucleus of the rat. Brain Res. 45, 617–621.

Machado, A.B.M. and Lemos, V.P.J. (1971). Histochemical evidence of a cholinergic sympathetic innervation of the rat pineal body. J. Neurovisc. Relat. 32, 104–111.

Machne, X. and Unna, K.R.W. (1963). Actions at the central nervous system. In: Cholinesterases and Anticholinesterase Agents, G.B. Koelle, ed. (Springer, Berlin) pp. 679–700.

MacIntosh, F.C. (1941). The distribution of acetylcholine in the peripheral and the central nervous system. J. Physiol. London 99, 436–442.

MacIntosh, F.C. and Oborin, P.E. (1953). Release of acetylcholine from intact cerebral cortex. XIX Intern. Physiol. Congr. Communic. pp. 580–581.

Maheshwari, U.R., Shirachi, D.Y. and Trevor, A.J. (1971). Adenosine triphosphate inhibition of ion activated microsomal acetylcholinesterase of ox caudate nucleus. Brain Res. 35, 437–445.

Maickel, R.P. (1968). Diverse central effects of chlorpromazine. Intern. J. Neuropharmacol. 7, 23–27.

Main, A.R. (1969). Kinetic evidence of multiple reversible cholinesterases based on inhibition by organophosphates. J. Biol. Chem. 244, 829–840.

Majno, G. and Karnovsky, M.L. (1961). A biochemical and morphologic study of myelination and demyelination. III. Effect of an organo-phosphorus compound (Mipafox) on the biosynthesis of lipid by nervous tissue of rats and hens. J. Neurochem. 8, 1–16.

Maletta, G.J. and Timiras, P.S. (1966). Acetyl- and butyrylcholinesterase activity of selected brain areas in developing rats after neonatal X-irradiation. J. Neurochem. 13, 75–84.

Maletta, G.J. and Timiras, P.S. (1967). Acetylcholinesterase activity in optic structures after complete light deprivation from birth. Exptl. Neurol. 19, 513–518.

Maletta, G.J. and Timiras, P.S. (1968). Choline acetyltransferase activity and total protein content in selected optic areas of the rat after complete light-deprivation during CNS development. J. Neurochem. 15, 787–793.

Maletta, G.J., Vernadakis, A. and Timiras, P.S. (1967). Acetylcholinesterase activity and protein content of brain and spinal cord in developing rats after pre-natal X-irradiation. J. Neurochem. 14, 647–652.

Malmgren, H. and Sylvén, B. (1955). On the chemistry of the thiocholine method of Koelle. J. Histochem. Cytochem. 3, 441–445.

Malmström, B.G., Levin, Ö. and Boman, H.G. (1956). Chromatography of human serum cholinesterase. Acta Chem. Scand. 10, 1077–1082.

Mandel'shtam, Yu-E. (1967). Penetration of anticholinesterase compounds into the nervous system of the Asiatic locust (in Russian). Byull. Eksp. Biol. Med. 63, 633–634.

Mann, K.H. (1962). Leeches (Hirudinea). Their Structure, Physiology, Ecology and Embryology (Pergamon Press, Oxford).

Mann, S.P. (1971). The innervation of mammalian bronchial smooth muscle: the localization of catecholamines and cholinesterases. Histochem. J. 3, 319–331.

Mann, T. (1954). The Biochemistry of Semen (Methuen, London) p. 174.

Mann, T. (1964). The Biochemistry of Semen and of the Male Reproductive Tract (Methuen, London) p. 208.

Manocha, S.L. and Shantha, T.R. (1970). *Macaca mulatta*. Enzyme Histochemistry of the Nervous System (Academic Press, New York).

Mansingh, A. (1967). The cholinergic system during diapause and development of *Rothschildia orizaba*. Indian J. Entomol. 29, 131–134.

Manukhin, B.N. and Buznikov, G.A. (1965). The physiological role of chemical transmitters in ontogenesis. In: Essays on Physiological Evolution, J.W.S. Pringle, ed. (Pergamon Press, Oxford) pp. 190–199.

Marchisio, P.C. and Consolo, S. (1968). Developmental changes of choline acetyltransferase (ChAc) activity in chick embryo spinal and sympathetic ganglia. J. Neurochem. 15, 759–764.

Markert, C.L. (1965). Epigenetic control of specific protein synthesis in differentiating cells. In: Molecular and Cellular Aspects of Development, E. Bell, ed. (Harper and Row, New York) pp. 267–280.

Markert, C.L. (1968). The molecular basis for isozymes. Ann. N.Y. Acad. Sci. 151, 14–40.

Markert, C.L. and Hunter, R.L. (1959). The distribution of esterases in mouse tissue. J. Histochem. Cytochem. 4, 42–49.

Marnay, A. and Nachmansohn, D. (1937a). Cholinesterase in voluntary frog's muscle. J. Physiol. London 89, 359–367.

Marnay, A. and Nachmansohn, D. (1937b). Cholinestérase dans le muscle strié. Compt. Rend. Soc. Biol. 124, 942–944.

Marnay, A. and Nachmansohn, D. (1937c). Cholinestérase dans le nerf de homard. Compt. Rend. Soc. Biol. 125, 1005–1006.

Marnay, A. and Nachmansohn, D. (1938). Choline esterase in voluntary muscle. J. Physiol. London 92, 37–47.

Marshall, W.H. (1940). An application of the frozen sectioning technic for cutting serial sections thru the brain. Stain Technol. 15, 133–138.

Martin, K. (1970). The effect of proteolytic enzymes on acetylcholinesterase activity, the sodium pump and choline transport in human erythrocytes. Biochim. Biophys. Acta 203, 182–184.

Martinez, A.J. and Friede, R.L. (1970). Accumulation of axoplasmic organelles in swollen nerve fibres. Brain Res. 19, 183–198.

Marx, G.L. and Carter, M.K. (1963). Histochemical and manometric studies of cholinesterases in mammalian kidney tissue. Am. J. Physiol. 204, 124–128.

Massoulié, J. and Rieger, F. (1969). L'acétylcholinestérase des organes électriques de poissons (torpille et gymnote); complexes membranaires. Eur. J. Biochem. 11, 441–455.

Massoulié, J., Rieger, F. and Bon, S. (1970a). Relations entre les complexes moléculaires de l'acétylcholinestérase. Compt. Rend. Acad. Sci. Paris 270 1837–1840.

Massoulié, J., Rieger, F. and Tsuji, S. (1970b). Solubilisation de l'acétylcholinestérase des organes électriques de gymnote. Action de la trypsine. Eur. J. Biochem. 14, 430–439.

Masters, C.J. and Holmes, R.S. (1972). Isoenzymes and ontogeny. Biol. Rev. 47, 309–361.

Matsuda, H. (1970). Ultrastructural localization of cholinesterase activity in the nervous system of the rabbit iris dilator. Jap. J. Ophthalmol. 14, 21–28.

Matsuda, T., Saito, K., Katsuki, S., Hata, F. and Yoshida, H. (1971). Studies on soluble proteins released from the synaptic vesicles of rat brain cortex. J. Neurochem. 18, 713–719.

Matsuura, H. (1967). Histochemical observation of bovine spinal ganglia. Histochemie 11, 152–160.

Mayer, O. and Michalek, H. (1971). Effects of DFP and Obidoxime on brain acetylcholine levels and on brain and peripheral cholinesterases. Biochem. Pharmacol. 20, 3029–3037.

Maynard, D.M. (1967). Organization of central ganglia. In: Invertebrate Nervous Systems, C.A.G. Wiersma, ed. (Univ. Chicago Press, Chicago) pp. 231–255.

Maynard, E.A. (1964). Esterases in crustacean nervous system. 1, Electrophoretic studies in lobsters. J. Exptl. Zool. 157, 251–266.

Maynard, E.A. (1966). Electrophoretic studies of cholinesterases in brain and muscle of the developing chicken. J. Exptl. Zool. 161, 319–336.

Maynard, E.A. and Maynard, D.M. (1960). Cholinesterase in the crustacean muscle receptor organ. J. Histochem. Cytochem. 8, 376–379.

Mazza, J.P., Hanker, J.S. and Dixon, A.D. (1973). Ultrastructural localization of cholinesterase activity in the trigeminal ganglion of the rat. J. Anat. 115, 65–78.

McCaman, R.E. and Dewhurst, S.A. (1970). Choline acetyltransferase in individual neurons of *Aplysia californica*. J. Neurochem. 17, 1421–1426.

McCaman, R.E. and Dewhurst, S.A. (1971). Metabolism of putative transmitters in individual neurons of *Aplysia californica*. Acetylcholinesterase and catechol-O-methyl transferase. J. Neurochem. 18, 1329–1335.

McCaman, R.E., Rodríguez de Lores Arnaiz, G. and De Robertis, E. (1965). Species differences in subcellular distribution of choline acetylase in the CNS. A study of choline acetylase, acetylcholinesterase, 5-hydroxytryptophan decarboxylase and monoamine oxidase in four species. J. Neurochem. 12, 927–935.

McCaman, M.W., Tomey, L.R. and McCaman, R.E. (1968). Radiometric assay of acetylcholinesterase activity in submicrogram amounts of tissue. Life Sci. Oxford 7, 233–244.

McCaman, R.E., Weinreich, D. and Borys, H. (1973). Endogenous levels of acetylcholine and choline in individual neurons of *Aplysia* J. Neurochem. 21, 473–476.

McCance, I. and Phillis, J.W. (1964a). The action of acetylcholine on cells in cat cerebellar cortex. Experientia 20, 217–218.

McCance, I. and Phillis, J.W. (1964b). Discharge patterns of elements in cat cerebellar cortex and their responses to iontophoretically applied drugs. Nature 204, 844–846.

McCance, I. and Phillis, J.W. (1968). Cholinergic mechanisms in the cerebellar cortex. Intern. J. Neuropharmacol. 7, 447–462.

McCance, I., Phillis, J.W. and Westerman, R.A. (1968a). Acetylcholine-sensitivity of thalamic neurones: its relationship to synaptic transmission. Brit. J. Pharmacol. 32, 635–651.

McCance, I., Phillis, J.W. and Westerman, R.A. (1968b). Physiological and pharmacological studies on the cerebellar projections to the feline thalamus. Exptl. Neurol. 21, 257–269.

McCance, R.A., Hutchinson, A.O., Dean, R.F.A. and Jones, P.E.H. (1949). The cholinesterase activity of the serum of newborn animals, and of colostrum. Biochem. J. 45, 493–496.

McCance, R.A., Brown, L.M., Comline, R.S. and Titchen, D.A. (1951). Cholinesterase activity in the secretions of the pancreas of the dog and parotid of the pig. Nature 168, 788–789.

McDonald, D.M. and Rasmussen, G.L. (1971). Ultrastructural characteristics of synaptic endings in cochlear nucleus having acetylcholinesterase activity. Brain Res. 28, 1–18.

McEnroe, W.D. (1971). Phenylthioacetate as a stain for cholinesterase. Science 171, 928.

McEwen, B.S. and Grafstein, B. (1968). Fast and slow components in axonal transport of protein. J. Cell Biol. 38, 494–508.

McEwen, B.S., Forman, D.S. and Grafstein, B. (1971). Components of fast and slow axonal transport in the goldfish optic nerve. J. Neurobiol. 2, 361–377.

McGeer, E.G., Ikeda, H., Asakura, T. and Wada, J.A. (1969). Lack of abnormality in brain aromatic amines in rats and mice susceptible to audiogenic seizure. J. Neurochem. 16, 945–950.

McGeer, E.G., Fibiger, H.C. and Wickson, V. (1971a). Differential development of caudate enzymes in the neonatal rat. Brain Res. 32, 433–440.

McGeer, P.L., McGeer, E.G., Fibiger, H.C. and Wickson, V. (1971b). Neostriatal choline ace-tylase and cholinesterase following selective brain lesions. Brain Res. 35, 308–314.

McGeer, E.G., Fibiger, H.C., McGeer, P.L. and Brooke, S. (1973). Temporal changes in amine synthesizing enzymes of rat extrapyramidal structures after hemitransections or 6-hydroxy-dopamine administration. Brain Res. 52, 289–300.

McIsaac, R.J. and Koelle, G.B. (1959). Comparison of the effects of inhibition of external, internal and total acetylcholinesterase upon ganglionic transmission. J. Pharmacol. Exp. Therap. 126, 9–20.

McKerracher, D.W., McGuire, W.A., Aronson, A. and Scott, J. (1966). Pseudocholinesterase and the prediction of stability in subnormal and psychopathic offenders. Brit. J. Psychiat. 112, 717–722.

McKinstry, D.N. and Koelle, G.B. (1967). Acetylcholine release from the cat superior cervical ganglion by carbachol. J. Pharmacol. Exp. Therap. 157, 319–327.

McLennan, H. (1954). Acetylcholine metabolism of normal and axotomized ganglia. J. Physiol., London 124, 113–116.

McLennan, H. (1964). The release of acetylcholine and of 3-hydroxytryptamine from the caudate nucleus. J. Physiol., London 174, 152–161.

McLennan, H. (1969). Cholinesterase in the feline red nucleus. Intern. J. Neuropharmacol. 8, 489–490.

McLennan, H. (1970). Synaptic Transmission, 2nd edn. (W.B. Saunders Co., Philadelphia).

McLennan, H. and York, D.H. (1966). Cholinergic mechanisms in the caudate nucleus. J. Physiol. London 187, 163–175.

McLennan, H. and York, D.H. (1967). The action of dopamine on neurones of the caudate nucleus. J. Physiol. London 189, 393–402.

McOsker, D.E. and Daniel, L.J. (1959). A colorimetric micro method for the determination of cholinesterase. Arch. Biochem. Biophys. 79, 1–7.

Mednick, M.L., Petrali, J.P., Thomas, N.C., Sternberger, L.A., Plapinger, R.E., Davis, D.A., Wasserkrug, H.L. and Seligman, A.M. (1971). Localization of acetylcholinesterase via pro-duction of osmiophilic polymers: new benzenediazonium salts with thiolacetate functions. J. Histochem. Cytochem. 19, 155–160.

Megazzini, P., Bernardi, G. and Ballotti, P.L. (1965). Regional changes of the cholinergic system in the guinea-pig's brain after physostigmine. Experientia 21, 406–408.

Méhes, I. and Décsi, L. (1958). Diminution de l'activité acétylcholinestérasique dans la vieil-lesse. Compt. Rend. Soc. Biol. 152, 688–689.

Mehrotra, K.N. and Dauterman, W.C. (1963). The specificity of rat brain acetylcholinesterase for N-alkyl analogues of acetylcholine. J. Neurochem. 10, 119–123.

Meizel, S., Boggs, D. and Cotham, J. (1971). Electrophoretic studies of esterases of bull sperma-tozoa, cytoplamic droplets and seminal plasma. J. Histochem. Cytochem. 19, 226–231.

Mellgren, S.I. (1973). Distribution of acetylcholinesterase in the hippocampal region of the rat during postnatal development. Z. Zellforsch. Mikroskop. Anat. 141, 375–400.

Mellgren, S.I. and Srebro, B. (1973). Changes in acetylcholinesterase and distribution of degen-erating fibres in the hippocampal region after septal lesions in the rat. Brain Res. 52, 19–36.

Mendel, B. and Myers, D.K. (1955). Identification of pseudocholinesterase in the tissues of ruminants. Nature 176, 783–784.

Mendel, B. and Rudney, H. (1943a). Studies on cholinesterase. 1. Cholinesterase and pseudo-cholinesterase. Biochem. J. 37, 59–63.

Mendel, B. and Rudney, H. (1943b). On the type of cholinesterase present in brain tissue. Science 98, 201–202.

Mendel, B. and Rudney, H. (1945). Some effects of salts on true cholinesterase. Science 102, 616–617.

Mendel, B., Mundell, D.B. and Rudney, H. (1943). Studies on cholinesterase 3. Specific tests for true cholinesterase and pseudo-cholinesterase. Biochem. J. 37, 473–476.

Mengle, D.C. and O'Brien, R.D. (1960). The spontaneous and induced recovery of fly-brain cholinesterase after inhibition by organophosphates. Biochem. J. 75, 201–207.

Menn, J.J. and McBain, J.B. (1968). Possible occurrence of natural cholinesterase inhibitors in the German cockroach, *Blattella germanica*. Ann. Entomol. Soc. Am. 61, 1578–1580.

Merck Index of Chemicals and Drugs (1968). P.G. Stecher ed. (Merck and Co. Inc., Rahway, N.J.).

Mészáros, T., Csuri, I.J., Házas, J. and Palkovits, M. (1969). Esterase activity in the hypothala-mus. Acta Morphol. Acad. Sci. Hung. 17, 201–215.

Metcalf, D.R. and Holmes, J.H. (1969). EEG, psychological and neurological alterations in humans with organophosphorus exposure. Ann. N.Y. Acad. Sci. 160, 357–365.

Metcalf, R.L. (1955). Organic Insecticides. Their Chemistry and Mode of Action (Interscience Publishers, New York).

Metcalf, R.L., Maxon, M., Fukuto, T.R., and March, R.B. (1956). Aromatic esterase in insects. Ann. Entomol. Soc. Am. 49, 274–279.

Metz, B. (1958). Brain acetylcholinesterase and a respiratory reflex. Am. J. Physiol. 192, 101–105.

Metzler, C.J. and Humm, D.G. (1951). The determination of cholinesterase activity in whole brains of developing rats. Science 113, 382–383.

Meynert, T. (1867). Der Bau der Grosshirnrinde und seine Örtlichen Verschiedenheiten nebst einem pathologisch-anatomischen Corallarium. Vierteljahrsschr.-Psychiat. pp. 198–252.

Michel, H.O. (1949). An electrometric method for the determination of red blood cell and plasma cholinesterase activity. J. Lab. Clin. Med. 34, 1564–1568.

Michel, H.O. and Krop, S. (1951). The reaction of cholinesterase with diisopropyl fluorophos-phate. J. Biol. Chem. 190, 119–125.

Miledi, R. (1964). Electron-microscopical localization of products from histochemical reactions used to detect cholinesterase in muscle. Nature 204, 293–295.

Miledi, R., Molinoff, P. and Potter, L.T. (1971). Isolation of the cholinergic receptor protein of *Torpedo* electric tissue. Nature 229, 554–557.

Millar, D.B. and Grafius, M.A. (1970). The subunit molecular weight of acetylcholinesterase. FEBS Letters 12, 61–64.

Miller, E., Heller, A. and Moore, R.Y. (1969a). Acetylcholine in rabbit visual system nuclei after enucleation and visual cortex ablation. J. Pharmacol. Exptl. Therap. 165, 117–125.

Miller, E., Reinwall, J., Brouwer J., Earl, F.L. and Curtis, J.M. (1969b). Regional distribution of cholinesterase in the central nervous system of miniature swine. Am. J. Vet. Res. 30, 2037–2039.

Mitchell, J.F. (1963). The spontaneous and evoked release of acetylcholine from the cerebral cortex. J. Physiol. London 165, 98–116.

Mitchell, J.F. and Phillis, J.W. (1962). Cholinergic transmission in the frog spinal cord. Brit. J. Pharmacol. 19, 534–543.

Mitchell, J.F. and Silver, A. (1963). The spontaneous release of acetylcholine from the denervated hemidiaphragm of the rat. J. Physiol. London 165, 117–129.

Mitchell, J.F. and Szerb, J.C. (1962). The spontaneous and evoked release of acetylcholine from the caudate nucleus. XXII Intern. Physiol. Congr. Communic. no. 819.

Mitropolitanskaya, R.L. (1941). On the presence of acetylcholine and cholinesterase in protozoa, spongia and coelenterata. C.R. (Doklady) Acad. Sci. USSR. 31, 717–718.

Mittag, T.W., Ehrenpreis, S., Detwiler, P. and Boyle, R. (1971a). Some properties of cholinesterases in intact rat diaphragm in vitro. Arch. Intern. Pharmacodyn. 191, 261–269.

Mittag, T.W., Ehrenpreis, S. and Patrick, P. (1971b). Some properties of cholinesterases in intact guinea-pig ileum in vitro. Arch. Intern. Pharmacodyn. 191, 270–278.

Mittag, T.W., Ehrenpreis, S. and Hehir, R.M. (1971c). Functional acetylcholinesterase of rat diaphragm muscle. Biochem. Pharmacol. 20, 2263–2273.

Mohamed, A.H., Kamel, A. and Ayobe, M.H. (1969). Enzymic activities of Egyptian snake venoms and a scorpion venom. Toxicon 7, 185–188.

Monod, J., Changeux, J.-P. and Jacob, F. (1963). Allosteric proteins and cellular control systems. J. Mol. Biol. 6, 306–329.

Monod, J., Wyman, J. and Changeux, J.-P. (1965). On the nature of allosteric transitions: a plausible model. J. Mol. Biol. 12, 88–118.

Montanini, I. and Porcellati, G. (1964). Protein metabolism of peripheral nerves during demyelination by organophosphorus compounds. Ital. J. Biochem. 13, 230–239.

Moore, E.J. and Petty, C.S. (1958). A note on cholinesterase activity in post-mortem tissues. J. Histochem. Cytochem. 6, 377–379.

Moore, R.A. (1929). Gelatin carmine injections. J. Tech. Meth. Bull. Intern. Assoc. Med. Museums 12, 55–58.

Morest, D.K. (1960). A study of the structure of the area postrema with Golgi methods. Am. J. Anat. 107, 291–303.

Morgan, I.G. and Austin, L. (1968). Synaptosomal protein synthesis in a cell-free system. J. Neurochem. 15, 41–51.

Morgan, I.G., Zanetta, J.-P., Breckenridge, W.C., Vincendon, G. and Gombos, G. (1973). The chemical structure of synaptic membranes. Brain Res. 62, 405–411.

Mori, S., Maeda, T. and Shimizu, N. (1964). Electron-microscopic histochemistry of cholinesterases in the rat brain. Histochemie 4, 65–72.

Morris, D., Maneckjee, A. and Hebb, C. (1971). The kinetic properties of human placental choline acetyltransferase. Biochem. J. 125, 857–863.

Morrison, J.F. (1965). Kinetic methods for the determination of enzyme reaction mechanisms. Aust. J. Biol. Sci. 27, 317–327.

Motoyama, N. and Saito, T. (1968). Substrate specificity of cholinesterases in mites. Bochu-Kagaku 33, 77–80.

Motulsky, A.G. and Morrow, A. (1968). Atypical cholinesterase gene $E_1{}^a$: rarity in Negroes and most orientals. Science 159, 202–203.

Mounter, L.A. and Whittaker, V.P. (1950). The esterases of horse blood. 2. The specificity of horse erythrocyte cholinesterase. Biochem. J. 47, 525–530.

Mounter, L.A. and Whittaker, V.P. (1953). The effect of thiol and other group-specific reagents on erythrocyte and plasma cholinesterases. Biochem. J. 53, 167–173.

Mršulja, B., Terzić, M., and Varagić, V.M. (1968). The effect of physostigmine and neostigmine on the concentration of glycogen in various brain structures of the rat. J. Neurochem. 15, 1329–1333.

Müller-Eberhard, H.J. (1972). The molecular basis of the biological activities of complement. Harvey Lectures 66, 75–104.

Mumenthaler, M. and Engel, W.K. (1961). Cytological localization of cholinesterase in developing chick embryo skeletal muscle. Acta Anat. 47, 274–299.

Mundell, D.B. (1944). Plasma cholinesterase in male and female rats. Nature 153, 557–558.

Murphy, S.D. (1967). Malathion inhibition of esterases as a determinant of malathion toxicity. J. Pharmacol. Exptl. Therap. 156, 352–365.

Mustea, I. (1969). Modifications de l'activité de la cholinestérase sérique dans le diagnostic différentiel des processus néoplastiques. Clin. Chim. Acta 24, 453–456.

Myers, D.K. (1953). Studies on cholinesterase. 9. Species variation in the specificity pattern of the pseudo cholinesterases. Biochem. J. 55, 67–79.

Myers, R.D. (1970). The role of hypothalamic transmitter factors in the control of body temperature. In: Physiological and Behavioral Temperature Regulation, J.D. Hardy, A.P. Gagge, and J.A. Stolwijk, eds. (Thomas, Springfield, Illinois), pp. 648–666.

Myers, R.D. and Yaksh, T.L. (1969). Control of temperature in the unanaesthetized monkey by cholinergic and aminergic systems in the hypothalamus. J. Physiol. London 202, 483–500.

Nachlas, M.M. and Seligman, A.M. (1949a). Histochemical demonstration of esterase. J. Natl. Cancer Inst. 9, 415–425.

Nachlas, M.M. and Seligman, A.M. (1949b). Evidence for the specificity of esterase and lipase by the use of three chromogenic substrates. J. Biol. Chem. 181, 343–355.

Nachmansohn, D. (1939). Cholinestérase dans le système nerveux central. Bull. Soc. Chim. Biol. 21, 761–796.

Nachmanshon, D. (1940a). Action of ions on choline esterase. Nature 145, 513–514.

Nachmansohn, D. (1940b). Choline esterase in brain and spinal cord of sheep embryos. J. Neurophysiol. 3, 396–402.

Nachmansohn, D. (1946). Chemical mechanism of nerve action. Ann. N.Y. Acad. Sci. 47, 395–425.

Nachmansohn, D. (1950). Studies on permeability in relation to nerve function. I. Axonal conduction and synaptic transmission. Biochim. Biophys. Acta 4, 78–95.

Nachmansohn, D. (1959). Chemical and Molecular Basis of Nerve Activity (Academic Press, New York).

Nachmansohn, D. (1962). Chemical and molecular basis of nerve activity. In: Neurochemistry, K.A.C. Elliott, I.H. Page and J.H. Quastel, eds. (Thomas, Springfield, Illinois), pp. 522–557.

Nachmansohn, D. (1963). Choline acetylase. In: Cholinesterases and Anticholinesterase Agents, G.B. Koelle, ed. (Springer, Berlin), pp. 40–54.

Nachmansohn, D. (1964). Chemical control of ion movements across conducting membranes. In: New Perspectives in Biology, M. Sela, ed. (Elsevier, Amsterdam) pp. 176–204.

Nachmansohn, D. (1970). Proteins in excitable membranes. Science 168, 1059–1066.

Nachmansohn, D. (1972). Biochemistry as part of my life. Ann. Rev. Biochem. 41, 1–28.

Nachmansohn, D. and Rothenberg, M.A. (1944). On the specificity of choline esterase in nervous tissue. Science 100, 454–455.

Nachmansohn, D. and Wilson, I.B. (1951). The enzymic hydrolysis and synthesis of acetylcholine. Advan. Enzymol. 12, 259–339.

Nachmansohn, D., Coates, C.W. and Cox, R.T. (1941). Electric potential and activity of choline esterase in the electric organ of *Electrophorus electricus* (Linnaeus). J. Gen. Physiol. 25, 75–88.

Nadol, J.B., Brzin, M. and De Lorenzo, A.J.D. (1970). Fine structural localization of acetylcholinesterase in sensory and motor neurons of the muscle receptor organ in *Homarus*. J. Comp. Neurol. 140, 399–420.

Naik, N.T. (1963). Technical variations in Koelle's histochemical method for demonstrating cholinesterase activity. Quart. J. Microsc. Sci. 104, 89–100.

Naik, N.T. and Cauna, N. (1971). Histochemical observations on cholinesterase activity in the autonomic ganglia of the human sympathetic trunk and vagus nerve. Histochem. J. 3, 47–53.

Nakajima, H. and Hatano, S. (1962). Acetylcholinesterase in the plasmodium of the myxomycete, *Physarum polycephalum*. J. Cell. Comp. Physiol. 59, 259–263.

Nakata, T. and Nishijima, S. (1971). Combined staining by thiolacetic acid and Bielschowsky methods. Stain Technol. 46, 151–153.

Namba, T. (1971). Cholinesterase activity of muscle fibers and motor end plates: comparative studies. Exptl. Neurol. 33, 322–328.

Namba, T. and Grob, D. (1967). Cholinergic receptors in skeletal muscle: isolation and properties of muscle ribonucleoprotein with affinity for d-tubocurarine and acetylcholine, and binding activity of the subneural apparatus of motor end plates with divalent metal ions. Ann. N.Y. Acad. Sci. 144, 772–800.

Namba, T. and Grob, D. (1968). Cholinesterase activity of the motor end plate in isolated muscle membrane. J. Neurochem. 15, 1445–1454.

Namba, T. and Grob, D. (1970). Cholinesterase activity of motor end plate in human skeletal muscle. J. Clin. Invest. 49, 936–942.

Namba, T., Nakamura, T. and Grob, D. (1967). Staining for nerve fiber and cholinesterase activity in fresh frozen sections. Am. J. Clin. Pathol. 47, 74–77.

Nandy, K. and Bourne, G.H. (1964). The effects of D-lysergic acid diethylamide tartrate (LSD - 25) on the cholinesterases and monoamine oxidase in the spinal cord: a possible factor in the mechanism of hallucination. J. Neurol. Neurosurg. Psychiatry 27, 259–267.

Nastuk, W.L. and Plescia, O.J. (1966). Current status of research on myasthenia gravis. Ann. N.Y. Acad. Sci. 135, 664–678.

Nathan, P. and Aprison, M.H. (1955). Cholinesterase activity in cytoplasmic particles from rabbit brain. Federation Proc. 14, 106–107.

Natoff, I.L. and Reiff, B. (1970). Quantitative studies of the effect of antagonists on the acute toxicity of organophosphates in rats. Brit. J. Pharmacol. 40, 124–134.

Navaratnam, V. (1965). The ontogenesis of cholinesterase activity within the heart and cardiac ganglia in man, rat, rabbit and guinea-pig. J. Anat. 99, 459–467.

Navaratnam, V. and Lewis, P.R. (1970). Cholinesterase-containing neurones in the spinal cord of the rat. Brain Res. 18, 411–425.

Navaratnam, V. and Palkama, A. (1966). Cholinesterases in the walls of the great arterial trunks and coronary arteries. Acta Anat. 63, 445–448.

Navaratnam, V., Lewis, P.R. and Shute, C.C.D. (1964). Effects of vagotomy on the cholinesterase content of the preganglionic innervation of the heart. J. Anat. 98, 287.

Navaratnam, V., Lewis, P.R. and Shute, C.C.D. (1968). Effects of vagotomy on the cholinesterase content of the preganglionic innervation of the rat heart. J. Anat. 103, 225–232.

Neal, R.A. and DuBois, K.P. (1965). Studies on the mechanism of detoxification of cholinergic phosphorothioates. J. Pharmacol. Exptl. Therap. 148, 185–192.

Needham, J. (1925). The metabolism of the developing egg. Physiol. Rev. 5, 1–62.

Neil, E. and O'Regan, R.G. (1969). Effects of sinus and aortic nerve efferents on arterial chemoreceptor functions. J. Physiol. London 200, 69–71P.

Nelson, L. (1964). Acetylcholinesterase in bull spermatozoa. J. Reprod. Fertility 7, 65–71.

Nelson, W.L. and Barnum, C.P. (1960). The effect of diisopropylphosphorofluoridate (DFP) on mouse brain phosphorus metabolism. J. Neurochem. 6, 43–50.

Neumann, S. and Walter, H. (1968). Frequencies of pseudocholinesterase variants in Icelanders, Greeks and Pakistanis. Nature 219, 950.

Newman, G., Kerkut, G.A. and Walker, R.J. (1968). The structure of the brain of *Helix aspera*. Electron microscope localization of cholinesterase and amines. Symp. Zool. Soc. London 22, 1–17.

Newton, L., Walker, R.J. and Woodruff, G.N. (1970). The effect of nicotinic and muscarinic agonists on single neurones of the leech, *Hirudo medicinalis*. J. Physiol. London 210, 54–55P.

Nichols, C.W. and Koelle, G.B. (1968). Comparison of the localization of acetylcholinesterase and non-specific cholinesterase activities in mammalian and avian retinas. J. Comp. Neurol. 133, 1–16.

Nichols, C.W., Hewitt, J. and Laties, A.M. (1972). Localization of acetylcholinesterase in the teleost retina. J. Histochem. Cytochem. 20, 130–136.

Nicholas, F., Madec, Y., Dubin, J.-C., Bainvel, C. and Bernard, S. (1967). Sur les variations de l'activité pseudo-cholinestérase plasmatique au cours de l'évolution du tétanos. Ann. Biol. Clin. Paris 25, 453–460.

Nomura, Y. and Schuknecht, H.F. (1965). The efferent fibers in the cochlea. Ann. Oto-laryngol. 74, 289–302.

Nordenfelt, I. (1965). Acetylcholine synthesis in salivary glands. Thesis: University of Lund.

Norris, C. and Guth, P. (1974). The release of acetylcholine (ACh) by the crossed olivocochlear (COCB) bundle. Acta Oto-laryngol. (in press)

Norvell, J.E., Harris, T.M. and Weitsen, H.A. (1971). The use of dark-field microscopy for the visualization of acetylcholinesterase in cholinergic neurons. Stain Technol. 46, 19–21.

Novikoff, A.B., Quintana, N., Villaverde, H. and Forschirm, R. (1966). Nucleoside phosphatase and cholinesterase activities in dorsal root ganglia and peripheral nerve. J. Cell Biol. 29, 525–545.

Nyberg-Hansen, R., Rinvik, E., Aarseth, P. and Barstad, J.A.B. (1969). Electronmicroscopic localization of cholinesterase at the neuromuscular junction by a quaternary carbon analogue of acetylthiocholine as substrate. Histochemie 20, 40–45.

O'Brien, R.D. (1953). Occurrence of cholinesterase in *Tenebrio* and *Tribolium*. Nature 172, 162–163.

O'Brien, R.D. (1963). Binding of organophosphates to cholinesterases. J. Agr. Food. Chem. 11, 163–166.

O'Brien, R.D. (1967). Insecticides: Action and Metabolism (Academic Press, New York).

O'Brien, R.D. (1968). Kinetics of the carbamylation of cholinesterase. Mol. Pharmacol. 4, 121–130.

O'Brien, R.D. (1969). Phosphorylation and carbamylation of cholinesterase. Ann. N.Y. Acad. Sci. 160, 204–214.

O'Brien, R.D., Eldefrawi, M.E., Eldefrawi, A.T. and Farrow, J.T. (1971). Binding of cholinergic ligands to electroplaxes and brain tissue. In: Cholinergic Ligand Interactions, D.J. Triggle, J.F. Moran and E.A. Barnard, eds. (Academic Press, New York) pp. 49–66.

O'Brien, R.D., Eldefrawi, M.E. and Eldefrawi, A.T. (1972). Isolation of acetylcholine receptors. Ann. Rev. Pharmacol. 12, 19–34.

Ochs, S. (1971). Characteristics and a model for fast axoplasmic transport in nerve. J. Neurobiol. 2, 331–345.

Ochs, S. (1972a). Fast transport of materials in mammalian nerve fibers. A fast transport mechanism for materials exists in nerve fibers which depends on oxidative metabolism. Science 176, 252–260.

Ochs, S. (1972b). Rate of fast axoplasmic transport in mammalian nerve fibres. J. Physiol. London 227, 627–645.

Ochs, S. and Smith, C.B. (1971). Fast axoplasmic transport in mammalian nerve in vitro after block of glycolysis with iodoacetic acid. J. Neurochem. 18, 833–843.

Odutola, A.B. (1970). The topographical localization of acetylcholinesterase in the adult rat cerebellum: a reappraisal. Histochemie 23, 98–106.

Odutola, A.B. (1972). The organization of cholinesterase-containing systems of the monkey spinal cord. Brain Res. 39, 353–368.

Ogston, A.G. (1955). Removal of acetylcholine from a limited volume by diffusion. J. Physiol. London 128, 222–223.

Oh, T.H. and Johnson, D.H. (1972). Effects of acetyl-β-methylcholine on development of acetylcholinesterase and butyrylcholinesterase activities in cultured chick embryonic skeletal muscle. Exptl. Neurol. 37, 360–370.

Okamoto, K., Tabei, R., Nosaka, S., Fukushima, A., Yamori, Y., Matsumoto, M., Yamabe, H., Morisawa, T., Suzuki, Y. and Tamegai, M. (1966). Enzyme histochemical studies on the hypothalamus of spontaneously hypertensive rats with special reference to that of rats subjected to various endocrine interferences. Jap. Circ. J. 30, 1483–1506.

Oki, Y., Oliver, W.T. and Funnell, H.S. (1964). Multiple forms of cholinesterase in horse plasma. Nature 203, 605–606.

Oki, Y., Oliver, W.T. and Funnell, H.S. (1965). Studies of esterases and multiple forms of cholinesterase in equine plasma. Can. J. Physiol. Pharmacol. 43, 147–156.

Okinaka, S., Yoshikawa, M., Vono, M., Muro, T., Mozai, T., Igata, A., Tanabe, H., Veda, S. and Tomonaga, M. (1961). Distribution of cholinesterase activity in the human cerebral cortex. Am. J. Phys. Med. 40, 135–145.

O'Leary, J.F., Kunkel, A.M. and Jones, A.H. (1961). Efficacy and limitations of oxime-atropine treatment of organophosphorus anticholinesterase poisoning. J. Pharmacol. Exptl. Therap. 132, 50–57.

Oliver, G.W.O., Taberner, P.V., Rick, J.T. and Kerkut, G.A. (1971). Changes in GABA level, GAD and CHE activity in CNS of an insect during learning. Comp. Biochem. Physiol. 38B, 529–535.

Olivier, A., Parent, A. and Poirier, L.J. (1970a). Identification of the thalamic nuclei on the basis of their cholinesterase content in the monkey. J. Anat. 106, 37–50.

Olivier, A., Parent, A., Simard, H. and Poirier, L.J. (1970b). Cholinesterasic striatopallidal and striatonigral efferents in the cat and the monkey. Brain Res. 18, 273–282.

Omole, A.A. (1972). Cholinesterase activity in the plasma and acid-secretory gastric mucosa of normal, castrated and castrated oestrogen-treated male rats. J. Pharm. Pharmacol. 24, 538–540.

Omoto, K. and Goedde, H.W. (1965). Pseudocholinesterase variants in Japan. Nature 205, 726.

Ord, M.G. and Thompson, R.H.S. (1950). Nature of placental cholinesterase. Nature 165, 927–928.

Ord, M.G. and Thompson, R.H.S. (1951). The preparation of soluble cholinesterases from mammalian heart and brain. Biochem. J. 49, 191–199.

Ord, M.G. and Thompson, R.H.S. (1952). Pseudo-cholinesterase activity in the central nervous system. Biochem. J. 51, 245–251.

Osen, K.K. and Roth, K. (1969). Histochemical localization of cholinesterases in the cochlear nuclei of the cat, with notes on the origin of acetylcholinesterase-positive afferents and the superior olive. Brain Res. 16, 165–185.

Ostrowski, K., Barnard, E.A., Stocka, Z. and Darzynkiewicz, Z. (1963). Autoradiographic methods in enzyme cytochemistry. I. Localisation of acetylcholinesterase activity using a ^{3}H-labelled irreversible inhibitor. Exptl. Cell Res. 31, 89–99.

Padykula, H.A. and Gauthier, G.F. (1970). The ultrastructure of the neuromuscular junctions of mammalian red, white and intermediate skeletal muscle fibers. J. Cell Biol. 46, 27–41.

Palek, J., Brabec, V., Vopatová, M., Michalec, Č. and Mircevová, L. (1969). The composition of red cell lipids and the activity of two enzymes of the erythrocyte membrane (ATPase and acetylcholinesterase) in primary refractory anemias. Clin. Chim. Acta 23, 133–138.

Palkama, A. (1967). Demonstration of adrenomedullary catecholamines and cholinesterases at electron microscopic level in the same tissue section. Ann. Med. Exptl. Biol. Fenniae 45, 295–306.

Palkama, A. and Sipponen, J. (1968). Light and electron histochemical observations on cholinesterase-containing centrifugal fibres in the optic nerve of the rat. Ann. Med. Exptl. Biol. Fenniae 46, 564–567.

Palmer, A.C. and Ellerker, A.R. (1961). Histochemical localization of cholinesterases in the brainstem of sheep. Quart. J. Exptl. Physiol. 46, 344–352.

Pandey, S.K., Uyas, B.K., Gupta, M.L. and Bhardwaj, S.D. (1970). Red blood cell enzyme activity of schizophrenia. Indian Med. Gaz. 9, 34–36.

Pannese, E., Luciano, L., Iurato, S. and Reale, E. (1971). Cholinesterase activity in spinal ganglia neuroblasts: a histochemical study at the electron microscope. J. Ultrastruct. Res. 36, 46–67.

Pantin, C.F.A. (1956). The origin of the nervous system. Pubbl. Staz. Zool. Napoli 28, 171–181.

Papp, M. (1968). Acetylcholinesterase activity of the human lower brain stem with special regard to the reticular formation. Acta Morphol. Acad. Sci. Hung. 16, 375–390.

Papp, M. and Bozsik, G. (1966). Comparison of the cholinesterase activity in the reticular formation of the lower brain stem of cat and rabbit. J. Neurochem. 13, 697–703.

Parpart, A.K. and Hoffman, J.F. (1952). Acidity vs. acetylcholine and cation permeability of red cells. Federation Proc. 11, 117.

Patočka, J. and Bajgar, J. (1971). Affinity of human brain acetylcholinesterase to some organophosphates and carbamates in vitro. J. Neurochem. 18, 2545–2546.

Patočka, J., Bielavský, J. and Ornst, F. (1970). Reactivating effect of $\alpha\omega$-bis-(4-pyridinealdoxime)-2-trans-butene dibromide on isopropylmethylphosphonylated acetylcholinesterase. FEBS Letters 10, 182–184.

Pauling, P. (1970). The conformation of molecules affecting cholinergic nervous systems. In: Drugs and Cholinergic Mechanisms in the CNS, E. Heilbronn and A. Winter, eds. (Research Institute for National Defence, Stockholm) pp. 457–464.

Pavlin, R. (1965). Cholinesterases in reticular nerve cells. J. Neurochem. 12, 515–518.

Pavlin, R. (1966). The activities of the cholinesterases in the reticular formation neurons investigated by micromanometric techniques. In: Biochemistry and Pharmacology of the Basal Ganglia, E. Costa, L.J. Côté and M.D. Yahr, eds. (Raven Press, New York) pp. 205–214.

Pearse, A.G.E. (1960). Histochemistry: Theoretical and Applied, 2nd edn. (Churchill, London).

Pearse, A.G.E. (1968). Histochemistry: Theoretical and Applied, Vol. 1, 3rd edn. (Churchill, London) pp. 475–494.

Pearson, C.K. (1963). A formalin-sucrose ammonia fixative for cholinesterases. J. Histochem. Cytochem. 11, 665–666.

Pecot-Dechavassine, M. (1962). Étude biochimique, pharmacologique et histochimique des cholinestérases des muscles striés chez les poissons, les batraciens et les mammifères. Thèse (Masson et Cie, Paris).

Pentreath, V.W. and Cottrell, G.A. (1968). Acetylcholine and cholinesterase in the radial nerve of *Asterias rubens*. Comp. Biochem. Physiol. 27, 775–785.

Pepeu, G. (1972). Cholinergic neurotransmission in the central nervous system. Arch. Intern. Pharmacodyn. 196 (Suppl), 229–243.

Pepler, W.J. and Pearse, A.G.E. (1957). The histochemistry of the esterases of rat brain, with special reference to those of the hypothalamic nuclei. J. Neurochem. 1, 193–202.

Perrotta, C.A. and Lewis, P.R. (1958). A histochemical study of placental esterases in the guinea-pig and in the human. J. Anat. 92, 110–117.

Peters, R.A. (1951). Lethal synthesis. Proc. Roy. Soc. London Ser. B 139, 143–170.

Petropoulos, E.A., Vernadakis, A. and Timiras, P.S. (1968). Developmental changes in the offspring of rats electroshocked during gestation. Exptl. Neurol. 20, 481–495.

Pham-Huu-Chanh and Plancade, Y. (1971). Étude comparée des effets du zinc et du cadmium sur les activités cholinestérasiques tissulaires du rat. Biochem. Pharmacol. 20, 729–736.

Phillis, J.W. (1965a). Cholinesterase in the cat cerebellar cortex, deep nuclei and peduncles. Experientia 21, 266–268.

Phillis, J.W. (1965b). Cholinergic mechanisms in the cerebellum. Brit. Med. Bull. 21, 26–29.

Phillis, J.W. (1968a). Acetylcholine release from the cerebral cortex: its role in cortical arousal. Brain Res. 7, 378–389.

Phillis, J.W. (1968b). Acetylcholinesterase in the feline cerebellum. J. Neurochem. 15, 691–698.

Phillis, J.W. (1971). The pharmacology of thalamic and geniculate neurons. Intern. Rev. Neurobiol. 14, 1–48.

Phillis, J.W. and Chong, G.C. (1965). Acetylcholine release from the cerebral and cerebellar cortices: its role in cortical arousal. Nature 207, 1253–1255.

Phillis, J.W. and Tebēcis, A.K. (1967). The effects of topically applied cholinomimetic drugs on the isolated spinal cord of the toad. Comp. Biochem. Physiol. 23, 541–552.

Phillis, J.W. and York, D.H. (1967). Cholinergic inhibition in the cerebral cortex. Brain Res. 5, 517–520.

Phillis, J.W. and York, D.H. (1968a). An intracortical cholinergic inhibitory synapse. Life Sci. Oxford 7, 65–69.

Phillis, J.W. and York, D.H. (1968b). Pharmacological studies on a cholinergic inhibition in the cerebral cortex. Brain Res. 10, 297–306.

Phillis, J.W., Tebēcis, A.K. and York, D.H. (1967). A study of cholinoceptive cells in the lateral geniculate nucleus. J. Physiol. London 192, 695–713.

Phillis, J.W., Tebēcis, A.K. and York, D.H. (1968). Acetylcholine release from the feline thalamus. J. Pharm. Pharmacol. 20, 476–478.

Pickford, M. (1939). The inhibitory effect of acetylcholine on water diuresis in the dog, and its pituitary transmission. J. Physiol. London 95, 226–238.

Pilgrim, R.L.C. (1954). The action of acetylcholine on the hearts of lamellibranch molluscs. J. Physiol. London 125, 208–214.

Piras, M.M., Szijan, I. and Gómez, C.J. (1970). Enzymatic and ultrastructural changes in subcellular fractions from developing rat brain. Acta Physiol. Latinoam. 20, 252–264.

Pisconti, G., De Santis, U., and Torelli, M. (1964). Serum cholinesterase activity in rheumatic disease of children (in Italian). Riv. Clin. Pediat. 73, 241–250.

Pitman, R.M. and Kerkut, G.A. (1970). Comparison of the actions of iontophoretically applied acetylcholine and gamma-aminobutyric acid with the EPSP and IPSP in cockroach central neurons. Comp. Gen. Pharmacol. 1, 221–230.

Plattner, F. and Hintner, H. (1930). Die Spaltung von Acetylcholin durch Organextrakte und Körperflüssigkeiten. Pflügers Arch. 225, 19–25.

Podleski, T.R. (1969). Molecular forces acting between ammonium ions and acetylcholine receptor. Biochem. Pharmacol. 18, 211–225.

Podleski, T.R. and Changeux, J.-P. (1970). On the excitability and cooperativity of the electroplax membrane. In: Fundamental Concepts in Drug Receptor Interactions, J.F. Danielli, J.F., Moran and D.J. Triggle, eds. (Academic Press, New York) pp. 93–119.

Pokrovskii, A.A. and Ponomareva, L.G. (1961). The distribution of cholinesterases in the brain of *Macacus rhesus* monkeys. Biochemistry USSR 26, 248–255.

Polak, R.L. (1969). The influence of drugs on the uptake of acetylcholine by slices of rat cerebral cortex. Brit. J. Pharmacol. 36, 144–152.

Polyakova, O.I. (1967). Cholinesterase activity in helminths (in Russian). Zh. Evol. Biokhim. Fiziol. 3, 124–127.

Pope, A. (1967). Microchemical architecture of human isocortex. Arch. Neurol. 16, 351–356.

Pope, A., Caveness, W. and Livingston, K.E. (1952). Architectonic distribution of acetylcholinesterase in the frontal isocortex of psychotic and nonpsychotic patients. Arch. Neurol. Psychiat. 68, 425–443.

Porcellati, G. (1965). Biochemical aspects of protein metabolism during nerve degeneration and regeneration; In: Protides of the Biological Fluids, H. Peeters, ed. (Elsevier, Amsterdam) pp. 115–126.

Porcellati, G. (1967). The effect of organo-phosphorus compounds on nerve phospholipid metabolism. Progr. Biochem. Pharmacol. 3, 49–58.

Porcellati, G. and Mastrantonio, M.A. (1964). Phospholipid metabolism of peripheral nerves during demyelination by organophosphorus compounds. Ital. J. Biochem. 13, 332–352.

Porcellati, G., Millo, A. and Manocchio, I. (1961). Proteinase activity of nervous tissues in organo-phosphorus compound poisoning. J. Neurochem. 7, 317–320.

Post, L.C. (1971). Inhibition of cholinesterase by carbamates. A new kinetic approach. Biochim. Biophys. Acta 250, 121–130.

Potter, L.T. (1967). A radiometric microassay of acetylcholinesterase. J. Pharmacol. Exptl. Therap. 156, 500–506.

Preston, E. and Heath, C. (1972a). Atropine-insensitive vasodilatation and hypotension in the organophosphate-poisoned rabbit. Arch. Intern. Pharmacodyn. 200, 231–244.

Preston, E. and Heath, C. (1972b). Depression of the vasomotor system in rabbits poisoned with an organophosphate anticholinesterase. Arch. Intern. Pharmacodyn. 200, 245–254.

Prince, A.K. (1966). A sensitive fluorometric procedure for the determination of small quantities of acetylcholinesterase. Biochem. Pharmacol. 15, 411–417.

Prosser, C.L. (1950). Comparative Animal Physiology (W.B. Saunders Co., Philadelphia) p. 333 and pp. 576–629.

Pryor, G.T. (1968). Postnatal development of cholinesterase, acetylcholinesterase, aromatic L-amino acid decarboxylase, and monoamine oxidase in C57 BL/6 and DBA/2 mice. Life Sci. Oxford 7, 867–874.

Pryor, G.T., Schlesinger, K. and Calhoun, W.H. (1966). Differences in brain enzymes among five inbred strains of mice. Life Sci. Oxford 5, 2105–2111.

Raab, W. (1969). Cholinesterase activity of rat serum following renal damage. Clin. Chim. Acta 24, 135–138.

Rabin, B.R. (1967). Co-operative effects in enzyme catalysis: a possible kinetic model based on substrate-induced conformation isomerization. Biochem. J. 102, 22–23c.

Ramon-Moliner, E. (1972). Acetylthiocholinesterase distribution in the brain stem of the cat. Ergeb. Anat. Entwichlungsgesch. 46, pt. 4, 7–53.

Rang, H.P. and Ritter, J.M. (1969). A new kind of drug antagonism:evidence that agonists cause a molecular change in acetylcholine receptors. Mol. Pharmacol. 5, 394–411.

Rang, H.P. and Ritter, J.M. (1970). On the mechanism of desensitization at cholinergic receptors. Mol. Pharmacol. 6, 357–382.

Ranish, N. and Ochs, S. (1972). Fast axoplasmic transport of acetylcholinesterase in mammalian nerve fibres. J. Neurochem. 19, 2641–2649.

Rasmussen, G.L. (1946). The olivary peduncle and other fiber projections of the superior olivary complex. J. Comp. Neurol. 84, 141–219.

Rasmussen, G.L. (1967). Efferent connections of the cochlear nucleus. In: Sensorineural Hearing Processes and Disorders, A.B. Graham, ed. (Churchill, London) pp. 61–75.

Ravin, H.A., Zacks, S.I. and Seligman, A.M. (1953). Histochemical localization of acetylcholinesterase in nervous tissue. J. Pharmacol. Exptl. Therap. 107, 37–53.

Raviola, E. and Raviola, G. (1961). Osservazioni istochimiche sulla retina di mammiferi adulti. Boll. Soc. Med-chir. Pavia 5–6, 611–632.

Raviola, E. and Raviola, G. (1962). Richerche istochimiche sulla retina di coniglo nel corso dello sviluppo post natale. Z. Zellforsch. Mikroskop. Anat. 56, 552–572.

Reale, E., Luciano, L. and Spitznas, M. (1971). The fine structural localization of acetylcholinesterase activity in the retina and optic nerve of rabbits. J. Histochem. Cytochem. 19, 85–96.

Reiff, B., Lambert, S.M. and Natoff, I.L. (1971). Inhibition of brain cholinesterase by organophosphorus compounds in rats. Arch. Intern. Pharmacodyn. 192, 48–60.

Reiner, E. (1971). Spontaneous reactivation of phosphorylated and carbamylated cholinesterases. Bull. World Health Organ. 44, 109–112.

Reiner, E. and Aldridge, W.N. (1967). Effect of pH on inhibition and spontaneous reactivation of acetylcholinesterase treated with esters of phosphorus acids and of carbamic acids. Biochem. J. 105, 171–179.

Reiner, E., Seuferth, W. and Hardegg, W. (1965). Occurrence of cholinesterase isoenzymes in horse serum. Nature 205, 1110–1111.

Renshaw, B. (1941). Influence of discharge of motoneurons upon excitation of neighboring motoneurons. J. Neurophysiol. 4, 167–183.

Rexed, B. (1954). A cytoarchitectonic atlas of the spinal cord in the cat. J. Comp. Neurol. 100, 297–379.

Reznik, S.R. and Kutovoi, A.I. (1970). Effect of metabolic products from *Bacillus subtilis* strain 110 on the activity of brain cholinesterase (in Russian). Mikrobiol. Zh. Kiev 32, 386–388.

Riov, J. and Jaffe, M.J. (1973). A cholinesterase from bean roots and its inhibition by plant growth retardants. Experientia 29, 264–265.

Ritchie, J.M. (1967). On the role of acetylcholine in conduction in mammalian non-myelinated nerve fibers. Ann. N.Y. Acad. Sci. 144, 504–516.

Robertson, G.S. (1967). Serum protein and cholinesterase changes in association with contraceptive pills. Lancet i, 232–235.

Robertson, J.D. (1959). The ultrastructure of cell membranes and their derivatives. Biochem. Soc. Symposia 16, 3–43.

Robins, E. and Smith, D.E. (1953). A quantitative histochemical study of eight enzymes of the cerebellar cortex and subjacent white matter in the monkey. Res. Publ. Ass. Nerv. Ment. Dis. 32, 305–327.

Robinson, N. (1966). Friedreich's ataxia: a histochemical and biochemical study. II. Hydrolytic enzymes. Acta Neuropathol. 6, 35–45.

Robinson, N. (1972). Enzyme histochemistry of the rat hypothalamus during early development. J. Neurochem. 19, 1577–1585.

Robinson, P.M. (1969). A cholinergic component in the innervation of the longitudinal smooth muscle of the guinea pig vas deferens. The fine structural localization of acetylcholinesterase. J. Cell Biol. 41, 462–476.

Robinson, P.M. and Bell, C. (1967). The localization of acetylcholinesterase at the autonomic neuromuscular junction. J. Cell Biol. 33, 93–102.

Rodríguez de Lores Arnaiz, G. and Pellegrino de Iraldi, A. (1972). Cholinesterase in cholinergic and adrenergic nerves: a study of the superior cervical ganglia and the pineal gland of the rat. Brain Res. 42, 230–233.

Rodríguez de Lores Arnaiz, G., Alberici, M. and De Robertis, E. (1967). Ultrastructural and enzymic studies of cholinergic and non-cholinergic synaptic membranes isolated from brain cortex. J. Neurochem. 14, 215–225.

Roepke, M.H. (1937). A study of choline esterase. J. Pharmacol. Exptl. Therap. 59, 264–276.

Roessmann, U. and Friede, R.L. (1966). Changes in butyryl cholinesterase activity in reactive glia. Neurology, 16, 123–129.

Roessmann, U. and Friede, R.L. (1967). The segmental distribution of acetyl cholinesterase in the cat spinal cord. J. Anat. 101, 27–32.

Rogers, A.W., Darzynkiewicz, Z., Salpeter, M.M., Ostrowski, K. and Barnard, E.A. (1969). Quantitative studies on enzymes in structures in striated muscles by labeled inhibitor methods. I. The number of acetylcholinesterase molecules and of other DFP-reactive sites at motor endplates, measured by radioautography. J. Cell Biol. 41, 665–685.

Rogers, D.C. (1965). The development of the rat carotid body. J. Anat. 99, 89–101.

Roşca, D.I. and Dordea, M. (1971). Studies on the functional role of acetylcholinesterases in osmoregulation in *H. medicinalis.* Studia Univ. Babes-Bolyai 16, 119–122.

Roşca, D.I., Wittenberg, C. and Ruşdea, D. (1958). Response to variations in salinity. XLV. Investigations on osmoregulation and the role of the nervous system in osmoregulation phenomena in *Hirudo medicinalis* (in Rumanian). Stud. Cercetări Biol. Cluj 9, 113–136.

Rosenberg, P. and Echlin, F.A. (1965). Cholinesterase activity of chronic partially isolated cortex. Neurology 15, 700–707.

Rosenzweig, M.R. (1966). Environmental complexity, cerebral change and behavior. Am. Psychologist 21, 321–332.

Rosenzweig, M.R., Krech, D. and Bennett, E.L. (1958). Brain enzymes and adaptive behaviour. In: Neurological Basis of Behaviour, Ciba Foundation Symp., G.E. Wolstenholme and C.M. O'Connor, eds. (Churchill, London) pp. 337–355.

Rosenzweig, M.R., Krech, D., Bennett, E.L. and Diamond, M.C. (1962). Effects of environmental complexity and training on brain chemistry and anatomy: a replication and extension. J. Comp. Physiol. Psychol. 55, 429–437.

Rosić, N. and Milošević, M.P. (1967). The relationship between cholinesterase activity and brain permeability to barbitone. Arch. Intern. Pharmacodyn. 165, 302–307.

Ross, D.M. (1965). Some problems of neuromuscular activity and behavior in the 'elementary nervous system'. In: Essays on Physiological Evolution, J.W.S. Pringle, ed. (Pergamon Press, Oxford) pp. 253–261.

Rossi, G. (1961). Acetylcholinesterase in the course of the development of the internal ear of the guinea pig. Acta Oto-laryngol. Suppl. 170, 1–91.

Rothenberg, M.A. and Nachmansohn, D. (1947). Studies on cholinesterase. III. Purification of the enzyme from electric tissue by fractional ammonium sulfate precipitation. J. Biol. Chem. 168, 223–231.

Rothschild, Lord (1961). A Classification of Living Animals (Longmans, Green & Co. Ltd. London)

Roufogalis, B.D. and Quist, E.E. (1972). Relative binding sites of pharmacologically active ligands on bovine erythrocyte acetylcholinesterase. Mol. Pharmacol. 8, 41–49.

Roulston, W.J., Schnitzerling, M.J. and Schuntner, C.A. (1968). Acetylcholinesterase insensitivity in the Biarra strain of the cattle tick *Boophilus microplus,* as a cause of resistance to organophosphorus and carbamate acaricides. Aust. J. Biol. Sci. 21, 759–769.

Rubinstein, H.M., Dietz, A.A., Hodges, L.K., Lubrano, T. and Czebotar, V. (1970). Silent cholinesterase gene: variations in the properties of serum enzyme in apparent homozygotes. J. Clin. Invest. 49, 479–486.

Ruckebusch, Y. and Ruckebusch, M. (1959). Mesure de l'activité cholinestérasique du sang, chez les animaux domestique. Rev. Méd. Vét. Toulouse 110, 627–637.

Russell, D.H. (1968). Acetylcholinesterase in the hypothalamo-hypophyseal axis of the white-crowned sparrow, *Zonotrichia leucophrys gambelii.* Gen. Comp. Endocrinol. 11, 51–63.

Russell, R.W. (1969). Behavioral aspects of cholinergic transmission. Federation Proc. 28, (Suppl.) 121–131.

Rutledge, L.T., Ranck, J.B. Jr., and Duncan, J.A. (1967). Prevention of supersensitivity in partially isolated cerebral cortex. Electroencephalog. Clin. Neurophysiol. 23, 256–262.

Rybicka, K. (1967). Embryogenesis in *Hymenolepis diminuta.* V. Acetylcholinesterase in embryos. Exptl. Parasitol. 20, 263–266.

Saba, S.R. and Mason, R.G. (1970). Acetylcholine, cholinesterase activity, and the aggregability of human platelets. Proc. Soc. Exptl. Biol. Med. 135, 104–107.

Sabatini, D.D., Bensch, K.G. and Barrnett, R.J. (1962). New means of fixation for electron microscopy and histochemistry. Anat. Record 142, 274.

Sakai, M. (1967). Hydrolysis of acetylthiocholine and butyrylthiocholine by cholinesterases of insects and a mite. Appl. Entomol. Zool. 2, 111–112.

Sakharov, D.A. (1970). Cellular aspects of invertebrate neuropharmacology. Ann. Rev. Pharmacol. 10, 335–352.

Sakharova, A.V. (1966). Ontogenetic dynamics of cholinesterase in cells of the cat's cerebellar cortex (in Russian).Tsitologiya, 8, 54–59.

Salánki, J., Hiripi, L. and Labos, E. (1966). Cholinesterase activity in the central nervous system of *Anodonta cygnea.* Ann. Inst. Biol. Tihany, Hung. Acad. Sci. 33, 143–150.

Salkeld, E.H. (1961). The distribution and identification of esterases in the developing embryo and young nymph of the large milkweed bug, *Oncopeltus fasciatus* (Dall.). Can. J. Zool. 39, 589–595.

Salmoiraghi, G.C. and Steiner, F.A. (1963). Acetylcholine sensitivity of cats' medullary neurons. J. Neurophysiol. 26, 581–597.

Salpeter, M.M. (1967). Electron microscope radioautography as a quantitative tool in enzyme cytochemistry. I. The distribution of acetylcholinesterase at motor end plates of a vertebrate twitch muscle. J. Cell Biol. 32, 379–389.

Salpeter, M.M. (1969). Electron microscope radioautography as a quantitative tool in enzyme cytochemistry. II. The distribution of DFP-reactive sites at motor endplates of a vertebrate twitch muscle. J. Cell Biol. 42, 122–134.

Salpeter, M.M., Plattner, H. and Rogers, A.W. (1972). Quantitative assay of esterases in end plates of mouse diaphragm by electron microscope autoradiography. J. Histochem. Cytochem. 20, 1059–1068.

Sams, W.M. Jr., and Carroll, N.V. (1966). Cholinesterase inhibitory property of dimethyl sulphoxide. Nature 212, 405.

Samuels, A.I., Moller, L. and Fischer, J.W. (1968). Effects of acetylcholinesterase on erythropoiesis in polycythemic and mildly plethoric mice. Ann. N.Y. Acad. Sci. 149, 406–408.

Sanderson, B.E. (1969). Acetylcholinesterase activity in *Nippostrongylus brasiliensis* (Nematoda). Comp. Biochem. Physiol. 29, 1207–1213.

Sanderson, B.E. and Ogilvie, B.M. (1971). A study of acetylcholinesterase throughout the life cycle of *Nippostrongylus brasiliensis.* Parasitology 62, 367–373.

Sanderson, B.E., Jenkins, D.C. and Phillipson, R.F. (1972). *Nippostrongylus brasiliensis:* relation between immune damage and acetylcholinesterase levels. Intern. J. Parasitol. 2, 227–232.

Saunders, B.C. (1957). Some Aspects of the Chemistry and Toxic Action of Organic Compounds containing Phosphorus and Fluorine (Cambridge Univ. Press, Cambridge).

Sawyer, C.H. (1943a). Cholinesterase and the behavior problem in *Amblystoma.* I. The relationship between the development of the enzyme and early motility. II. The effects of inhibiting cholinesterase. J. Exptl. Zool. 92, 1–29.

Sawyer, C.H. (1943b). Cholinesterase and the behavior problem in *Amblystoma*. III. The distribution of cholinesterase in nerve and muscle throughout development. IV. Cholinesterase in nerveless muscle. J. Exptl. Zool. 94, 1–31.

Sawyer, C.H. (1945). Hydrolysis of choline esters by liver. Science 101, 385–386.

Sawyer, C.H. (1946). Cholinesterases in degenerating and regenerating peripheral nerves. Am. J. Physiol. 146, 246–253.

Sawyer, C.H. (1955). Further experiments on cholinesterase and reflex activity in *Amblystoma* larvae. J. Exptl. Zool. 129, 561–578.

Sawyer, C.H. and Everett, J.W. (1946). Effects of various hormonal conditions in the intact rat on the synthesis of serum cholinesterase. Endocrinology 39, 307–322.

Sawyer, C.H. and Hollinshead, W.H. (1945). Cholinesterases in sympathetic fibers and ganglia. J. Neurophysiol. 8, 135–153.

Sawyer, C.H., Markee, J.E. and Townsend, B.F. (1949). Cholinergic and adrenergic components in the neurohumoral control of the release of LH in the rabbit. Endocrinology 44, 18–37.

Schaumann, W. and Job, C. (1958). Differential effects of a quaternary cholinesterase inhibitor, phospholine, and its tertiary analogue, compound 217-AO, on central control of respiration and on neuromuscular transmission. The antagonism by 217–AO of the respiratory arrest caused by morphine. J. Pharmac. Pharmacol. 23, 114–120.

Scheibel, M.E. and Scheibel, A.B. (1969). A structural analysis of spinal interneurons and Renshaw cells. In: The Interneuron, UCLA Forum on medical sciences 11, M.A.B. Brazier, ed. (Univ. Chicago Press, Berkeley) pp. 159–208.

Schlaepfer, W.W. (1968). Acetylcholinesterase activity of motor and sensory nerve fibers in the spinal roots of the rat. Z. Zellforsch. Mikroskop. Anat. 88, 441–456.

Schlaepfer, W.W. and Torack, R.M. (1966). The ultrastructural localization of cholinesterase activity in the sciatic nerve of the rat. J. Histochem. Cytochem. 14, 369–378.

Schlinke, J.C. (1970). Chronic toxicity of Dursban in chickens. J. Econ. Entomol. 63, 319.

Schmitt, F.O. (1968). The molecular biology of neuronal fibrous proteins. Neurosciences Res. Prog. Bull. 6, 119–144.

Schonbach, J. and Cuénod, M. (1971). Axoplasmic migration of protein. A light microscopic autoradiographic study in the avian retino-tectal pathway. Exptl. Brain Res. 12, 275–282.

Schonbach, J., Schonbach, C. and Cuénod, M. (1971). Rapid phase of axoplasmic flow and synaptic proteins: an electron microscopical autoradiographic study. J. Comp. Neurol. 141, 485–498.

Schuster, F.L. and Hershenov, B. (1969). Ultrastructural localization of cholinesterase in the pellicle of *Tetrahymena pyriformis* (W). Exptl. Cell Res. 55, 385–392.

Schwab, A. (1949). Über die nerven- und muskelphysiologie des pferde Egels *Haemopis sanguisuga*. Z. Vergleich. Physiol. 31, 506–526.

Schwabe, C.W. (1959). Host-parasite relationships in echinococcosis. I. Observations on the permeability of the hydatid cyst wall. Am. J. Trop. Med. Hyg. 8, 20–28.

Schwabe, C.W., Koussa, M. and Acra, A.N. (1961). Host-parasite relationships in echinococcosis. IV. Acetylcholinesterase and permeability regulation in the hydatid cyst wall. Comp. Biochem. Physiol. 2, 161–172.

Schwarzacher, H.G. (1958). Der Cholinesterasegehalt motorischer Nervenzellen während der axonalen Reaktion. Acta Anat. 32, 51–65.

Schwarzacher, H.G. (1961). Acetylcholinesterase in mammalian myotendinous junctions. Bibliotheca Anat. 2, 220–227.

Scott, F.H. (1905). On the metabolism and action of nerve cells. Brain 28, 506–526.

Scott, F.H. (1906). On the relation of nerve cells to fatigue of their nerve fibres. J. Physiol. London 34, 145–162.

Seaman, G.R. and Houlihan, R.K. (1951). Enzyme systems in *Tetrahymena geleii* S. III. Acetylcholinesterase activity. Its relation to motility of the organism and to co-ordinated ciliary action in general. J. Cell. Comp. Physiol. 37, 309–321.

Segal, H.L. (1959). The development of enzyme kinetics. In: The Enzymes, Vol. 1, P.D. Boyer, H. Lardy and K. Myrbäck, eds. (Academic Press, New York) pp. 1–48.

Sekine, T. (1951). Choline esterase in pig spermatozoa. J. Biochem., Tokyo 38, 171–179.

Serafini-Fracassini, A., and Frasson, P. (1966). Histochemical observations on the carotid body of the dog. Acta Anat. 63, 240–248.

Sereni, F., Principi, N., Perletti, L. and Piceni, S.L. (1966). Undernutrition and the developing rat brain. I. Influence on acetylcholinesterase and succinic acid dehydrogenase activities and on norephinephrine and 5-OH-tryptamine tissue concentrations. Biol. Neonatorum 10, 254–265.

Shafai, T. and Cortner, J.A. (1971a). Human erythrocyte acetylcholinesterase. I. Resolution of activity into two components. Biochim. Biophys. Acta 236, 612–618.

Shafai, T. and Cortner, J.A. (1971b). Human erythrocyte acetylcholinesterase. II. Evidence for the modification of the enzyme by ion-exchange chromatography. Biochim. Biophys. Acta 250, 117–120.

Shantha, T.R., Iijima, K. and Bourne, G.H. (1967). Histochemical studies on the cerebellum of squirrel monkey (*Saimiri sciurea*). Acta Histochem. 27, 129–162.

Shanthaveerappa, T.R. and Bourne, G.H. (1965). Histochemical studies on distribution of dephosphorylating and oxidative enzymes and esterases in olfactory bulb of the squirrel monkey. J. Natl. Cancer Inst. 35, 153–165.

Sharma, N.N. (1968). Studies on the histochemical distribution of simple esterases and cholinesterases in the olfactory bulb of the rat. Acta Anat. 69, 168–175.

Sharma, S., Savithramma, M. and Sharma, K.N. (1970). Changes in cholinesterase activity in the genital tract: effect of olfactory lobe lesions. Indian J. Physiol. Pharmacol. 14, *P*.29.

Shaw, C.R. (1969). Isozymes: classification, frequency and significance. Intern. Rev. Cytol. 25, 297–332.

Shen, S.C., Greenfield, P. and Boell, E.J. (1955). The distribution of cholinesterase in the frog brain. J. Comp. Neurol. 102, 717–743.

Shen, S.C., Greenfield, P. and Boell, E.J. (1956). Localization of acetylcholinesterase in chick retina during histogenesis. J. Comp. Neurol. 106, 433–462.

Shimizu, N. and Ishii, S. (1966). Electron microscopic histochemistry of acetylcholinesterase of rat brain by Karnovsky's method. Histochemie 6, 24–33.

Shukuya, R. (1953a). Kinetics of human blood cholinesterase. II. The temperature effect upon cholinesterase activity. J. Biochem. Tokyo 40, 135–140.

Shukuya, R. (1953b). On the kinetics of the human blood cholinesterase. III. Inhibition of cholinesterase by urethane. J. Biochem. Tokyo 40, 535–545.

Shukuya, R. and Shinoda, M. (1956). Kinetics of the human blood cholinesterase. V. The inhibition of acetylcholinesterase and cholinesterase by hydrogen ion and tetraethylammonium bromide. J. Biochem. Tokyo 43, 315–326.

Shute, C.C.D. and Lewis, P.R. (1960). The salivatory centre in the rat. J. Anat. 94, 59–73.

Shute, C.C.D. and Lewis, P.R. (1961). The use of cholinesterase techniques combined with operative procedures to follow nervous pathways in the brain. Bibliotheca Anat. 2, 34–49.

Shute, C.C.D. and Lewis, P.R. (1963). Cholinesterase - containing systems of the brain of the rat. Nature 199, 1160–1164.

Shute, C.C.D. and Lewis, P.R. (1965). Cholinesterase-containing pathways of the hindbrain: afferent cerebellar and centrifugal cochlear fibres. Nature 205, 242–246.

Shute, C.C.D. and Lewis, P.R. (1966a). Electron microscopy of cholinergic terminals and acetylcholinesterase-containing neurones in the hippocampal formation of the rat. Z. Zelforsch. Mikroskop. Anat. 69, 334–343.

Shute, C.C.D. and Lewis, P.R. (1966b). Cholinergic and monoaminergic pathways in the hypothalmus. Brit. Med. Bull. 22, 221–226.

Shute, C.C.D. and Lewis, P.R. (1967a). Ultrastructural changes in cholinergic hippocampal afferent fibres after section of the fornix. J. Anat. 101, 604–606.

Shute, C.C.D. and Lewis, P.R. (1967b). The ascending cholinergic reticular system: neocortical, olfactory and subcortical projections. Brain 90, 497–520.

Shute, C.C.D. and Lewis, P.R. (1969). Localization of cholinesterases in monkey brain. J. Anat. 104, 186–187.

Sideman, M.B., Pearse, J.J. and Suwalski, R.T. (1965). Cholinesterase levels in spastic children. Intern. J. Neuropsychiat. 1, 183–184.

Silman, I. (1969). Microenvironmental effects on enzyme activity. J. Gen. Physiol. 54, 50s–57s.

Silver, A. (1960). Some aspects of nerve degeneration with special reference to changes in enzyme composition. Ph.D. Thesis: University of Edinburgh.

Silver, A. (1963). A histochemical investigation of cholinesterases at neuromuscular junctions in mammalian and avian muscle. J. Physiol. London 169, 386–393.

Silver, À. (1967). Cholinesterases of the central nervous system with special reference to the cerebellum. Intern. Rev. Neurobiol. 10, 57–109.

Silver, A. (1969). A method for demonstrating acetylcholinesterase and choline acetylase in foetal brain. Experientia 25, 63–64.

Silver, A. (1971). The significance of cholinesterase in the developing nervous system. In: Progress in Brain Research, Vol. 34, O. Eränkö, ed. (Elsevier, Amsterdam) pp. 346–355.

Silver, A. (1972). The localization of cholinesterases in the leech and some other invertebrates. J. Physiol. London 226, 27P.

Silver, A. and Wolstencroft, J.H. (1971). The distribution of cholinesterases in relation to the structure of the spinal cord in the cat. Brain Res. 34, 205–227.

Simpson, N.E. (1966). Factors influencing cholinesterase activity in a Brazilian population. Am. J. Human Genet. 18, 243–252.

Singer, J.J. and Goldberg, A.L. (1970). Cyclic AMP and transmission at the neuromuscular junction. In: Role of Cyclic AMP in Cell Function, Advances in Biochemical Psychopharmacology, Vol. 3, P. Greengard and E. Costa, eds. (Raven Press, New York) pp. 335–348.

Singh, I. (1964). Seasonal variations in the nature of neurotransmitters in the frog vagus-stomach muscle preparation. Arch. Intern. Physiol. Biochim. 72, 843–851.

Singh, D., Johnston, R.J. and Klosterman, H.J. (1967). Effect on brain enzyme and behaviour in the rat of visual pattern restriction in early life. Nature 216, 1337–1338.

Sinha, A.K. and Rose, S.P.R. (1972). Monoamineoxidase and cholinesterase activity in neurons and neuropil from the rat cerebral cortex. J. Neurochem. 19, 1607–1610.

Sirchia, G., Ferrone, S., Mercuriali, F. and Zanella, A. (1970). Red cell acetylcholinesterase activity in autoimmune haemolytic anaemias. Brit. J. Haematol. 19, 411–415.

Sjöqvist, F. (1962). Cholinergic sympathetic ganglion cells: a histochemical, biochemical and pharmacological analysis of acetylcholinesterase-rich nerve cells in the cat. Thesis: Karolinska Institutet, Stockholm.

Sjöstrand, J. (1970). Fast and slow components of axoplasmic transport in the hypoglossal and vagus nerves of the rabbit. Brain Res. 18, 461–467.

Skaer, R.J. (1973). Acetylcholinesterase in human erythroid cells. J. Cell Sci. 12, 911–923.

Skangiel-Kramska, J. and Niemierko, S. (1971). Isoenzymes of acetylcholinesterase in the sciatic nerve of rabbit and their molecular weights. Bull. Pol. Acad. Sci., Ser. Biol. 19, 389–393.

Skramstad, K.H. (1956). Cholinesterases of mice in relation to sex and age. Acta Physiol. Scand. 36, 383–388.

Smallman, B.N. and Pal, R. (1957). The activity and intra-cellular distribution of choline acetylase in insect nervous tissue. Bull. Entomol. Soc. Am. 3, 25.

Smallman, B.N. and Wolfe, L.S. (1954). The effect of salts on the estimation of cholinesterase activity. Enzymologia 17, 133–144.

Smallman, B.N. and Wolfe, L.S. (1956). Soluble and particulate cholinesterase in insects. J. Cell. Comp. Physiol. 48, 197–213.

Smelik, P.G. and Ernst, A.M. (1966). Role of nigro-neostriatal dopaminergic fibers in compulsive gnawing behavior in rats. Life Sci. Oxford 5, 1485–1488.

Smissaert, H.R. (1964). Cholinesterase inhibition in spider mites susceptible and resistant to organophosphate. Science 143, 129–131.

Smith, C.A. and Sjöstrand, F.S. (1961). Structure of the nerve endings on the external hair cells of the guinea pig cochlea as studied by serial sections. J. Ultrastruct. Res. 5, 523–556.

Smith, C.C. and Glick, D. (1939). Some observations on cholinesterase in invertebrates. Biol. Bull. 77, 321–322.

Smith, C.C., Jackson, B. and Prosser, C.L. (1940). Cholinesterase content, ACh sensitivity of *Cerebratulus*. Biol. Bull. 79, 377.

Smith, D.P., Fourman, J.M. and Haase, P. (1971a). The effect of an intravenous injection of isotonic saline or sucrose on the secretory activity and butyrylcholinesterase content of the salt glands of *Anas domesticus*. Cytobios. 3, 49–55.

Smith, D.P., Fourman, J.M. and Haase, P. (1971b). Secretory activity and butyrylcholinesterase content of the salt glands of *Anas domesticus* in response to an intravenous injection of hypertonic saline. Cytobios. 3, 57–64.

Smith, D.S. and Treherne, J.E. (1965). The electron microscopic localization of cholinesterase activity in the central nervous system of an insect, *Periplaneta americana* L. J. Cell Biol. 26, 445–465.

Smith, H.V. and Spalding, J.M.K. (1959). Outbreak of paralysis in Morocco due to *ortho*-cresyl phosphate poisoning. Lancet ii, 1019–1021.

Smith, M.I., Elvove, E., Valaer, P.J. Jr., Frazier, W.H. and Mallory, G.E. (1930). Pharmacological and chemical studies of the cause of so-called ginger paralysis. Preliminary report. U.S. Pub. Health Repts. 45, 1703–1716.

Smyth, R.D. and Beck, H. (1969). The effect of time and concentration of ethanol administration on brain acetylcholine metabolism. Arch. Intern. Pharmacodyn. 182, 295–299.

Smythies, J.R. (1970). The chemical nature of the receptor site. A study in the stereochemistry of synaptic mechanisms. Intern. Rev. Neurobiol. 13, 181–222.

Smythies, J.R. (1974). The molecular structure of acetylcholine and adrenergic receptors: an all protein model. Intern. Rev. Neurobiol. 17, (in press).

Snell, R.S. (1958a). The histochemical appearances of cholinesterase in the superior cervical sympathetic ganglion and the changes which occur after preganglionic nerve section. J. Anat. 92, 408–418.

Snell, R.S. (1958b). The histochemical appearances of cholinesterase in the parasympathetic nerves supplying the submandibular and sublingual salivary glands of the rat. J. Anat. 92, 534–543.

Snell, R.S. (1959). The histochemical appearances of cholinesterase in the parotid salivary gland of the rat. Z. Zellforsch. Mikroskop. Anat. 49, 330–334.

Snell, R.S. (1961). The histochemical localization of cholinesterase in the central nervous system, Bibliotheca Anat. 2, 50–58.

Snell, R.S. and Garrett, J.R. (1957). The distribution of cholinesterase in the submaxillary and sublingual salivary glands of the rat. J. Histochem. Cytochem. 5, 236–245.

Snell, R.S. and Garrett, J.R. (1958). The effect of postganglionic sympathectomy on the histochemical appearances of cholinesterase in the nerves supplying the submandibular and sublingual salivary glands of the rat. Z. Zellforsch. Mikroskop. Anat. Abt. Histochem. 48, 201–214.

Solly, S.R.B., Harrison, D.L., Hunnego, J.N. and Shanks, V. (1971a). Fensulfothion II. The effects of grazing sheep on Fensulfothion-treated pasture or on pasture grown in Fensulfothion-treated soil. N.Z. J. Agr. Res. 14, 79–87.

Solly, S.R.B., Harrison, D.L. and Ritchie, A.R. (1971b). Fensulfothion III. Effects of grazing dairy cows on Fensulfothion-treated pastures. N.Z.J. Agr. Res. 14, 88–96.

Souza, J.A., Barreto, H.E. and Martins, E.O. (1970). Alterações da colinesterasemia e proteinemia provocadas em ratos por alguns agentes anestésicos gerais (I). O Hospital (Rio de Janeiro) 78, 1167–1183.

Sovershaev, P.F. (1968). Content and localization of cholinesterase in pine tree seeds and vegetative buds (in Russian). Izv. Vyssh. Ucheb. Zaved. Les. Zh. 11, 30–33.

Spehlmann, R. (1971). Acetylcholine and the synaptic transmission of nonspecific impulses to the visual cortex. Brain 94, 139–150.

Spehlmann, R., Daniels, J.C. and Smathers, C.C. Jr. (1971). Acetylcholine and the synaptic transmission of specific impulses to the visual cortex. Brain 94, 125–138.

Sperti, L. and Sperti, S. (1959a). Effects of chronic lesions of the peduncles on cerebellum cholinesterase activity in the albino rat. Experientia 15, 441–442.

Sperti, L. and Sperti, S. (1959b). Effects of midline cerebellar splitting and of lesions in cerebral cortex on cerebellum cholinesterase activity in the albino rat. Experientia 15, 442.

Sperti, L., Sperti, S. and Zatti, P. (1960a). Cholinesterases distribution in cerebellum. Arch. Ital. Biol. 98, 41–52.

Sperti, L., Sperti, S. and Zatti, P. (1960b). Development of cholinesterase activity in the cerebellum and optic lobes of the chick embryo. Arch. Ital. Biol. 98, 53–59.

Srinivasan, R., Karczmar, A. and Bernsohn, J. (1972). Activation of acetylcholinesterase by Triton X-100. Biochim. Biophys. Acta 284, 349–354.

Štalc, A. and Župančič, A.O. (1972). Effect of α-bungarotoxin on acetylcholinesterase bound to mouse diaphragm end-plates. Nature New Biol. 239, 91–92.

Starzl, T.E., Taylor, C.W. and Magoun, H.W. (1951). Ascending conduction in reticular activating system, with special reference to the diencephalon. J. Neurophysiol. 14, 461–477.

Stedman, E. and Stedman, E. (1935). The relative choline-esterase activities of serum and corpuscles from the blood of certain species. Biochem. J. 29, 2107–2111.

Stedman, E., Stedman, E. and Easson, L.H. (1932). Choline-esterase. An enzyme present in the blood-serum of the horse. Biochem. J. 26, 2056–2066.

Stefani, E. and Gerschenfeld, H.M. (1969). Comparative study of acetylcholine and 5-hydroxytryptamine receptors on single snail neurons. J. Neurophysiol. 32, 64–74.

Stegwee, D. (1951). Studies on cholinesterase in insects. Physiol. Comparata et Oecol. 2, 241–247.

Steiner, F.A. and Meyer, M. (1966). Actions of L-glutamate, acetylcholine and dopamine on single neurones in the nuclei cuneatus and gracilis of the cat. Experientia 22, 58–59.

Steiner, F.A. and Pieri, L. (1969). Comparative microelectrophoretic studies of invertebrate and vertebrate neurones. In: Progress in Brain Research, Vol. 31, K. Akert and P.G. Waser, eds. (Elsevier, Amsterdam) pp. 191–199.

Steiner, F.A. and Weber, G. (1965). Die Beeinflussung labyrinthär erregbarer Neurone des Hirnstammes durch Acetylcholin. Helv. Physiol. Pharmacol. Acta 23, 82–89.

Stevens, J.T., Stitzel, R.E. and McPhillips, J.J. (1972). Effects of anticholinesterase insecticides on hepatic microsomal metabolism. J. Pharmacol. Exptl. Therap. 181, 576–583.

Storm-Mathisen, J. (1970). Quantitative histochemistry of acetylcholinesterase in rat hippocampal region correlated to histochemical staining. J. Neurochem. 17, 739–750.

Storm-Mathisen, J. and Blackstad, T.W. (1964). Cholinesterase in the hippocampal region. Distribution and relation to architectonics and afferent systems. Acta Anat. 56, 216–253.

Straschill, M. and Perwein, J. (1971). Effect of iontophoretically applied biogenic amines and of cholinomimetic substances upon the activity of neurons in the superior colliculus and mesencephalic reticular formation of the cat. Pfluger's Arch. 324, 43–55.

Straughan, D.W. and Legge, K.F. (1965). The pharmacology of amygdaloid neurones. J. Pharm. Pharmacol. 17, 675–677.

Strickland, K.P. and Thompson, R.H.S. (1954). The effect of cholinesterase inhibitors on the leakage of potassium from brain slices. Biochem. J. 58, xx.

Strickland, K.P. and Thompson, R.H.S. (1955). On the mechanism of potassium loss from brain slices induced by cholinesterase inhibitors. Biochem. J. 60, 468–475.

Strizhak, E.K. (1967). Hygienic significance of methacrylamides in the problem of the sanitary protection of reservoirs (in Russian). Mater. Nauch-Prakt. Konf. Molodykh. Gig. Sanit. Vrachei 11th, 79–82.

Stromblad, B.C.R. (1957). Supersensitivity caused by denervation and by cholinesterase inhibitors. Acta Physiol. Scand. 41, 118–138.

Strumia, E. and Baima-Bollone, P.L. (1964). AChE activity in the spinal ganglia of the chick embryo during development. Acta Anat. 57, 281–293.

Strumwasser, F. (1962). Post-synaptic inhibition and excitation produced by different branches of a single neuron and the common transmitter involved. XXII Intern. Physiol. Congr. 2, Abst. 801.

Sturge, L.M. and Whittaker, V.P. (1950). The esterases of horse blood. 1. The specificity of horse plasma cholinesterase and ali-esterase. Biochem. J. 47, 518–525.

Surgenor, D.M. and Ellis, D. (1954). Preparation and properties of serum and plasma proteins. Plasma cholinesterase. J. Am. Chem. Soc. 76, 6049–6051.

Surgenor, D.M., Strong, L.E., Taylor, H.L., Gordon, R.S. Jr., and Gibson, D.M. (1949). The separation of choline esterase, mucoprotein, and metal-combining protein into subfractions of human plasma. J. Am. Chem. Soc. 71, 1223–1229.

Sushko, E.P. (1970). Activity of blood acetylcholinesterase in children with Botkin's disease (infectious hepatitis) (in Russian). Dokl. Akad. Nauk. Beloruss. SSR 14, 279–282.

Svensmark, O. (1963a). Electrophoretic properties of atypical human serum cholinesterase. Acta Chem. Scand. 17, 876.

Svensmark, O. (1963b). Enzymatic and molecular properties of cholinesterases in human liver. Acta Physiol. Scand. 59, 378–389.

Svensmark, O. (1965). Molecular properties of cholinesterases. Acta Physiol. Scand. 64, Suppl. 245, 1–74.

Szabolcs, M., Kővér, A. and Kovács, L. (1968). Study of the physical chemical and enzyme chemical properties of contractile proteins in postnatal life. Acta Biochim. Biophys. Acad. Sci. Hung. 3, 141–151.

Szász, G. (1968). Cholinesterase bestimmung im Serum mit Acetyl- und Butyrylthiocholin als Substrat. Clin. Chim. Acta 19, 191–204.

Szász, G., Czirbesz, Zs., Király, L. and Csáki, P. (1964). Enzymaktivitatsbestimmungen im Kindesalter. V. Serumcholinesterase beim nephrotischen Syndrom. Acta Paediat. Acad. Sci. Hung. 5, 367–390.

Szentágothai, J. (1958). The anatomical basis of synaptic transmission of excitation and inhibition in motoneurons. Acta Morphol. Acad. Sci. Hung. 8, 287–309.

Szentágothai, J. (1964). Neuronal and synaptic arrangement in the substantia gelatinosa Rolandi. J. Comp. Neurol. 122, 219–240.

Szentágothai, J., Flerkó, B., Mess, B., and Halász, B. (1968). Hypothalamic control of the anterior pituitary. An experimental morphological study (Akadémiai Kiadó, Budapest).

Szerb, J.C. (1964). The effect of tertiary and quaternary atropine on cortical acetylcholine output and on the electroencephalogram in cats. Can. J. Physiol. Pharmacol. 42, 303–314.

Szerb, J.C. (1967). Cortical acetylcholine release and electroencephalographic arousal. J. Physiol. London 192, 329–343.

Szerb, J.C. and Somogyi, G.T. (1973). Depression of acetylcholine release from cerebral cortical slices by cholinesterase inhibition and by oxotremorine. Nature New Biol. 241, 121–122.

Szerb, J.C., Malik, H. and Hunter, E.G. (1970). Relationship between acetylcholine content and release in the cat's cerebral cortex. Can. J. Physiol. Pharmacol. 48, 780–790.

Taber, E. (1961). The cytoarchitecture of the brain stem of the cat. I. Brain stem nuclei of cat. J. Comp. Neurol. 116, 27–69.

Takagi, K. and Takahashi, A. (1968). Studies of separation and characterization of acetylcholine receptor labeled with tritiated dibenamine. Biochem. Pharmacol. 17, 1609–1618.

Takeuchi, A. and Takeuchi, N. (1964). The effect on crayfish muscle of iontophoretically applied glutamate. J. Physiol. London 170, 296–317.

Tammelin, L.-E. (1953). An electrometric method for the determination of cholinesterase activity. 1. Apparatus and cholinesterase in human blood. Scand. J. Clin. Lab. Invest. 5, 267–270.

Tammelin, L.-E. (1957). Dialkoxy-phosphorylthiocholines, alkoxy-methyl-phosphorylthiocholines, and analogous choline esters. Synthesis, pK_a of tertiary homologs, and cholinesterase inhibition. Acta Chem. Scand. 11, 1340–1349.

Tammelin, L.-E. (1958a). Organophosphorylcholines and cholinesterases. Arkiv. Kemi 12, 287–298.

Tammelin, L.-E. (1958b). Choline esters: substrates and inhibitors of cholinesterases. Svensk. Kemisk. Tidskrift. 70, 157–181.

Tammelin, L.-E. and Löw, H. (1951). Calibration of an electrometric method for the determination of cholinesterase activity. Acta Chem. Scand. 5, 322–323.

Tammelin, L.-E. and Strindberg, B. (1952). Cholinesterase activity determined with an electrometric method. Acta Chem. Scand. 6, 1041–1047.

Tanaka, Y. and Katsuki, Y. (1966). Pharmacological investigations of cochlear responses and of olivo-cochlear inhibition. J. Neurophysiol. 29, 94–108.

Tapp, J.T. and Markowitz, H. (1963). Infant handling: effects on avoidance learning, brain weight and cholinesterase activity. Science 140, 486–487.

Tauc, L. and Gerschenfeld, H.M. (1962). A cholinergic mechanism of inhibitory synaptic transmission in a molluscan nervous system. J. Neurophysiol. 25, 236–262.

Taxi, J. (1952). Action du formol sur l'activité de diverses préparations de cholinestérases. J. Physiol. Paris 44, 595–599.

Taxi, J. (1961). La distribution des cholinestérases dans divers ganglions du système nerveux autonome des Vertébrés. Bibliotheca Anat. 2, 73–89.

Taylor, I.M. and Anderson, R.H. (1973). Cholinesterase and the atrioventricular node and bundle in the human fetus up to midterm. J. Histochem. Cytochem. 21, 464-468.

Taylor, I.M. and Smith, R.B. (1971). Cholinesterase activity in the human fetal heart between the 35- and 160-millimeter crown-rump length stages. J. Histochem. Cytochem. 19, 498–503.

Tebēcis, A.K. (1970a). Properties of cholinoceptive neurones in the medial geniculate nucleus. Brit. J. Pharmacol. 38, 117–137.

Tebēcis, A.K. (1970b). Studies on cholinergic transmission in the medial geniculate nucleus. Brit. J. Pharmacol. 38, 138–147.

Tebēcis, A.K. (1972). Cholinergic and non-cholinergic transmission in the medial geniculate nucleus of the cat. J. Physiol. London 226, 153–172.

Ten Cate, J. (1955). Contribution à la question de l'innervation cholinergique du coeur de l'*Anodonta cygnea*. Pubbl. Staz. Zool. Nap. 27, 199–203.

Tennyson, V.M. and Brzin, M. (1970). The appearance of acetylcholinesterase in the dorsal root neuroblast of the rabbit embryo. A study by electron microscope cytochemistry and microgasometric analysis with the magnetic diver. J. Cell Biol. 46, 64–80.

Tennyson, V.M., Brzin, M. and Duffy, P. (1968). Electron microscopic cytochemistry and microgasometric analysis of cholinesterase in the nervous system. In: Progress in Brain Research, Vol. 29, A. Lajtha and D.H. Ford, eds. (Elsevier, Amsterdam) pp. 41–62.

Tennyson, V.M., Brzin, M. and Kremzner, L.T. (1973). Acetylcholinesterase activity in the myotubule and muscle satellite cell of the fetal rabbit. An electron microscopic-cytochemical and biochemical study. J. Histochem. Cytochem. 21, 634–652.

Teräväinen, H. (1967). Electron microscopic localization of cholinesterases in the rat myoneural junction. Histochemie 10, 266–271.

Teräväinen, H. (1969a). Localization of acetylcholinesterase in the rat myoneural junction. Histochemie 17, 162–169.

Teräväinen, H. (1969b). Ultrastructural distribution of cholinesterase activity in the ventral nerve cord of the earthworm *Lumbricus terrestris.* Histochemie 18, 177–190.

Terzani, G., Natalizi, G. and Marinucci, G. (1968). Blood cholinesterase activity in acute and chronic hepatopathy. Lab. Diagn. Med. 13, 139–151.

Tewari, H.B. and Bourne, G.H. (1960). Histochemical localization of specific and non-specific cholinesterases and simple esterase in myelinated nerves. Exptl. Cell Res. 21, 245–248.

Tewari, H.B. and Bourne, G.H. (1962). Histochemical studies on the distribution of specific and non-specific cholinesterases and simple esterase in the cerebellum of the rat. Acta Anat. 51, 349–368.

Tewari, H.B. and Bourne, G.H. (1963). Histochemical studies on the distribution of simple esterase, specific and non-specific cholinesterase in trigeminal ganglion cells of rat. Acta Anat. 53, 319–332.

Thomas, J.F. and Hopkins, T.L. (1966). Unpublished observations cited by Menn and McBain (1968) q.v.

Thompson, E.H. and Whittaker, V.P. (1952). Cholinesterase activity and sodium transport in the human red cell. Biochim. Biophys. Acta 9, 700–701.

Tibbs, J. (1960). Acetylcholinesterase in flagellated systems. Biochim. Biophys. Acta 41, 115–122.

Tingari, M.D. and Lake, P.E. (1972a). The intrinsic innervation of the reproductive tract of the male fowl (*Gallus domesticus*). A histochemical and fine structural study. J. Anat. 112, 251–271.

Tingari, M.D. and Lake, P.E. (1972b). Histochemical localization of glycogen, mucopolysaccharides, lipids, some oxidative enzymes and cholinesterases in the reproductive tract of the male fowl (*Gallus domesticus*). J. Anat. 112, 273–287.

Toman, J.E.P., Woodbury, J.W. and Woodbury, L.A. (1947). Mechanism of nerve conduction block produced by anticholinesterases. J. Neurophysiol. 10, 429–441.

Torack, R.M. and Barrnett, R.J. (1962). Fine structural localization of cholinesterase activity in the rat brainstem. Exptl. Neurol. 6, 224–244.

Toschi, G. (1959). A biochemical study of brain microsomes. Exptl. Cell Res. 16, 232–255.

Tranzer, J.P. and Thoenen, H. (1967). Significance of 'empty vesicles' in post-ganglionic sympathetic nerve terminals. Experientia 23, 123–124.

Tranzer, J.P., Thoenen, H., Snipes, R.L. and Richards, J.G. (1969). Recent developments on the ultrastructural aspect of adrenergic nerve endings in various experimental conditions. In: Progress in Brain Research, Vol. 31, K. Akert and P.G. Waser, eds. (Elsevier, Amsterdam) pp. 33–46.

Treherne, J.E. (1966). The Neurochemistry of Arthropods (Cambridge Univ. Press, Cambridge).

Treherne, J.E. and Smith, D.S. (1965). The penetration of acetylcholine into the central nervous tissues of an insect (*Periplaneta americana* L.) J. Exptl. Biol. 43, 13–21.

Trotter, J.L. and Burton, R.M. (1969). Acetylcholine esterase activity of synaptic vesicle fractions and membrane fractions prepared from rat brain tissue. J. Neurochem. 16, 805–811.

Trotter, M.D. and Fairburn, E.A. (1966). Serum pseudocholinesterase in eczema. Brit. J. Dermatol. 78, 469–471.

Trueman, T. and Herbert, J. (1970). Monoamines and acetyl-cholinesterase in the pineal gland and habenula of the ferret. Z. Zellforsch. Mikroskop. Anat. 109, 83–100.

Tsuji, S. (1964). Sur l'innervation cholinergique du système conducteur du coeur de la Grenouille. Compt. Rend. Soc. Biol. 158, 1769–1771.

Tsuji, S., Rieger, F., Peltre, G., Massoulié, J. and Benda, P. (1972). Acetylcholinestérase du muscle, de la moelle épinière et du cerveau de Gymnote. J. Neurochem. 19, 989–997.

Tucci, A.F. and Seifter, S. (1969). Preparation and properties of porcine parotid butyrylcholinesterase. J. Biol. Chem. 244, 841–849.

Tuček, S. (1966). On subcellular localization and binding of choline acetyltransferase in the cholinergic nerve endings of the brain. J. Neurochem. 13, 1317–1327.

Tucker, E.M. (1974). A shortened life span of sheep red cells with a glutathione deficiency. Res. Vet. Sci. 16, 19–22.

Turbow, M.M. and Burkhalter, A. (1968). Acetylcholinesterase activity in the chick embryo spinal cords. Develop. Biol. 17, 233–244.

Turpaev, T.M., Abashkina, L.I., Brestkin, A.P., Brick, I.L., Grigorjeva, G.M., Pevzner, D.L., Rozengart, V.J., Rozengart, E.V. and Sakharov, D.A. (1968). Cholinesterase of squid optical ganglia Ommatostrephes sloanei-pacificus. Eur. J. Biochem. 6, 55–59.

Twarog, B.M. (1954). Responses of a molluscan smooth muscle to acetylcholine and 5-hydroxytryptamine. J. Cell. Comp. Physiol. 44, 141–163.

Tyshtchenko, V.P. and Mandelstam, J.E. (1965). A study of spontaneous electrical activity and localization of cholinesterase in the nerve ganglia of Antheraea pernyi Guer. at different stages of metamorphosis and in pupal diapause. J. Insect Physiol. 11, 1233–1239.

Tzagóloff, A. (1963). Metabolism of sinapine in mustard plants. II. Purification and some properties of sinapine esterase. Plant Physiol. 38, 207–213.

Usdin, E. (1970). Reactions of cholinesterases with substrates, inhibitors and reactivators. In: International Encyclopedia of Pharmacology and Therapeutics, Section 13, Vol. I, Anticholinesterase Agents, A.G. Karczmar, ed. (Pergamon Press, Oxford) pp. 47–354.

Utley, J.D. (1966). Acetylcholinesterase and pseudocholinesterase in neural and nonneural tissue in the medial geniculate body of the cat. Biochem. Pharmacol. 15, 1–6.

Vaccarezza, J.-R. and Peltz, L. (1960). Action de la corticotropine sur l'activité cholinestérasique sanguine chez des sujets normaux et chez des malades allergiques respiratoires. Presse Méd. 68, 723–724.

Vaccarezza, J.R. and Willson, J.A. (1965a). Blood cholinesterase in adrenalectomized rats. Experientia 21, 205.

Vaccarezza, J.R. and Willson, J.A. (1965b). The relationship between blood cholinesterase activity and neoplastic diseases. Experientia 21, 405–406.

Vaccarezza, J.R., Ruiz, D.C. and Domínguez, A.C. (1969). Colorimetric method for measuring pseudocholinesterase, acetylcholinesterase and whole blood cholinesterase. Its use in the presumptive diagnosis of cancer. Experientia 25, 808–809.

Valverde, F. (1962). Reticular formation of the albino rat's brain stem cytoarchitecture and corticofugal connections. J. Comp. Neurol. 119, 25–53.

Van Asperen, K. (1962). A study of housefly esterases by means of a sensitive colorimetric method. J. Insect Physiol. 8, 401–416.

Van der Kloot, W.G. (1955). The control of neurosecretion and diapause by physiological changes in the brain of cecropia silkworm. Biol. Bull. 109, 276–294.

Van der Kloot, W.G. (1956). Cholinesterase and sodium transport in frog muscle. Nature 178, 366–367.

Vanha-Perttula, T. (1966). Esterases of the rat adenohypophysis. Acta Physiol. Scand. 69, Suppl. 283, 1–104.

Van Hooidonk, C., De Borst, C., Mitzka, F.A. and Groos, C.C. (1972). Chromogenic substrates for cholinesterases: 1-[2-thiazolylazo]-2-acetoxybenzene derivatives. Anal. Biochem. 48, 33–44.

Van Ros, G. and Druet, R. (1966). Uncommon electrophoretic patterns of serum cholinesterase (pseudocholinesterase). Nature 212, 543–544.

Van Rossum, J.M. (1964). Receptor theory in enzymology. In: Molecular Pharmacology, Vol. II, E.J. Ariens, ed. (Academic Press, New York) pp. 199–255.

Varagić, V. and Krstić, M. (1966). Adrenergic activation by anticholinesterases. Pharmacol. Rev. 18, 799–800.

Varanka, I. (1968a). Effect of some water-miscible organic solvents on the activity of acetyl-cholinesterase in nervous- and muscular tissues of *Anodonta cygnea* (Pelecypoda). Ann. Inst. Biol. Tihany. Hung. Acad. Sci. 35, 83–91.

Varanka, I. (1968b). Biochemical investigation of cholinesterase in the central nervous system of *Lymnaea stagnalis* L. (Gastropoda). Ann. Inst. Biol. Tihany. Hung. Acad. Sci. 35, 93–107.

Varga, E. and Szöör, Á. (1971). Cholinesterase activity in myosin. In: Contractile Proteins and Muscle, E. Laki, ed. (Marcel Dekker, New York) pp. 263–271.

Varga, E., Kővér, A., Kovács, T. and Hetényi, E. (1957a). Changes in cholinesterase activity of striated muscle after denervation. Acta Physiol. Acad. Sci. Hung. 11, 235–242.

Varga, E., Kővér, A., Kovács, T. and Hetényi, E. (1957b). Changes in the acetylcholine-sensitivity and cholinesterase activity of skeletal muscles in the course of ontogenesis. Acta Physiol. Acad. Sci. Hung. 11, 243–251.

Venkatachari, S.A.T. and Dass, P.M. (1968). Cholinesterase activity rhythm in the ventral nerve cord of scorpion. Life Sci. Oxford 7, 617–621.

Venkatachari, S.A.T. and Naidu, V.D. (1969). Choline esterase activity in the nervous system and the innervated organs of the scorpion *Heterometrus fulvipes*. Experientia 25, 821–822.

Vernadakis, A. and Rutledge, C.O. (1973). Effects of ether and pentobarbital anaesthesia on the activities of brain acetylcholinesterase and butyrylcholinesterase in young adult rats. J. Neurochem. 20, 1503–1504.

Viale, G. and Apponi, G. (1961). Histochemie Untersuchungen über die Cholinesterasen in der menschlichen Netzhaut. Z. Zellforsch. Mikroskop. Anat. 55, 673–678.

Vigh-Teichmann, I. and Goslar, H.-G. (1968). Enzymhistochemische Studien am Nervensystem. III. Das Verhalten einiger Hydrolasen im Nervensystem des Regenwurmes (*Eisenia foetida*). Histochemie 14, 352–365.

Vitale, V. and Blincoe, C.B. (1971). Rapid potentiometric method for the determination of erythrocyte and plasma acetylcholinesterase activity. Lab. Pract. 20, 716–718.

Vincent, D., Roux, G., Ghiloni, J. and Dumas, J.C. (1965). Sur les estérases du sang de Hamster (*Mesocricetus auratus*). Compt. Rend. Soc. Biol. 159, 2064–2066.

Vincent, D., Moskovtchenko, J.F. and Traeger, J. (1970a). Les cholinestérases (erythrocytaire et plasmatique) dans les anémies renales. Clin. Chim. Acta 28, 97–102.

Vincent, D., Perrier, H. and Rouzioux, J.M. (1970b). Sur les isoenzymes de la cholinestérase sérique chez espèces animales. Compt. Rend. Soc. Biol. 164, 1767–1769.

Vlk, J. and Tuček, S. (1962a). The distribution of cholinesterases in the mammalian heart. Physiol. Bohemoslov. 11, 46–52.

Vlk, J. and Tuček, S. (1962b). Changes in the acetylcholine content and cholinesterase activity in dog atria during the early postnatal period. Physiol. Bohemoslov. 11, 53–57.

Vogt, O. and Vogt, C. (1902). Neurobiologische Arbeiten, Vol. I, O. Vogt, ed. (Fischer, Jena).

Volle, R.L. and Koelle, G.B. (1961). The physiological role of acetylcholinesterase (AChE) in sympathetic ganglia. J. Pharmacol. Exptl. Therap. 133, 223–240.

Voorhorst, C.D. (1971). Use of 1-methyl-acylquinolium iodides for assaying acetylcholinesterase in tissue homogenates. Differentiation between acetylcholinesterase and cholinesterase. Biochem. Pharmacol. 20, 1321–1323.

Voss, G. and Matsumura, F. (1965). Biochemical studies on a modified and normal cholinesterase found in the Leverkusen strains of the two-spotted spider mite *Tetranychus urticae*. Can. J. Biochem. 43, 63–72.

Wachtel, H. and Kandel, E.R. (1967). A direct synaptic connection mediating both excitation and inhibition. Science 158, 1206–1208.

Wachtel, H. and Kandel, E.R. (1971). Conversion of synaptic excitation to inhibition at a dual chemical synapse. J. Neurophysiol. 34, 56–68.

Wagner, W.R. (1848). Lohengrin i, iii (Novello, Ewer and Co.) Octavo ed.

Waldron, H.A. (1967). Cholinergic fibres in the spinal cord of the rat. Brain Res. 4, 113–116.

Waldron, H.A. (1969). The morphology of the lateral cervical nucleus in the hedgehog. Brain Res. 16, 301–306.

Waldron, H.A. (1970). Acetylcholinesterase activity in the red nucleus of the guinea pig and the cat. Brain Res. 20, 251–257.

Waldron, H.A. and Gwyn, D.G. (1969). Acetylcholinesterase activity in the red nucleus of the rat and its response to axotomy. Brain Res. 13, 146–154.

Walker, R.J., Hedges, A. and Woodruff, G.N. (1968). The pharmacology of the neurones of *Helix aspera*. Symp. Zool. Soc. London 22, 33–74.

Waller, A. (1850). Experiments on the section of the glossopharyngeal nerve and hypoglossal nerves of the frog and observations on the alteration produced thereby in the structure of their primitive fibres. Phil. Trans. Roy. Soc. London 141, 423–429.

Wang, E.I.C. and Braid, P.E. (1967). Oxime reactivation of diethylphosphoryl human serum cholinesterase. J. Biol. Chem. 242, 2683–2687.

Waser, P.G. and Nickel, E. (1969). Electronmicroscopic and autoradiographic studies of normal and denervated endplates. In: Progress in Brain Research, Vol. 31, K. Akert and P.G. Waser, eds. (Elsevier, Amsterdam) pp. 157–169.

Waterson, J.G., Hume, W.R. and De La Lande, I.S. (1970). The distribution of cholinesterase in the rabbit ear artery. J. Histochem. Cytochem. 18, 211–216.

Watkins, J.C. (1965). Pharmacological receptors and general permeability phenomena of cell membranes. J. Theoret. Biol. 9, 37–50.

Watson, W.E. (1968). Centripetal passage of labelled molecules along mammalian motor axons. J. Physiol. London 196, 122–123*P*.

Watts, D.C. (1968). The origin and evolution of the phosphagen phosphotransferases. In: Homologous Enzymes and Biochemical Evolution, N. van Thoai and J. Roche, eds. (Gordon and Breach, New York) pp. 279–296.

Waymire, J.C., Weiner, N. and Prasad, K.N. (1972). Regulation of tyrosine hydroxylase activity in cultured mouse neuroblastoma cells: elevation induced by analogs of adenosine 3′:5′ cyclic monophosphate Proc. Natl. Acad. Sci. U.S. 69, 2241–2245.

Weight, F.F. (1968). Cholinergic mechanisms in recurrent inhibition of motoneurons. In: Psychopharmacology, a Review of Progress 1957–1967, D.H. Efron, ed. (U.S. Printing Office, P.H.S. Publ. No. 1836) pp. 69–75.

Weinstock, M. (1971). Acetylcholine and cholinesterases. In: Narcotic Drugs: Biochemical Pharmacology, D.H. Clouet, ed. (Plenum Press, New York) pp. 254–261.

Weiss, P. (1947). Protoplasm synthesis and substance transfer in neurons. XVII Intern. Physiol. Congr. Abstr. p. 101.

Weiss, P. and Hiscoe, H.B. (1948). Experiments on the mechanism of nerve growth. J. Exptl. Zool. 107, 315–395.

Welsch, F. and Dettbarn, W.-D. (1970). The subcellular distribution of acetylcholine, choline acetyltransferase and cholinesterases in lobster walking leg nerves. J. Neurochem. 17, 927–940.

Welsch, F. and Dettbarn, W.-D. (1971). Protein synthesis in lobster walking leg nerves. Comp. Biochem. Physiol. 38B, 393–403.

Welsch, F. and Dettbarn, W.-D. (1972a). The subcellular distribution of acetylcholine, cholinesterases and choline acetyltransferase in optic lobes of the squid *Loligo pealei*. Brain Res. 39, 467–482.

Welsch, F. and Dettbarn, W.-D. (1972b). Inhibition of cholinesterases of rat diaphragm muscle by organophosphates and spontaneous recovery of enzyme activity in vitro. Biochem. Pharmacol. 21, 1039–1049.

Welsch, U. and Pearse, A.G.E. (1969). Electron cytochemistry of BuChE and AChE in thyroid and parathyroid C cells, under normal and experimental conditions. Histochemie 17, 1–10.

Welsh, J.H. (1966). Neurohumors and neurosecretion. In: Physiology of *Echinodermata*, R.A. Boolootian, ed. (Interscience Publishers, New York) pp. 545–560.

Wender, M. and Kozik, M. (1970). Studies of the histoenzymatic architecture of the Ammon's horn region in the developing rabbit brain. Acta Anat. 75, 248–262.

Werman, R. (1966). Criteria for identification of a central nervous transmitter. Comp. Biochem. Physiol. 18, 745–766.

Wermuth, B. and Brodbeck, U. (1972). Oligomeric forms of acetylcholinesterase from *Electrophorus electricus*. Experientia 28, 740.

Werner, G. and Kuperman, A.S. (1963). Actions at the neuromuscular junction. In: Cholinesterases and Anticholinesterase Agents, G.B. Koelle, ed. (Springer, Berlin) pp. 570–678.

Werner, I., Peterson, G.R. and Shuster, L. (1971). Choline acetyltransferase and acetylcholinesterase in cultured brain cells from chick embryos. J. Neurochem. 18, 141–151.

Whitfield, I.C. (1968). Centrifugal control mechanisms of the auditory pathway. In: Hearing Mechanisms in Vertebrates, A.V.S. De Reuck and J. Knight, eds. (Churchill, London) pp. 246–258.

Whitfield, I.C. and Comis, S.D. (1968). A reciprocal gating mechanism in the auditory pathway. In: Cybernetic Problems in Bionics, H.L. Oestreicher and D.R. Moore, eds. (Gordon and Breach, New York) pp. 301–312.

Whittaker, M. (1968). The pseudocholinesterase variants. Differentiation by means of alkyl alcohols. Acta Genet. Statist. Med. 18, 325–334.

Whittaker, M. (1969). The serum cholinesterase variants. Differentiation by means of formaldehyde. Clin. Chim. Acta 26, 141–145.

Whittaker, M. and Hardisty, C.A. (1969). Properties of 'unusual' and 'atypical' serum cholinesterases using O-nitrophenyl butyrate as substrate. Clin. Chem. 15, 445–451.

Whittaker, V.P. (1951). Specificity, mode of action and distribution of cholinesterases. Physiol. Rev. 31, 312–343.

Whittaker, V.P. (1953). The specificity of pigeon brain aceto-cholinesterase. Biochem. J. 54, 660–664.

Whittaker, V.P. (1965). The application of subcellular fractionation techniques to the study of brain function. Progr. Biophys. Mol. Biol. 15, 39–96.

Whittaker, V.P., Michaelson, I.A. and Kirkland, R.J.A. (1964). The separation of synaptic vesicles from nerve-ending ('synaptosomes'). Biochem. J. 90, 293–303.

Whittaker, V.P., Dowe, G. and Scotto, J. (1971). Vesiculin: a possible counter-ion for acetylcholine in the cholinergic synaptic vesicle. Proc. IIIrd Intern. Meeting Intern. Soc. Neurochem. p. 266.

Widlund, L. and Heilbronn, E. (1973). Turnover of brain-cortex phospholipids and its relation to endogenous acetylcholine. Biochem. Soc. Trans. 1, 463.

Wieland, Th. (1968). Multiple forms of enzymes. In: Homologous Enzymes and Biochemical Evolution, N. van Thoai and J. Roche, eds. (Gordon and Breach, New York) pp. 3–18.

Wieth, J.O. (1970a). Paradoxical temperature dependence of sodium and potassium fluxes in human red cells. J. Physiol. London 207, 563–580.

Wieth, J.O. (1970b). Effect of some monovalent anions on chloride and sulphate permeability of human red cells. J. Physiol. London 207, 581–609.

Wigglesworth, V.B. (1958). The distribution of esterase in the nervous system and other tissues of the insect *Rhodnius prolixus*. Quart. J. Microsc. Sci. 99, 441–450.

Williams, J.D. and Cooper, J.R. (1965). Acetylcholine in bovine corneal epithelium. Biochem. Pharmacol. 14, 1286–1289.

Willmer, E.N. (1960). Cytology and Evolution (Academic Press, New York).

Wills, J.H. (1970). Toxicity of anticholinesterases and treatment of poisoning. In: International Encyclopedia of Pharmacology and Therapeutics, Section 13, Vol. I, Anticholinesterase Agents, A.G. Karczmar, ed. (Pergamon Press, Oxford) pp. 357–469.

Wilson, B.W., Montgomery, M.A. and Asmundson, R.V. (1968). Cholinesterase activity and inherited muscular dystrophy of the chicken. Proc. Soc. Exptl. Biol. Med. 129, 199–206.

Wilson, I.B. (1951). Mechanism of hydrolysis. II. New evidence for an acylated enzyme as an intermediate. Biochim. Biophys. Acta 7, 520–525.

Wilson, I.B. (1952). Acetylcholinesterase. XIII. Reactivation of alkyl phosphate-inhibited enzyme. J. Biol. Chem. 199, 113–120.

Wilson, I.B. (1967). Conformation changes in acetylcholinesterase. Ann. N.Y. Acad. Sci. 144, 664–674.

Wilson, I.B. (1971). The possibility of conformational changes in acetylcholinesterase. In: Cholinergic Ligand Interactions, D.J. Triggle, J.F. Moran and E.A. Barnard, eds. (Academic Press, New York) pp. 1–18.

Wilson, I.B. and Bergmann, F. (1950). Acetylcholinesterase. VIII. Dissociation constants of the active groups. J. Biol. Chem. 186, 683–692.

Wilson, I.B. and Cabib, E. (1956). Acetylcholinesterase: enthalpies and entropies of activation. J. Am. Chem. Soc. 78, 202–207.

Wilson, I.B. and Harrison, M.A. (1961). Turnover number of acetylcholinesterase. J. Biol. Chem. 236, 2292–2295.

Wilson, I.B., Bergmann, F. and Nachmansohn, D. (1950). Acetylcholinesterase. X. Mechanism of the catalysis of acylation reactions. J. Biol. Chem. 186, 781–790.

Wilson, I.B., Hatch, M.A. and Ginsburg, S. (1960). Carbamylation of acetylcholinesterase. J. Biol. Chem. 235, 2312–2315.

Winckler, J. (1970a). Zum Einfrieren von Gewebe in Stickstoff-gekühlten Propan. Histochemie 23, 44–50.

Winckler, J. (1970b). Verwendung gefriergetrockneter Kryostatschnitte für histologische und histochemische Untersuchungen. Histochemie 24, 168–186.

Winkelmann, R.K. (1966). Some unusual histochemical properties of kangaroo skin. J. Invest. Dermatol. 46, 446–452.

Winkelmann, R.K., Wolff, K. and Pierce, J. (1967). Removal of epidermis with sodium bromide for cholinesterase staining of dermal nerves. Stain Technol. 42, 214–215.

Wins, P., Schoffeniels, E. and Foidart, J.-M. (1970). Inhibition of membrane-bound acetylcholinesterase by d-tubocurarine and its reversal by bivalent cations. Life Sci. Oxford 9, 259–267.

Winson, J. and Miller, N.E. (1970). Comparison of drinking elicited by eserine or DFP injected into the preoptic area of rat brain. J. Comp. Physiol. Psychol. 73, 233–237.

Winteringham, F.P.W. and Disney, R.W. (1964). A radiometric study of cholinesterase and its inhibition. Biochem. J. 91, 506–514.

Winton, M.Y., Metcalf, R.L. and Fukuto, T.R. (1958). The use of acetylthiocholine in the histochemical study of the action of organophosphorus insecticides. Ann. Entomol. Soc. Am. 51, 436–441.

Wislocki, G.B. and Leduc, E.H. (1952). Vital staining of the hematoencephalic barrier by silver nitrate and trypan blue, and cytological comparisons of the neurohypophysis, pineal body, area postrema, intercolumnar tubercle and supraoptic crest. J. Comp. Neurol. 96, 371–413.

Witter, R.F. and Gaines, T.B. (1963). Relationship between depression of brain or plasma cholinesterase and paralysis in chickens caused by certain organic phosphorous compounds. Biochem. Pharmacol. 12, 1377–1386.

Wofsy, L. and Michaeli, D. (1967). Affinity labeling of the anionic site of acetylcholinesterase. Proc. Natl. Acad. Sci. U.S. 58, 2296–2298.

Wofsy, L., Metzger, H. and Singer, S.J. (1962). Affinity labeling – a general method for labeling the active sites of antibody and enzyme molecules. Biochemistry 1, 1031–1039.

Wolfe, L.S. and Smallman, B.N. (1956). The properties of cholinesterase from insects. J. Cell. Comp. Physiol. 48, 215–235.

Wollemann, M. and Zoltan, L. (1962). Cholinesterase activity of cerebral tumors and tumorous cysts. Arch. Neurol. Psychiat. Chicago 6, 161–167.

Wolter, J.R. (1964). Thin nerves with simple endings containing cholinesterase .in striated human eye muscle. Neurology 14, 283–286.

Wood, W.G. and Ballantyne, B. (1968). Sodium ion transport and β-glucuronidase activity in the nasal glands of *Anas domesticus*. J. Anat. 103, 277–287.

Woodson, P.B.J., Schlapfer, W.T. and Barondes, S.H. (1972). Postural avoidance learning in the headless cockroach without detectable changes in ganglionic cholinesterase. Brain Res. 37, 348–352.

Woolley, D.E. (1963). Sex differences in brain pseudo-cholinesterase activity in the rat. J. Neurochem. 10, 447–452.

Wright, C.I. and Sabine, J.C. (1948). Cholinesterases of human erythrocytes and plasma and their inhibition by antimalarial drugs. J. Pharmacol. Exp. Therap. 93, 230–239.

Yagihara, Y. and Hawthorne, J.N. (1972). Effects of acetylcholine on the incorporation of [^{32}P] orthophosphate in vitro into the phospholipids of nerve-ending particles from guinea pig brain. J. Neurochem. 19, 355–367.

Yamauchi, A. and Lever, J.D. (1971). Correlations between formol fluorescence and acetylcholinesterase (AChE) staining in the superior cervical ganglion of normal rat, pig and sheep. J. Anat. 110, 435–443.

Young, W. (1973). The influence of hydrostatic pressure on the rate of hydrolysis of acetylcholine and contractility in the vagal heart system. Experientia 29, 802–804.

Youngstrom, K.A. (1938). On the relationship between choline esterase and the development of behavior in amphibia. J. Neurophysiol. 1, 357–363.

Youngstrom, K.A. (1941). Acetylcholine esterase concentration during development of the human fetus. J. Neurophysiol. 4, 473–477.

Zacks, S.I. (1954). Esterases in the early chick embryo. Anat. Record 118, 509–537.

Zacks, S.I. and Blumberg, J.M. (1961). The histochemical localization of acetylcholinesterase in the fine structure of neuromuscular junctions of mouse and human intercostal muscle. J. Histochem. Cytochem. 9, 317–324.

Zaimis, E. (1963). Actions at autonomic ganglia. In: Cholinesterases and Anticholinesterase Agents, G.B. Koelle, ed. (Springer, Berlin) pp. 530–569.

Zajicek, J. (1956). Studies on the histogenesis of blood platelets. II. Quantitative determination of acetylcholinesterase activity in single megakaryocytes from various mammals. Acta Haematol. 15, 296–302.

Zajicek, J. (1957). Studies on the histogenesis of blood platelets and megakaryocytes. Acta Physiol. Scand. 40, Suppl. 138, 1–32.

Zajicek, J., Sylvén, B. and Datta, N. (1954). Attempts to demonstrate acetylcholinesterase activity in blood and bone-marrow cells by a modified thiocholine technique. J. Histochem. Cytochem. 2, 115–121.

Zelená, J. (1970). Ribosome-like particles in myelinated axons of the rat. Brain Res. 24, 359–363.

Zelená, J. (1972). Ribosomes in myelinated axons of dorsal root ganglia. Z. Zellforsch. Mikroskop. Anat. 124, 217–229.

Zelená, J. and Lubińska, L. (1962). Early changes of acetylcholinesterase activity near the lesion in crushed nerves. Physiol. Bohemoslov. 11, 261–268.

Zeller, E.A. and Bissegger, A. (1943). Über die Cholin-esterase des Gehirns und der Erythrocyten. Helv. Chim. Acta 26, 1619–1630.

Zeller, E.A. and Joël, C.A. (1941). Beiträge zur Fermentchemie des männlichen Geschlechtsapparates. Über das Vorkommen der Cholinesterase, der Mono- und Diaminoxydase in Sperma und Prostata, und über die Beeinflussung der Spermien-Beweglichkeit durch Fermentinhibitoren. Helv. Chim. Acta 24, 968–976.

Zeller, E.A., Birkhäuser, H., Wattenwyl, H.V. and Wenner, R. (1941). Geschlechtsfunktion und Serum-Cholinesterase des Menschen. Helv. Chim. Acta 24, 962–968.

Zimmerman, J.J. and Goyan, J.E. (1971). Linear free energy relationships in the enzymatic hydrolysis of substituted benzoylcholine esters. J. Med. Chem. 14, 1206–1211.

Zsigmond, E.K. (1972). The effect of epinephrine and its congeners on human cholinesterases. Arch. Intern. Pharmacodyn. 197, 102–107.

Zsoltan-Nagy, I. and Salánki, J. (1965). Histochemical investigations of cholinesterase in different molluscs with reference to functional conditions. Nature 206, 842–843.

Župančič, A.O. (1953). The mode of action of acetylcholine. A theory extended to a hypothesis on the mode of action of other biologically active substances. Acta Physiol. Scand. 29, 63–71.

Župančič, A.O. (1964a). The equilibrium constant for the reaction between acetylcholine and the anionic centres of horse-plasma cholinesterase. Biochim. Biophys. Acta 81, 411–412.

Župančič, A.O. (1964b). An attempt to determine the equilibrium constant for the reaction between acetylcholine and the anionic centres of horse plasma cholinesterase. Intern. J. Neuropharmacol. 3, 333–339.

Župančič, A.O. (1967). Evidence for the identity of anionic centers of cholinesterases with cholinoreceptors. Ann. N.Y. Acad. Sci. 144, 689–693.

Župančič, A.O. (1970). Kinetic response of a membrane-bound acetylcholinesterase to cholinergic activating and blocking agents. FEBS Letters 11, 277–280.

Župančič, A.O., Majcen, Z. and Štalc, A. (1972). Kinetics of muscarinic cholinoreceptors and of acetylcholinesterase of frog heart ventricle under the influence of atropine. Life Sci. Oxford 11 pt. 1, 135–140.

Subject index

The following abbreviations are used: ACh, acetylcholine; AChE, acetylcholinesterase; AcThCh, acetylthiocholine; antiChE(s), anticholinesterase(s); BuCh, butyrylcholine; BuChE, butyrylcholinesterase; BuThCh, butyrylthiocholine; BzCh, benzoylcholine; BzChE, benzoylcholinesterase; ChAc, choline acetyltransferase; ChE(s), cholinesterase(s); CNS, central nervous system; MeCh, acetyl-β-methylcholine; PrCh, propionylcholine; RNA, ribonucleic acid. Note: re ch. 6, individual page numbers have been given for species but not for tissues (except in a few cases) listed in the tables.